Encyclopedia of

Architectural and Engineering Feats

Encyclopedia of

Architectural and Engineering Feats

Donald Langmead and Christine Garnaut

ABC-CLIO

Santa Barbara, California
Denver, Colorado
Oxford, England

Library of Congress Cataloging-in-Publication Data
    Langmead, Donald.
      Encyclopedia of architectural and engineering feats / Donald Langmead
    and Christine Garnaut.
        p. cm.
    Includes bibliographical references and index.
     ISBN 1-57607-112-X (hardcover)—ISBN 1-57607-569-9 (eBook)
     1. Architecture. 2. Engineering—History. I. Garnaut, Christine.
  II. Title.
  NA200 .L32 2001
  721'.03—dc21

                                          2001004229

1-57607-112-X (hardcover)
1-57607-569-9 (eBook)

07 06 05 04 03 02 01   10 9 8 7 6 5 4 3 2 1 (cloth)

ABC-CLIO, Inc.
130 Cremona Drive, P.O. Box 1911
Santa Barbara, California 93116-1911

*This book is also available on the World Wide Web as an eBook. Visit
www.abc-clio.com for details.*
This book is printed on acid-free paper ∞.
Manufactured in the United States of America

Contents

Preface

The engineer, inspired by the law of economy and governed by mathematical calculation, puts us in accord with universal law. He achieves harmony. The architect, by his arrangement of forms, realizes an order which is pure creation of his spirit. By the relationships which he creates he awakes profound echoes in us, he gives us the measure of an order which we feel to be in accordance with that of our world.

—Le Corbusier, *Vers une Architecture* (1923)

This book offers an overview of the architectural and engineering works that represent major steps, as well as significant innovations, in the creation of the built environment. Its scope is wide in both time and space, presenting achievements from prehistory until the present, including work in progress, in all the inhabited continents—one entry even describes a building in outer space.

Not all these feats were performed by trained architects or engineers, or by others acting in any professional capacity. We have concluded that some of the more awesome accomplishments and exciting responses to what seemed to be insurmountable challenges have been the work of communities that many people might regard as underdeveloped. That leads us to the observation that the greatest achievements have depended, not on the sophistication of a culture's technology, but on the social imperatives that stimulated people to push the available technology to its limits. In the ancient and preliterate world, corporate will, brilliant social organization, and shared commitment to a spiritual ideal were as powerful influences upon the creation of the great

works of humankind as any of the inventions of the Industrial Age.

Therefore, the reader will find in these pages many references to feats that were in effect social products. It is generally true of the ancient and medieval worlds that architects and engineers stood deep in the shadow of their patrons, although a few individuals like Hippodamos, Iktinus, and Mnesikles are identifiable. But the patrons are usually better known: for example, Perikles, Hadrian, and Charlemagne. The closer we come to the present, the more often particular personalities stand out; architects and engineers are identifiable, and some are even famous. The watershed seems to have been the fifteenth-century Italian movement known as the Renaissance, defined 400 years later by the Swiss historian Jacob Burckhardt as "an affirmation of the individual, who emerged from the anonymous crowd of the Middle Ages."

Had we restricted ourselves to the *major* steps forward—new structural systems and new building materials—this would be a slim volume indeed. They have been few and far between. Throughout its entire time on Earth, humanity has discovered just three essential ways of building: post-and-beam, or trabeated, construction; the arch and its extensions, the vault and the dome, known as arcuated construction; and the use of stretched filaments and membranes to produce a tensile structure. The ways in which each has been applied at different times and in different places have depended not only upon what materials were available but also upon beliefs about how long a building should last. Until about 150 years ago the most durable material was stone, whether naturally occurring or replaced by synthetic variations—that is, brick or concrete. Each develops great compressive strength (resistance to crushing) but readily fails in tension (when it is stretched). For millennia, that limiting physical property confined builders to the trabeated or arcuated structural systems. Durable tensile materials (steel and reinforced concrete) were not developed until the late nineteenth century A.D.; synthetic membranes did not appear until a whole century later. Before that, tensile systems were used only in structures like Bedouin tents, which were nonetheless remarkable feats for their economy of means, functionality, and environmental sustainability.

Naturally, there were developments, often born of necessity, within the narrow range of appropriate structural systems. The prehistoric peoples of the stone-poor Egyptian, Mesopotamian, and Pakistani river valleys invented the brick, later making it more durable by baking it in kilns. In the second millennium B.C. the warlike Mycenaeans, having invented the "relieving triangle" to reduce loads over openings, extended and refined it to form corbeled vaults and domes. Over a thousand years later, the Romans developed the arch—known to, but rarely used by, the ancient Egyptians—and exploited its space-forming potential in all manner of variations built in concrete. Around A.D. 500 Byzantine engineers found a way to build domes above square compartments by inventing the pendentive. After that, at least in the West, no advances were made for another five centuries.

Then, in a thrilling burst of creative activity, the itinerant master masons of France, England, and Germany produced the great Gothic churches. Ironically, those revolutionary structures were made possible by the adoption of an Islamic invention, the pointed arch; together with rib-and-panel vaulting and the flying buttress; that simple device formed the structural essence of a new framed architecture, giving a hint of things to come. From about A.D. 1500 building technology suffered another 300-year hiatus. The Renaissance—whose architecture Victor Hugo called "the sunset that all of Europe mistook for dawn"—saw the mason-architect replaced, first with the artist-courtier-architect and finally with the abstruse academician. The structural adventure all but ceased.

Late in the 1700s, metal building frames were developed, first in iron, a material that also worked best in compression. A century later the technological revolution produced steel, and subsequently reinforced, prestressed, and shell concrete. Since then, structural innovation has accelerated, culminating in the creation of buildings beyond the earth.

The works and ideas in this book have been chosen for any of several reasons. Some represent those erratic major leaps in the use of materials or technology but many exemplify the creative refinement of existing systems. Others, like Frank Lloyd Wright's Frederick C. Robie house in Chicago, are socially significant. Still others are inspired responses to economic needs: for example, the architecture associated with the railroads or the building of the Panama Canal. Some, like the Deltaworks in the Netherlands or China's Three Gorges Dam, are works that have been undertaken on an almost unimaginably ambitious scale. A few others are not structures at all, but individuals, ideas, or institutions that have generated new directions in engineering or architectural thought, leaving their mark on the built environment. No entry is included on the basis of a single factor. Although inevitably there are certain emphases, each feat is considered within a cultural, social, artistic, and technological context. Wherever possible, we have included information about the condition of the structures at the time of writing: a fact borne out poignantly in the final entry—the World Trade Center Towers in New York. This book was due to go to press when the twin skyscrapers were destroyed on 11 September 2001 by a terrorist attack that claimed a multitude of innocent lives. The apparent ease and swiftness of their destruction is a sobering reminder, not only of the fragility of life but also of the structures that we build.

We are grateful to the University of South Australia for practical support and encouragement during the compilation of this book, and especially to the library staff at the City West Campus. Thanks are also due to Caryl Bosman, who found and photocopied reference material for us. We have appreciated and enjoyed the opportunity to collaborate with ABC-CLIO editorial staff in the preparation of this encyclopedia and thank them warmly for their encouragement and gracious assistance.

Of course there will be objections to our choice of subjects, especially over what has been omitted. Our only answer is that we could not include everything. We have tried to avoid embracing structures merely because they are the biggest of their kind, because records for height, span, or capacity are regularly broken. Rather, we have sought to identify those achievements that demonstrate discovery, creativity, and innovation, as well as those that manifest humanity's propensity to "just do it," especially against the odds. As Arnold Toynbee once pointed out, societies and individuals perform at their best, not when life is easy but when they are faced with a challenge. As the reader will discover, at those times the result has been a remarkable and exciting architectural or engineering feat.

Donald Langmead
Christine Garnaut
Adelaide, South Australia
October 2001

Abomey Royal Palaces
Benin, Africa

The Royal Palaces of Abomey in the West African Republic of Benin (formerly the Kingdom of Dahomey), on the Gulf of Guinea, are a substantial reminder of a vanished kingdom. From 1625 to 1900 Abomey was ruled by a succession of twelve kings. With the exception of Akaba, who created a separate enclosure, each built a lavish cob-wall palace with a high, wide-eaved thatched roof in the 190-acre (44-hectare) royal grounds, surrounded by a wall about 20 feet (6 meters) high. There are fourteen palaces in all, standing in a series of defensible courtyards joined by what were once closely guarded passages. Over centuries, the complex—really a "a city within a city"—was filled with nearly 200 square or rectangular single-story houses, circular religious buildings, and auxiliary structures, all made of unbaked earth and decorated with colorful bas-reliefs, murals, and sculpture; it was a major and quite unexpected feat of contextual architecture in a preliterate society.

According to tradition, in the twelfth or thirteenth century A.D., Adja people migrated from near the Mono River in what is now Togo and founded a village that became the capital of Great Ardra, a kingdom that reached the zenith of its power about 400 years later. Around 1625 a dispute over which of three brothers should be king resulted in one, Kokpon, re-

taining Great Ardra. Another, Te-Agdanlin, founded Little Ardra (known to the Portuguese as Porto-Novo). The third, Do-Aklin, established his capital at Abomey and built a powerful centralized kingdom with a permanent army and a complex bureaucracy. Intermarriage with the local people gradually formed the largest of modern Benin's ethnic groups, the Fon, or Dahomey, who occupy the southern coastal region. Abomey is their principal town.

The irresistible Fon armies—they included female warriors—carried out slave raids on their neighbors, setting up a trade with Europeans. By 1700 about 20,000 slaves were sold each year, and the trade became the kingdom's main source of wealth. Despite British efforts to stamp it out, it persisted, and Dahomey continued to expand northward well into the nineteenth century. King Agadja (1708–1732) subjugated much of the south, provoking the neighboring Yoruba kingdom to a war, during which Abomey was captured. The Fon were under Yoruba domination for eighty years from 1738. In 1863, in a bid to balance Fon power, Little Ardra (the only southern town not annexed by Agadja) accepted a French protectorate. France, fearing other European imperialists, tried to secure its hold on the Dahomey coast. King Behanzin (1889–1893) resisted, but France established a protectorate over Abomey, exiled him, and made his brother, Agoli-Agbo, puppet king under a colonial government. By 1904 the

French had seized the rest of present-day Benin, absorbing it into French West Africa.

Tradition has it that the first palace was built for King Dakodonou in 1645 and that his successors followed with structures of the same materials and similar design—in architectural jargon, each palace was contextual. King Agadja was the first to incorporate 40-inch-square (1-meter) panels of brightly painted bas-relief in niches in his palace facade. After that they proliferated as an integral decorative device; for example, King Glélé's (1858–1889) palace had fifty-six of them. As esthetically delightful as they were, the main purpose of the panels was not pleasure but propaganda. An important record of the preliterate Fon society, many documented key events in its rise to supremacy, rehearsing in images the (probably exaggerated) deeds of the kings. Just as history books might do in another society, they held for posterity the Fon's cultural heritage, customs, mythology, and liturgy.

When French forces advanced on Abomey in 1892, King Behanzin commanded that the royal palaces were to be burned rather than fall into their hands. Under Agoli-Agbo I, the buildings were restored. Although contemporary documents describe the compound as a "vast camp of ruins," the exact extent of both the damage and the reconstruction is unclear. The palace of King Glélé (known as the Hall of the Jewels) was among the buildings to survive. Although there are doubts about the age of the existing bas-reliefs, which may be reproductions, those from that palace are probably original and the oldest of the remaining works. In 1911 the French made an ill-informed attempt at architectural restoration, particularly in the palaces of Guezo and Glélé. Further inappropriate work in the early 1930s included replacing some of the thatched roofs with low-pitched corrugated steel. Denied the protection of the traditional wide eaves, the earthen bas-reliefs were badly damaged.

The palaces seem to have been under continual threat. After damage from torrential rain in April 1977, the Benin government sought UNESCO's advice on conserving and restoring them. In 1984 the complex was inscribed on the World Heritage List and simultaneously on the List of the World Heritage in Danger because of the effects of a tornado. The royal compound, the Guezo Portico, King Glélé's tomb, and the Hall of the Jewels were badly damaged. Several conservation programs have been initiated subsequently. In 1988 fifty of the fragile reliefs from the latter building, battered by weather and insect attack, were removed before reconstruction was initiated. After removal, they were remounted as individual panels in stabilized earth casings, and between 1993 and 1997 an international team of experts from the Benin government and the Getty Conservation Institute worked on their conservation. The Italian government has financed other projects.

Today the glory of the royal city of Abomey has passed. Most of the palaces are gone; only those of Guezo (1818–1858) and Glélé tenuously stand. Their size gives a glimpse of their splendid past: together they cover 10 acres (4 hectares) and comprise 18 buildings. They were converted into a historical museum in 1944. Apart from them, the enclosure of the Royal Palaces is abandoned. Many buildings, including the Queen Mother's palace, the royal tombs, and the so-called priestesses' house remain in imminent danger of collapse.

Further reading

Ben-Amos, Pauline G. 1999. *Art, Innovation, and Politics in Eighteenth Century Benin.* Blommington: Indiana University Press.

Piqué, Francesca, Leslie H. Rainer, et al. 1999. *Palace Sculptures of Abomey: History Told on Walls.* Los Angeles: Getty Conservation Institute and the J. Paul Getty Museum.

The Acropolis
Athens, Greece

The Acropolis of Athens, as it was rebuilt in the fifth century B.C., is a wonderful example of unified civic design in a theocentric society. Under the general control of Pheidias, the greatest artist of his day, the popular *strategos* (elected general) Perikles (ca. 495–429 B.C.) initiated a fifty-year plan comprising architectural and artistic projects that included the Parthenon, the great western Propylaea, the precious little Temple of Athena Nike, and the Erechtheion. The costly works were paid for by misappropriating

The Acropolis, Athens, Greece; Pheidias, principal designer, after 447 B.C.. Reconstruction as the site appeared from the west in the first century A.D.

funds from the treasury of the Delian League, an action that led to the eventual overthrow of Athens.

In Greek, *acropolis* signifies "highest city," and it is clear that the original purpose of the 8-acre (3-hectare) steep-sided limestone outcrop 500 feet (150 meters) above the neighboring city was for defense in a region dotted with rivalrous city-states. The first inhabitants of Athens, the Palasgians, fortified the Acropolis with 20-foot-thick (6-meter) walls, protecting the king's palace and his courtiers' houses. The lower city *(asty)* in its shadow accommodated mundane urban functions in times of peace, and was the center of an agricultural hinterland. Other noteworthy *acropolae*, with even more dramatic topog-

raphy, included Acrocorinth, High Nafplion, and the Cadmea of Thebes. When the lower parts of Athens were walled, its acropolis assumed a religious rather than military role.

Archeologists continue to debate the history of the Acropolis, differing over the disposition of its older buildings. What is certain is that it was the site of successive shrines built over centuries for Athena, the city's patron goddess. Early among them was the Hekatompedon, which may well have been a *temenos* (sacred enclosure) and not a temple. The tyrant Peisistratos (602–527 B.C.) championed the Athena cult and commissioned the "old temple of Athena." Completed by his sons, it was the only temple on the

Acropolis when the ragtag Persian armies sacked the abandoned city of Athens in 480, destroying it only five years after it was finished. Within thirteen years of the defeat of the Persians, Cimon and Themistokles had rebuilt the perimeter walls of the Acropolis and cleared the huge site of debris.

After subsequent disputes between the Greek city-states had been resolved in a truce, the fourteen peaceful years of the Delian League gave Perikles an opportunity to propagandize the power of Athens among the participating city-states. In 447 B.C., he appointed Pheidias (ca. 493–430 B.C.) to oversee the restoration of the Acropolis. That Athenian golden age witnessed the creation of the greatest Hellenic architecture: the Parthenon (447–432 B.C.), the Erechtheion (421–406 B.C.), and the Temple of Athena Nike (427–424 B.C.). Grouped at the western end of the plateau, they were reached through the magnificent ceremonial Propylaea (437–432 B.C.), the only opening in the perimeter wall.

Amongst the earliest additions to Perikles' Acropolis was Pheidias's 30-foot-high (9.2-meter) bronze figure of a seated Athena Promachos, directly inside the Propylaea. According to the ancient Greek traveler Pausanius, her gleaming helmet and spear could be seen by sailors off Cape Sounion, 30 miles (50 kilometers) away. The Byzantine emperor Justinian had the statue removed to Constantinople in the sixth century A.D.

The Propylaea, never completed, was designed by the architect Mnesikles to replace a gateway, commenced about fifty years before, that had remained unfinished when the Persians took Athens. It seems that Mnesikles designed his gateway in geometrical relationship with the Parthenon, or even (as Constantine Doxiadis claimed) with the salient corners of all buildings on the Acropolis. The Propylaea had Doric porches and a central hall flanked by two wings, one of which contained a famous picture gallery with works by Polygnotos and others. The paintings have long since gone, but Pausanius describes legendary and historical figures as the subject matter. Intriguingly, the interior of the gateway had Ionic columns supporting its roof, as did the Parthenon's smaller chamber, perhaps as subtle references to the supremacy of Athens over the Ionic

colonies, which, after all, had been overrun by the Persians.

South of the gateway, perched on the west wall of the Acropolis, is the perfect little Ionic Temple of Athena Nike (Athena, Giver of Victory), sometimes known as Nike Apteros (the Stay-at-Home Victory). Only 27 feet long and less than 19 wide (8.62 by 5.8 meters), it celebrated the Battle of Plataea (479 B.C.), when the Greeks soundly routed the Persians. The sanctity of the site is emphasized by the succession of buildings upon it: there had evidently been a small Helladic shrine, over which Peisistratos had built a more substantial edifice, destroyed by the Persians in 480. The Periklean building replaced it about fifty years later. Other statues and small shrines proliferated in the high city, some by Pheidias, others by Myron, and a huge bronze figure of the Trojan horse. There were also a number of altars including, between the Parthenon and the Erechtheion, the great altar of Athena.

The Romans made few changes to the Acropolis after they conquered Athens in 86 B.C. A *quadriga* (four-horse chariot) was placed before the Propylaea, and a small circular temple (albeit in the Greek style) was later built for the worship of the emperor Augustus. Around A.D. 50 Claudius tried to improve access by building a monumental stair to the Propylaea, and a monument to Agrippa was erected on the approach path. Other buildings were erected outside the walls on the sides of the Acropolis: the Stoici of Eumenes and the huge theater financed by Herodes Atticus around A.D. 161. Under Flavius Septimius Marcellinus, the Acropolis again became a fortress.

Subsequent invaders desecrated the architectural and artistic treasures. As noted, Justinian's armies looted them in the fifth century (only to be looted in turn when the Fourth Crusade sacked Constantinople in 1204). The Ottoman Turks occupied Athens in 1456. In 1645 lightning struck their gunpowder magazine in the Propylaea, causing extensive damage. In 1686 they quarried the Temple of Athena Nike to build defenses against Venetian invaders; a year later, Venetian bombards destroyed what was left of the Propylaea's west facade. The largely intact Parthenon, also used as a powder magazine, was hit and left in ruins. When the Turks were expelled after nearly four centuries, they left the Acropolis covered with gardens and hundreds of small huts, its monuments in ruin.

More depredations were inflicted by a supposed friend: in 1801 the British ambassador, Thomas Bruce, the seventh earl of Elgin, arrived in Athens with a Turkish decree permitting him to look for fragments of sculpture on the Acropolis. Among the fifty pieces he plundered was much of Pheidias's surviving Parthenon sculpture, later sold to the British Museum. He also took a caryatid from the Erechtheion, replacing it with a plaster cast. The moral debate over ownership has become an international issue and still rages.

Greece won its independence in 1836. Under Otho, first king of the Hellenes, every postclassical structure on the Acropolis was removed as he set about to restore "the glory that was Greece." The Athena Nike temple was rebuilt in 1835–1836. The Acropolis Museum was opened in 1878. In the twentieth century the American School of Classical Studies partly rebuilt the Erechtheion, which had been destroyed by a combination of wars and weathering. Some fallen columns of the Parthenon were restored, but the building suffered more damage in World War II and remains empty and roofless, a noble ruin. At the beginning of the Acropolis's fifth millennium, the worst threats to its survival are atmospheric pollution, the vibration set up by passing aircraft, and most paradoxical of all, tourism. All that one can hope for today is a mere glimpse of the original splendor of the Athenian Acropolis. In the Golden Age, its buildings were bathed in the clarity of Aegean sunshine and glowed with color. Reds, blues, and greens, not just painted but patterned, picked out the structural elements, all hung with swags of gold and silver and punctuated with glinting bronze rosettes. The sculpted stone friezes were rendered in realistic hues, and the brilliant yellow columns of the Parthenon added their fire to what must have been a breathtaking scene.

See also Erechtheion; Parthenon

Further reading

Economakis, Richard, ed. 1994. *Acropolis Restoration: The CCAM Interventions*. London/New York: Academy Editions.

Rhodes, Robin Francis. 1995. *Architecture and Meaning on the Athenian Acropolis*. Cambridge, UK: University Press.

Takahashi, Bin. 1969. *The Acropolis*. Tokyo and Palo Alto, CA: Kodansha International.

Afsluitdijk
The Netherlands

The 20-mile-long (32-kilometer) Afsluitdijk (literally, "closing-off dike"), constructed from 1927 to 1932 between Wieringen (now Den Oever) and the west coast of Friesland, enabled the resourceful Dutch to turn the saltwater Zuider Zee (South Sea) into the freshwater IJsselmeer and eventually to create an entire new province, Flevoland. Like their successful responses to similar challenges before and since, it was an audacious and farsighted feat of planning, hydraulic engineering, and reclamation.

Throughout their history, the Netherlanders have fought a battle against the water. Much of their tiny country is well below average sea level, in places up to 22 feet (7 meters). The threat of inundation comes not only from the sea but also from the great river systems whose deltas dominate the geography of Holland. Over centuries, literally thousands of miles of dikes and levees have been built to win agricultural land back from the water, and having gained it, to protect it. From the seventeenth century Amsterdam merchants invested their profits in building the North Holland polders— Beemstermeer, the Purmer, the Wormer, the Wijde Wormer, and the Schermer—reclaimed through the ingenious use of the ubiquitous windmill.

In 1250 the 79-mile-long (126-kilometer) Omringdijk was built along Friesland's west coast to protect the land from the sea, and as early as 1667 the hydraulic engineer Hendric Stevin bravely proposed to close off the North Sea and reclaim the land under the Zuider Zee. His scheme was then technologically impossible. The idea was revived in 1891 by the civil engineer and statesman Cornelis Lely. Based on research undertaken over five years, his plan was straightforward: a closing dike across the neck of the Zuider Zee would create a freshwater lake fed by the River IJssel and allow the reclamation of 555,000 acres (225,000 hectares) of polder land— in the event, 407,000 acres (165,000 hectares) were won. Despite Lely's assurances about the feasibility of the plan, his parliamentary colleagues were unenthusiastic. But attitudes changed after the region around the Zuider Zee was disastrously flooded in 1916; moreover, World War I (in which Holland remained neutral) convinced the Dutch government that internal transportation links needed to be improved. The Zuiderzee Act was passed in 1918.

The Zuiderzeeproject commenced in 1920 with the construction of the Amsteldiepdijk, also known as the Short Afsluitdijk, between Van Ewijcksluis, North Holland, and the westernmost point of the island of Wieringen. There were some initial foundation problems and a financial calamity for the contractor, but the dike was completed in 1926. There followed the construction of the small test polder Andijk (1927) and the Wieringermeer (1927–1930).

The key element in the daring plan was the construction of the Afsluitdijk across the Waddenzee, an arm of the North Sea. The project was undertaken by a consortium of Holland's largest dredging firms, known as N. V. Maatschappij tot Uitvoering van de Zuiderzeewerken. All the work, involving moving millions of tons of earth and rock, was carried out manually by armies of laborers working from each end of the structure. Built during the Great Depression, the Afsluitdijk was a welcome source of employment. It was completed on 28 May 1932. It was intended later to build a railroad over the broad dike, but as the volume of road traffic increased in Holland, priority was given to a four-lane motorway. The railroad was never built, although adequate space remains for it.

The closure of the Afsluitdijk enabled the eventual reclamation of three huge tracts of land formerly under the sea: the Noordoostpolder (1927–1942), East Flevoland (1950–1957), and South Flevoland (1959–1968). They were later combined to become a new province, Flevoland, with a total area of over 500 square miles (1,400 square kilometers). Its rich agricultural land supports two cities, Lelystad and Almere, although the latter is more properly a dormitory for Amsterdam. Flevoland is on average 16 feet (5 meters) below sea level. The great freshwater body south of the Afsluitdijk was renamed IJsselmeer.

Its balance, carefully controlled through the use of sluices and pumps, is determined by inflow and outflow rates, rainfall and evaporation, and storage level changes. With a surface of nearly 500 square miles (131,000 hectares), it is the largest inland lake in the Netherlands. A proposal to reclaim a fifth polder, the 230-square-mile (60,300-hectare) Markerwaard, behind a 66-mile-long (106-kilometer) dike between Enkhuizen and Lelystad was not pursued, mainly because of ecological concerns.

In February 1998 the Dutch Ministry of Transport, Waterways, and Communication published the *Waterkader* report, setting out national water-management policies until 2006. Aiming to keep the Netherlands safe from flooding, it presents a case for reserving temporary water-storage areas—"controlled flooding"—against times of high river discharge or rainfall. The government, recognizing that raising the dikes and increasing pumping capacity cannot continue forever, has adopted the motto "Give water more space." The document *Long-Range Plan Infrastructure and Transport* of October 1998 promised to invest 26 billion guilders (approximately U.S.$13 billion) in the nation's infrastructure before 2006. Part of the money is earmarked for waterways, including links between Amsterdam and Friesland across the IJsselmeer.

See also Deltaworks; Storm Surge Barrier

Further reading
Groenman, Sjoerd, et al. 1958. *Land Out of the Sea: The Building of the Dikes*. Meppel, the Netherlands: Roelofs van Goor.
Robert, Paul, and Rolf Bos. 1982. *50 Jaar Afsluitdijk: Herinneringen van Dijkers, Denkers en Drammers*. Bussum, the Netherlands: De Boer Maritiem.
Terpstra, Pieter. 1982. *50 Jaar IJsselmeer: Afsluiting van de Zuiderzee en de Gevolgen Daarvan*. Bussum, the Netherlands: De Boer Maritiem.

Airplane hangars
Orvieto, Italy

The Italian engineer and architect Pier Luigi Nervi (1891–1979) was among the most innovative builders of the twentieth century and a pioneer in the application of reinforced concrete. In 1932 he produced some unrealized designs for circular aircraft hangars in steel and reinforced concrete that heralded the remarkable hangars he built for the Italian Air Force at Orvieto. None have survived but they are well documented: more than enough to demonstrate that they were a tour de force, both as engineering and architecture.

Nervi had graduated from the University of Bologna in 1913. Following World War I service in the Italian Engineers Corps he established an engineering practice in Florence and Bologna before moving to Rome, where he formed a partnership with one Nebbiosi. Nervi's first major work, the 30,000-seat Giovanni Berta Stadium at Florence (1930–1932), was internationally acclaimed for its graceful, daring cantilevered concrete roof and stairs. The revolutionary hangars followed soon after.

There were three types, all with parabolic arches and elegant vaulted roofs that paradoxically conveyed a sense of both strength and lightness. The first type, of which two were built at Orvieto in 1935, had a reinforced concrete roof made up of a lattice of diagonal bow beams, 6 inches (15 centimeters) thick and 3.7 feet (1.1 meters) deep, intersecting at about 17-foot (5-meter) centers. They supported a deck of reinforced, hollow terra-cotta blocks covered with corrugated asbestos-cement. The single-span roof measured 133 by 333 feet (40 by 100 meters), and its weight was carried to the ground through concrete equivalents of medieval flying buttresses. The 30-foot-high (9-meter) doors that accounted for half of one of the long sides of the hangar were carried on a continuous reinforced concrete frame.

In the other types Nervi's fondness for structural economy led to the prefabrication of parts, saving time and money. Type two was his first experiment with parallel bow trusses assembled from open-web load-bearing elements, spanning the 150-foot (45-meter) width of the hangar. A reinforced-concrete roof covering provided stiffening. The third type combined the diagonal configuration of the first and the prefabrication techniques of the second. He built examples of it six times between 1939 and 1941 for air bases at Orvieto, Orbetello, and Torre del Lago. The massive roofs, covered with corrugated asbestos cement on a prefabricated concrete deck, were supported on only six sloping columns—at each corner and the midpoints of the long sides—that carried the weight and thrust beyond the perimeter of the hangars. All the components were cast on-site in simple wooden forms.

The Germans bombed these amazing structures as they retreated from Italy toward the end of World War II. Nervi was delighted to learn that, even in the face of such a tragedy, the prefabricated joints had held together despite the destruction of his hangars. He later included them amongst his most "interesting" works, observing that their innovative forms would have been impossible to achieve by the conventional concrete technology of the day. In the early 1940s Nervi extended his experiments to *ferro-cimento*—a very thin membrane of dense concrete reinforced with a steel grid—which he used to build a number of boats.

He next combined that material with the prefabrication techniques he had developed for the hangars. For Salone B at the Turin Exhibition of 1949–1950, he designed a 309-by-240-foot (93-by-72-meter) vaulted rectangular hall with a 132-foot-diameter (40-meter) semicircular apse at one end. The main hall roof and the hemidome over the apse consisted of corrugated, precast *ferro-cimento* units less than 2 inches (5 centimeters) thick, supported on in situ buttresses, creating one of the most wonderful interior spaces of the twentieth century.

Nervi's designs were too complex to be calculated by orthodox mathematical analysis, and he developed a design methodology that used polarized light to identify the stress patterns in transparent acrylic models. A few unbuilt projects were followed by three structures for the 1960 Rome Olympic Games. He built the Palazzo dello Sport (1959, with Marcello Piacentini), the Flaminio Stadium (1959, with Antonio Nervi), and the Palazzetto dello Sport (1957, with Annibale Vitellozzi). The last is a gem of a building whose rational structure is so transparently expressed that the observer can almost *see* the loads being shepherded to the ground in a way redolent of late English Gothic fan vaulting.

See also Reinforced concrete; Shell concrete

Further reading

Huxtable, Ada Louise. 1960. *Pier Luigi Nervi*. New York: Braziller.

Nervi, Pier Luigi. 1957. *The Works of Pier Luigi Nervi*. New York: Praeger.

Airship hangars
Orly, France

The French dominated the early history of human flight. In September 1783 the Montgolfier brothers launched a hot-air balloon carrying farm animals to show that it was safe to travel in the sky, and a few weeks later Pilatre de Rozier and the Marquis d'Arlandes took to the air for a 5.5-mile (9-kilometer) trip over Paris. In 1852 another Frenchman, the engineer Henri Giffard, built the first successful airship—a steam-powered, 143-foot-long (44-meter), cigar-shaped affair that flew at about 6 mph (10 kph). About thirty years later Charles Renard and Arthur Krebs constructed an electrically powered airship that was maneuverable even in light winds. By 1914 the French military had built a fleet of semirigid airships, but they proved ineffective as weapons in the Great War. On the other hand, nonrigid airships were widely used for aerial observation, coastal patrol, and submarine spotting. Their advent generated a different type of very large building: the airship hangar. The first zeppelin shed at Friedrichshafen, Germany (1908–1909), had been 603.5 feet long, 151 wide, and 66 high (184 by 46 by 20 meters). Like most others built Europe, it was a steel-lattice structure with a light cladding. Much more inventive and spectacular were the parabolic reinforced concrete hangars built in from 1922 to 1923 on a small military airfield among farmlands at Orly, near Paris. They were a major achievement of engineering and architecture.

The French engineer-architect Marie Eugène Léon Freyssinet (1879–1962) studied at the École Polytechnique and the École Nationale des Ponts et Chaussées in Paris. After serving in the army in World War I he became director of the Société des Enterprises Limousin and later established his own practice. A great innovator, he worked mainly with reinforced concrete, building several bridges. By 1928

he was to patent a new technique, prestressing, that eliminated tension cracking in reinforced concrete and solved many of the problems encountered with curved shapes. Simply, steel reinforcing cables were stretched and the concrete poured around them; when it set the cables were released and (because it was in compression) the structural member acquired an upward deflection. When it was loaded in situ the resulting downward deflection brought it back to the flat position while remaining in compression.

At Orly, Freyssinet was presented with a brief that called for two sheds that could each contain a sphere with a radius of 82 feet (25 meters), to be built at reasonable cost. He responded by designing prestressed reinforced concrete buildings consisting of a series of parallel tapering parabolic arches that formed vaults about 985 feet long, 300 wide, and 195 high (300 by 90 by 60 meters). The internal span was about 266 feet (80 meters), and each arch was assembled from 25-foot-wide (7.5-meter) stacked, profiled sections only 3.5 inches (9 centimeters) thick; those at the base of the arch were 18 feet (5.4 meters) deep and those at the crown 11 feet (3.4 meters). Placed side by side, they formed a very stiff corrugated enclosure. Starting at a height of 65 feet (20 meters), reinforced yellow glass windows were cast in the outer flanges of the arches.

Freyssinet specified an easily compactable concrete to ensure that the hangars would be waterproof. It was reinforced with steel bars and poured into reusable pine formwork that was itself stressed with tension rods to create prestressed concrete. The concrete was also designed to flow into every corner of the complicated molds, and it was fast-setting so that formwork could be quickly stripped and reused. The structure was temporarily supported on timber centering, and a network of cables held the formwork in tension until the concrete developed its full strength. In other structures lateral wind loading could be resisted by cross bracing, but because clear spans were imperative, Freyssinet provided the necessary stiffening by "folding" the concrete on the component arches. The selfweight of the massive structure was accommodated by increasing the cross-sectional area of the arches as they approached the ground, where the foundations consisted of deep horizontal concrete pads laid with an inward slope toward the center of the hangars. Tragically, in 1944, U.S. aircraft bombed these revolutionary and beautiful structures.

See also Shell concrete

Further reading

Fernández Ordóñez, José A. 1978. *Eugène Freyssinet*. Barcelona: 2c Ediciones.

Nicolaou, Stéphane. 1997. *Les premiers dirigeables français*. Boulogne, France: E.T.A.I. Le Bourget: Musée de l'air et de l'espace.

Akashi-Kaikyo Bridge
Kobe, Japan

The graceful Akashi-Kaikyo Bridge, linking Kobe City and Awajishima Island across the deep straits at the entrance to Osaka Bay, was opened to traffic on 5 April 1998. Exploiting state-of-the-art technology, it formed the longest part of the bridge route between Kobe and Naruto in the Tokushima Prefecture, completing the expressway that connects the islands of Honshu and Shikoku. With a main span of 1.25 miles (1.99 kilometers) and a total length of nearly 2.5 miles (3.91 kilometers), it was then the longest suspension bridge ever built.

With the growing demand for faster land travel, more convenient links over water obstacles become necessary. If long-span—say, over 1,100 yards (1,000 meters)—bridges are to be politically, economically, and structurally viable, design must be optimized. Because a bridge's selfweight increases in direct proportion to its span, the structure must be as light as possible while achieving minimum deformation and maximum stiffness under combined dead, wind, and traffic loads. A cable-supported suspension bridge is an ideal way to achieve that.

Alternative designs were developed for the Akashi-Kaikyo Bridge, considering a range of main span lengths. The most economical length was between 6,500 and 6,830 feet (1,950 and 2,050 meters); the final choice of 6,633 feet (1,990 meters) was constrained by geological and topographical factors. The length of the side spans was fixed at 3,200 feet (960 meters), enabling the cable anchorages to be located near the original shorelines. The clients insisted that, because of its immense span, the form of bridge had to assure the public that it would withstand all kinds of loads, including typhoons and earthquakes. Also, it had to express the essential beauty of the Seto-Inland Sea region and evoke a bright future for the Hyogo Prefecture. The Akashi-Kaikyo Bridge would be painted green-gray because it was redolent of the forests of Japan.

Construction began in May 1988. The reinforced concrete anchorages for the cables on the respective shores are of different sizes, because of different soil conditions. As an indication, the one at the Kobe end has a diameter of 283 feet (85 meters) and is 203 feet (61 meters) deep. It is the largest bridge foundation in the world.

Huge cylindrical steel chambers (caissons) form the foundation of the main towers. Fabricated off-site, they are 217 feet (65 meters) high—more than a 30-story building—and 267 feet (80 meters) in diameter; each weighs 15,000 tons (15,240 tonnes). To provide a level base, an area of seabed about as big as a baseball field was excavated under each of them. They were floated into position, and their exterior compartments were flooded to carefully sink them in 200 feet (60 meters) of water. This was achieved to within a 1-inch (2.54-centimeter) tolerance. Each was then filled with 350,000 cubic yards (270,000 cubic meters) of submarine concrete. The foundations of the bridge were seismically designed to withstand an earthquake of Richter magnitude 8.5, with an epicenter 95 miles (150 kilometers) away. On 17 January 1995 the Great Hanshin Earthquake (magnitude 7.2) devastated nearby Kobe; its epicenter was just 2.5 miles (4 kilometers) from the unfinished bridge. A careful postquake investigation showed that, although the quake had lengthened the bridge by about 3.25 feet (1 meter), neither the foundations nor the anchorages were damaged. As the builders boasted, it was "a testament to the project's advanced design and construction techniques."

The towers rise to 990 feet (297 meters) above the waters of the bay (for comparison, those on the Golden Gate Bridge are 750 feet [230 meters] high). They have steel shafts, each assembled in thirty tiers, generally made up of three prefabricated blocks that were hoisted into place and fixed with high-tensile bolts. The shafts are cruciform in cross section,

Akashi-Kaikyo Bridge; Kobe, Japan, 1988–1998.

designed to resist oscillation induced by the wind. The main cables, fixed in the massive anchorages and passing through the tops of towers, were spun from 290 strands of galvanized steel wire—a newly developed technology—each containing 127 filaments about 0.2 inch (5 millimeters) in diameter. Their high strength does away with the need for double cables, and because they achieve a sag:span ratio of 1:10, the height of the main towers could be reduced. To prevent corrosion of the cables in the salt atmosphere, dehumidified air flows through a hollow inside them, removing moisture. The towers and the suspended structure are all finished with high-performance anticorrosive coatings to suit the demanding marine environment.

From the main cables, polyethylene-encased, parallel-wire-strand suspension cables support the truss-stiffened girder that carries a six-lane highway with a traffic speed of 60 mph (100 kph). The preassembled truss members were hoisted to the deck level at the main towers, carried to their location by a travel crane, and connected; then the suspension cables were attached. This construction technique was chosen because it did not disrupt activity on the water, where 1,400 ships daily pass through the straits.

Further reading

Thomas, Mark. 2001. *The Akashi-Kaikyo Bridge: World's Longest Bridge*. New York: Rosen Publishing Group.

Alberobello trulli
Italy

The *Murgia dei Trulli*, with its communes of Martina Franca, Locorotondo, Cisternino, and Alberobello, is located in the Apulian interior at the upper part of the heel of Italy. Although *trulli* are scattered throughout the region, more than 1,500 of them are in the Monti and Aja Piccola quarters, on the western hill of Alberobello. This unique conical house form is significant in the history of architecture because it perpetuated well into the twentieth century a construction technique practiced throughout the northern Mediterranean since prehistoric times.

The name derives from *truddu*, Greek for "cupola." The clustered stone dwellings of Alberobello, small by modern Italian housing standards, are built by roofing almost square or rectangular bases (although some tend toward a circle) with approximately conical cupolas of roughly worked flat limestone slabs, stacked without mortar in corbeled courses. These gray roofs, no two of which are quite the same, are normally crowned with a whitewashed pinnacle in the form of a sphere standing on a truncated inverted cone. Some are painted with symbols: astrological signs or Christian ones, and even some of older pre-Christian significance. As is often the case with vernacular architecture, geometrical precision is not a priority: nothing is truly right-angled, nothing truly plumb. Bernard Rudofsky describes the roof as a retrocedent wall, because it also encloses habitable space that is traditionally used for storage. Typically, the inside of the roof is a parabolic dome, formed by packing the gaps between the larger structural stones. The walls of the ground floor are thick enough—they can be up to 10 feet (3.27 meters) in older houses—to include alcoves for a hearth or cupboards, or even a curtained-off recess for a bed. Doorways are low, and the interior, though whitewashed, is usually quite dingy because the windows are small, possibly for structural reasons. Curved walls make furnishing difficult. More recent trulli, the last of which were built in the 1950s, are interconnected with others to gain more living space.

The oldest documented Alberobello examples date from the fifteenth century, coinciding with the foundation of a permanent agricultural community centered in the town. However, the essential building technique and the consequent house form are much older. The type, clearly related to the prehistoric *nuraghi* of Sardinia and the rather more sophisticated Mycenaean *tholos*, has been archeologically linked to both the nomadic pastoral Early Bronze culture and permanent agrarian communities in the Apennine region. Remarkably, similar constructions can be found in the middle of Scotland and on the west coast of Sweden.

A plausible and somewhat romantic tradition dates the development of trulli as the house form of Alberobello to a single historical event. It is said that in the eighteenth century the local ruler Count Girolamo II of Acquaviva compelled the peasant farmers to build their houses with mortarless stone roofs. Because drywall structures were tax-exempt, and because they could be (relatively) easily dismantled before the regular visits of inspectors from Naples, he chose this method of tax avoidance. Although the people were freed from his regulation by a decree from Ferdinando IV of Bourbons in May 1797, the house form persisted, perhaps because of rural conservatism. Trulli are no longer built by the traditional technique and in the traditional style, but some of the master builders are still living, and the craft skills have not yet been lost. After the mid-1950s the "romantic" trulli were noticed by tourists and real-estate agents, and that has been to the detriment of many of them. Since the inclusion of the Alberobello precinct on UNESCO's World Heritage List in 1996, serious archeological study has been undertaken, and the old craft skills have been applied to an extensive restoration program.

Further reading

Allen, Edward. 1969. *Stone Shelters*. Cambridge, MA: MIT Press.

Esposito, Gabriella. 1983. *Architettura e storia dei trulli: Alberobello, un paese da conservare*. Rome: Casa del libro.

Rudofsky, Bernard. 1977. *The Prodigious Builders*. London: Seker and Warburg.

La Alhambra palace
Granada, Spain

Built between 1238 and 1391, the most outstanding reminder of Granada's glorious Moorish epoch is La Alhambra (the Red Castle), a complex of fortresses, palaces, and gardens for the Nasrid kings on a high plateau called the Cerro del Sol. Granada lies beneath it on the southeast, and beyond the city the fertile

Andalusian plain stretches toward the mighty Sierra Nevada. It has been justifiably claimed that in La Alhambra "all the refinement, wealth and delicacy of Islamic art and architecture reached its last climax in the West."

Following the Arab conquest of the Berbers in the seventh century A.D., intermarriage between the two peoples produced the ethnic group now known as the Moors. In 711 a Moorish army led by Tariq ibn-Ziyad swarmed across the Straits of Gibraltar, and within a little over two decades they had conquered much of Spain. They made Córdoba the center of al-Andalus (Andalusia), part of an Islamic empire extending from the borders of China and India to the Atlantic. Seville, Jaén, and Granada were soon established as seats of Islamic culture and commerce. The Visigoths were expelled from Granada in 711 by the Moors, who governed it from Córdoba until the fall of the caliphate in 1031, after which it was ruled for two centuries by the successive Berber dynasties of the Almoravides and the Almohades. When Córdoba was taken by Christian armies in 1236, Moorish Granada grew in importance, reaching its apogee under the Nasrid kings, beginning with Ibn al-Ahmar, called Mohammed I, in 1241. Granada was the last Islamic outpost in Spain until the Treaty of Santa Fé consigned it to Ferdinand and Isabella 250 years later.

In 1238 Mohammed I repaired an irrigation channel from the Darro River to the top of the Red Hill and reinforced the ninth-century fortress known as La Alcazaba with 90-foot-high (27-meter) towers and five fortified gates. The stronghold became the kernel of La Alhambra. Mohammed II (1273–1302) extended the fortifications, and La Alcazaba was again modified as a luxurious residence for Mohammed III. In 1318 the architect Aben Walid Ismail was commissioned to design El Generalife, the Nasrids' summer palace, among beautiful irrigated gardens on an adjoining hilltop. Although the majority of the buildings of La Alhambra cannot be as accurately dated as that, it is known that most were initiated by Yusuf I (1333–1354) and Mohammed V (1354–1391). After the surrender of Boabdil, successive Catholic kings refurbished the palace, carefully retaining the Moorish style, an approach that

comments upon its sublime beauty. In the sixteenth century the Holy Roman Emperor Charles V had some of the older buildings demolished to make way for his own, designed by the architect Pedro Machuca.

The confluence of cultures in Andalusia generated the unique Moorish architecture that continues to be influential in Spain and has made an impact elsewhere, especially upon garden design. Because La Alhambra is such an accumulation of sequential elements, many of them starkly contrasting (like massive towers and delicate arcades), the paradoxical fortress-palace is almost impossible to describe. Within the forbidding utilitarian curtain wall of the fortress there are the inviting and surprising delights of the palace, built around secluded courtyards: sumptuous halls and chambers, arcaded internal patios with pools and fountains, wooded plazas, and peaceful gardens with streams of tinkling or chattering water. All is laid out with symmetrical geometry and careful proportion, the various buildings placed in a composition of studied informality. And, as would never be suspected from looking upon the austere outer defenses, all within is profusely decorated with restrained taste in the finest materials and finishes: glazed tile skirtings; walls, friezes, and arcades replete with stucco plant motifs; and ceilings ornamented with bows and *mocarae* (designs of several prisms on a concave base), sometimes picked out with gold or lapis lazuli, and sometimes bearing verses from the Koran, inscribed with exquisite calligraphy.

In *The Alhambra* (1832) the American writer Washington Irving wistfully remarked: "A few broken monuments are all that remain to bear witness to [Moorish] power and dominion. Such is the Alhambra—a Moslem pile in the midst of a Christian land; an Oriental palace amidst the Gothic edifices of the West; an elegant memento of a brave, intelligent, and graceful people, who conquered, ruled, flourished, and passed away." La Alhambra and the gardens of El Generalife (whose buildings are all but gone) were added to UNESCO's World Heritage List in 1984.

Further reading

Grabar, Oleg. 1992. *The Alhambra*. Sebastopol, Calif.: Solipsist Press.

Irving, Washington. 1983 (1832). *The Alhambra*. Boston: Twayne.

Villa-Real, Ricardo. 1978. *The Alhambra and the Generalife*. Granada: Miguel Sánchez.

Alpine railroad tunnels
Switzerland

Switzerland's government-owned, 3,100-mile (5,000-kilometer) railroad network is world renowned for its efficiency, despite the difficulties imposed by the mountainous terrain. Two of the four major rail links that pass through the small, land-locked country to connect northern Europe and Italy cross the 13,000-foot-high (4,000-meter) Swiss Alps. That access was made possible only by the remarkable engineering feats embodied in the construction, between 1872 and 1922, of the St. Gotthard, Simplon, and Lötschberg Tunnels, drilled through the rock thousands of feet underground. However, the Swiss were not the first to conquer the mountains.

The earliest European alpine railroad tunnel, the Frejus Tunnel, was drilled through Mont Cenis to connect Bardonecchia in the Italian province of Savoy (north of the Alps), through Switzerland, with Modena on the Italian peninsula. King Carlo Alberta of Sardinia championed the scheme in 1845, and his successor Victor Emmanuel II took it up in 1849. Drilling did not begin on the 8-mile (13-kilometer) double-track tunnel—over twice the length of any before attempted—until late 1857, supervised by the engineer Germain Sommeiller (1815–1871), assisted by Sebastiano Grandis and Severino Grattoni. Sommeiller patented the first industrial pneumatic drill, which greatly expedited the work. Finished in 1870, the tunnel was opened in 1871, just two months after his death.

The following year, work began on a 100-mile (160-kilometer) railroad, the Gotthardbahn, which crossed the Lepontine Alps in south-central Switzerland to link Zurich, at the heart of the country's northern commercial centers, with Chiasso at the Italian frontier. Before then the way across the Alps, used for 800 years, was over the 6,935-foot (2,114-meter) St. Gotthard Pass. A road was built in the 1820s. Alfred Escher, the founder of Credit Suisse, was the initiator of the Gotthardbahn, and as its president, with Emil Welti he negotiated German and Italian cooperation for the project in 1869–1871. Two feeder lines meet at Arth-Goldau; from there the mountain section runs through Brunner, Fluelen, and Altdorf to Erstfeld. There it commences the steep climb to Goeshenen at the northern end of the St. Gotthard Tunnel. Designed by the Geneva engineer Louis Favre, the double-track tunnel is 9.25 miles (15 kilometers) long, passing through the mountain 5,500 feet (1,700 meters) below the surface. The southern ramp is even steeper, and at Giornico more loops take the line to Chiasso. The tunnel was drilled from both ends, and the bores joined in 1880. The railroad was opened in 1882, when the difficult approach lines were completed. Favre had accepted punishingly tight schedules for the contract. He drove his force of 4,000 immigrant laborers to cut almost 18 feet (5.4 meters) a day—over twice that achieved in the Frejus Tunnel—in horrifying working conditions: water inrushes, rock falls, dust, and (because of the great depth) temperatures up to 102°F (39°C). About 1,000 men suffered serious injury; 310 were killed.

Twenty years later, the safety record on the Simplon Tunnel, although far from perfect, was much better. From the thirteenth century, the 6,590-foot (2,009-meter) Simplon Pass near the Swiss-Italian border was a key to trade between northern and southern Europe; and in the beginning of the nineteenth century, probably for military reasons, Napoléon I ordered a road built over it. Begun around 1898, the Simplon Railroad connects the Swiss town of Brig with Iselle, Italy. Its 12.3-mile (19.8-kilometer) tunnel—in reality two tunnels—under Monte Leone was conceived as a twin-tube single-track system by the German engineer Alfred Brandt; separate galleries 55 feet (17 meters) apart were linked with cross-hatches. Until the completion of Japan's Seikan Tunnel in 1988, the Simplon Tunnel was the world's longest railroad tunnel. Because of its depth—up to 7,000 feet (2,140 meters) below ground—temperatures exceeding 120°F (49°C) were faced during construction. The first gallery, Simplon I, was completed by January 1905 and traffic commenced the following year. Various problems, including the intervention of World War I, delayed Simplon II until 1921; it was opened in 1922.

The Lötschberg Tunnel, opened in 1913, is a 9-mile (14.6-kilometer) double-track railroad tunnel between Kandersteg and Goppenstein in south-central Switzerland's Bernese Alps. It is part of the 46-mile (74-kilometer) standard-gauge Bern-Lötschberg-Simplon Railway connecting Spietz and Brig. The branch lines from Thun and Interlaken meet at Spietz, where the main trunk leads to Frutigen and begins a steep mountain section, much like the Gotthardbahn's, to the Lötschberg Tunnel at Kandersteg. South of the tunnel the line descends from Goppenstein to the Rhone valley, where it reaches Brig and the line to the Simplon Tunnel and Domodossola, Italy. Together, Lötschberg and Simplon completed a through-route from Germany and France to Italy.

In 1987, the Swiss government initiated further investment in its railroad network. The major part of the plan, estimated to cost EUR10 billion (U.S.$8.8 billion), is the largest construction project in Europe. Known as NEAT (for Neue Eisenbahn-Alpen Transversale, i.e., New Alpine Railroad Crossing), it involves the creation of two new 30-foot-diameter (9-meter) twin-tube alpine tunnels, suitable for high-speed trains, through the St. Gotthard and Lötschberg Mountains, respectively. Built at lower altitudes than their predecessors, they will double rail-transit capacity and significantly reduce journey times between northern and southern Europe. The first axis is expected to be in service by 2006.

Further reading

Allen, Cecil John. 1965. *Switzerland's Amazing Railways*. London: Nelson.

Marti, Franz, and Walter Trüb. 1971. *The Gotthard Railway*. London: Allan.

Amsterdam Central Station
The Netherlands

Amsterdam Central Station is in fact geographically central in the city. Although it conformed to the general pattern of many metropolitan railroad stations before and after, it was an architectural and engi-neering achievement in that it was built on three artificial islands in the River IJ, supported by no fewer than 26,000 timber piles driven into the soft river bottom. That was a feat perhaps remarkable to the rest of the world but quite commonplace to the Dutch, who for centuries had coped with too much water and too little land.

Economic activity in Amsterdam revived with the railroads in the second half of the nineteenth century. New shipyards and docks were built. Extravagant public buildings such as P. J. H. Cuypers's National Museum (1876–1915) and H. P. Berlage's famous Stock Exchange (1884–1903) celebrated both the financial boom and awakening nationalism. In 1876 Cuypers and A. L. van Gendt were commissioned to design the Amsterdam Central Station. It was the first time that such work had been trusted to an architect rather than to engineers, a decision taken because the building would hold an important place in the nation's image. Indeed, the brief jingoistically demanded that it should be in the Oud-Hollandsche (Old Dutch) style.

That qualification presented little difficulty to Cuypers, who had developed a personal historical-revivalist manner based on late Gothic and early Renaissance forms and ideas. His abundantly decorated National Museum was already under construction. Eclectically drawing on a wide variety of styles, it did not readily expose his rationalist architectural philosophy, gleaned from E. E. Viollet-le-Duc's theories. Cuypers wanted to restore the crafts to a place of honor and insisted on the honest application of traditional materials. He was responsible for the appearance of the station; van Gendt, thoroughly experienced as mechanical engineer for the railroad, would take care of constructional aspects.

Work commenced in 1882. The station was built on the artificial islands in the Open Havenfront of Amsterdam's original harbor, which had been cut off from the River IJ by the railroad. Special engineering skill was needed to create a solid foundation for the massive building and the rolling loads imposed by trains. As noted, 26,000 timber piles support the structure. The four-story station building, of red brick with stone dressings, is unmistakably Dutch. It is 1,020 feet (312 meters) long and 100 feet (30.6

Amsterdam Central Station, The Netherlands; P. J. H. Cuypers, architect; L. J. Eijmer, engineer, 1884–1889. Exterior view of platform sheds; the roof on the left was added in 1922.

meters) deep. On the axis of Damrak—the main street leading to the dam in the downtown area—a central pavilion flanked with clock towers houses the main entrance to the concourse. Its facade is resplendent with ornament: the clock faces; the arms of those European cities to which the railroad gave access; and an assortment of allegorical relief sculptures wherever they could fit, aptly representing such themes as "Steam," "Cooperation," and "Progress." Convinced that the building process needed the collaboration of all the arts, Cuypers sought the artistic advice and skill of others, especially J. A. Alberdingk Thijm and V. de Steurs, who had worked on the National Museum.

Late in 1884 the architect produced two sketches for the platform roof; they have been characterized as "unassuming." But that part of the design was not in his contract, and the structure—anything *but* unassuming—was designed by the railroad's own civil engineer, L. J. Eijmer. Carried on a frame of fifty

semicircular, open-web trusses of wrought iron, spanning 150 feet (49 meters), the original station shed covered about 3.75 acres (1.5 hectares). During construction, problems arose over anchoring the arches, no doubt due to the foundation soil, but rejecting a suggestion to build several smaller, lighter roofs, it was resolved to proceed with the monumental design "on a scale that could compare with that of the great examples abroad." Cuypers designed the decorative elements of the rafters and the glazed gable end. The roof was completed in October 1889. In 1922, to cover new platforms, another similar arch was added beside the IJ.

The final phase of construction was the King's Pavilion at the station's eastern end in 1889—in the event, an ironic title, since the kingdom of the Netherlands was to be ruled only by queens for more than a century. Coaches could be driven inside, where a stair led to the royal waiting room, all in Cuypers's individualistic neo-Gothic style and enriched with a color scheme by the Austrian G. Sturm and executed by G. H. Heinen. The room was restored in 1995.

The building of Amsterdam Central Station, "a palace for the traveler," clearly demonstrates two issues that confronted architects and engineers late in the nineteenth century. First, after sixty years of building railway stations, they were no closer to finding an esthetic that suited the building type, fitted the new materials and technology, and removed the unnecessary tension between utility and beauty. Second, and related to the first, the nature of architectural practice was changing as increased knowledge called for specialization and the eventual replacement of the omniscient, not to say omnipotent, architect by a design *team*: architect, yes, but also mechanical engineer, structural engineer, interior designer, and consultant artist. That idea would not be enunciated until Walter Gropius wrote the Bauhaus manifesto in 1919.

Further reading

Hoogewoud, Guido, Janjaap Kugt, and Aart Oxenaar. 1985. *P. J. H. Cuypers en Amsterdam: Gebouwen en Ontwerpen, 1860–1898*. The Hague: Staatsuitgeverij.

Oxenaar, Aart. 1989. *Centraal Station Amsterdam: Het Paleis Voor de Reiziger*. The Hague: SDU.

Angkor Wat
Cambodia

Angkor Wat, a temple complex dedicated to the Hindu deity Vishnu, was built in the twelfth century A.D. in the ancient city of Angkor, 192 miles (310 kilometers) northwest of Phnom Penh. It is probably the largest (and, as many have claimed, the most beautiful) religious monument ever constructed. Certainly it is the most famous of all Khmer temples.

Angkor served as the capital of the Khmer Empire of Cambodia from A.D. 802 until 1295. Evidence uncovered since 1996 has led some scholars to assert that the site may have been occupied some 300 years earlier than first thought, obviously affecting accepted chronologies. Whatever the case, its powerful kings held sway from what is now southern Vietnam to Yunnan, China, and westward from Vietnam to the Bay of Bengal. The city site was probably chosen for strategic reasons and for the agricultural potential of the region. The Khmer civilization was at its height between 879 and 1191, and as a result of several ambitious construction projects, Angkor eventually grew into a huge administrative and social center stretching north to south for 8 miles (13 kilometers) and east to west for 15 miles (24 kilometers). The population possibly reached 1 million.

Apart from the hundreds of buildings—temples, schools, hospitals, and houses—there was an extensive system of reservoirs and waterways. The public and domestic buildings, all of timber, have long since decayed. But because they were the only structures in which masonry was permitted, over 100 temple sites survive. Earlier examples were mostly of brick, but later, the porous, iron-bearing material known as laterite was used, and still later sandstone, quarried about 25 miles (40 kilometers) away.

The city of Angkor was the cult center of Devaraja, the "god-king," and an important pilgrimage destination. The Khmer kings themselves, from Jayavarman II (802–850) onward, had come to be worshiped as gods, and the temples they built were regarded as not only earthly but also as symbols of Mount Meru, the cosmological home of the Hindu deities. The official state religion was worship of the Siva Lingam, which signified the king's divine authority. Jayavarman II had identified the kingship with Siva, and acting upon that precedent, King Suryavarman II (1113–ca. 1150) presented himself as an incarnation of Vishnu. He built Angkor Wat as a temple and administrative center for his empire and as his own sepulcher (which is why it faces west); to celebrate his status, he dedicated it to Vishnu.

Financed by the spoils of war and taking over thirty years to finish, the sandstone-and-laterite Angkor Wat occupies a 2,800-by-3,800-foot (850-by-1,000-meter) rectangular site. Its layout provides an architectural allegory of the Hindu cosmology. The temple is surrounded by a 590-foot-wide (180-meter) moat, over 3 miles (5 kilometers) long, which represents the primordial ocean. A causeway decorated with carvings of the divine serpents leads to a 617-foot-long (188-meter) bridge that gives access to the most important of four gates. The temple is reached by passing through three galleries separated by paved walkways. It is an approximately pyramidal series

Angkor Wat, Cambodia; architect(s) unknown, early twelfth century A.D. Aerial view of main shrine. Restored 1986–1999.

of terraces and small buildings arranged in three ascending stories—they stand for the mountains that encompass the world—and surmounted at the center by a temple "mountain" of five lotus-shaped towers, symbolizing the five peaks of Mount Meru. Four of the original nine towers have succumbed to time and weather. The temple walls are replete with wonderfully crafted bas-reliefs, many of which were once painted and gilded, including about 1,700 heavenly nymphs and others that depict scenes of Khmer daily life, episodes from the epics *Ramayana* and *Mahabharata*, the exploits of Vishnu and Siva, and (of course) the heroic deeds of King Suryavarman II.

In 1177 Angkor fell to the Cham army from northern Cambodia, who held it until it was retaken early in the reign of the Khmer King Jayavarman VII (1181–ca. 1215). When he built Angkor Thom nearby he dedicated his new capital to Buddhism, and Angkor Wat became a Buddhist shrine. Many of its carvings and statues of Hindu deities were replaced by Buddhist art. The Thais sacked Angkor in 1431. The following year the Khmers abandoned the city, and it was left to the encroaching jungle for a few centuries. However, Theravada Buddhist monks kept Angkor Wat as intact as possible until the late nineteenth century, making it one of the most important pilgrimage destinations in Southeast Asia.

The French explorer Henri Mouhot "discovered" Angkor in 1860. After French imperialism imposed itself in Indochina in 1863, the site attracted the scholarly interest of westerners. In 1907, when Cambodia had been made a French protectorate and Thailand returned Angkor to its control, L'École Française d'Extreme Orient established the Angkor Conservation Board. It seems that for forty years the European colonizers were more interested in reconstructing Angkor Wat than in undertaking scholarly restoration. The prodigal use of reinforced concrete made many of the buildings unrecognizable. The vandalism was mercifully halted when Khmer Rouge guerrillas occupied the site, followed by the Vietnamese army. When an uneasy peace was restored in 1986, the Archaeological Survey of India took up the project, replacing much of the French work with more modern and less intrusive techniques. At the invitation of the Cambodian government, the Japanese Government Team for Safeguarding Angkor began a four-year preservation and restoration project in November 1994, initially focused on the Bayon temple in Angkor Thom but extending to the outer buildings of Angkor Wat. Because of delays caused by the July 1997 conflicts in Cambodia, the program was extended into 1999.

Further reading

Fujioka, Michio Tsuenari Kazumori, and Chikao Mori. 1972. *Angkor Wat*. Tokyo: Kodansha International.

Mannikka, Eleanor. 1996. *Angkor Wat: Time, Space, and Kingship*. Honolulu: University of Hawaii Press.

Narasimhaiah, B. 1994. *Angkor Wat, India's Contribution in Conservation, 1986–1993*. New Delhi: Archaeological Survey of India.

Appian Way
Italy

The Appian Way *(Via Appia)*, the oldest and perhaps most famous Roman road, was built by the Censor Appius Claudius Caecus in 312 B.C. Enlarging a track between Rome and the Alban Hills and forming the main route to Greece and the eastern colonies, this so-called queen of roads *(regina viarumeters)* ran south from the Porta Capena in Rome's Servian Wall to Capua. It passed through the Appii Forum to the coastal town of Anxur (now Terracina), 60 miles (100 kilometers) from Rome, to which point it was almost straight, despite crossing the steep Alban Hills and the swampy Pontine Marshes. In 190 B.C. it was extended to Brundisium (modern Brindisi) on the Adriatic coast—more than 350 miles (560 kilometers) from the capital and eighteen days' march for a legion. Parts of it—now called the *Via Appia Antica*—remain in use after more than 2,000 years.

The medieval proverb "A thousand roads lead man forever toward Rome" was popularized in William Black's *Strange Adventures of a Phaeton* (1872) as "All roads lead to Rome." That was probably once true: the Romans built about 50,000 miles (80,000 kilometers) of paved roads throughout their empire, mainly to expedite movements of the legions. Inevitably, the system was put to wider use and eventually served all kinds of travelers: dignitaries, politicians, commercial traffic of all kinds, and even an official postal service.

Roman engineers efficiently developed road-building techniques to create enduring structures. Usually (but not always), roads were laid upon a carefully constructed embankment *(agger)* to provide a foundation—rubble laid in such a way as to provide proper drainage—for the base. The dimensions of the agger varied according to the importance of the road. Sometimes it may have been just a small ridge, but on major routes it could be up to 5 feet high and 50 wide (1.5 by 15 meters). For very minor roads no embankment was built, but two rows of curbstones defined the carriageway; the excavation between them was layered with stones and graded material, the topmost sometimes forming the pavement. Overall, the depth of a Roman road from the surface to the bottom of the base was up to 5 feet. It seems that road width varied according to function, importance, and topography. The widest *(decumanus maximus)* was 40 feet (12.2 meters) wide, while a minor road might be only 8 feet (2.4 meters). Rural thoroughfares were generally 20 feet (6 meters), but all roads became narrower over difficult terrain: some mountain passes, at less than 10 feet, were too narrow (and often too steep) for carts.

Although stone was sometimes transported from a few miles away, local material was normally used. Of course, that practice gave rise to differences in construction along the length of a road, as is evident in the *Via Appia*. At one place a 3-foot-thick (1-meter) bottom layer of earth and gravel from the neighboring mountains was consolidated between the curbs and covered by a thinner layer of gravel and crushed limestone, also contained by parallel rows of closely placed large stones. Elsewhere, a base layer of sand was covered with another of crushed limestone into which slabs of lava up to 15 inches (50 centimeters) thick were fixed. Stone surfaces were mandatory for urban streets after 174 B.C., but other roads were not always stone-paved, especially in difficult terrain. Like the substructure, surfaces varied according to what materials were locally available: gravel, flint, small broken stones, iron slag, rough concrete, or sometimes fitted flat stones were used. The pavement thickness varied from a couple of inches on some roads to 2 feet (0.6 meter) at the crown of others. Surfaces sloped down—as steeply as 1 in 15—from the center, to allow rainwater run-off into flanking ditches.

Roman roads were strong enough to support half-ton metal-wheeled wagons, and many were wide enough to accommodate two chariots abreast. Some roads were provided with intentional ruts, intended to guide wagons on difficult stretches. Under normal traffic a paved Roman road lasted up to 100 years. Beginning with the Appian Way, the ancient Roman engineers flung an all-weather communication network across Italy and eventually their empire. The poet Publius Papinius Statius wrote late in the first century A.D.:

> How is it that a journey that once took till sunset
>
> Now is completed in scarcely two hours?
>
> Not through the heavens, you fliers, more swiftly
>
> Wing you, nor cleave you the waters, you vessels.

Further reading

Chevallier, Raymond. 1976. *Roman Roads*. London: Batsford.

Hagen, Victor von, and Adolfo Tomeucci. 1967. *The Roads That Led to Rome*. Cleveland: World Publishing.

White, K. D. 1984. *Greek and Roman Technology*. London: Thames and Hudson.

Archigram

The Archigram group was established in 1961 by a few young British architects "united by common interests and antipathies." Its founders were Peter Cook, Michael Webb, and David Greene, who were soon joined by Dennis Crompton, Ron Herron, and Warren Chalk. Archigram's international impact—its architectural feat, so to speak—was significant. Other architects would give form to its notions. The Centre Pompidou, Paris, by Renzo Piano and Richard Rogers, and Arata Isozaki's buildings at the 1970 Osaka World's Fair are redolent of the fantastic schemes drawn, but never built, by Archigram. The

Austrian architect Hans Hollein, too, admits his debt to them after 1964. It is in the realm of ideas about living in an advanced industrial civilization that they offered most.

All the founders had been students at the Architecture Association school in London, where they had learned, in the face of a then-reactionary architectural profession, to apply democratic principles to the art. The members who came later assimilated those ideas and blended them with other influences, notably the futuristic urban visions of Friedrich Kiesler and Bruno Taut and the technological notions of Richard Buckminster Fuller, whom they heroized. They also formed a symbiotic intellectual association with the exactly contemporary Japanese Metabolist group, in which Isozaki was preeminent. The Japanese applauded their efforts to "dismantle the apparatus of Modern Architecture."

Like the Dutch De Stijl group around 1920, Archigram's cooperation was mainly through a polemical journal; and like the Hollanders, it drew its name from the title of the journal. *Archigram* (derived from "architecture" and "telegram" or "aerogram") was published (almost) annually between 1961 and 1974. *Archigram*, more like a polemical broadsheet than a journal, directed an attack on the smugness of modernist architectural conservatism, reinforced by what can best be called Britishness. The powerful publication ran to ten annual issues, preaching an urgent message about architecture that has been described as "esthetic technocratic idealism." Possibly the most significant architectural publication of the decade, its "pop" format, including beautifully drawn comic strips, declared the group's "optimism and possibilities of technology and the counterculture of the pop generation."

The 1964 issue, after a controversial *"Living City"* exhibition at London's Institute of Contemporary Arts, attracted the critic Reyner Banham, who became the group's champion. There followed a succession of (perhaps) outlandish architectural proposals. Archigram's direction was urban, technological, autocratic—and some have said inhumane. The members believed that technology was the hope of the world, so traditional means of building houses and cities must be superseded. Their favorite words were *change, adaptability, flexibility, metamorphosis, impermanence,* and *ephemerality.* Accordingly, they designed a living environment that incorporated all kinds of gadgetry. They proposed an inflatable bodysuit containing food, radio, and television, and the "suitaloon," a house carried on the back. These eccentric ideas extended from the individual to the communal: Chalk's *Capsule Homes* (1964) were projected alongside Cook's *Plug-in City* (1964–1966), in which self-contained living units could be temporarily fitted into towering structural frames, and Herron's nomadic *Walking City,* in which skyscrapers could move on giant telescoping legs. The group published its *Instant City* in 1968.

It has been suggested that in the 1960s Archigram was to modern architecture what the Beatles were to modern music. But in the early 1970s they more or less dispersed, Greene and Herron (for a while) becoming teachers in the United States. Crompton, Cook, and Herron formed Archigram Architects (1970–1974). Herron and Cook then established independent practices in various partnerships. Crompton maintained links with the Architectural Association, and Greene turned to writing poetry and practicing architecture. Webb moved permanently to the United States and after 1975 taught at Cornell and Columbia Universities in New York. Chalk continued writing and teaching in the United States and England, mostly at the Architectural Association, until he died in 1987.

See also Industrialized building; Pompidou Center (Beaubourg)

Further reading

Archigram. (1961–1974). 10 issues (numbers 1–9½). London: Archigram.
Cook, Peter, ed. 1972. *Archigram.* London: Studio Vista.
Crompton, Dennis ed. 1994. *A Guide to Archigram, 1961–1974.* London: Architectural Press.

Artemiseion
Ephesus, Turkey

The Artemiseion, a huge Ionic temple dedicated to the goddess Artemis, stood in the city of Ephesus on the Aegean coast of what was then Asia, near the modern town of Selcuk, about 30 miles (50 kilome-

Artemiseion in Ephesus, Turkey; Demetrios and Paeonios, architects, 356 B.C.–ca. 240 B.C. Artist's reconstruction.

ters) south of Izmir, Turkey. The splendid building was acclaimed as one of the seven wonders of the world, as attested by Antipater of Sidon: "When I saw the sacred house of Artemis … the [other wonders] were placed in the shade, for the Sun himself has never looked upon its equal outside Olympus." Among several attempts to identify the architectural and sculptural wonders of the ancient world, the seven best known are those listed by Antipater in the second century B.C. and confirmed soon after by one Philo of Byzantium.

Artemis was the Greek moon goddess, daughter of Zeus and Leto. Whatever form she was given, it was always linked with wild nature. On the Greek mainland she was usually portrayed as a beautiful young virgin, a goddess in human form. In Ephesus and the other Ionic colonies of Asia, where ancient ideas of the Earth Mother and associated fertility cults persisted, she was linked with Cybele, the mother goddess of Anatolia, and her appearance was dramatically different, even grotesque. The original cult statue has long since disappeared, but copies survive. That is hardly surprising, because the trade in them flourished in Ephesus at least until the first century A.D. They portray a standing figure, her arms outstretched like those of the earlier décolleté figurines common in Minoan Crete. Artemis was fully dressed except for her many breasts, symbolizing her fertility (although some recent scholars have suggested that the bulbous forms are bulls' scrotums). The lower part of her body was covered with a tight-fitting skirt, decorated with plant motifs and carved in relief with griffins and sphinxes. She wore a head scarf decorated in the same way and held in place with a four-tiered cylindrical crown. Ancient sources say that the original statue was made of black stone, enriched with gold, silver, and ebony.

The Artemis shrines at Ephesus had a checkered history. The earliest was established on marshy land near the river, probably around 800 B.C.; it was later rebuilt and twice enlarged. The sanctuary housed a sacred stone—perhaps a meteorite—believed to have fallen from Zeus. By 600 B.C. Ephesus had become a major port, and in the first half of the fifth century, its citizens commissioned the Cretan architect Chersiphron and his son Metagenes to build a larger temple in stone to replace the timber structure. In 550 B.C. it also was destroyed when the Lydian king, Croesus, invaded the region. Croesus, whose name has passed into legend for his fabulous wealth, contributed generously to a new temple, the immediate predecessor to the "wonder of the world." It was four times the area of Chersiphron's temple, and over 100 columns supported its roof. In 356 B.C. one Herostratos, a young man "who wanted his name to go down in history," started a fire that burned the temple to the ground.

The Ephesian architects Demetrios and Paeonios (and possibly Deinocrates) were commissioned to design a more magnificent temple, built to the same plan and on the same site. The first main difference was that the new building stood on a 9-foot-high (2.7-meter) stepped rectangular platform measuring 260 by 430 feet (80 by 130 meters), rather than a lower *crepidoma* like the earlier stone building. Another departure from the normally austere and reserved Greek architectural tradition was the opulence of the temple, which went beyond even its great size. Its porch *(pronaos)* was very deep: eight bays across and four deep. The Ionic columns towered to 48 feet (17.7 meters); each had, in place of the usual Ionic base, a 14-foot-high (3.5-meter) lower section, carved with narrative decorations in deep relief. The other difference was in the quality of the detail. The wonder of the world was decorated with bronze statues by the most famous contemporary artists, including Scopas of Paros. Their detail can only be guessed at, as can the overall appearance of the great temple. Attempts have been made at graphical reconstruction, but they vary widely in their interpretation of the sparse archeological evidence. Antipater described the Artemiseion as "towering to the clouds," and Pliny the Elder called it a "wonderful monument of Grecian magnificence, and one that merits our genuine admiration." Pliny also asserted that it took 120 years to build, but it may have taken only half that time. It was unfinished in 334 B.C. when Alexander the Great arrived in Ephesus.

By the time the Artemiseion was vandalized by raiding Goths in A.D. 262—it was partly rebuilt—both the city of Ephesus and Artemis-worship, once flaunted as universal, were in decline. When the Roman emperor Constantine redeveloped elements of the city in the fourth century A.D., he declined to restore the temple. By then, with most Ephesians converted to Christianity, it had lost its reason for being. In A.D. 401 it was completely torn down on the instructions of John Chrysostom. The harbor of Ephesus silted up, and the sea retreated, leaving barely habitable swamplands. As has so often happened, the ruined temple was reduced to being a quarry, and its stone sculptures were broken up to make lime for plaster. The old city of Ephesus, once the administrative center of the Roman province of Asia, was eventually deserted.

The temple site was not excavated until the nineteenth century. In 1863 the English architect John Turtle Wood set out to find the legendary building, under the auspices of the British Museum. He persisted through six expeditions and in 1869 discovered the base under 20 feet (6 meters) of mud. He ordered an excavation that exposed the whole platform. Some remains are now in the British Museum, others in the Istanbul Archeological Museum. In 1904 and 1905 another British expedition, led by David Hogarth, found evidence of the five temples, each built on top of the former. Today the site is a marshy field, a solitary column the only reminder that in that place once stood one of the seven wonders of the ancient world.

Further reading

Clayton, Peter, and Martin Price. 1988. *The Seven Wonders of the Ancient World*. London: Routledge.

Cox, Reg, and Neil Morris. 1996. *The Seven Wonders of the Ancient World*. Parsippany, NJ: Silver Burdett.

Aswan High Dam
Egypt

The Aswan High Dam, replacing earlier dams, contains the River Nile nearly 600 miles (1,000 kilometers) upstream from Cairo by a massive embankment 375 feet (114 meters) high and 3,280 feet (1,003 meters) long, built of earth and rock fill with a clay and concrete core. It impounds Lake Nasser, one of the largest reservoirs in the world, covering an area more than 300 miles (480 kilometers) long and 10 miles (16 kilometers) wide, that holds enough water to irrigate over 7 million acres (2.8 million hectares) of farmland for many years. Its economic and social impact on the lower reaches of the Nile (that is, in the north of Egypt) makes it an engineering feat of some importance, although not necessarily always beneficial.

The annual flooding of the Nile has been the historical life source of Egypt, in what is almost a rainless region. Almost all the population lives within 12 miles (20 kilometers) of the river. The flooding—13 billion to 169 billion cubic yards (12 billion to 155 billion cubic meters)—is caused by late-summer rains on Ethiopia's plateaus that find their way into the Nile's tributaries. Late in the nineteenth century, regional population growth was outstripping agricultural production, and the river had to be controlled to recover stability. The first Aswan Dam was built from 1899 to 1902 and raised in 1907–1912 and again in 1929–1934. When its potential to generate

Aswan High Dam, Egypt, 1952–1968.

the power needed in industrializing economies was realized, hydroelectric installations were added in 1960. But its inadequate storage capacity meant that in years of extreme flooding, sluices would have to be opened to protect the structure, thereby, ironically, inundating the very areas the dam was meant to protect. Designs for the High Dam, 4 miles (6.4 kilometers) south of the existing structure, were put in hand in 1952. Egypt and Sudan signed the Nile Water Agreement in November 1959.

President Gamal Abdel Nasser initiated the Aswan High Dam project. Because of his connections with the Communist bloc, the United States and Britain refused to secure loans, so the Soviet Union provided civil engineers to design the earthen dam and supplied the equipment and 400 technicians and electrical engineers to build the hydroelectric power station. Moreover, almost a third of the estimated U.S.$1 billion cost was met by the Soviet Union, and the remainder was funded by Egypt's controversial nationalization of the Suez Canal. Construction, commenced in 1960, was complete by mid-1968. The last of the twelve turbines was installed in 1970, and President Anwar Sadat officially inaugurated the High Dam.

Despite causing some ecological problems, the dam brought good outcomes for the people of the Nile valley. For example, with a regulated agricultural system in place, through multiple cropping the nation's agricultural income has increased by 200 percent. For all that, the poor drainage of the newly irrigated lands has led to increased salinity, and more than half of Egypt's arable soils are now medium to poor in quality. Village communities were provided with water and electricity. A fishing industry was established, with an annual production target of 112,000 tons (101,600 tonnes) by the year 2000. The hydroelectric scheme generates 2,100 megawatts, about half of Egypt's annual needs at the time of construction; demand has since increased, and the Aswan High Dam now provides a little over 20 percent. Egypt was untouched by the drought over much of Africa in the late 1980s, and during the following decade the land was saved from several unusually high floods. In 1996 Lake Nasser rose above the spill level for the first time, and plans are in hand to open

up more irrigated farmland, even recovering parts of the Sahara Desert.

Nevertheless, the construction of the Aswan High Dam has caused some problems, mostly as cumulative effects of impeding the flow of the river. Farmers have had to turn to artificial fertilizers to replace the nutrients that no longer reach the floodplain. Before the High Dam was built, half the water flowing in the Nile reached the Mediterranean; the millions of tons of silt it once carried to the sea are now mostly trapped behind the dam; consequently the ocean is eroding the coastline. The now-stagnant waters in the Delta destroyed long-standing ecological systems, and the loss of nutrients from the river has drastically damaged the sardine industry in the Mediterranean: 20,000 tons (18,288 tonnes) were caught in 1962; that figure was reduced to only 670 tons (609 tonnes) in 1969. Other branches of the Mediterranean fishing industry were similarly affected. Thankfully, for whatever reasons, there has since been a complete recovery.

The Aswan High Dam also had social and cultural implications. Not least traumatic of these was the displacement of population. More than 90,000 Nubians had to be relocated; those living in Egypt were moved about 28 miles (45 kilometers), but Sudanese Nubians had to move to new homes 370 miles (600 kilometers) away. Much of Lower Nubia was submerged under Lake Nasser, including archeological sites between the First and Third Cataracts of the Nile. At the urging of UNESCO, a rescue program named the Nubia Salvage Project was started in 1960 by the Oriental Institute, University of Chicago. As a result, twenty monuments from the Egyptian part of Nubia (including the front of the rock-hewn tomb at Abu Simbel) and four others from the Sudan were dismantled, relocated, and reerected. Others were documented before their inundation, but some were lost forever without being recorded.

Further Reading

Fahim, Hussein M. 1981. *Dams, People, and Development: The Aswan High Dam Case*. New York: Pergamon.

Little, Tom. 1965. *High Dam at Aswan: The Subjugation of the Nile*. New York: John Day.

Shibl, Yusuf A. 1971. *The Aswan High Dam*. Beirut: Arab Institute for Research and Publishing.

Avebury Stone Circle, Wiltshire, England, ca. 2600–2000 B.C. Aerial view.

Avebury Stone Circle
England

The Avebury Stone Circle, covering around 28 acres (11 hectares), is the largest known stone circle in the world. It partly embraces the linear village of Avebury, 90 miles (145 kilometers) west of London in a part of England that is replete with prehistoric remains: Silbury Hill; the Sanctuary; and the long barrows of East Kennet, West Kennet, and Beckhampton. John Aubury, who accidentally discovered it while foxhunting in the winter of 1648, wrote that Avebury "does as much exceed in greatness the so renowned Stonehenge as a Cathedral doeth a parish Church." Indeed, it is sixteen times the size of Stonehenge.

When the Avebury circle was intact, its complex, if rather irregular, geometry comprised a 30-foot-deep (9.2-meter) ditch inside a 20-foot-high (6-meter) grass-covered chalk bank 1,396 feet (427 meters) in diameter. One observer describes it as "a curiously amorphous 'D' shape." The ditch, possibly

once filled with water, enclosed an outer circle of about 100 enormous, irregular standing stones that varied in height from 9 to 20 feet (2.7 to 6 meters). Within the large circle, there were two inner circles, each about 340 feet (104 meters) in diameter. The northern one (now largely destroyed) seems to have comprised two concentric rings, one of twenty-seven stones and one of twelve; at their center stood three larger stones. The southern circle had a single 20-foot-high (6-meter) stone at its center. The inmost circles are thought to have been set up about 2600 B.C.; the outer ring and enclosing earthworks have been dated at a century later.

Its construction called for colossal effort on the part of the builders. The standing stones were quarried and dressed 2 miles (3.2 kilometers) from their final position, dragged or perhaps sledded to the site—some weighed 45 tons (41 tonnes)—and set upright. The excavation of the vast surrounding ditch with rudimentary stone tools yielded an estimated

200,000 tons (203,200 tonnes) of spoil, mostly chalk stone. Some of the spare material may have been carried 1 mile (1.6 kilometers) to construct the mysterious 130-foot-high (39.6-meter) chalk mound known as Silbury Hill, just outside the village of Avebury. Many of the stones are now missing, possibly "quarried" by farmers or cleared for agricultural and even religious reasons since about the fourteenth century A.D., when villagers actually buried some of them. Only 36 of the original 154 megaliths remain standing.

The outer circle was broken to form four 50-foot-wide (15.3-meter) entrances, facing approximately north, south, east, and west. Two were the terminations of avenues of the same width, defined by standing stones and extending up to 1.5 miles (2.5 kilometers) across the surrounding countryside. According to the eighteenth-century antiquarian William Stukeley, the so-called West Kennet Avenue ran south to the Sanctuary, another stone circle on Overton Hill; the one named Beckhampton Avenue ran west to end at the neolithic tomb known as Beckhampton Long Barrow. Stukeley's measured drawings, made before 1743, are the only surviving record of the former condition of the site. He interpreted the ground plan of Avebury as the body of a serpent passing through a circle—a traditional alchemical symbol—and whose head and tail were marked by the avenues.

Recent investigations have led some scholars to speculate that the Avebury circle was part of a network of sacred places that stretched 200 miles (360 kilometers) across southern England. Similar to Stonehenge and many other megalithic monuments in Britain, the Avebury Stone Circle formed part of at least a local complex of megalithic works. The whole complex probably continued to be used for around 2,300 years. That persistence and the very size of the Avebury Stone Circle give weight to the suggestion that it was "perhaps the most significant sacred site in all of Britain, if not the entire continent of Europe." The renaissance of paganism in the West at the end of the twentieth century excited new interest in its elusive mysteries.

Further reading

Burl, Aubrey, and Edward Piper. 1980. *Rings of Stone: The Prehistoric Stone Circles of Britain and Ireland.* New Haven, CT: Ticknor and Fields.

Grundy, Alan H. 1994. *Britain's Prehistoric Achievements.* Lewes, UK: Book Guild.

Hadingham, Evan. 1975. *Circles and Standing Stones.* London: Heinemann.

B

Babylon: Nebuchadnezzar's city
Iraq

The city of Babylon ("Gate of God") once stood on the banks of the Euphrates River, 56 miles (90 kilometers) south of Baghdad, Iraq. It was the capital of Babylonia in the second and first millennia B.C. In A.D. 1897 the German archeologist Robert Koldewey commenced a major excavation. During the next twenty years he unearthed, among many other structures, a processional avenue to the temple of Marduk and the legendary fortified city wall, which once enjoyed a place among the seven wonders of the ancient world. It was not until the sixth century A.D. that its place was usurped by the so-called Hanging Gardens.

Babylon entered the pages of history as the site of a temple around 2200 B.C. At first it was subject to Ur, an adjacent city-state, but gained its independence in 1894 B.C., when the Sumu-abum established the dynasty that reached its zenith under Hammurabi, known as "the Lawgiver." The Hittites overran the city 330 years later. It was governed by the Kassite dynasty, which extended its borders and made it the capital of the country of Babylonia, with southern Mesopotamia under its control. When the Kassites yielded to pressure from the Elamites in 1155 B.C., Babylon was governed by a succession of ephemeral dynasties and became part of the Assyrian Empire in the late eighth century B.C. In turn, the Assyrians

were driven out by Nabopolassar, who founded the Neo-Babylonian dynasty around 615 B.C. His son Nebuchadnezzar II (ca. 604–561 B.C.) built the kingdom into an empire that covered most of southwest Asia.

Babylon, now Nebuchadnezzar's imperial capital, underwent a huge rebuilding program—new temples and palace buildings, defensive walls and gates, and a splendid processional way—to make it the largest city in the known world, covering some 2,500 acres (1,000 hectares). It must have impressed visitors, because the myth sprang up, perhaps from the assertion of the Greek historian Herodotus, that it was 200 square miles (510 square kilometers) in area, with 330-foot-high (99-meter) walls, 80 feet (25 meters) thick. Of his achievement, Nebuchadnezzar boasted, "Is not this great Babylon that I have built for the house of my kingdom by the might of my power, and for the honor of my majesty?"

The Euphrates River divided the city into two unequal sectors. The "old" quarter, including most of the palaces and temples, stood on the east bank; Nebuchadnezzar's new city was on the west. The whole was surrounded by an 11-mile-long (17-kilometer) outer wall enclosing suburbs and the king's summer palace. The inner wall, penetrated by eight fortified gates leading to the outlying regions of Babylonia, was wide enough to allow two chariots to be driven abreast on its top. Most prominent among the portals was the northern Ishtar Gate, dedicated to the queen of heaven: a defensible turreted building with double towers and a barbican, faced with blue glazed brick and richly ornamented with 500 bulls, dragons, and other animals in colored brick relief.

Through the Ishtar Gate passed the north-south processional way, which ran past the royal palace and was used in the New Year festival. It was paved with limestone slabs, about 3.5 feet (1 meter) square; the flanking footpaths were of breccia stones about 2 feet (600 millimeters) square. Joints were beveled and the gaps filled with asphalt. The road was contained by 27-foot-thick (8-meter) turreted walls, behind which citadels were strategically placed. The faces of the walls were decorated with lions in low relief. Much of the significance of the road lies in the exotic and doubtless expensive materials employed. The land between the rivers had little naturally occurring stone, and except for their faces, the city walls and gatehouses and even the king's palace were constructed of sun-dried brick.

Inside the Ishtar Gate, at the northwest corner of the old city, stood Nebuchadnezzar's extensive palace with its huge throne room, and the fabled Hanging Gardens of Babylon. It is more likely that they were "overhanging" gardens. Described by one first century B.C. visitor as "vaulted terraces raised one above another," they were irrigated with water pumped from the Euphrates. Another early description says that this 400-foot-square (122-meter) artificial mountain was more than 80 feet (25 meters) high and built of stone. It was planted with all manner of vegetation, including large trees. There is a romantic legend that the Hanging Gardens were built for Nebuchadnezzar's wife, Amytis, a Mede who missed the green mountains of her motherland. Beside the palace stood the rebuilt temple of the city's patron god, Marduk, replete with gold ornament. In a sacred precinct north of the temple stood a seven-story ziggurat (stepped pyramid); some descriptions put its height at 300 feet (90 meters).

Nebuchadnezzar was Babylon's last great ruler. Because his successors were comparatively weak, the Neo-Babylonian Empire quickly passed. In 539 B.C. the Persian Cyrus II took the city by stealth, over-threw Nebuchadnezzar's grandson Belshazzar, and subsumed Babylon into his empire. The city became the official residence of the crown prince, but following a revolt in 482 B.C., Xerxes I demolished the temples and ziggurat, thoroughly destroying the statue of Marduk. Alexander the Great captured the city in 330 B.C. but he died before he could carry out his intention to refurbish it as the capital of his empire. For a few years after 312 B.C., the Seleucid dynasty used Babylon as a capital until the seat of government was moved (with most of the population) to the new city of Seleucia on the Tigris. Babylon the Great became insignificant, and by the foundation of Islam in the seventh century A.D., it had almost disappeared.

Now Babylon is being rebuilt. In April 1989 the *New York Times International* reported that, under Iraqi President Saddam Hussein, "walls of yellow brick, 40 feet [12 meters] high and topped with pointed crenellations, have replaced the mounds that once marked [Nebuchadnezzar's] Palace foundations. And as Babylon's walls rise again, the builders insert inscribed bricks recording how [it] was 'rebuilt in the era of the leader Saddam Hussein.'" An annual International Babylon Festival—one was subtitled "From Nebuchadnezzar to Saddam Hussein"—is part of the megalomaniac dictator's projection of himself as the ancient king's successor. Portraits of the two hang side by side on a restored wall in Babylon.

See also Hanging Gardens of Babylon

Further reading

Reade, Julian. 1991. *Mesopotamia*. Cambridge, MA: Harvard University Press.

Saggs, Henry William Frederick. 1988. *The Greatness That Was Babylon*. London: Sidgwick and Jackson.

Service, Pamela F. 1998. *Ancient Mesopotamia*. New York: Benchmark Books.

Banaue rice terraces
Ifugao Province, Philippines

In the Banaue municipality of the northern Ifugao Province on the Philippine island of Luzon, the indigenous Igorot people have constructed 49,500 acres (20,000 hectares) of agricultural land upon the inhospitable bedrock of the steep Cordillera Central Mountain Range. For millennia, succeeding generations of farmers built and maintained 12,500 miles (20,000 kilometers) of dikes and retaining walls—enough to stretch halfway around the equator—creating a unique, irregular patchwork of terraced rice paddies. The American anthropologist Roy Barton called these terraces and others in the region "a modification by man of the earth's surface on a scale unparalleled elsewhere."

The Cordillera rice terraces were added to UNESCO's World Heritage List in December 1995, a decision justified in the following terms: "The fruit of knowledge passed on from one generation to the next, of sacred traditions and a delicate social balance, they helped form a landscape of great beauty that expresses conquered and conserved harmony between humankind and the environment." Moreover, they were cited as "outstanding examples of living cultural landscapes."

The tiers rise to about 4,900 feet (1,500 meters) above sea level. Each is defined by a stone or clay retaining wall, snaking along the contours of the steep mountainside. Stone walls are up to 50 feet high (15 meters); some of the clay walls are more than 80 feet (25.5 meters) high. Some garden terraces have been backfilled with soil, ash, and composted vegetable material, while others have been simply carved from the rock and overlaid with soil washed down from the higher levels. Rice cannot be grown without large quantities of water, and the terraces are served by an elaborate irrigation system, comprising canals cut through the rock and bamboo and wooden aqueducts. Once the highest terraces are flooded, water spills over the descending walls until the whole hillside is irrigated.

What of their builders? Igorot (literally, "the mountain people") is a broad ethnic classification applied to a number of groups bound by common sociocultural and religious characteristics—Ibaloy, Kankanay, Ifugao, Kalinga, Apayao, and Bontoc—who occupy the Cordilleras. They originate from the warlike immigrants who reached the northern islands of the Philippines from Vietnam and China, some scholars believe 10,000 years ago. Their descendants eventually became rice farmers and, against the dif-

Banaue rice terraces, Ifugao Province, Luzon, Philippines; possibly commenced 200 B.C.

ficulties presented by the hostile topography, built their amazing tiers of rice fields on the precipitous mountainsides. The true age of the terraces remains in question: some sources suggest that the Igorot commenced them between 200 B.C. and A.D. 100, others that they date from at least 1000 B.C. As late as the 1990s rising nationalism had not permeated their tribal highlands, and the Igorots, while regarded as citizens, did not think of themselves as Filipinos. They were further alienated by the Marcos administration's dam-building schemes, which included flooding the mountain valleys in their Cordillera homelands. They continue to resist integration into Filipino society.

The rice culture of the Igorot, central to their way of life, inevitably had a spiritual dimension. As Joaquin Palencia remarks, "the adversarial nature of the geography of this region and the tremendous odds faced by the Ifugao to assure access to food ... set the stage for the *bul-ul*, the rice god figures that came to be a mechanism through which superhuman restraint became central to the production of a basic need." Indeed, the Igorot embrace no fewer than 1,500 gods, each type fulfilling a different function. The *bul-ul* is a large-headed, seated or standing humanoid figure, ritually carved, usually from sacred narra wood. The sizes of the rice field and its guardian *bul-ul* are directly related: the Banaue terraces have large, thickset *bul-ul*. Once the ceremonies and feasts are completed, the figure is installed in a granary in the attic of a house, from which it is believed to protect crops and ensure abundant harvests. But the forces threatening the Banaue rice terraces, and others like them at Hungduan, Kiangan, Mayoyao, and Bontok, are other than spiritual.

Rice farming is labor intensive—and hard labor at that—and yields low financial returns. The main threat to the terraces is the departure of young Ifugaos, who seek better work opportunities in the cities. Water shortage is also a problem: the lack of rain in the dry season is exacerbated by systematic deforestation and illegal logging. Because of such poor forestry management there is no longer enough water for irrigation, and recent harvests have been unable to sustain even the terrace owners, much less provide a cash crop. Moreover, with only one crop a year, mountainside farming compares poorly with the lowland paddies, where there are two. Many Ifugao farmers, encouraged by the Rice Terraces Commission (RTC), established under President Fidel Ramos in 1994, are now planting vegetables that can be harvested after six weeks, a quarter of the time needed for rice.

In 1998, when these combined problems were exacerbated by accelerated erosion caused by introduced giant earthworms, the RTC introduced a plan to maintain the terraces, focused on the preservation of Ifugao culture, diversification of the regional economic base, and the application of appropriate current agricultural technology. Its success has yet to be proven. Promoted by the Philippines tourism authority as "the eighth wonder of the world," the Banaue rice terraces are among the country's major attractions. Already under threat from cultural change, neglect, and inadequate irrigation, if they are not maintained they will be in ruins within a couple of decades.

Further reading

Barton, Roy Franklin. 1922. *Ifugao Economics*. Berkeley: University of California Press.

Palencia, Joaquin G. 1998. The Ifugao Bul-ul. *Tribal Arts Online*. May. http://www.tribalarts.com.

Reyes, Angelo J., and de los Aloma, M., eds. 1987. *Igorot: A People Who Daily Touch the Earth and the Sky*. Baguio, Philippines: CSG.

Scott, William Henry. 1966. *On the Cordillera: A Look at the Peoples and Cultures of the Mountain Province*. Manila: MCS Enterprises.

BART (Bay Area Rapid Transit)
San Francisco, California

BART (Bay Area Rapid Transit) is a 95-mile (152-kilometer) automated rapid-transit system, the first of the "new generation" of such systems in the United States. By the end of the twentieth century there were thirteen in operation, including Washington, D.C. (opened 1976), Atlanta (1979), and Miami (1986). BART has thirty-nine stations on five lines radiating out from San Francisco to serve Contra Costa and Alameda Counties in the eastern Bay Area of northern California.

In 1947 a joint Army-Navy review board predicted that another connecting link between San Francisco and Oakland would be needed to prevent intolerable traffic congestion on the Bay Bridge. It proposed the construction of a tube to carry high-speed electric trains under the waters of the bay. Four years later the California State Legislature created the San Francisco Bay Area Rapid Transit Commission and charged it with finding a long-term transportation solution in the context of environmental problems, not least among them the danger from earthquakes. After six years of investigation, the commission concluded that any transportation plan would have to be part of a total regional development plan. Because no such plan existed, the commission prepared a coordinated master strategy, later adopted by the Association of Bay Area Governments.

The commission's most economical transportation solution was to establish a five-county rapid-transit district, with the task of building and operating a high-speed rapid rail network linking major commercial centers with suburban nodes. The San Francisco BART District was formed, comprising the counties of Alameda, Contra Costa, Marin, San Francisco, and San Mateo. Plans were made for a revolutionary rapid-transit system. Electric trains would run on grade-separated corridors at maximum speeds of 80 mph (128 kph) and averaging around 45 mph (72 kph). Sophisticated, well-appointed vehicles would compete with private automobiles in the Bay Area, and well-designed, conveniently located stations would be built.

By mid-1961, after extensive public consultation, the final plan was submitted to the five counties for approval. San Mateo County was unconvinced and withdrew from the scheme in December. Marin County also withdrew a few months later, not only because it could not sustain its share of the cost but also because there were questions about the feasibility of running trains across the Golden Gate Bridge. The original proposal was therefore revised as a three-county plan, providing links across the bay between San Francisco and Contra Costa and Alameda. Those counties accepted the BART Composite Report in July 1962.

As part of the following November's general election, voters approved a $792 million bond issue to finance the high-speed transit system and to rebuild 3.5 miles (5.6 kilometers) of the San Francisco Municipal Railway. The estimated $133 million cost of the Transbay Tube was to be funded by Bay Bridge tolls. The rolling stock, which would run on 1,000-volt direct current, was estimated to cost another $71 million, and the total cost of the system was projected at $996 million—the largest public works project ever undertaken by local residents in the United States. There were to be many delays, and costs would inevitably rise, eventually totaling $1.62 billion.

Parsons-Brinckerhoff-Tudor-Bechtel was the consortium appointed to manage the project, consisting of Parsons-Brinckerhoff-Quade and Douglas (the New York originators of the first plan); and from San Francisco, Tudor Engineering Company and the Bechtel Corporation. BART construction began on 19 June 1964, on the Diablo Test Track in Contra Costa County; completed ten months later, it was used to develop and test the vehicular system.

The Oakland subway was commenced in January 1966. In the following November the first of the fifty-seven, 24-foot-high-by-28-foot-wide (7.4-by-14.8-meter) steel-and-concrete sections of the Transbay Tube, almost 4 miles (6.4 kilometers) long in total, was submerged in the bay. A 3-mile-long (4.8-kilometer) drilled rock tunnel through the Berkeley Hills was completed four months later. The Transbay Tube structure was completed in August 1969. Lying as much as 135 feet (41.3 meters) underwater, it took six years to design (seismic studies were an integral part of the process), and under three to build. The tunnel (indeed the entire BART system) would survive intact the Loma Prieta earthquake of 1989. The final cost of the tunnel was $180 million. Before the tube was closed to visitors so the rail tracks could be installed, thousands of pedestrians passed through.

In July 1967 construction began on the two-level Market Street subway, 100 feet (30.6 meters) below San Francisco. The work was complicated by a difficult mud-and-water environment and the century-old network of underground utilities. The first tunneling on the west coast was carried out entirely under compressed-air conditions; this section of the project brought the BART workforce to 5,000 in 1969. On 27 January 1971 the bore into the west end of Montgomery Street Station marked the completion of that phase of the project.

Although delays and inflation were eroding capital, public and governmental pressure groups forced the relocation of 15 miles of line and 15 stations, and a general improvement of station designs. They were also substantially altered during construction to improve access. Discussion of BART's financial problems is not the purpose of this essay: suffice it to say that an increasing input of federal money was needed to support the constant variations and improvements to the original plan. BART's linear park was constructed to demonstrate how functionality need not spoil the amenity of the environment, and major landscaping was partly funded by federal money.

When the first 250 vehicles were eventually ordered from Rohr Industries of California, the price had reached $80 million—$18 million above the estimate for the whole 450-car fleet. The first car was delivered in August 1970, and within months, 10 test cars operated on the Fremont Line. The paid service began operation on 11 September 1972 on the 28 miles (45 kilometers) between Fremont and MacArthur Stations. Heavily subsidized by federal grants, 200 more cars were bought by July 1975. In the late 1980s, BART purchased another 150 from SOFERVAL, an American subsidiary of Alsthom Atlantique of France, and 80 more from Morrison-Knudsen a few years later.

A central control room, installed in 1972 in the Lake Merritt Administration Building, was replaced in 1979 by an Operations Control Center, from which train operations and remote control of electrification, ventilation, and emergency-response systems are supervised.

In 1991, the BART Extensions Program launched a $2.6 billion plan to expand services in Alameda, Contra Costa, and San Mateo Counties. Since then 5 stations and 21 miles (33 kilometers) of double track have been added, including the Pittsburg-Antioch Extension, whose North Concord/Martinez Station opened in December 1995, the first new one in over 20 years. The $517 million Dublin/Pleasanton Extension opened in May 1997. A proposal to connect BART to San Francisco International Airport (SFO) was first considered in 1972, just as the inaugural service was opened. The first stage opened in February 1996. During the next phase, BART will move further down the San Francisco peninsula, adding 9 miles (14.4 kilometers) of track and 4 new stations, including one inside the new International Terminal. Work on the final leg started in 1997, and the line was scheduled for completion early in the twenty-first century. In 1995, BART launched a ten-year program, costing $1.1 billion, to overhaul the system infrastructure and the original fleet of cars.

Further reading

Anderson, Robert M., et al. 1980. *Divided Loyalties: Whistle-Blowing at BART*. West Lafayette, IN: Purdue University Press.

Grant, Howard. 1976. *Notes from Underground: An Architect's View of BART*. San Francisco: Reid and Tarics Associates.

Grefe, Richard, and Richard Smart. 1976. *A History of the Key Decisions in the Development of Bay Area Rapid Transit*. San Francisco: McDonald and Smart.

Baths of Caracalla
Rome, Italy

The Baths of Caracalla *(Thermae Antoninianae)* were built between A.D. 212 and 216 by the emperor Marcus Aurelius Antoninus (A.D. 188–217), usually known as Caracalla. Although in layout the Baths of Caracalla largely emulated the model established about a century before in the Baths of Trajan, their massive scale and opulent internal finishes were without precedent. Their fully integrated plan and imposing scale and grandeur amply demonstrated the Romans' design skills. Significantly, the baths demonstrated the structural advances made possible through the masterful use of concrete to span vast spaces using barrel and groin vaults, domes, and half-domes, as well as the sophisticated mechanical engineering services developed by the Romans.

Public baths *(thermae)* were an essential part of all Roman towns. The majority of citizens lived in crowded tenements *(insulae)* without running water or sanitary facilities, so communal baths were constructed and made available to both sexes of all social classes. Entry was free. Generally, mixed bathing was not favored, so the baths were open to women in the mornings and men in the afternoons and evenings. The thermae were the center of Roman social life—people could meet friends there and engage in any number of leisure and cultural pursuits on offer. As well as changing rooms, gymnasia, saunas, and pools of various temperatures, there were libraries, museums, restaurants, bars, shops, lecture theaters, concert halls, playing fields, gardens, and courtyards, all richly furnished with mosaics, fountains, and statues. Although extremely costly to build, the baths were a political investment—a means for the emperor to demonstrate his concern for the well-being of the community.

The Baths of Caracalla occupied a 50-acre (20.25-hectare) site. The complex was divided into three parts: the rectangular main building, approximately 750 by 380 feet (225 by 115 meters) and large enough to accommodate 1,600 bathers; encircling landscaped parks and gardens; and a perimeter ring of shops, lecture halls, and pavilions. Laid out symmetrically, the compactly planned baths offered identical bathing circuits on either side of the central (and shorter) axis. The sequence of bathing spaces on that axis comprised the hot bath *(caldarium)*, warm bath *(tepidarium)*, and the cold bath *(frigidarium)* in a large unheated central hall. The last, which also served as a foyer, was open on one side, allowing easy access to the open-air swimming pool *(natatio)*. Changing rooms *(apodyteria)*, gymnasia, or exercise yards *(palaestrae)*, with terraced porticoes, and sauna *(laconica)* were arranged symmetrically on the transverse axis. Rooms for massage, manicure, and other services associated with the bathing routine were featured on either side of the baths. Decorative interior finishes—colored marble veneers on walls, marble, basalt and granite columns and arches, and coarsely textured black-and-white mosaic floors—created a rich and sumptuous character.

Since the baths were public facilities that attracted large numbers of people, the gathering spaces needed to be vast and uncluttered with structural elements. In the absence of structural impediments, bathers were afforded extended views to various parts of the thermae. The Romans achieved these objectives by exploiting the semicircular arch. The rectangular central hall of the Baths of Caracalla demonstrated their structural method. It was roofed with an enormous semicircular intersecting concrete vault divided into three compartments. Each was 108 feet (30 meters) high and rested at the corners on enormous piers. Clerestory windows adequately lit the hall.

Water for the Baths of Caracalla flowed from a branch of the Aqua Marcia aqueduct into a huge reservoir, divided into eighteen chambers with a total capacity of about 2.2 million gallons (10 million liters). The water was carried through pipes laid underneath the gardens to the main building, where it was distributed directly to the cold pools, or to wood-fired boilers, where it was heated for the warm and hot baths. For ease of inspection and maintenance, distribution pipes and waste drains were located in separate tunnels. A separate network of tunnels was used to store wood for about fifty furnaces *(praefurnia)* that heated the saunas *(laconica)* and other rooms via a hot-air system *(hypocausta)* beneath the floors. The heated rooms were on the southwestern side of the complex to gain maximum benefit from the sun; all had large windows. The hottest room, the circular, protruding caldarium, was covered by a 115-foot-diameter (35-meter) dome, higher than the Pantheon's and only slightly less in span.

The Baths of Caracalla are now in ruins, but their soaring height and impressive scale allow visitors to appreciate their size and massiveness.

See also Roman concrete construction

Further reading

Piranomonte, Marina. 1998. *Le Terme di Caracalla*. Milan: Electra.
Sear, Frank. 1989. *Roman Architecture*. London: Batsford.
Ward-Perkins, J. B. 1981. *Roman Imperial Architecture*. Harmondsworth, UK: Penguin.

The Bauhaus
Germany

The German design school known as the Bauhaus (literally, house of building), that functioned between 1919 and 1932, laid the foundation of a different kind of architectural education, one that was eventually adopted throughout the world. It restored the links between design and making that had been undermined during the Renaissance and virtually destroyed by the European academies. Much of the Bauhaus's significance lies in the fact that some of its leaders migrated to the United States in the 1930s to head up the schools of architecture at Harvard and the Illinois Institute of Technology; other members also became teachers and practitioners in America.

The Bauhaus was conceived by Walter Gropius (1883–1969). After reluctantly commencing architectural studies at Berlin-Charlottenberg in 1905, between 1907 and 1910 he worked in the office of Peter Behrens before forming a partnership with a fellow employee, Adolf Meyer. During World War I Gropius served as a cavalry officer, and following the November 1918 armistice he was appointed director of two separate institutions in Weimar, Saxony, Germany:

The Bauhaus, Dessau, Germany; Walter Gropius, architect, 1925–1926. Exterior of workshops.

the Grand Ducal Academy of Arts and the Grand Ducal Academy of Crafts. He immediately proposed that they should be combined, and in April 1919 courses started at Das Staatliches Bauhaus Weimar. Gropius's 1919 *Manifesto* called for "the unification of all the creative arts under the leadership of architecture"; building on the doctrines of nineteenth-century English reformers, Gropius sought to improve design standards by combining art and production.

Architecture was not in the curriculum of the Bauhaus's first phase at Weimar (1919–1923). Because he believed that good art, architecture, and design were more the result of collaboration than of individual virtuosity, Gropius's formal program was based upon the proposition that one cannot design without understanding the process by which the design is realized. The designed object must be "by systematic practical and theoretical research into formal, technical, and economic fields" derived from "natural functions and relationships"—in short, the Bau-

haus provided an applied design education based on Marxist materialism. Under Johannes Itten students were introduced to elements of design—shape, line, color, pattern, texture, rhythm, and density. There were also workshops for stone, wood, metal, pottery, glass, painting, and textiles. Every course was conducted by a team: a craftsperson and an artist.

The aims of the Bauhaus were maintained through the three phases of its existence in three different places and despite several changes in its direction. They were: first, "rescue all of the arts from the isolation in which each then found itself"; second, raise the status of craft to that of the so-called fine arts; and third, link the designer with emerging industrial production. Those ideas are taken for granted now, but they were first spelled out by the Bauhaus.

In spite of Gropius's ostensible nonpolitical stance, the unfamiliar ideas, left-wing beliefs, and eccentric ways evident at the Bauhaus unsettled the government and brought opposition. Objecting to official insistence upon an exhibition *Art and Technics* in 1923, immediately afterward the staff resigned. Gropius was swamped with offers to relocate, and accepted one from Dessau. To house the school he designed a group of connected blocks (1925–1926): administration, classrooms, studios, workshops, and accommodations for staff and students. Although Gropius often denied any such intention, the need for modern architecture—a tangible expression of the spirit of the age—meant that the Dessau complex would be adopted as a model internationally.

Architecture was introduced into the curriculum at Dessau. Just then, groups of European architects, mostly socialists, were searching for a pure form of architecture, liberated from the historical styles that they associated with a decadent aristocracy or (worse in their eyes) with the rising industrial bourgeoisie. The architects included English Arts and Crafts, Italian Futurists, Dutch De Stijl, and German Expressionists. Buildings inevitably became expressions of their beliefs, and their response to Europe's widespread housing crisis of the 1920s was an austere form of workers' housing with open floor plans, white interiors, and furniture that "worked," whatever that meant. For them, a building must have a flat roof and flat walls, devoid of all ornament and decoration. And because color was bourgeois, the exteriors of houses must be white, gray, or black—in fact, just like the Dessau Bauhaus. It is not surprising that by 1932 the Americans Henry-Russell Hitchcock and Philip Johnson recognized in all this what they (inaccurately) dubbed an International Style. It was soon imitated throughout the world, frequently with no heed to the underlying sociopolitical theory.

Gropius resigned the Bauhaus directorship in April 1928, not only to concentrate upon his architectural practice but also in an attempt—futile, as it happened—to stem the growing National Socialist (Nazi) Party's propaganda attacks upon the school. He recommended the Swiss architect Hannes Meyer as successor. But because Meyer was overtly Communist, the mayor of Dessau dismissed him in 1930, appointing in his place a German architect, Ludwig Mies van der Rohe. Under mounting pressure to close altogether, Mies moved the Bauhaus to Berlin in 1932. A year later he disbanded it.

Although its ideas were spread internationally by many publications—not least the *Bauhausbüche* series after 1925—and exhibitions, the Bauhaus became more influential through the diaspora of staff and students: for example, Gropius went to head up the Graduate School of Design at Harvard, Mies became dean of architecture at the Illinois Institute of Technology, and László Moholy-Nagy established the "New Bauhaus" in Chicago. A Bauhaus archive, originally at Darmstadt, moved to Berlin in the 1970s; another is housed at Harvard. The design philosophy and the educational philosophy of the Bauhaus continue to have impact on the teaching and practice of architecture and design.

Further reading

Gropius, Walter. 1965. *The New Architecture and the Bauhaus.* Cambridge, MA: MIT Press.

Hochman, Elaine S. 1997. *Bauhaus: Crucible of Modernism.* New York: Fromm.

Wingler, Hans Maria. 1969. *The Bauhaus: Weimar, Dessau, Berlin, Chicago.* Cambridge, MA: MIT Press.

Bedouin tents

There are only three essential structural systems in architecture: the post and beam (trabeated), the arch and its extensions (arcuated), and those that employ stretched filaments and membranes (tensile). Because durable tensile materials like steel and reinforced concrete were not developed until after 1865, and synthetic membranes, like fiberglass-Teflon laminate and Kevlar, until more than a century later, tensile technology was limited to buildings not considered "proper" architecture. But despite the denial of *means*, the *method* of creating them has been understood, refined, and applied from ancient times. Purest among such applications are the tents of the Bedouin. Their origins are lost, but they are indeed architectural feats for their structural economy, functionality, and environmental sustainability.

The nomadic Arabs known as Bedouin (*badawi*, for "desert dwellers") inhabited Arabia from sometime in the second millennium B.C. With the expansion of Islam in the seventh century A.D., they spread into the Syrian and Egyptian deserts and invaded northern Africa, where their flocks, allowed to overgraze, soon turned much of the coastal pasture into semi-desert. The Bedouin, who now comprise about 10 percent of the population of the Middle East, continue to herd camels, sheep, goats, and sometimes cattle. Their patterns of migration depend on availability of pasture: in winter, if there is rain, they move farther into the desert; in summer, they locate near assured water supplies and build simple mud-and-stone temporary houses. While on the move, the Bedouin live in a *beit al-sha'r* ("house of hair"). The dwelling, little changed for about 4,000 years, consists of short wooden posts supporting a framework of tightly stretched goat-hair ropes, over which a loosely woven goat-hair cloth membrane *(fala'if)* is stretched to serve as walls and roof.

The goat-hair yarn is spun on a drop spindle by the older women and woven into cloth strips on a horizontal loom. The breadth of the strips approximates the ancient cubit (about 20 inches or 50 centimeters); they vary in length from 23 to 65 feet (7 to 20 meters), depending on the size of the tent for which they are made. Because the women work only in "spare" time, even short strips of cloth may take several months to produce. The portable loom allows any unfinished work to be rolled up when the group moves on. The finished strips are sewn together with black goat-hair thread to make up a single roof membrane. That is a social occasion, with women working together. The goat hair's natural color, usually black, is retained, although sometimes the addition of sheep's wool yields a streaked cloth. The black fabric absorbs heat, but it also provides deep shade, so that temperatures inside can be considerably lower than outside. The coarse weave allows heat to disperse, and the covering provides good insulation in the cold desert night. When it rains, the loosely woven fabric swells, stopping most leaks. A tent cloth lasts an average of five years, and its maintenance and replacement depend upon a renewable resource, as they have for centuries.

When the Bedouin make camp, the leader of the band directs the women in pitching the tent. Before the poles are raised, the roof is spread on the ground, with one of its long sides facing windward, and stretched by tightening lines attached to pegs. Once it has been lifted on the pole and rope frame, the goat-hair flaps that form the walls—long enough to enclose the entire tent at night—are hung and pegged down, with the entrance facing away from the prevailing wind. The low profile of the roof and very long guy ropes are designed to maximize wind resistance.

Traditionally, brightly decorated curtains divide the interior. The men's area, always at the end toward Mecca, also incorporates the *majlis*, where guests are received around a hearth. The private family area *(mahram)*, or women's section, is much larger and barred to all men except the head of the family. The third space is the kitchen. Of necessity for a nomadic lifestyle, furnishings are sparse. Carpets and mattresses cover the desert floor; pillows stacked around a camel saddle may provide seating for guests.

The Tuaregs, descendants of the Berbers, whom the Arabs displaced from North African coastal regions, also live in tents. About 800,000 strong, the seven major Tuareg confederations inhabit an area from the western Sahara to western Sudan. Although some have permanent settlements, most prefer small nomadic groups. Believing that "houses are the graves of the living," they set up rectangular tents about 10 feet (3 meters) long and 10 to 15 feet (3 to 4.6 meters) wide, covered with up to forty tanned goatskins, dyed red and sewn together, or mats of palm fiber. In this matriarchal society, when a woman marries, her family makes a tent for her, and it remains her property. In about two hours, she can put her household on pack animals, ready to move on.

Two other examples will demonstrate that not all transportable houses are tensile structures. The nomadic lifestyle of some Amerindian tribes was constrained by the migration of the great buffalo herds. Their houses needed to be strong enough to withstand the prairie winds while lending themselves to easy dismantling, carrying, and reerection. Possibly derived from the Inuit's Arctic summer dwellings, the tepee was adopted about two hundred years ago as the year-round house of the Plains nations. A conical skeleton frame of up to thirty wooden poles was lashed together near the top and covered with a fitted membrane of tanned buffalo hides. Although it was transportable, it did not share all the tensile characteristics of the Bedouin tent. The same is true of the *ger* (or yurt), the traditional house of Mongolian herdspeople, still in use all year-round. Its self-supporting framed structure—a cylinder roofed with a dome—applies a dynamic arrangement, refined over centuries, of leather-lashed saplings, a roof ring, and tensioning bands. The covering, traditionally felt, is secured with ropes. The ger can be dismantled and carried by pack animals, although sometimes it is transported intact on a wagon.

Many Middle Eastern governments are attempting to impose a permanent sedentary lifestyle on the Bedouin. Modernization, if not altogether desirable, is probably inevitable. Trucks are displacing camels as the principal means of transportation; some camps have refrigerators and television sets powered by portable generators whose noise disturbs the quiet of the desert. Coffee is brewed for guests on gas stoves rather than the traditional hearth, and "off-the-hook" canvas tents are appearing among the "houses of hair."

See also Tension and suspension buildings

Further reading

Chatty, Dawn. 1986. *From Camel to Truck: The Bedouin in the Modern World*. New York: Vantage.

Crociani, Paola. 1994. *Bedouin of the Sinai*. Reading, UK: Garnet.

Hobbs, Joseph J. 1989. *Bedouin Life in the Egyptian Wilderness*. Austin: University of Texas Press.

Weir, Shelagh. 1976. *The Bedouin: Aspects of the Material Culture of the Bedouin of Jordan*. London: World of Islam.

Beijing-Hangzhou Canal
China

The Grand Canal (Chinese, *Da Yunhe*) in China is the world's longest artificial waterway and the oldest canal still in existence. The 1,121-mile-long (1,794-kilometer) series of linked channels extends from Hangzhou on the southeast coast to the capital, Beijing, in the north. As an engineering achievement of the ancient Chinese, the canal compares with the more familiar Great Wall. It passes through twenty-four sophisticated locks and is crossed by sixty bridges. Most of China's large rivers, including the Huai, the Huang Ho, the Wei, and the Yangtze flow from the west to the Pacific Ocean in the east, and the north-south Grand Canal provides a vital connector between their systems. That fact in itself presented a challenge to which the ancient builders were equal: the gradient of the canal was carefully designed and maintained by dredging to ensure that the seasonal flooding of the rivers did not inundate agricultural land along the artificial waterway. In places, dikes and levees provided further protection.

The Grand Canal—once known as the Grand Imperial Canal—had a simple reason for being. Successive emperors wanted to secure communication between the heavily populated politico-military centers of North China and the rice-producing regions of the south. This meant constructing a link that enabled the rapid deployment of troops and provided a faster, safer corridor for transporting grain and freight, free from the threat of the pirates who preyed on coastal shipping. During the Song dynasty (A.D. 960–1279), the annual grain traffic on the canal exceeded 340,000 tons (345,440 tonnes), carried by fleets of up to forty barges, lashed together up to four abreast and towed by water buffalo.

Suggested dates for the commencement of the canal vary from the fourth to the sixth century B.C. The 140-mile (225-kilometer) section traditionally known as the Shanyang Canal, from Qingjiang in northern Jiangsu to the Yangtze, probably was constructed sometime in that period and extended almost a thousand years later, during the Northern Qi dynasty (A.D. 550–576), when existing waterways were linked to form a single system. The second Sui emperor, Yang Di, launched an intensive building program between 605 and 610. He is said to have employed 6 million peasants constructing links between the Huang Ho and Yangtze Rivers. By thus joining the north and south of China, the canal allowed for the development of an integrated national economy and reestablished the power of the imperial civil service. Therefore, it is not surprising that it retained its importance during the Tang dynasty (618–907), when China was at the height of its power.

The canal was a key to trade expansion under the Tang and Song, and around 800 the center of political and economic activity slowly began to move to the south. By the twelfth century, Jiangsu and Zhejiang Provinces had become the heart of China, and the Southern Song dynasty (1127–1279) established a capital at Hangzhou in 1138. In 1282, under the Mongols, another canal was built between the Huang Ho and the Ta-ch'ing River in northern Shantung, but several attempts to join it to the sea proved unsuccessful. Eventually the Hui-t'ung Canal was built to join the Huang Ho and the Wei Rivers. The Ming dynasty (1368–1644) reigned from Yingtian until 1421, after which the capital was returned to Beijing. The whole Grand Canal, comprising six main sections, was dredged and repaired. Since then it has been widened repeatedly, the last changes being made at the beginning of the Ch'ing dynasty in the middle of the seventeenth century.

Early in the twentieth century the Grand Canal began to fall into disuse for reasons that included the frequent flooding of the Huang Ho; the move to coastal shipping; the construction of major north-south railroads; and not least, general neglect as a result of political turmoil. However, the Communist regime started rehabilitation in 1958, and over the next eight years the canal was dredged, straightened, and widened, and a new 40-mile (64-kilometer) section was built. But it was not until the late 1980s that plans were put in hand to dredge the entire Grand Canal, reinstating it as an important highway for local and medium-distance freight vessels, especially in the south. Shallow-draft vessels—mostly barges and tourist boats—can now navigate the stretch south of the Yangtze all year-round. The section north of the Yangtze is seasonally navigable, and major works are in progress to allow bulk carriers to reach Xuzhou; beyond that, the canal remains impassable.

Further reading

Bishop, Kevin, and Annabel Roberts. 1997. *China's Imperial Way: Retracing an Historical Trade and Communications Route from Beijing to Hong Kong*. Hong Kong: Odyssey.

The Grand Canal: An Odyssey. 1987. Beijing: Foreign Languages Press.

Harrington, Lyn. 1974. *The Grand Canal of China*. Folkestone, UK: Bailey and Swinfen.

Borobudur Temple
Indonesia

Borobudur Temple stands on the plain of Kedu, about 25 miles (40 kilometers) northwest of the modern city of Yogyakarta on the Indonesian island of Java. Its name is derived from Hhumtcambharabudara ("the mountain of the accumulation of virtue in the ten stages of Bodhisatva"). Crowning a 150-foot (46-meter) hill, this largest of all Buddhist buildings is a masterpiece of religious architecture. One of the world's best-preserved ancient monuments, it was built about 300 years before many of the great Christian cathedrals of western Europe and the famous Angkor Wat in Cambodia, some of whose temples are thought to have been influenced by it.

Sometime before the fifth century A.D., Hinduism and Buddhism spread along maritime trade routes between the Asian mainland and Java, Sumatra, and Bali. By about the seventh century Mahayana teachings dominated Buddhist thought in East Asia, and Java eventually became an important center of monastic scholarship. Mahayana Buddhist precepts constrained the form of such edifices as Borobudur.

Built from more than 1 million carved blocks of gray andesite lava quarried at nearby Mount Merapi, Borobudur was initiated as a Hindu precinct, probably a Siva temple, around A.D. 775. The lower two terraces had been completed when a shift in power to the Buddhist Sailendra dynasty brought the project to a halt. Naturally, the finished stages were unsuited to the liturgical needs of Buddhism; on the other hand, such a huge structure—its several levels are 17,800 square yards (15,000 square meters) in total area—was a powerful evocation of Hinduism, so after about fifteen years work resumed to convert the building into the largest stupa ever built. The stupa as a building type is almost exclusive to Buddhism: in essence it is a square base surmounted by an inverted circular bowl and capped with a spire. It was almost complete in 832 when the Hindu Sanjaya dynasty set out to reunify central Java and took over all religious buildings. Because most of the population was Buddhist, Borobudur remained a focus of that religion.

Influenced by the Gupta architecture of fourth-century India, the Borobudur Temple modeled the

Borobudur Temple, Java, Indonesia; architect(s) unknown, ca. 775–835. Aerial view. Restored 1971–1983.

Buddhist concept of the cosmos, organized around the mythical Mount Meru, the "Axis of the World," which rose from the Waters of Chaos. The whole precinct, standing on a 670-foot-square (200-meter) platform, represented a lotus flower, sacred to Buddha. Its three stages represented the major divisions of the universe: the material world, the world of thought, and the world of cosmic order and balance. From the eastern gateway, 3 miles (5 kilometers) of open galleries bore pilgrims through 10 levels of clockwise ascent to the top—symbolically, from the physical world to nirvana, the sought-after annihilation. Much of that processional way was lined with more than 24,000 square feet (2,500 square meters) of relief panels.

The lower five terraces—"the world of desire"—were square in plan, with 160 richly ornamented relief panels providing cautionary Buddhist tales, stories of Buddha's journey toward enlightenment, and a lively documentation of daily life in ninth-century Java. The next three terraces, circular in plan, had no wall reliefs, symbolizing the world of thought. A total of seventy-two bell-shaped, stone-latticework stupas was spaced evenly along them, each containing a statue of Buddha. The uppermost stage of the temple, originally rising to a height of 140 feet (42 meters), was a large central stupa crowned with a spire. Representing nirvana, it was empty.

Borobudur remained the spiritual center of Javanese Buddhism for about 150 years, until about 1,000 years ago, when it was suddenly abandoned. The reasons are probably complex. Its demise could have been due to natural disaster: soon after the building was finished, Mount Merapi erupted, depositing thick layers of ash over a large region and partially burying Borobudur. And at least in part, the departure from the site must be linked with the gradual transfer of power from central Java to the east, through the tenth and eleventh centuries. The jungle quickly reclaimed the great temple.

In 1814, Thomas Stamford Raffles, British lieutenant-governor of Java, hearing reports of the ruins, sent the Dutch engineer H. C. Cornelius to investigate. Cornelius found Borobudur so long neglected that his large work team took six weeks to clear vegetation and dirt enough to uncover only its outline. Spasmodic recovery work continued until the 1870s, when the last reliefs were exposed. But once the protective layer of soil was removed, the stone face began to deteriorate rapidly. Dr. Theodore van Erp began serious restoration in 1907, but it was discontinued after only four years. The newly independent Indonesian government took responsibility for preservation in the late 1940s, and a few years later it asked UNESCO for assistance. Consequently a major rescue project—costing U.S.$21 million and funded by the Indonesian government, UNESCO, private citizens, and foreign governments—was initiated in 1971. The restoration of the monument was completed by February 1983.

Further reading

Dumarçay, Jacques. 1991. *Borobudur*. Singapore/New York: Oxford University Press.

Frédéric, Louis, and Jean-Louis Nou. 1996. *Borobudur*. New York: Abbeville Press.

Miksic, John, and Marcello Tranchini. 1990. *Borobudur: Golden Tales of the Buddhas*. New York: Random House.

Brasília
Brazil

Brasília, the inland capital of Brazil, stands in a largely isolated region nearly 750 miles (1,200 kilometers) northwest of Rio de Janeiro. The design and construction of the city in such a remote place, uninhabited before 1956, was a major logistical achievement in planning and urban design. Conceived on the scale and in the grand manner of L'Enfant's Washington, D.C., of 1789–1791, it followed in the tradition of such cities as New Delhi, India (Lutyens and Baker, 1911–1931), and Canberra, Australia (Walter and Marion Griffin, 1913–1920). With its tall blocks in expansive landscaped parks, Brasília translated into reality for the first time the radical urban theories only envisioned in H. Th. Wijdeveld's *Amsterdam 2000* (1919–1920) and a little later in Le Corbusier's *Ville Radieuse*.

The plan to move Brazil's capital from Rio de Janeiro to an inland site, secure from naval attack, had been mooted first around 1789, and it was continually revived for the next thirty turbulent years. In 1823, soon after independence from Portugal was proclaimed, José Bonifácio presented the Constituent Assembly with a bill to fulfill the intention and to name the new city Brasília. Social and political upheavals dotted the rest of the century: burgeoning population; rapid economic growth; the spread of railroads; revolts and insurrections; civil and foreign war; the rise and fall of the Brazilian Empire; and, over thirty-five years, the abolition of slavery. The republic was proclaimed at the end of 1889, and the constitution of the United States of Brazil was adopted in February 1891.

That document defined the general location of the future Federal District: somewhere within the state of Goias on the sparsely inhabited 3,609-foot-high (1,200-meter) Central Plateau. The Exploring Commission of the Brazilian Central Upland was appointed, and it selected a 5,700-square-mile (14,400-square-kilometer) area—the "Cruls Quadrilateral" (named for the commission's Belgian leader, Louis Cruls). In 1953 a 2,300-square-mile (5,800-square-kilometer) section of it was chosen as the general site for the new capital. The announcement was expected to encourage a population movement westward into what was largely unused land, relieving urban congestion in Rio de Janeiro.

In September 1956 President Juscelino Kubitschek de Oliveira, promising Brazilians an economic development plan that he ambitiously called "Fifty Years in Five," initiated the foundation of Brasília. A design competition for a *Plano Piloto* (pilot plan) attracted forty-one entries from twenty-six architects and urbanists, and in March 1957 that of the Brazilian Lúcio Costa was announced as winner. His design was described by the president of the competition jury, British architect-planner William Holford, as "a work of genius and one of the greatest contributions to contemporary urbanism."

The importance of Costa's plan has been largely eclipsed by the beautiful, even spectacular, public

Brasília, Brazil, Lúcio Costa, planner; Oscar Niemeyer, architect, 1957–1960. View south along the major axis of the central city.

architecture of another Brazilian, Oscar Niemeyer, who had been his student at the Escola Nacional de Belas Artes early in the 1930s. They had collaborated before, and Niemeyer had also worked on urban design commissions for Kubitschek, when the latter was mayor of Belo Horizonte. For Brasília, Niemeyer designed the Congress Building; the law courts; the cathedral; the university; the National Theater; the Palácio do Planalto; the Palácio dos Arcos; and the president's residence, Palácio da Alvorada (Palace of the Dawn). It is interesting to note that construction of this presidential residence, and the airport, began in 1956, before Costa's success became public. The internationally reputed Brazilian landscape designer Roberto Burle Marx, who had previously worked with both Costa and Niemeyer, planned the major landscaping elements, a critical aspect of the capital.

Despite the general popularity of the vision, partly whipped up by the media, there was also strong dissension. But Kubitschek was determined to continue. Under the direction of Novacap, the corporation created to manage the project, the center of the city was built in the remarkably short period of three years. On 21 April 1960, Brasília was officially inaugurated as the capital. Soon after, Kubitschek was briefly replaced by Jânio da Silva Quadros, who solved national economic problems with draconian spending cuts, including projects at Brasília. That hiatus continued under the next president, reformer João Goulart. Then in March 1964 Goulart was overthrown in an army coup that brought military rule for the next twenty years. Although pressure would persist through most of the decade to return the seat of government to Rio, Brasília was confirmed as the national capital during the 1964–1966 presidency of General Humberto Castelo Branco.

The public cost of building the city remains unknown; some sources put it as high as U.S.$100 billion. The ways in which the money was raised and the efficiency with which it was spent are also under a cloud. It is claimed, for example, that the Banco do Brasil simply printed money for Novacap, almost on demand, and there were rumors that, at the start of the project, Brazilian air force transport airplanes carried equipment and building materials for the Palácio da Alvorada. Soon, a massive road-building program was initiated and highways were constructed to São Paulo and Belo Horizonte in the south, Belém in the north, and eventually westward to the Mato Grosso.

What of the urban form? In presenting his *Plano Piloto*, Costa explained that he intended to make a city that was monumental yet comfortable, efficient yet welcoming and intimate, spacious yet neat, rustic yet urban, and lyrical yet functional. The cruciform layout—some critics have compared it to a swept-wing aircraft, an analogy accepted by the planner—has its framework defined by "two axes, two terraces, one platform, two broad highways running in one direction, one super highway in the other."

The Monumental Axis runs east-west. At its eastern end, on the shores of Lake Paranoá (formed by damming the Paraná River), is the Plaza of the Three Powers. Around it are located the Supreme Court and the Congress Building with its twin twenty-eight-story towers and two striking hemispheres housing the Senate in a dome and the Chamber of Deputies in a bowl. The group is completed by the Palácio da Alvorada, surrounded by an inverted colonnade of white marble. The startling cathedral, redolent of a crown of thorns, and the university, are nearby. The lake wraps around the Plano Piloto, its shores dotted with embassies, private clubs, and sports facilities. From this grand focus, the broad Esplanade of the Ministries, flanked with buildings housing the bureaucracy, leads west to the central business district at the intersection of the main axes.

Each arm of the sweeping north-south Residential Axis is surrounded by nine bands of subdivision flanking an elevated highway. Those closer to the city core accommodate 780-foot-square (240-meter) residential *superquadras* (superblocks), most of which contain between eight and sixteen rectangular concrete-and-glass apartment buildings, usually six (but sometimes three) stories high, set in traffic-free parks. Each group was designed as a self-contained, middle-class neighborhood unit for an average of 3,000 residents, with shops, churches, schools, and playgrounds. Other recreational facilities serve a number of adjacent superblocks. The taller apartment buildings are raised on *pilotis*, so that at ground level the parks are uninterrupted. Open green space makes up about 60 percent of Brasília's total area—about five times as much per capita as, say, São Paulo. As elsewhere in the world, the imposition of an international modernist ideal on house form has not been socially successful; while doubtless well intentioned, it is not well received because it denies the tradition of household organization developed over centuries. The extensive, more upmarket residential developments, mostly one-family houses, are on the peninsulas known as Lago Norte and Lago Sud, across the lake.

Most of the people who work in support industries—domestic servants and others—live in one of the fifteen nearby satellite towns within the Federal District and commute by bus to the Plano Piloto. Some of the satellites are planned developments; others have grown laissez-faire. They have very little open space, and some have social problems stemming from high unemployment. Of course, government is Brasília's primary function, but it was inevitable that banking and commerce would flourish. Mainly because of the famous plan and architecture, tourism has also developed. Construction is an important part of the industrial infrastructure, but apart from that, only light industry is permitted.

Originally designed for 500,000 people, Brasília has grown rapidly. The 1960 population was around 90,000, and by 1980 it had increased to more than 411,000. A 1996 census showed that it had reached just over 1.8 million, and it probably rose to 2 million—mostly civil servants and businesspeople—by the turn of the century. Since about 1990 traffic problems such as gridlock and inadequate parking space have arisen in Brasília. A Y-shaped, partly underground rail system was started in 1992. Linking the south wing of the Plano Piloto with five of the satellite towns and with a total length of 26 miles (42 kilometers), it was designed to cater to two-thirds of the population. Commercial operation has been promised several times, but it still had not begun by 2001.

In 1987, Brasília was inscribed on UNESCO's World Heritage List. According to some residents, that was a mixed blessing for a living city: while it certainly increased tourist revenue and helps preserve the quality of life (for some), at the same time it inhibits the character of future expansion.

See also Chandigarh

Further reading

Epstein, David G. 1973. *Brasília, Plan and Reality: A Study of Planned and Spontaneous Urban Development*. Berkeley and Los Angeles: University of California Press.

Shoumatoff, Alex. 1987. *The Capital of Hope: Brasília and Its People*. Albuquerque: University of New Mexico Press.

Underwood, David. 1994. *Oscar Niemeyer and the Architecture of Brazil*. New York: Rizzoli.

Bricks

The humble brick literally shaped the face of world architecture. The Nile, Tigris-Euphrates, and Indus River valleys were the locations of what has been called "the urban implosion," the sudden emergence of cities from the neolithic villages that lined the waterways. The alluvial expanses on whose agricultural produce the new urban centers burgeoned had little naturally occurring stone, so the city walls, the buildings within, and even the royal palaces were built of brick. Packed clay had been used for centuries, and as it does in parts of the Arab world today, it yielded soft, curvilinear free forms. The advantage of the brick was that it was a prefabricated modular building unit, made easy to handle by its size and weight. Its shape and standard size—functions of the manufacturing process—inevitably generated a rectilinear architecture and affected the way people built (by assembling units rather than allowing the building to grow) as well as limiting such details as proportion and the subdivision of surfaces. Those causes and effects persist until this day.

Sun-dried bricks were made from puddled clay, perhaps containing a little sand or gravel, reinforced with a fibrous material (usually straw) that minimized cracking as the bricks dried. The mixture was packed into wooden molds, without tops or bottoms, that were removed once initial hardening had occurred. The bricks were then stacked and left to dry in the sun, sometimes for as long as two years, before being used in buildings. They were usually set in beds of wet mud, although the ancient Egyptians are known to have used gypsum-based mortar. The Babylonians employed hot natural bitumen, imported from lakes at Id on the Iranian Plateau; every several courses, the bed joints were reinforced with woven reeds. The dry climates of the river valleys presented few problems with weathering, but sometimes walls were plastered over with mud.

The Indus valley culture employed kiln-fired bricks long before its contemporaries, in buildings, pavements, and drains. Fired bricks also appeared a little later in Mesopotamia, where they were employed only in such special situations as decorative facings (with colored glazes) of public buildings or copings on walls. Timber for building was in short enough supply, and it was unreasonable to use it unnecessarily to fuel kilns. In the land between the rivers, sophisticated brick technology was early applied to massive structures like King Ur-Nammu's ziggurat at Ur (ca. 2100 B.C.). It was mainly of sun-dried brick, with thick facings of fired brick. Sixteen centuries later the Babylonian king Nebuchadnezzar II built his new city, with an 11-mile-long (17-kilometer) outer wall and an inner wall wide enough to allow two chariots to be driven abreast on its top. Both of these huge structures were of sun-dried brick, and the northern Ishtar Gate was faced with blue glazed brick, ornamented with colored brick relief figures of bulls, dragons, and other beasts. Nebuchadnezzar also refaced the Marduk ziggurat—thought by some to be the Tower of Babel—with a 50-foot-thick (15-meter) fired brick casing. Because the successive cultures that later dominated the region were builders in stone, the value of brick architecture was overlooked for centuries, to appear again in the Roman world.

For the pragmatic Romans, brick construction was more economical than stone, so the material was widely used. Brick making became a major industry that eventually was nationalized. To maintain quality control, brick makers were obliged to stamp their products with the brick maker's name and the place and date of manufacture. Flat Roman bricks, laid in thick beds of lime mortar, were used to build arches and principally as "lost formwork" in the ubiquitous concrete structures, in which they were covered with marble, mosaic, or stucco. Although it was maintained as a decorative material in the Byzantine Empire, with the decline of the Western Roman Empire, brick again went out of fashion. For several centuries after about A.D. 400, the only bricks used in western Europe were recycled from Roman buildings. It was only when those supplies were exhausted by about the beginning of the twelfth century that brick was again revived. As had been the case in the protohistoric river civilizations, necessity gave birth to invention, and brick architecture reappeared in the stone-poor Low Countries. Trade routes through Flanders were integral to the spreading use of bricks and clay roof tiles as building materials, and they moved as trade goods or as ballast in ships. Even toward the end of the Middle Ages, English architects and their clients regarded the brick as an exotic, luxurious, and somewhat suspicious building material.

Further reading

Davey, Norman. 1971. *A History of Building Materials*. New York: Drake.

Knevitt, Charles. 1995. *The Story of Brick*. Bedford, UK: Hanson Brick/Polymath.

Teutonico, Jeanne Marie, ed. 1996. *Architectural Ceramics: Their History, Manufacture, and Conservation*. London: James and James.

Brihadisvara Temple
Thanjavur, India

The so-called Big Temple, the Brihadisvara at Thanjavur (Tanjore) in the Indian state of Tamil Nadu, was built between A.D. 1003 and 1010. It is the epitome of Dravidian temple architecture and a wonderful gallery of South Indian art and craft. Vijayalaya Cholan (A.D. 846–871), founder of the Chola dynasty, chose the well-established settlement beside the River Kaveri as his capital, and for four centuries Chola influence on Indian religion, culture, art, and architecture spread from the royal city. Thanjavur is now a country town of about 200,000 people.

Chola dominion was extended under Vijayalaya's son Aditya I, the beginning of an empire that reached its apogee of power and prosperity under the greatest Chola ruler, Rajaraja (Arunmozhivarma), soon

Brihadisvara Temple, Thanjavur, India; architect(s) unknown, 1003–1010. Inner and outer eastern gatehouses.

after he became king in A.D. 985. By about 1005 he subjugated much of southern India, and the dynasty eventually controlled the Malay Peninsula, Sumatra, and parts of Sri Lanka. His conquests complete, until his death in 1014 Rajaraja turned his thoughts to religion and the arts, initiating many temples and replacing older brick shrines with ones of stone.

Although remarkably open-minded in religious matters, Rajaraja was a pious devotee of Siva, and the greatest of his cultural achievements was a masterpiece of South Indian art and architecture, the Brihadisvara at Thanjavur (also known as Peruvudaiyar Koil). He named it Rajarajeswaram. The temple precinct occupies most of the Sivaganga Fort, which was enclosed within perimeter walls in the sixteenth century. The fort is now bounded on the east and west by moats, and on the south by the Grand Anaicut Channel, part of an extensive eleventh-century irrigation system. To the north is the Sivaganga Garden, which also postdates the temple.

The Brihadisvara Temple stands within two concentric rectangular spaces. The outer court, measuring 793 by 397 feet (238 by 119 meters), is entered at its eastern end through a magnificent towered gatehouse, flanked by shrines dedicated to Ganapathi and Mrurgan. Facing the outer gatehouse is a similar structure, 90 feet (27 meters) high, that gives access to the carefully planned 500-by-250-foot (150-by-75-meter) stone- and brick-paved inner courtyard, at whose western end stands the main shrine of Sri Brihadisvara.

The temple proper, so to speak, is a complex suite of several elements. Although they are separate structures, they form an entity, the *garbhagriham*—that is, the holy of holies—surrounded by a 1,500-foot (450-meter) colonnaded cloister *(prakaram)*. The cloister's two levels, although dimly lit, are decorated with brilliantly colored frescoes from the Chola period, depicting the lives of the sixty-four Nayanmars (Saivite saints), the sacred bull *(nandi)*, and the ceremonial mount of Siva. There are also sculptured panels showing the Bharata Natyam dance postures *(karanas)* and the manifestations of Siva known as *Sivalingams*. Altogether, there are about 250 Sivalingams throughout the temple complex: the largest, 29 feet (8.7 meters) high, is set in a two-story sanctum. The main shrine also includes a large hall with open aisles *(Maha-mandapam)*, intended for religious discourse; another terraced hall enclosing the shrine of Sri Thyagarajar; and various ancillary halls for storing religious trappings and housing musicians. Standing in an elaborately decorated open hall in the inner court is a massive monolithic nandi, 12 feet (3.6 meters) high and nearly 20 feet (6 meters) long. There are several subshrines in the complex, but only one seems to have been built at the same time as the main temple.

The inner court is dominated by the 96-foot-square (29-meter) granite base of the *vimana*, a tower that rises through fourteen diminishing stories to a height of almost 220 feet (66 meters). Its facades are encrusted with hundreds of stucco figures of the myriad Hindu gods, standing in niches between carefully wrought pilasters. The vimana is crowned with an octagonal dome *(sikaram)* resting on an 80-ton (7.3-tonne) granite structure (contrary to popular accounts, it is not a single block), enriched with nandis at each corner. Rising above the dome is a 12.5-foot (3.8-meter) finial *(kalasam)* ending in a copper pot overlaid with gold plate, a gift of King Rajaraja. The generous endowments of the devout king and his sister Kundavai to the temple are recorded in inscriptions on the walls of the vimana. Everywhere, the surfaces of the building provide a vehicle for Chola art, making the Brihadisvara more than just an architectural masterpiece—it is also a magnificent repository of the highest artistic and craft skills of a golden age. Someone has said that the Chola artists conceived like giants and finished like jewelers. The temple was added to UNESCO's World Heritage List in 1987.

Beginning with Rajaraja the Great, the piety of the Chola dynasty is evidenced by more than seventy temples built in and near Thanjavur over the next two centuries. Noteworthy among them were Gangaikondacholisvaram Temple at Gangaikonda Cholapuram, whose vimana was a little shorter than that at Thanjavur; the more diminutive Airavateswarar Temple at Darasuram, described by one critic as "a sculptor's dream re-lived in stone"; and the Kampahareswarar Temple at Tribhuvanam. In each of the four, the vimana was taller—usually much taller—than the towers of the entrance gates; after them, Chola architects returned to their traditional forms, in which the relative heights were reversed.

Further reading

Balasubrahmanyam, S. R. 1977. *Middle Chola Temples: Rajaraja I to Kulottunga I (A.D. 985–1070)*. Amsterdam: Oriental Press.

Barrett, Douglas E. 1974. *Early Chola Architecture and Sculpture, 866–1014 A.D.* London: Faber.

Moorthy, K. K. 1991. *The Temples of Tamil Nadu*. Tirupathi, Andhra Pradesh, India: Message Publications.

Brooklyn Bridge
New York City, New York

When it was opened on 24 May 1883, the Brooklyn Bridge, joining the boroughs of urban Manhattan and semirural Brooklyn across New York's East River, was the longest suspension bridge in the world—twice as long as any previously built. More significantly, it was the first structure of its kind to be supported by cables of galvanized steel wire instead of the usual iron.

From the early seventeenth century through most of the nineteenth, the only transport link between Manhattan and Brooklyn was a ferry service, latterly the Fulton Street Ferry. As early as 1802 the New York State Legislature had been petitioned to build a bridge between Long Island and Manhattan Island, but it was not until 1857—a decade before the enabling legislation was passed—that serious consideration was given to the project. The German-born engineer John Augustus Roebling had been thinking about an East River bridge since 1852. Supported by influential local politicians Abram Hewitt and William C. Murray, he proposed a suspension bridge, composed of two 800-foot (247-meter) spans linked by a 500-foot (153-meter) cantilever section over Blackwells Island (now Roosevelt Island), close to the site of the present-day Queensboro Bridge. But the economic depression was followed in 1861 by civil war, two events that delayed the project until 1866.

Then the New York civil engineer Julius W. Adams proposed a suspension bridge, also on a different site from the final structure. In April 1867, under the entrepreneurship of William C. Kingsley, the New York Bridge Company was founded to build Adams's version of "The New York and Brooklyn Bridge." His design was superseded a month later by Roebling's scheme, prepared in collaboration with Wilhelm Hildenbrand. Their respective contributions are unknown, but most of the credit has gone to Roebling. The Bridge Company was a private corporation, but legislation provided that the city of New York might subscribe $1.5 million of the total capital, the city of Brooklyn $3 million, and private stockholders $500,000; in the event, more than 60 percent of the private funding came from Kingsley and his connections.

The total length of Roebling's bridge, including approaches and land spans, was to be 5,989 feet (1,796 meters); as built, it was some 800 feet (245 meters) longer. The 1,595-foot-6-inch (479-meter) span across the river would enter the tower arches 119 feet (36 meters) above the shore. A clearance of 135 feet (over 40 meters) at midspan would allow even the tallest ships to sail under the graceful arch. Roebling proposed to run extensions of the New York and Brooklyn elevated railroad tracks down the center of the bridge; they were to be flanked by vehicular carriageways. Above the railroad he designed an elevated pedestrian path.

In June 1869 the U.S. Army Corps of Engineers approved (over the signature of Ulysses S. Grant) the construction of Roebling's bridge. Surveying began without delay. Tragedy was just as immediate: while Roebling was locating the Brooklyn tower, a ferry collided with the Fulton slip on which he stood, crushing his foot. Within about a month, his masterpiece barely started, he died of tetanus poisoning. His son, Washington Augustus Roebling, was appointed chief engineer, and William Kingsley assumed superintendence of construction.

In 1870, with the necessary surveying and dredging completed, the foundations of the tower arches

were commenced. The two massive timber caissons were built at Webb & Bell's Greenpoint shipyard and towed 4 miles (6.4 kilometers) to the bridge site. As an indication of size, the *smaller* measured 168 by 102 feet (50 by 30 meters); built of foot-square (300-by-300-millimeter) flitches of yellow pine, its roof was 15 feet (4.5 meters) and its walls 9 feet (2.7 meters) thick. The hollow structures were sunk to the riverbed, piled with the granite blocks of the tower bases, as workers inside them removed the spoil, laboring in very uncomfortable and extremely dangerous conditions. After fourteen months' digging, the Brooklyn caisson reached bedrock in March 1871, at just over 40 feet (13 meters) under the riverbed. The caisson on the Manhattan side reached firm soil at almost twice that depth, although it was still thirty feet short of bedrock, by May 1872. The pressures experienced at such depths killed some workers, prompting Roebling's decision to go no deeper.

Having spent a lot of time in the Manhattan caisson, he also suffered "caisson disease" (commonly known as the bends). His whole body was crippled, and by the end of 1872 he was barely able to speak and beginning to go blind. With his wife, Emily Roebling, he sought treatment at a spa in Weisbaden, Germany, remaining for several months. Then they lived for three years in Trenton, New Jersey, the location of the Roebling wire works. In 1877 they returned to New York and took a house with a view of the bridge. From there, with Emily's help, Washington Roebling would supervise the remaining phases of construction.

Meanwhile, work continued on the anchorages, towers, and cables. The anchorage at either end of the bridge comprises thousands of tons of masonry, into which are embedded four huge anchor plates, from which 152 anchor bars in each plate take up the enormous tensile loads imposed by the four huge cables that carry the bridge's superstructure. The granite neo-Gothic towers, designed to resist the compressive loads exerted on them by the cables, took three years to build. The Brooklyn tower was finished in May 1875, and the Manhattan tower the following July. Rising over 276 feet (83 meters) above the river—equivalent to about twenty-eight stories—they were higher than any building in New York City except the spire of Trinity Church.

In August 1876 the two anchorages were linked across the East River by a wire rope. The spinning of the four bridge cables in situ—two outer ones and two near the middle of the 85-foot-wide (25-meter) bridge—began in February 1877 and was completed on 5 October 1878. The process combined 278 galvanized steel wires into a strand, and nineteen strands were bound into an iron-wire-wrapped cable, almost 16 inches (40 centimeters) in diameter; each cable could support 11,200 tons (10,200 tonnes). Secured at the anchorages and passing over the towers, they hung in a natural curve, or catenary. At the bottom they were attached to the center of the main span of the bridge deck. From the cables, vertical "suspenders," about as thick as a man's wrist, supported the deck along its length, assisted by a system of heavy wire ropes radiating in both directions from the towers. Construction of the understructure, the stiffening trusses, and the roadway began in March 1879.

The Brooklyn Bridge, also then known as the Great East River Bridge, was opened on 24 May 1883 when Hewitt formally presented it to the mayors of New York and Brooklyn. He boasted, "The cities of New York and Brooklyn have constructed, and today rejoice in the possession of, the crowning glory of an age memorable for great industrial achievements." The New York Bridge Company had been wound up in 1874, when the project was taken over by the cities of Brooklyn and New York. Instead of the promised three, the bridge took thirteen years to build. And, including a little under $4 million for land acquisition, it cost $15 million—at present values, around $1.5 billion—instead of the estimated $7 million. Until the marginally longer Williamsburg Bridge over the East River was completed in 1903, the Brooklyn Bridge remained the longest bridge in the world.

Major reconstruction was undertaken in 1954, when the engineer David Steinman modified the inner and outer trusses and removed the railroad tracks to widen the roadways. New approach ramps were built, and augmented in 1969. There has been a major rehabilitation of the main span and the approaches since 1979, and the latest renovation involved emergency redecking completed at a cost of $33.5 million in October 1999. The bridge was designated a National Historic Landmark by the U.S. Government in 1964 and a National Historic Civil Engineering Landmark by the American Society of Civil Engineers in 1972.

Further reading

Roebling, Karl. 1983. *The Age of Individuality: America's Kinship with the Brooklyn Bridge.* Fern Park, FL: Paragon Press.

Trachtenberg, Alan. 1979. *Brooklyn Bridge: Fact and Symbol.* Chicago: University of Chicago Press.

Weigold, Marilyn E. 1984. *Silent Builder: Emily Warren Roebling and the Brooklyn Bridge.* Port Washington, NY: Associated Faculty Press.

C

Cahokia mounds
Illinois

At a time when settlements in the Americas rarely exceeded 400 or 500 inhabitants, the Native American center of Cahokia was as large as contemporary London, a size that no other city in the United States would attain until the nineteenth century. The well-organized aggregation of mounds and residential districts had a population estimated at 10,000 to 30,000—some sources claim 40,000. Cahokia's distinctive earth mounds (there were 120 of them) took three forms: conical, "ridge top," and, most commonly, platforms, often crowned with ceremonial buildings or the houses of the powerful. At the heart of the city stood the huge ceremonial embankment (now known as Monks Mound) that was in itself a stupendous feat of planning and engineering.

The indigenous American civilization known as Mississippian—no one knows what they called themselves—sprang up in the American Bottom, an extensive fertile floodplain near the confluence of the Mississippi, Missouri, Illinois, Kaskaskia, and Meramec Rivers. Between about A.D. 1000 and 1250, they lived near what is now central and East St. Louis and where the Illinois towns of Fairmont City, Dupo, Lebanon, and Mitchell now stand. This suburban concentration was eclipsed by their greatest achievement: Cahokia, dubbed "America's lost metropolis." Cahokia was named for the branch of the Illinois people who occupied the region in the seventeenth century, long after the builders had departed.

In terms of both agriculture and trade, Cahokia was perfectly located. The predictable annual flooding of farmland enabled planning and replenished the soil so that maize and other crops were sustainable for centuries. The river systems reaching out to much of North America facilitated trade, and there is evidence of commercial traffic over a network that extended from Minnesota in the north to Mississippi in the south; Cahokian traders reached west as far as Kansas and east to Tennessee. Raw materials such as copper, seashells, and mica were imported and processed in Cahokia to be exported as copper ornaments and shell beads—indications of a sophisticated manufacturing industry. It was once believed that this productive economic environment led to population growth, as Cahokian civilization slowly flowered.

Recently, archeologist Timothy Pauketat has questioned this conclusion, claiming that there is no evidence for it. Although not all his peers agree, he suggests that Cahokia experienced an urban implosion in little more than a decade early in the eleventh century A.D., growing from a village of only 1,000 into a city ten times that size. Based on studies of wider Native American beliefs, that event may have been due to the emergence of a charismatic chief whose arrival prompted villagers to abandon their

Cahokia mounds, Illinois, 1000–1250. Modern stairway, Monks Mound.

settlements throughout eastern Missouri and southern Illinois and migrate to Cahokia.

It is now widely accepted that the Middle Mississippian area of which Cahokia forms a large part was under some kind of chiefdom government. Each chief—a Brother of the Sun—seems to have ruled a territory that depended upon a specific floodplain, and he managed food distribution between the central place and outlying settlements. Perhaps he had other roles, including matters of trade, administration of a civil service, and most probably religio-political duties. Little more is known.

However it came into being, the *fact* of Cahokia is staggering. Its earthen mounds extended over 6 square miles (15 square kilometers). At the heart of the city, defended by a wooden stockade, was the 200-acre (81-hectare) precinct of the ruling class, with the great ceremonial flat-topped mound at its center. The engineers and architects built to a master plan that was almost certainly based upon Mis-

sissippian cosmology—a sort of model of the universe. Cahokians viewed their universe as Father Sky and Mother Earth, and the layout of streets and structures mirrored that. The northern half of the city represented Sky, the southern half, Earth. They were defined by a long east-west street; another, running northeast, formed a cross symbolizing north, south, east, and west, its center point just in front of the central mound and at the end of a grand plaza. Archeologists have uncovered four circular solar calendars built of large, evenly spaced red cedar posts at the outer limits of the two streets. These "woodhenges," so called because they had the same purpose as Stonehenge in England, were essential to the Cahokians' agriculture-based economy, both in a practical and a ceremonial sense.

From about 1100 the central precinct, containing 17 earth mounds, was protected by a 2-mile-long (3.2-kilometer) stockade, constructed from some 15,000–20,000 1-foot-thick (30-centimeter) oak and hickory logs. The wall was about 12 feet (3.6 meters) high, with projecting bastions every 70 feet (21 meters) along its length. Outside it, thousands of single-family houses clustered, organized in small groups around ceremonial poles. Although it may have served as a social barrier between the Cahokian elite and the general population, it is clear from its form and the evidence of some hastily built parts that the palisade's main purpose was defense. It was rebuilt three times before 1300.

The inner city of Cahokia was dominated by an enormous platform mound, identified as the largest prehistoric earthwork in the Americas. Surviving today, Monks Mound was named after a Trappist monastery in the vicinity. Its base, measuring 1,037 by 790 feet (291 by 236 meters), extends over 14 acres (5.25 hectares), and the structure rises through four sloping-sided rectangular terraces to a height of 100 feet (30.6 meters). It contains 820,000 cubic yards (692,000 cubic meters) of earth, all of which was hand-excavated from large "borrow" pits and carried in woven baskets to the site. Monks Mound was built in several stages over about 200 years, with carefully designed strata of sand and clay, and drains to deal with water saturation. Long ago, it was crowned with a 50-foot-high (15-meter) thatched-roof building of timber-pole construction, 105 by 48 feet (31 by 14 meters). Some scholars identify it as a temple. It was certainly the chief's residence, in which the political and religious observances were conducted that ensured the nation's continuing prosperity. In effect, the mound was a means of lifting Mother Earth to Father Sky, bringing male and female together. That these ancient builders could set out their city with its streets aligned to the cardinal compass points and construct such a durable monument over generations, without having a written language or the wheel, makes their accomplishment the more marvelous.

Around 1200, for reasons that may only be guessed, Cahokia began to decline. Perhaps growth had placed too much burden upon the agricultural hinterland or overloaded the urban infrastructure; perhaps deforestation had changed the local ecology. Or perhaps there was civil war over dwindling resources. Other scholars attribute the demise of the city to a mud slide on the great mound, which may have been construed as an omen. No one really knows. And no one knows where the Cahokians went. By 1400 their remarkable metropolis was abandoned. Arriving much later in the area, the first Europeans mistook the mounds, overgrown by then, for natural hillocks. Monks Mound was not discovered until the beginning of the nineteenth century.

Modern farming, expanding towns, highways, and pollution continue to threaten those smaller communities around Cahokia that have not already been destroyed. The 2,200-acre (890-hectare) Cahokia Mounds State Historic Site is administered by the Illinois Historic Preservation Agency. It was added to UNESCO's World Heritage List in 1982. Archeological investigation continues. Following major slumps on the east and west sides of Monks Mound in the mid-1980s, attempts were made to reduce internal waterlogging. In January 1998 construction workers, drilling horizontally into the west side, struck a deep layer of limestone or sandstone cobbles 40 feet (12 meters) beneath the surface. Further tests were hampered by groundwater, but the find has excited scientists because stone does not naturally occur in the region. There is much more to be revealed at Cahokia.

Further reading

Pauketat, Timothy R., and Thomas E. Emerson, eds. 1997. *Cahokia: Domination and Ideology in the Mississippian World*. Lincoln: University of Nebraska Press.
Young, Biloine Whiting, and Melvin L. Fowler. 1999. *Cahokia, the Great Native American Metropolis*. Champaign: University of Illinois Press.

Canal system
England

The creation of England's inland water-transport network during the 1700s was among the most important contributors to the Industrial Revolution. In the second half of the century, manufacturing, already transformed by entrepreneurial labor management, was shifting from cottage industry to factories, where machines mass-produced goods. A cheap, efficient transport infrastructure was vital to gather raw materials and distribute products. Because

England's disjointed road network was inadequate, and because new industrial areas in the north were not always served by navigable rivers, the initiative (and money) of industrialists and merchants combined with engineering skill and a great deal of hard work to develop a national system capable of moving bulk goods. England's so-called Canal Age opened the country to the Industrial Revolution as the itinerant canal builders—they were known as "navigators"—changed the face of Britain. Between 1700 and 1835 some 4,000 miles (6,400 kilometers) of waterways were added to the 1,000 miles (1,600 kilometers) of navigable rivers.

England's first modern canal was the Sankey Brook Navigation, engineered by Henry Berry and Thomas Steers. Authorized by Parliament in 1755, two years later it was carrying coal to the industries of Liverpool on the River Mersey. In 1759 the third Duke of Bridgewater, Francis Egerton, impressed by a recent visit to France's Canal du Midi (1667–1694), proposed building a 10-mile (16-kilometer) waterway to link his Worsley coal mines with the River Irwell and thus with industrial Manchester. The millwright James Brindley (1716–1762) was employed to work on the project with the duke's land agent, John Gilbert. In the event, the Bridgewater Canal, which became operational by 1765, bypassed the Irwell, taking the coal directly to Manchester and Liverpool. It was more important than the Sankey Canal because it began a national network of waterways that would eventually join the manufacturing center of Birmingham to Britain's major rivers: the Mersey, the Severn, the Trent, and the Thames.

Over the next seventy years those rivers were connected by 2,000 miles (3,200 kilometers) of canals, and industrial regions like the Staffordshire Potteries and the Midlands Black Country prospered because of their access to national and world markets. Many of the most successful canals were built between 1760 and 1770, the first authorized to be built similar in size to the river navigations. But the construction cost of canal locks constrained developers to reduce their size, and as trade increased the narrow waterways could no longer meet demand. In the early 1780s an economic depression practically halted canal building, but recovery a decade later led to what has been called "canal mania," and there was a great deal of speculative promotion and ill-advised investment. Although final construction costs often exceeded estimates, most proposals were oversubscribed, often with ruinous results. Some schemes were profitable; others were abandoned during construction. Few showed much profit.

The construction of the canal system was an awesome enterprise. Some names appear often and they are generally interrelated. In many ways, James Brindley set the standards for those who followed. The success of the Bridgewater Canal gave him impetus for other canal projects: the Grand Trunk, the Staffordshire and Worcestershire, the Coventry, the Oxford, the old Birmingham, and the Chesterfield—altogether, a 360-mile (580-kilometer) network—were designed and constructed by this self-educated engineer. Another self-styled "civil engineer," John Smeaton (1724 –1792), built the Forth and Clyde Canal in Scotland and the Grand Canal in Ireland (with William Jessup, whom he trained). Jessup (1745–1814) worked on several river navigations and canals, mostly in the Midlands and eastern England; he was engineer on the Grand Junction, Ellesmere (later the Llangollen), and Rochdale Canals. Under Jessup, the famous Thomas Telford (1757–1834) was an engineer on the Ellesmere Canal. He became chief engineer on the Liverpool and Birmingham (Shropshire Union) Canal, where, unlike his predecessors who chose to follow land contours, he built embankments and made cuttings to follow a more direct route. He also made improvements to the Birmingham Canal systems. John Rennie (1761–1821) was a university-trained engineer who became surveyor and engineer on the Kennet and Avon Canal and on the Rochdale and Lancaster Canals.

Around 1830, investors began to turn to the new railroads. For a while canals and railroads were complementary, the canals carrying bulk cargoes while the railroads conveyed passengers and light goods. But by the mid–nineteenth century a national network of standard-gauge railroads had developed, and canal tolls were forced down. Most could no longer compete economically. Some railroad companies bought up canals and closed or abandoned them. But that was not the only reason for decline;

the other was that, although they were interlinked and covered large areas of Britain, the canals were never *conceived* as an integrated system. The canals for the most part were built piecemeal for local traffic using traditional regional vessels that often varied in size. Because there was no standard canal lock, a fragmented, inefficient transport system resulted. They gradually went out of use as commercial thoroughfares.

World War II witnessed a temporary revival. Following years of neglect and war damage, the canals were soon regarded as derelict. They were nationalized under the aegis of British Waterways in 1947 and over the next couple of decades their leisure and recreation value began to be recognized. The Inland Waterways Association was formed to "rescue" them, and volunteer restoration projects continue. Millions of pounds are being spent on maintenance projects, and there are now more craft using British canals than at the height of their commercial success.

See also Hydraulic boat lifts

Further reading

Boughey, Joseph. 1994. *Hadfield's British Canals*. Stroud, UK: Alan Sutton.

Chrimes, Mike, ed. 1997. *The Civil Engineering of Canals and Railways before 1850*. Aldershot, UK: Ashgate.

Paget-Tomlinson, E. W. 1993. *The Illustrated History of Canal and River Navigations*. Sheffield, UK: Sheffield Academic Press.

Cappadocia: underground cities
Turkey

Cappadocia, a region of central Anatolia in Turkey, lies within the triangle of Nevsehir, Aksaray, and Kayseri. It is bounded by the now dormant Mount Erciyes in the east and Mount Hasandag in the south. Prehistoric eruptions of these volcanoes blanketed a wide area with a 1,500-foot (450-meter) layer of ash and detritus. The hardening tufa was carved by nature into thousands of distinctive pyramidal rock formations known as "fairy chimneys," within which generations of settlers have created astounding subterranean cities. Guesses at the total number vary from 30 to 200. Carved from the living rock to a depth of at least twenty stories, and each able to house tens of thousands of people, the underground cities result from 3,000 years of continual adaptation and extension. Derinkuyu and Kaymakli, described below, are only two of such architectural feats in the region.

Who were these intrepid constructors, who built downward instead of upward, and whose houses were framed with shafts and corridors rather than columns and beams? Over millennia Cappadocia has been occupied in turn by invading Lycians, Phrygians, Persians, Greeks, Romans, Arabs, Byzantines, and Seljuk and Ottoman Turks. The indigenous Hittites were probably first to build underground. In the fourteenth century B.C., retreating from Phrygian invaders, they made excavations, normally of no more than two levels. The next major wave of building was not until the fourth century A.D. Always strategically vital, fertile Cappadocia became a Roman province in A.D. 17, and its towns flourished under stable Roman rule. Within about 200 years it became a center of eastern Christianity and when the persecution reached its final peak around A.D. 305, the Christians withdrew to the mountain fastnesses, building secure subterranean places in which to live and worship. The peril passed with the Edict of Toleration (A.D. 313) but reemerged for different reasons under the excesses of iconoclasm (726–843), as well as the incursions of Arabs. The Christian response to renewed threats was to build rock-cut churches and monasteries, often adapting and extending much older underground houses. The Göreme Valley abounds with well-hidden churches and monastic buildings—the number has been estimated at 600 to 3,000—carved out of the soft tufa. Most were built in the tenth century. The Seljuk Turks defeated the Byzantine army at the Battle of Manzikert (A.D. 1071) and then spread over Anatolia. They were followed in the fourteenth century by the Muslim Ottoman Turks. None of these changes put the Christian communities of Anatolia under threat, but by then rock-hewn architecture had become an established cultural expression.

At the beginning of the twentieth century, a Jesuit named Guillaume de Jerphanion began a long study of the well-preserved wall paintings that adorned many of the churches. International interest in Cappadocia was awakened when he published his research in 1925, but the great underground cities were not discovered until the 1960s. Two of the largest so far unearthed are Derinkuyu, located in 1963, and Kaymakli, 6 miles (10 kilometers) to the south, a year later. They were once joined by a well-ventilated tunnel, almost certainly wide enough to allow three people to walk abreast.

Derinkuyu, probably dating from the eighth century and capable of housing a population of between 10,000 and 20,000 inhabitants, was built around a 280-foot-deep (84-meter) main air shaft. The ventilation system had at least fifty smaller vertical vents linked by narrow horizontal corridors. This network formed a multistory building "frame," so to speak, and rooms—very comfortable living spaces, community kitchens, meeting rooms, chapels, stores, and even cemeteries—were cut to open from it. To date, eight levels have been excavated to a depth of 165 feet (55 meters), with twelve or more still buried. The top three levels appear to have been used as private and communal living quarters. Some scholars believe that each family unit had its own living room, bedroom, kitchen, toilet, and assorted storerooms. The lower levels housed storerooms and churches, and the lowest was a last resort of retreat in times of danger. It is possible that Derinkuyu was not permanently inhabited but served as a refuge at such times. Security was thus the main determinant in its planning: entrances were small and defensible, the ventilation outlets were carefully hidden, and there were several wells and a large cistern at the lowest level. Each section of the city could be isolated by large stone gates. Kaymakli was much the same, but only four of the eight levels remain accessible. The cities were last occupied during an Egyptian invasion in 1839.

Because of its unique geomorphic and cultural features, the entire region was added to UNESCO's World Heritage List in 1985. At the beginning of the twenty-first century, the underground cities, constructed as they are of soft tufa, are under threat from two main sources. Increasing tourism is exposing them to accidental and, sadly, deliberate damage.

More significant, climatic changes are turning the once-fertile surrounding agricultural land to desert. As farmers leave, the ecology changes: rainwater, once absorbed by vegetation, now permeates the soil, damaging the subterranean structures. Although appropriate technology is available to at least reduce deterioration, the severity of the problem and the fragility of the stone limit its application to the fascinating underground cities of Cappadocia.

Further reading

Demir, Ömer. 1990. *Cappadocia: Cradle of History: Göreme*. Ankara: International Society for the Investigation of Ancient Civilizations.

Kostof, Spiro. 1989. *Caves of God: Cappadocia and Its Churches*. New York: Oxford University Press.

Central Artery/Tunnel
Boston, Massachusetts

Toward the end of the twentieth century, Boston had traffic problems as severe as any city in the world. When the elevated six-lane Central Artery Highway, which ran through the downtown area, was opened in 1959, it quite easily coped with 75,000 vehicles a day; by the early 1990s the traffic load had increased to 190,000—effectively more cars per lane than any other urban interstate road in the United States. Movement was slowed to a snail's pace for over ten hours each day, and the accident rate was four times the national average for similar thoroughfares. Moreover, the urban area was divided by the elevated road so that access between the north and south sectors was greatly restricted. Naturally, the same congestion characterized the two tunnels under Boston Harbor that joined downtown Boston with East Boston and Logan Airport; the airport, only 1 mile (1.6 kilometers) from the central business district, was an hour away by road!

The $10.8 billion Central Artery/Tunnel Project (CA/T), conceived in 1981 and under construction as of 2001 by the Massachusetts Turnpike Authority, deserves a place among the engineering marvels of the modern world. Referred to by Bostonians as the "Big Dig," it is the largest, most complex highway project ever initiated in a U.S. city—indeed, the largest public works project of any kind in the United

States. Scheduled for completion in 2004, the project faces all the challenges associated with building in the heart of a busy city: that is, to meet the continuing demands of traffic capacity, to make sure that life and business are not unduly disrupted over a construction period lasting thirteen years, and to satisfy environmental and esthetic standards.

The spine of the multifaceted project is an eight-lane underground expressway directly under the existing road; in places its roof is at ground level, and at its deepest point it is 120 feet (36 meters) below ground, resting on bedrock. Tunneling was made especially difficult by the fact that there are four distinct soil types beneath Boston. Much of the downtown area is built on landfill placed at various times between the late eighteenth and the mid–twentieth centuries. Under the fill is a layer of mixed silt, sand, and peat, and below that the marine clay known as Boston blue on the bedrock. The demolition of the old elevated road releases about 27 acres (10 hectares) of open space for a linear park in the center of the downtown and for the construction of new city streets connecting North and South Stations; and existing cross streets, cut off since 1959, will be reconnected. Other advantages spring from the project, including a predicted 12 percent reduction in carbon monoxide levels and the creation (by using the spoil from excavations) of 105 acres (42 hectares) of open space at Spectacle Island in Boston Harbor and 40 acres (16 hectares) of new parks on the riverbanks below two new bridges.

The Central Artery rises to the surface at Causeway Street on the northern edge of Boston and crosses the Charles River on a 1,407-foot (42-meter), ten-lane, asymmetrical, cable-stayed bridge designed by the Swiss engineer Christian Menn. The bridge, constructed at a cost of $87 million, is the widest of its kind in the world. The Charles River Bridge links with National Route 1 and local access roads. The project also included a parallel, 830-foot (250-meter) four-lane bridge, also for local traffic, which was opened in October 1999. The Massachusetts Turnpike has been extended to Boston's international airport via a new tunnel connected to the four-lane Ted Williams Tunnel under Boston Harbor; the new tunnel was opened to commercial traffic in December 1996. Four highway interchanges will eventually connect the new roads with the regional system.

The part of the program that caused most local concern was the crossing of the Charles River. A proposal in August 1989, after construction had started on the Central Artery North Area Project, included three bridges, with a large area of the north shore being occupied by connecting ramps. Marshaled by the press (the *Boston Globe* dubbed the scheme a "grotesque monstrosity"), local residents, environmentalists, and even public servants opposed this design—to the point of litigation—on the grounds that it would "overwhelm their neighborhood with visual blight, shadows, noise and air pollution." A Bridge Design Review Committee (BDRC), appointed in January 1991, next produced an alternative plan that was not finalized until September 1992. Although this proposal won an Urban Design Award from the American Institute of Architects, the state rejected it, doubtless under political pressure. Instead, it prepared its own new plan, with two bridges and an underground ramp on the south shore; it was approved by state and federal environmental agencies in June 1994. Objections continued until 1997, when the U.S. District Court found that all necessary conservation measures had been taken. Construction began on the bridges in 1999.

The project has been under construction since late 1991. The extension of Interstate 90 through South Boston to the Ted Williams Tunnel and the airport was scheduled to open in 2001. The northbound lanes of the underground highway through downtown Boston will follow in 2002, and the southbound lanes in the following year, allowing the demolition of the elevated highway and the creation of the landscaping to be carried out by 2004. The federal government will meet about 70 percent of the cost, and the Commonwealth of Massachusetts the remainder. The Central Artery/Tunnel Project is owned and managed by the Massachusetts Turnpike Authority. Design and construction management was provided by the Bechtel-Parsons Brinckerhoff consortium.

Further reading

Hughes, Thomas P. 1998. *Rescuing Prometheus*. New York: Pantheon.

Luberoff, David, Alan Altshuler, and Christie Baxter. 1994. *Mega-Project: A Political History of Boston's Multi-Billion Dollar Artery/Tunnel Project*. Cambridge, MA: Harvard University Press.

Chandigarh
Punjab, India

When India won independence from the British in 1947, Pakistan and India were partitioned. The Punjab was divided and its capital, Lahore, was lost to Pakistan. Soon, East Punjab's population was quickly doubled by the flood of refugees from Pakistan. In March 1948 the provincial government, in consultation with the Indian central government and the enthusiastic support of Prime Minister Pandit Nehru, approved a new 45-square-mile (114-square-kilometer) capital site on a sloping plain near the Shivalik foothills. Designed by an international team under the leadership of Le Corbusier—it was his only realized urban planning scheme—the new city introduced India to a modern architectural and urbanistic idiom. Named for one of the two dozen existing villages in the area, Chandigarh, about 150 miles (240 kilometers) north of New Delhi, has been called "one of the most significant urban planning experiments of the twentieth century" and a "symbol of planned urbanism."

The Punjab government, on the crest of a wave of nationalism, probably would have preferred to commission Indian professionals, but none was suitably qualified. In December 1949 it approached the New York architect-planner Albert Mayer, who was then engaged on master plans for Greater Bombay and Kanpur. He accepted the Chandigarh brief: a master plan for a city of 500,000, detailed designs for selected buildings, and planning controls for adjacent areas. He assembled an expert consultancy team and involved Matthew Nowicki as codesigner. Their fan-shaped plan sat between two seasonal riverbeds that crossed the site. The seat of the state government was at its head, and the city center was located at its heart. Two linear parklands ran from the northeast head of the plan to its southwest tip, and a curving road network defined "superblock" neighborhood units like those of Brasília. The Americans also pro-

Chandigarh, Punjab, India; Le Corbusier, architect, 1951–1965. The building in the foreground is the Palace of Justice; the Secretariat is in the background.

vided concept sketches for the capitol, the city center, and a typical superblock. But when Nowicki died in a plane crash in August 1950, Mayer withdrew from the project.

The Punjab government then engaged the Swiss architect and urban theorist Le Corbusier, whose ideas were quite different from the Americans'. His team comprised the European modernists Pierre Jeanneret (his cousin), Maxwell Fry, and Jane Drew, as well as several Indian professionals. In four days of February 1951 Le Corbusier and his colleagues redesigned the city, covering approximately the same site but exchanging Mayer and Nowicki's garden city–influenced plan for an orthogonal grid. Nevertheless, many of their ideas, including the basic framework and such elements as the capitol, city center, university, industrial area, and a linear parkland known as Leisure Valley, were retained, in most cases in or at least close to their original locations. The "super-block" was replaced by the "sector," but the neighborhood unit remained as important in Le Corbusier's plan, although Mayer's curving streets were replaced with a rectangular layout. Indeed, the whole city plan was constrained by a system of seven types of roads, establishing a traffic hierarchy that provided quiet and safe residential areas. The Periphery Control Act of 1952 created a green belt and regulated development within 10 miles (16 kilometers), ensuring that Chandigarh would be surrounded by countryside. Sadly, in the event, that did not happen.

The 250-acre (100-hectare) city center was laid out immediately southeast of the intersection of the main axes of the plan. Its northern zone is reserved for commercial and civic purposes, and the southern zone is the district administration center, with courthouses, police headquarters, fire station, and the interstate bus terminal. The State Bank of India's branch office was designed by J. K. Chowdhury, and Pierre Jeanneret designed the Town Hall and the State Library.

Altogether, about 2,000 acres (800 hectares) of planted open space dot the city area. Le Corbusier included a chain of linear parks with pedestrian paths and cycle tracks from one end of Chandigarh to the other. A number of gardens transformed the eroded bed of a seasonal stream into the charmingly named Leisure Valley, where Chandigarh's cultural institutions—the Government Museum and Art Gallery, the Museum of Evolution of Life, the College of Fine Arts, and the College of Architecture—were built.

The capitol is Le Corbusier's pièce de résistance, and he began designing its buildings on his first visit to India, early in 1951. They were completed by 1965. The geometric concrete masses—the Secretariat, the Assembly, and the Palace of Justice—are grouped around a piazza and dominate the city. He also proposed a Governor's Palace, later changed in the proposals to a Museum of Knowledge, but it was never built. The Secretariat is the largest building in the complex, a huge 10-story linear slab that houses 4,000 civil servants. Its flat roofline is interrupted only by a sculptural composition containing a restaurant, ramp, and terraced garden. Inside, each floor is planned as a long central corridor with flanking offices, a layout expressed in the arrays of repetitive balconies on the long facades. An asymmetrical sculptural *brise-soleil* (sun-shading screen) provides visual relief to the sameness at about the middle of the building. The vast square Assembly Building is said to have been inspired by the geometrical forms of Maharaja Jai Singh's observatory, the Jantar Mantar, which Le Corbusier saw in Delhi, and (somewhat less romantically) by the cooling towers of a power station. A huge hyperbolic drum and a connected pyramidal skylight rise from the flat roof; beneath, in the base of the drum, the upper and lower parliamentary chambers receive dramatic shafts of light. A basement-level entrance is provided for daily use, and another huge ceremonial portal leads from the piazza through a massive enameled door decorated with a Le Corbusier mural. The external facades are protected from the sun on three sides with another sculptural *brise-soleil*. The Palace of Justice is a linear block whose main facade looks toward the piazza, continuous with the great central hall from which the courtrooms open. Wanting to express the majesty of the law, Le Corbusier designed for the judges a tall portico resting on three immense pylons painted in primary colors. The pylons serve to visually relieve the repetitive pattern of the arcaded facade. The Palace of Justice has a gigantic folded-concrete roof that shades the whole building. Le Corbusier also designed several monuments for the main piazza of the capitol complex. The most striking is the giant sheet-metal *Open Hand*, Chandigarh's official emblem; there is also a Tower of Shadows and a memorial to the martyrs of the Punjab partition.

Chandigarh is universally held to be an icon of European Modernism, and one Indian scholar has observed that it was "a pacesetter for post-independence architecture in India." He continues, "In contrast to the undifferentiated sprawl of contemporary Indian towns, Chandigarh is endowed with a specific identity—given by its picturesque setting, the well-ordered, orthogonal matrix and, above all, a distinctive architectural vocabulary" (Joshi 1999, introduction). For all that, and despite the fact that it provides (by Indian standards) good services and utilities, it wants for the sense of enclosure, density, and coherence that characterize the traditional Indian environment. Therefore, it fails in many nonmaterial respects to meet its inhabitants' needs.

See also Brasília

Further reading

Bhatnager, V. S. 1996. *Chandigarh—The City Beautiful.* New Delhi: A. P. H. Publishing.

Joshi, Kiran. 1999. *Documenting Chandigarh—Volume I.* Ahmedabad, India: Mapin Publishing.

Le Corbusier. 1952. *1946–1952.* Vol. 5 of *Le Corbusier, Oeuvre complete.* Edited by Willy Boesinger. Zurich: Artemis.

Channel Tunnel
England and France

The English Channel, known to the French as *la Manche* (the Sleeve), is a narrow strip of the Atlantic Ocean that separates England from the rest of Europe. It is at its narrowest at the hazardous Dover Strait, notorious for its strong tides, dense fogs, and frequent gale-force winds. The Channel Tunnel—popularly called "the Chunnel"—provides a railroad connection between Britain and France *under* the Dover Strait and is one of the most ambitious infrastructure projects ever undertaken in Europe, among the great engineering feats of the twentieth century. The complicated, visionary project was dogged by financial, logistical, and safety problems, exacerbated by two languages, two governments, two sets of legal requirements, ten contractors, and 220 financial backers in twenty-six countries.

The idea of the fixed link has a long history, the earliest recorded proposal, by a French engineer named Nicolas Desmaret, dating from 1751. About fifty years later Albert Mathieu Favier suggested a horse-drawn railroad under the Channel; his scheme included an artificial island that would serve as a staging post. Another French engineer, Aimé Thomé de Gamond, worked on several plans for almost forty years from 1830, making careful geological surveys of the seabed. In 1856 he proposed a railroad tunnel between Folkestone in the southern English county of Kent and Cap Gris-Nez on the French coast. A modified version was supported by the British engineers William Low and Sir John Clarke Hawkshaw in 1867, and a report was published the following year. For several reasons, largely political, the project went no further.

In the meantime the development of a pneumatic boring machine revolutionized tunneling techniques. In the mid-1870s Channel Tunnel companies were formed in England and France. In 1881 the South Eastern Railway acquired land near Folkestone, and the Submarine Railway Company bored 2,100 yards (2,000 meters) of pilot tunnel under the English Channel at Shakespeare Cliff. In France, 1,800 yards (1,600 meters) were drilled at Sangatte, southwest of Calais. Work stopped in May 1882 when the security-conscious British Parliament, afraid of undersea invasion, opposed the project. It remained in abeyance until after the Great War.

Work on a trial bore at Folkestone Warren in 1922 was aborted after only 140 yards (128 meters), again because of political antagonism in England. Despite support from eminent politicians, the Channel Tunnel was shelved until the Great Depression and another World War had passed. In 1948 the South Eastern Railway (by then Southern Railways) assigned its plans on to the nationalized British Railways, but it was not until 1956 that the French/British Channel Tunnel Study Group was formed to investigate the economic and engineering aspects of a fixed link. Four years later, it recommended a tunnel—in fact, two single-track railway tunnels and a service tunnel—between Folkestone and Sangatte. The two governments agreed to proceed with the project.

Years of surveys and research yielded a scheme the cost of which would be divided equally between Britain and France, and work began on both sides of the Channel in 1974. Only a year later Britain withdrew from the project when the estimated cost was increased by 200 percent. A pilot tunnel at Shakespeare Cliff was abandoned, and the project again lapsed. In 1984 it was once more agreed to in principle at an Anglo-French summit, and applications were invited from the private sector to build the tunnel. The successful tenderer for the design, planning, and construction, announced in January 1986, was Transmanche Link (TML), a consortium

Channel Tunnel, England and France; Transmanche Link consortium, engineers, 1986–1993. Diagram showing the relationship of the tubes.

of British and French corporations. The British Channel Tunnel Group (Balfour Beatty Construction, Costain UK, George Wimpey International, Taylor Woodrow Construction, Tarmac Construction, Midland Bank, and National Westminster Bank) was to build the English terminal and 14 miles (22.3 kilometers) of tunnels from Shakespeare Cliff. France-Manche, the French group (Bouygues, Dumez, Societé Auxiliaire d'Enterprises, Societé Generale d'Enterprises Sainrapt et Brice, Spie Batignolles, Banque Nationale de Paris, Credit Lyonnais, and Banque Indosuez) was responsible for the French terminal and the remainder of the tunnels from Sangatte. In order to finance the work, a private Anglo-French organization, Eurotunnel, was established and given a fifty-five year concession agreement to build and operate the link. Construction was under way, with three tunnel-boring machines at Shakespeare Cliff and three more at Sangatte by November 1987. The excavators met on 1 December 1990.

The 31-mile-long (50-kilometer) Channel Tunnel connects the terminals at Folkestone, England,

and Coquelles, near Calais, France. The submarine section is nearly 24 miles (38 kilometers) long. The two concrete-lined, single-track railroad tunnels, 25 feet (7.6 meters) in diameter, are spaced 98 feet (30 meters) apart, and a 16-foot-diameter (4.8-meter) tunnel between them is used for maintenance and ventilation. Two huge crossover chambers allow trains to switch tunnels. Maintenance-access cross passages every 1,230 feet (375 meters) link the central service tunnel and the rail tunnels. At 820-foot (250-meter) intervals, piston ducts arch above the service tunnel to link the others and relieve the pressure created by speeding trains. The tunnels are drilled through the rock at an average of 150 feet (45 meters) beneath the seabed. Electrical power for drainage pumps, lighting, and trains is fed from the national supply grids in England and France.

The Chunnel was officially opened on 10 December 1993, and Eurotunnel commenced its commercial operations six months later. At the time of completion, the project had cost U.S.$13.5 billion. Four different services pass through the tunnel: Le Shuttle carries tourist vehicles; Le Shuttle Freight handles commercial vehicles such as vans, trucks, and semitrailers; Eurostar transports pedestrian passengers; and other freight trains travel between Britain and mainland Europe. The journey between Paris and London takes just three and a half hours; the actual Channel crossing only thirty-five minutes. The Channel Tunnel is only one element of the European Community's plan for a 12,500-mile (20,000-kilometer) high-speed rail network linking cities across the continent.

See also Thames Tunnel

Further reading

Fetherston, Drew. 1997. *The Chunnel: The Amazing Story of the Undersea Crossing of the English Channel.* New York: Times Books.

Hunt, Donald. 1994. *The Tunnel: The Story of the Channel Tunnel, 1802–1994.* Upton-upon-Severn, UK: Images Publishing.

Charlemagne's Palatine Chapel
Aachen, Germany

The city of Aachen stands 40 miles (64 kilometers) southwest of Cologne on the River Wurm, a tributary of the Roer, in the German state of North Rhine-Westphalia. The Romans knew the place as Aquisgranum, famous for its health spas since the first century A.D. The Merovingian kings, who ruled the Franks from A.D. 481 to 751, held court there, but the town enjoyed great eminence during the Carolingian dynasty, especially under Charlemagne (reigned 768–814). His Palatine Chapel, now the central element of Aachen Cathedral, is the finest surviving example of Carolingian architecture. This architectural jewel copied the centrally planned Byzantine church of San Vitale at Ravenna, Italy (525–548), clearly demonstrating one way in which building ideas are transmitted between cultures. The ability of its northern builders to assimilate a southern European style was in itself a considerable achievement.

Charlemagne succeeded his father, Pepin the Short, as king of the Franks in 768. The first strong secular ruler in Europe since the ancient Roman Empire, he was in theory—but only in theory—subordinate to the pope, a relationship symbolized by his coronation by Pope Leo III as Holy Roman Emperor on Christmas Day 800. Six years earlier he had established his residence and Court at Aachen, the town where he was born. In 792, he commissioned Bishop Odo of Metz to design and build the royal complex, 50 acres (20 hectares) in area: the palace, law court, and, of course, the Palatine Chapel. Einhard (who was also Charlemagne's biographer) was appointed as works supervisor. Wanting to imitate the grandeur of the imperial Roman rulers, the king had looked for precedents. Historians have suggested that his palace was based on several models, Constantine's palatine court (ca. 310) in Trier, Germany, among them. Charlemagne also had been to Ravenna on Italy's northern Adriatic coast, where he had been dazzled by the glorious Byzantine buildings. Kenneth Clark opines that, when the Frankish king saw the scintillating mosaics in San Vitale, he "realized how magnificent an emperor could be." Returning to Aachen, Charlemagne gave instructions for a replica to be built as his private chapel.

Constructed at the southern end of the palace complex on the site of an earlier church, the domed octagonal Palatine Chapel was built between 796 and 804. It was consecrated by Pope Leo III in 805 to serve as Charlemagne's chapel, a reliquarium for his collection, and a church for members of the royal court. It is 54 feet (16.5 meters) in diameter and 124 feet (38 meters) high—at the time the largest dome north of the Alps. Of course, beautiful as it is, in the circumstances Odo's building could never have been a perfect replica. Architectural ideas are transmitted by several means: traveling architects, craftsmen, or patrons; images of buildings; and published theories. None is ideal. Images cannot convey the spatial aspects of buildings, and a visit to a building, no matter how perceptive and prolonged, leaves the visitor with mere impressions only. For those reasons, San Vitale lost a good deal in the translation, so to speak, even if Charlemagne imported columns and marbles from Ravenna and Rome and Byzantine craftsmen to assist with the work. Moreover, the refinement of the Italian church had been achieved after years of experiment with indigenous structural and decorative systems. Nevertheless, the Palatine

Charlemagne's Palatine Chapel, Aachen, Germany; Odo of Metz, architect, 796–804. Interior.

Chapel at Aachen is an extraordinary advance upon preceding Carolingian buildings.

It is much sturdier than San Vitale, having an unmistakably Roman structure. Like early Roman churches, it was approached from the west through a huge symmetrical atrium (said to have held 7,000 people), the well-defined entrance to the octagon flanked by towers with turret staircases leading to an upper level. Above the entrance was a place from which the emperor could appear to his people. None of the atrium survives. The octagonal central space of the original chapel is crowned with a lofty mosaic-faced dome constructed as a series of groin vaults; opposite the entrance, on both levels, was the sanctuary. The octagon is surrounded at the lower level by an ambulatory with a groin-vaulted dark sandstone ceiling. Those vaults, remarkable for the absence of transverse arches—Odo's own innovation—are supported at the angles of the octagon on large piers that also carry a semicircular dividing arcade. The upper level of the ambulatory is roofed with an annular barrel vault and separated from the octagon by a screen of two pairs of superimposed marble, porphyry, and granite columns within wide arched openings. At right angles to the main axis of the chapel, and reached at both levels through the sanctuary, were once mirrored north and south annexes.

On the decision of the members of the court, although he wished to be buried at St. Denis, Charlemagne's remains were interred in the Palatine Chapel in 814. Thereafter, until 1531, it became the imperial coronation church. From 1355, to accommodate the enormous traffic of pilgrims, the choir was rebuilt in the Gothic style, several chapels and a narthex were added, and the building became Aachen Cathedral. It was dedicated in 1414. The original mosaic on the interior of the dome was replaced by one Salviati, a Venetian, between 1870 and 1873. The cathedral was designated a UNESCO World Heritage site in 1978. A restoration program began in 1995.

Further reading

Collins, Roger. 1998. *Charlemagne*. Toronto: University of Toronto Press.

Conant, Kenneth John. 1973. *Carolingian and Romanesque Architecture, 800 to 1200*. Harmondsworth, UK: Penguin.

Hubert, Jean, Jean Percher, and Wolfgang Volbach. 1970. *The Carolingian Renaissance*. New York: Braziller.

Chartres Cathedral (Cathedral of the Assumption of Our Lady)
France

Chartres, capital of France's Department of Eure-et-Loir, stands on the Eure River, about 60 miles (100 kilometers) southwest of Paris. An important center in pre-Roman Gaul, it was one of the sacred places of the Druids. Overrun by the Normans, the region later settled down, and late in the thirteenth century it became the appanage of Charles de Valois, who was briefly (1284–1290) king of Aragon and Sicily. François I made it a Duchy in 1528. Louis XIV granted the Duchy of Chartres to the House of Orléans, an arrangement that lasted until about 1850.

Chartres prospered in the Middle Ages because it possessed a precious relic—a piece of oriental silk believed to be the veil worn by the Virgin Mary during the birth of Christ. Chartres therefore became an important pilgrimage site, and the chapter of the cathedral established trade fairs to coincide with the four annual feasts of the Virgin. The late-twelfth-century Cathedral of the Assumption of Our Lady at Chartres, recognized as "a reference point of French Gothic art," is a milestone in the development of Western architecture because it employs all the elements of a new structural system: the pointed arch; the rib-and-panel vault; and, most significantly, the flying buttress.

Only the Royal Portal on the west front (1150–1175) and the crypt remain of the Romanesque cathedral commenced on the site of an earlier church in 1145. The remainder was destroyed by fire in 1194, and, not least because religious fervor was running high in France, construction immediately commenced on a new cathedral, a "turning point in Gothic architecture" that rose upon the foundations of the old. Financed by all levels of local society (which also provided much of the voluntary labor), most of the building was finished by 1223. The cathedral was consecrated in 1260.

Chartres Cathedral, Eure-et-Loir, France; architect(s) unknown, 1145–1223. West front.

This Chartres was architecturally radical because the upper part of the walls above the arcades that separated aisles from nave was essentially stone frames for the expansive colored stained-glass windows, punctuated by the piers that carried the 112-foot-high (34-meter) quadripartite vaults. The lateral stability of earlier churches had depended upon massive masonry walls with frugal openings; here, there were diaphanous, luminous walls because the stability was provided by flying buttresses, used in a way previously unseen.

Flying buttresses, derived by the master masons by persistent experiment, are masonry arches that transmit the sideways thrusts of the stone roof vaults to vertical buttresses—in effect, very thick but very narrow walls at right angles to the building—constructed against the outside walls of the aisles. The resultant force of the thrust and the tremendous selfweight of the towering buttresses created a stable structural system. Once hidden beneath the roofs, at Chartres the flying buttresses were exposed and decorated as a feature of the architecture. There was now available a construction system in which rib-and-panel vaulting (employing the pointed arch) was carried by piers and buttresses whose stability was ensured by the dynamic balance of thrust and counterthrust. These had been used in Durham Cathedral, completed around 1133, and the pointed arch had been exploited in Suger's St. Denis a decade later. They reached a mature synthesis at Chartres in 1194, where, as one historian has observed, the master mason—sadly, he remains anonymous—"outlined new principles which would inspire all the great architects of the thirteenth century." After Chartres, the builder-architects of northern Europe further developed the structural skeleton whose columns, arches, and flying buttresses liberated the wall from its load-bearing function. The inevitable result was that the interiors of the Gothic cathedrals became loftier and lighter, illuminated by vast expanses of stained glass. Of those transcendent spaces, Chartres was the forerunner.

The cathedral is celebrated for its 152—originally there were 186—stained-glass windows, dating from about 1200 to 1235, with a total area exceeding 21,500 square feet (2,000 square meters). Used to instruct the illiterate masses, most are replete with figures from Bible stories and religious legends; others propagandize the trade guilds and organizations that paid for them. The architecture of Chartres is also enriched with sculpture; in all, there are about 2,000 figures, some dating from the Romanesque church. These figures, too, are remarkably innovative, because they are among the earliest medieval carvings to depart from the iconographic renderings of human beings to impart individual features, the reawakening of a naturalism that foreshadows the rise of Christian humanism in Europe.

Chartres has altered only a little in its 800-year lifetime. Another fire damaged it in the twelfth century, and the northwest spire was hit by lightning and replaced between 1507 and 1513. The church survived the political and religious conflicts of the sixteenth century and, remarkably, those of the French Revolution (1787–1799). The roof was damaged by fire in 1836, necessitating replacement. The current problem is more insidious, and preservation programs are in hand to guard against air-pollution damage to the historic stained-glass windows.

See also St. Pierre Cathedral

Further reading
Branner, Robert, ed. 1996. *Chartres Cathedral: Illustrations, Introductory Essay, Documents, Analysis, Criticism.* New York: W. W. Norton.
James, John. 1990. *The Master Masons of Chartres.* Sydney and New York: West Grinstead Publishing.
Miller, Malcolm. 1997. *Chartres Cathedral.* New York: Riverside Book Co.

Chek Lap Kok International Airport
Hong Kong

Hong Kong's new international airport at Chek Lap Kok is the product of what was at the time the world's largest engineering and architectural project—a logistical marvel that developed designs in only twenty-one months and managed a workforce of up to 21,000 to build the airport facilities as well as the island on which they stand and the extensive ground transport links, in only five years. In 1999, a convention of U.S. construction executives and editors named it one of the top ten architectural and engineering achievements of the twentieth century.

Anyone who flew into Hong Kong before mid-1998 will always remember the unnerving experience of looking directly into apartment buildings that seemed almost to touch the wingtips as the plane descended to Kai Tak Airport—a dubious thrill that is no longer part of a visit to the crowded island. Kai Tak airfield commenced operations around 1924, becoming a Royal Air Force base three years later. In 1935 it was upgraded to suit growing commercial traffic, and two more runways were added over the next twenty-five years. It was renamed Hong Kong International Airport in 1958 and underwent continual extensions and improvements as the number of flights increased at a dizzying rate. Shortly before it closed in 1998, Kai Tak was processing nearly 30 million international passengers and over 1.5 million tons (1.36 million tonnes) of international cargo every year.

There had been discussions about an out-of-town airport since the 1960s, within an international transport strategy that also included shipping; a plan to construct a new airport was announced in October 1988. Although well down the government's list of preferred sites (after Nim Wan, Lamma Island, and Clearwater Bay), Chek Lap Kok was chosen, but not unanimously. When it opened on 6 July 1998 the new airport had an annual capacity of 2.76 million tons (2.50 million tonnes) of cargo and 35 million passengers, planned to rise to 87 million by the year 2040. The Provisional Airport Authority, charged with planning and realizing the facility, was established in April 1990. The contract, estimated at almost HK$50 billion (then equivalent to U.S.$6.4 billion), was awarded to the Mott Consortium, comprising Mott Connell, Ove Arup and Partners; Fisher Marantz, Renfro Stone, O'Brien Kreitzberg and Associates; Wilbur Smith Associates; and the architectural firm of Norman Foster and Partners, which undertook the design of the terminal building.

The first construction stage project was the re-creation of the site. In 1992 Chek Lap Kok was a 330-foot (100-meter) hilltop rising from the sea; by June 1995 dredging and reclamation had reshaped it into a 3.7-by-2.2-mile (6-by-3.5-kilometer) flat platform—about four times its original area—23 feet (7 meters) above sea level. For the first year the airport operated with a single runway. Now known as the South Runway, it is used mostly for landings; the North Runway, put into service late in August 1999, is used principally for departures. Handling an average of 450 flights a day, Chek Lap Kok has forty-eight frontal aircraft gates at the terminal, twenty-seven on the apron, and thirteen cargo gates.

The 1,400-yard-long (1.27-kilometer), nine-level terminal building, under 45 acres (18 hectares) of 120-foot-wide (36-meter) steel barrel vaults, is the largest enclosed public space ever built. An indicator of the logistical achievement of the entire project, the superstructure of the vast Y-shaped building was completed in only three years. Its design was constrained by off-site fabrication of components that could be site-assembled, in much the same way as Joseph Paxton's Crystal Palace 150 years earlier. The air-cooled central terminal space, over 1,000 feet (300 meters) wide, houses the usual airport functions. More than 1 mile (1.6 kilometers) of moving walkways carry incoming passengers along the 2,400-foot (720-meter) concourse, through the baggage hall, to 124 immigration desks and seventy-six custom positions. Departing passengers are served by 288 check-in desks. Dimensions are difficult to convey; suffice it to say that the baggage hall alone is as big as New York's Yankee Stadium, and the fully automatic baggage-handling system can process 19,000 items an hour. There is also the inevitable shopping area—the "Hong Kong Sky Mall"—in five zones and comprising 154 specialist retail, food, and drink outlets. Nearby, the twelve-story Regal Airport Hotel, with 1,100 rooms and connected to the passenger terminal by a covered walkway, completes the facility. Internal shuttle trains run through a 20-foot-high (6-meter) tunnel, 106 feet (32 meters) wide, beneath the building. The design of Chek Lap Kok allows for expansion that will include an additional concourse and passenger terminal, as well as additional air cargo, catering, and maintenance facilities.

Chek Lap Kok was complemented by a complex Airport Core Project involving several elements and costing HK$155.3 billion (about U.S.$20 billion). The high-speed Airport Express Railway, part of Hong Kong's mass-transit rail link, and 21 miles (34 kilometers) of 3-lane highway across the Tsing Ma Bridge (the world's longest road-rail suspension bridge) provide alternative routes between the airport and Kowloon and further through the new Western Tunnel to Hong Kong Island and the central business district. The scheme also includes a new town for 150,000 people, because height restrictions, so necessary for Kai Tak Airport, have now been lifted. And, of course, the 2,350-acre (940-hectare) Kai Tak site became free for redevelopment. Plans are in hand for mixed commercial and recreational uses among residential towers accommodating 300,000 people. Work should be completed by 2003.

Ch'in Shi Huangdi's tomb
X'ian, China

In 1974, peasants digging a well in a field about 25 miles (40 kilometers) east of X'ian unearthed pits containing thousands of life-size, carefully detailed terra-cotta warriors, horses, and chariots. The soldiers were poised to defend the tomb of Ch'in Shi Huangdi (259–210 B.C.). Among the greatest archeological finds of the twentieth century, the ceramic army is but a small part of the great funerary monument—a necropolis with huge underground rooms around a gigantic burial mound—that the despotic ruler commissioned for himself many years before his death. The imperial tomb itself has not yet been uncovered.

In 246 B.C., when he was thirteen years old, Ying Zheng ascended the throne of Ch'in, the strongest of China's seven surviving territories. Unifying the divided states into a single nation, in 221 B.C., he took the title Ch'in Shi Huangdi (literally, "Ch'in, the First Emperor"). Great changes ensued in his short, tyrannical reign. The feudal system was abolished, and China was divided into about forty provinces, all controlled by a centralized bureaucracy. To ensure its efficiency over such a vast area, Ch'in Shi Huangdi commissioned the construction of over 6,000 miles (10,000 kilometers) of roads and more than 1,000 miles (1,600 kilometers) of canals, which also served for irrigation and flood mitigation. Southward, his empire extended to Vietnam's Red River Delta, encompassing most of what are now Yunnan, Guizhou,

and Sichuan Provinces; to the north, it reached as far as Lanzhou in Gansu Province and into parts of modern Korea. To defend his domain against nomad incursions, the first emperor commissioned the building of the Great Wall of China. He also initiated census taking, as well as the compulsory standardization of currency, weights and measures, writing, and even axle widths. As another means of control, in 213 B.C. he decreed that history and philosophy books, especially those contradicting Ch'in theories, should be burned. His despotism was resented by the common people. The foreign wars, the construction of the Wall, and other extravagant, self-indulgent public works (including his tomb), supported by policies of military conscription, heavy taxation, and forced labor, had imposed a terrible financial and social cost. Toward the end of his life, fearing assassination, Ch'in Shi Huangdi became reclusive. He died in 210 B.C., and his empire collapsed. After eight years of widespread rebellions, Liu Pang founded the Han dynasty.

The first-century-B.C. historian Sima Qian described Ch'in Shi Huangdi's tomb as a microcosm of the universe. Ironically, the first emperor's obsessive quest for an elixir of life had probably caused his madness and death; he had ingested mercury as a means to immortality. Because it was intended to serve as Ch'in Shi Huangdi's capital in the afterlife, the necropolis has many of the elements of a living city: encircling walls, parks and gardens, buildings for officials and the army, cemetery walls, and, of course, a palace. It was built mainly underground by (according to historical records) a labor force of 700,000 conscripts from all over China, over a period of thirty-six years. The 7,500-strong terra-cotta army stood guard in three vaults, about 0.75 mile (1.2 kilometers) to the east. Their weapons were looted, possibly during the uprising after Ch'in Shi Huangdi's death. The tomb complex proper, oriented perfectly to the cardinal points of the compass, was surrounded by a 65-foot-high (20-meter) wall that enclosed the rectangular imperial tomb gardens, covering an area of about 1.3 by 0.6 miles (2.17 by 0.97 kilometers). In the center of the precinct stood the building in which funerary rituals were performed. Close to it on one side were three blocks housing the

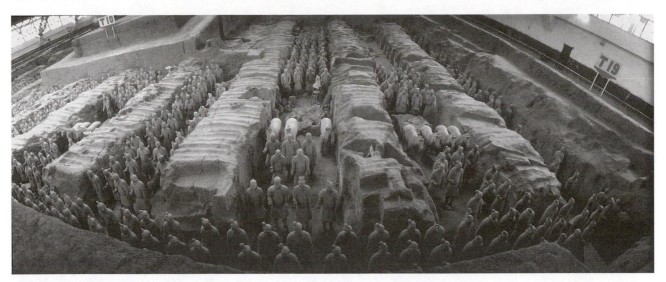

Ch'in Shi Huangdi's Tomb, X'ian, China, 221–210 B.C. Pit of the terra-cotta warriors.

Residence of the Garden and Temple Officials; on the other side were twenty-seven graves of Ch'in Shi Huangdi's high-ranking counselors and bureaucrats, buried with him so they could continue to serve him. Nearly 100 other pits (now containing the skeletons of horses and terra-cotta grooms) were the emperor's eternal stables. It is thought that other pits containing clay models of plants and birds were evocations of his parks and gardens.

The building known as the Main House, a sort of servery for Ch'in Shi Huangdi's food, stood near the 164-foot (50-meter) pyramidal grave mound, axially located at the southern end of the complex, within a second walled enclosure, measuring 749 by 632 yards (685 by 578 meters). There was a wide gate on each side. The burial chamber was lined with a waterproofing layer of bronze sheets. The tomb is believed to have been an opulent palace that accommodated all the emperor's needs, based on his accustomed extravagant lifestyle. According to reports, it was rich with "fine utensils, precious stones and rarities." There were scale models of palaces, towers, and official buildings, and a mechanically circulated system in which rivers of mercury represented the rivers of China and the Pacific Ocean, under a ceiling studded with pearls describing the constellations. Lamps burned whale oil to illuminate the space, and crossbow booby traps were installed to kill graverobbers. An official account reads, "Once the First Emperor was placed in the burial chamber and the treasures were sealed up, the middle and outer gates were shut to imprison all those who had worked on the tomb. No one came out. Trees and grass were then planted over the mausoleum to make it look like a hill" (cited in Cotterell 1981, 17). Archeological excavations continue at the site.

Yuan Zhongyi, leader of the team of archeologists working on the grave site, believes that the burial ground extends over an area of about 20 square miles (50 square kilometers); only a fifth of it has been uncovered. Work is funded by proceeds from the museum at the terra-cotta warriors' site; most of the money is used to maintain that site, but in 1997, Yuan Zhongyi's annual budget was only U.S.$25,000, about a tenth of what is needed. Consequently, the dig at the tomb was temporarily suspended. The team also lacked the special conservation skills needed to handle the 2,000-year-old artifacts of silk and wood. Work resumed in 1999, and new discoveries continue.

Further reading

Ch'in Shi Huangdi Pottery Figures of Warriors and Horses. 1981. Compiled by the Museum of Ch'in Shi Huangdi. Shanghai: The Museum.

Cotterell, Arthur. 1981. *The First Emperor of China.* London: Macmillan.

Wu, Bolun, and Zhang Wenli. 1990. *The Qi Shihuang Mausoleum.* Shanghai: The Museum.

CIAM (International Congresses of Modern Architecture)

Founded in 1928, the International Congresses of Modern Architecture (in French, Congrès Internationaux d'Architecture Moderne—CIAM) was the chief propagandist of avant-garde notions of architecture and urbanism—the voice of the Modern Movement—from 1930 to 1934 and again from 1950 to 1955. CIAM contended that architecture was inextricably linked with politics and economics and encouraged architects to turn from purely artistic endeavors to engage in social-engineering experiments with new urban and architectural forms—especially in housing. It was a principal milestone in the evolution of Western architectural thought.

The Swiss-French architect Le Corbusier unsuccessfully took part in a 1927 design competition for a new League of Nations center in Geneva and submitted a design that was not in a historical revivalist style. Rather, it responded to function and *zeitgeist*— the spirit of the age. Although his entry was rejected, ostensibly because it was not drawn in ink, it is most probable that conservative jury members conspired against the modernist proposal. The consequent scandal propelled Le Corbusier into the limelight, identifying him with avant-garde architecture. Some historians believe that an immediate outcome of the incident was the birth of CIAM. More positive impetus was given by the international acclaim for the Deutscher Werkbund's Weissenhofsiedlung (1927) in Stuttgart, Germany.

In Europe, the second half of the 1920s witnessed an interchange of the radical notions of contemporary architecture, largely effected by the modernist control of journals. Through publications and conferences (and by their contributions to Weissenhof), many German, Russian, Dutch, and French architects showed themselves eager to meet the "demands of industrialization" as great changes occurred in social structure. Acting together, architects could apply unified pressure to bring about the urbanistic and housing reforms they all believed to be urgently necessary.

In 1928 F. T. Gubler, secretary of the Swiss chapter of the Deutscher Werkbund, suggested to Madame Hélène de Mandrot that she offer her chateau at La Sarraz, Switzerland, for a meeting of twenty-five of Europe's leading architects. Austria, Belgium, France, Germany, the Netherlands, Spain, and Switzerland were represented. At a three-day gathering in June, facilitated by Le Corbusier and the Swiss critic Siegfried Giedion, CIAM was formed. The group was unanimous that rationalization and standardization must be priorities if the urbanistic and housing problems that each faced at home were to be humanely solved. The creation of CIAM, in an attempt to impose an international order on the varying aspects of the "new" architecture, established Modernism as a unified movement, complete with a manifesto and statutes. It even had a committee and an official address in Zurich—that of Giedion, who was elected as the founding secretary. Another Swiss, the architect Karl Moser, was CIAM's first president.

The La Sarraz meeting, really a clearinghouse for ideas, was dominated by Le Corbusier. But the Dutchman Mart Stam and the Swiss Hannes Meyer composed the closing declaration, simply restating the "best aspirations" and "fashionable fetishes" of the day and railing against academic conservatism. The second congress at Frankfurt (1929) dealt with more substantial issues, and discussion centered around Giedion's notion of *existenzenminimum*—low-cost residential units. As its deliberations were focused on urbanism and housing policies, CIAM was obliged to enter the political lists. Giedion argued that, in the same way that the individual living unit leads to the organization of construction methods, those methods lead to the organization of the entire city—a materialistic doctrine that ignored the complex social interactions, especially of the industrial city. City planning was therefore simply "architecture writ large." CIAM formed the Committee for Resolving the Problems of Contemporary Architecture (French acronym CIRPAC).

At the Brussels congress of November 1930, the Dutch architect-planner Cornelis van Eesteren was elected president, an office he held until 1947. The appointment flagged CIAM's shift toward rationalist urban planning policies, and the theme for the 1933 congress—the first of a planned series—was "The Functional City." After a conference planned for Moscow was canceled, members took a "working cruise" between Marseilles and Athens aboard *Patris II*. The outcome was the provocative *Athens Charter*, published anonymously in 1943, which reviewed earlier discussions, restated the capitalistic barriers to acceptable urban renewal or design, and identified the new problems of regional planning and urban contextuality. The charter was the closest CIAM ever came to a definitive credo. But it offered no specific solutions except the familiar generic one: modern technology. It called for balance between individual and community requirements; for dominance of the landscape over buildings, including generous urban green areas; for due consideration of physical environmental factors; for the conservation of historic buildings; and for separation of the main urban functions (living, working, recreation, and a carefully designed transport infrastructure). Moreover, housing should take priority among the urban planning. Legislation should ensure the provision of all these qualities. In it can be seen a legacy that persists in present land-use planning and zoning. The Italian historians Manfredo Tafuri and Francesco Dal Co offer the following criticism:

> To it probably belongs the credit for having founded a large measure of the predominant ideology of modern architecture, endowing architects with a model of action as flexible as it was already out of date.... It was also the most extreme demonstration of the radical diversities and the profound fragmentation of experiences that marked those early heroic years of contemporary architecture. Attempting to synthesize experience in large measure mutually contradictory, the Charter flattened out their originality, ignored their defeats, befuddled their tracks (Tafuri and Dal Co 1980, 219).

Reviews and revisions would be incorporated during the 1951 conference. The *Athens Charter* was republished, signed by Le Corbusier, in 1957.

The fifth CIAM congress, held in Paris in 1937, dealt with housing and leisure. When CIAM was overtaken by war in Europe, Giedion, Walter Gropius, Jose Luis Sert, Richard Neutra, and Stamos Papadaki sustained the group in the United States under the name CIAM, Chapter for Relief and Postwar Planning. Following World War II, some younger members envisioned other ways of considering the role of building within the context of urban design. The first postwar conference, organized by the British section of CIAM, the Modern Architectural Research Group (MARS), at Bridgewater in 1947, was followed by others at Bergamo, Italy (1949), Hoddeston, England (1951), and Aix-en-Provence, France (1953). The Dutch architect Aldo van Eyck had been a delegate at the Bridgewater meeting. He responded to what he heard there with the "Statement against Rationalism" and in 1956 at Dubrovnik, Yugoslavia, where CIAM was dissolved. He became a founding member of an international group—"a loose association of friends" of the Modern Movement—calling itself Team Ten, independent of CIAM and with a new agenda. Other rebels included Jan Bakema, Shadrach Woods, Giancarlo de Carlo, Georges Candilis, Alexis Josic, and Alison and Peter Smithson. Team Ten set its "own goals for a new, more humane system of public housing" and demanded a fresh comprehension of architecture, particularly within urban social life, in the context of rebuilding European cities. At the 1959 conference, the old regime was replaced by the new.

CIAM's urbanist ideals were realized on a massive scale in much of Europe's emergency reconstruction following World War II, when industrialized prefabricated building systems were applied to urgent needs for housing. The new dwelling type was also widely adopted in developing countries where revolutionary governments claimed to establish socialist societies. Almost without exception, these vast housing estates, wherever they are, evidence the failure of the Modern Movement's well-meaning ideas about mass housing. They have become concrete jungles, insecure, vandalized ghettos fraught with crime. Urban plans are sterile because of the strict adherence to functional zoning, and the massive housing blocks lack individuality and character. The overwhelming task now facing architects and urbanists is their rehabilitation.

See also Unité d'Habitation; Weissenhofsiedlung

Further reading

Mumford, Eric Paul. 2000. *CIAM Discourse on Urbanism, 1928–1960*. Cambridge, MA: MIT Press.

Tafuri, Manfredo, and Francesco Dal Co. 1980. *Modern Architecture*. London: Academy Editions.

Wilson, Colin St. John. 1995. *The Other Tradition of Modern Architecture*. London: Academy Editions.

Circus Maximus
Rome, Italy

The Circus Maximus stood in the Murcia Valley, between the Palatine and the Aventine Hills, the largest and oldest of the four chariot-racing tracks in ancient Rome. It was extended under various administrations until the time of Julius Caesar (100–44 B.C.). His alterations, and those ordered by his nephew, the emperor Augustus (reigned 27 B.C.–A.D. 14), created a building about 2,035 feet long by 460 wide (620 by 140 meters), with an arena measuring 1,850 by 280 feet (564 by 85 meters). On each side concrete vaults supported tiers of seats that accommodated at least 150,000 spectators: some sources put the number above 200,000, and others even more. For the purposes of comparison, the Houston Astrodome has a capacity of around 62,000, and Australia's Melbourne Cricket Ground holds only 100,000 spectators. Like many Roman public edifices, the circus, while not entirely a new building type (it was based on the Greek hippodrome), was built on a scale that the world had not seen before.

Founded when the city was part of the Etruscan kingdom (ca. 600 B.C.), the Circus Maximus remained the major site of diversions for the Roman populace for over a thousand years. The brook that ran through the Murcia Valley was diverted to a culvert, over which the central barrier *(spina)* of the hairpin track was constructed. The original circus was built of wood, but it was rebuilt and enlarged several times. In 196 B.C., Lucius Stertinius built an arch facing the starting gate, and a year or so later the censors for the games ordered the seating changed so that senators were separated from the plebeians. About thirty years later a vast stage was built for musicians and dancers, and the starting gate was altered. Julius Caesar commissioned a major reconstruction and extension in the first century B.C., and Augustus constructed a shrine that also served as an imperial box from which he could watch the races. In 10 B.C. he

erected an obelisk on the spina to commemorate his conquest of Egypt, bringing the Circus Maximus to its greatest glory. Dionysius of Halicarnassus described the Augustan arena as "one of the most beautiful and admirable structures in Rome."

Following disastrous fires in the wooden parts of the structure in A.D. 103, Trajan again restored the Circus. Each of the three stories of seats was divided by aisles. Marble seats in the first tier were reserved for senators—and for the equestrian class behind them. Senators were also allowed to sit along the podium that defined the track. The plebeians occupied the rows above the select seats. Unlike in other places of public entertainment, the sexes were allowed to sit together—a degree of permissiveness that some Romans considered scandalous.

Events other than chariot racing—animal hunts, gladiatorial games, athletic competitions, and processions—were held in the Circus Maximus. In order to display wild beasts, Julius Caesar had a water-filled moat 10 feet wide and 10 feet deep (3 by 3 meters) made around the arena. About a century later it was filled in to gain more seating space; for safety reasons animal fights were discontinued and eventually staged at the Colosseum. Although all kinds of entertainment were popular, chariot races remained the Romans' favorite spectator sport, probably for the excitement and the vicarious danger of the reckless races. The crowds fanatically supported the various professional racing factions, named for the colors worn by the charioteers: the red, green, blue, and white. The chariots—usually drawn by four horses—started from twelve gates leading from the Forum Boarium, near the starter's position. At the far end, where the track entered its sharp 180-degree turn, stood the triumphal arch (built in A.D. 80) through which processions entered the arena. The spina was adorned with gilded shrines, including one to Consus, a god of the harvest, and another to Murcia; at either end were the turning posts. Run under very strict rules, races comprised thirteen turns around those posts, a distance of approximately 4 miles (6.4 kilometers).

During the reign of Augustus, Rome gave no fewer than seventy-seven days a year to public spectacles; seventeen of those were for chariot races. Usually,

Circus Maximus, Rome, Italy; architect(s) unknown, ca. 600 B.C.–A.D. 103. Detail of model in the Museo della Civita, Rome.

twelve races were run each day, although the infamous emperor Gaius Caligula had the number doubled. It is reported that Domitian once had 100 races in a day but was forced, simply for the sake of time, to reduce the thirteen laps to five. By the fourth century A.D. the annual number of race days had risen to sixty-six. Convinced, possibly with good reason, that the circus was the devil's playground, the church fathers later condemned it. Nevertheless, events continued to be organized well into the Christian era, and the last race was recorded in A.D. 549, seventy-five years after Rome had fallen to the barbarians.

Now, the only visible remains of the Circus Maximus are at the semicircular end. The vaulted brick-and-concrete substructures of the seats on the Palatine side were uncovered by archeologists in the 1930s, and those excavations were extended in 1976.

A few years later, work began on the Aventine side of the same end. Every spring, and sometimes in the fall, the Roseto Comunale, Rome's municipal rose garden on the lower slopes of the Aventine, is opened to the public. Located about halfway along the southwestern side of the Circus Maximus, it presents a spectacle of a less exciting kind.

See also Colosseum (Flavian Amphitheater)

Further reading

Auguet, Roland. 1994. *Cruelty and Civilization: The Roman Games*. London: Routledge.

Humphrey, John H. 1986. *Roman Circuses: Arenas for Chariot Racing*. Berkeley and Los Angeles: University of California Press.

Plass, Paul. 1995. *The Game of Death in Ancient Rome: Arena Sport and Political Suicide*. Madison: University of Wisconsin Press.

Clifton Suspension Bridge
Bristol, England

The River Avon rises in the Cotswolds and falls about 500 feet (150 meters) in its 75-mile (120-kilometer) course to the Severn Estuary at Avonmouth. Near Bristol it passes through a channel that was cut in the nineteenth century to give access to oceangoing vessels, and then through the steep Clifton Gorge, where it is daringly crossed by the Clifton Suspension Bridge, 245 feet (75 meters) above the water. The iron structure, with a main span of 702 feet (214 meters), challenged conventional wisdom and pushed the new material and contemporary technology beyond the theoretical limits.

Bristol's port of Avonmouth was a well-established center for coastwise and international shipping. As the nineteenth century saw accelerating growth in trade and economic prosperity, Bristol's wealthier citizens wished to secure a market share for their city, and the renown that went with it, in the face of intense competition from such rivals as Liverpool. Perhaps they envied the prestigious bridge at Conwy, Wales, and the Menai Suspension Bridge, both designed by the Scots engineer Thomas Telford. Funds were in hand to start the project: the Bristol wine merchant William Vick, who died in 1754, had bequeathed £1,000 to build a bridge across Clifton Gorge; the money had been accruing interest while held in trust.

A design competition, announced in May 1830, attracted twenty-two entries, including four from the brilliant engineer Isambard Kingdom Brunel, who was then only twenty-four years old. The spans he proposed varied between 879 and 916 feet (267 and 279 meters); all were longer than any existing suspension bridge. The jury short-listed four designs (one of Brunel's among them), before seeking Telford's opinion. In an arrogant gesture he rejected all the

Clifton Suspension Bridge, Bristol, England; Isambard Kingdom Brunel, engineer, 1830–1864.

schemes. His given reason was pragmatic enough: his Menai bridge (1819–1826) had almost been destroyed by crosswinds; it was nearly 579 feet (175 meters) long, and Telford believed that nothing over 600 feet (184 meters) was feasible—the 700 feet across the exposed Clifton Gorge was out of the question. The committee then asked him to submit an alternative design, but the three-span bridge carried on soaring Gothic spires that he produced was unsuitable, even comical. A second competition followed in October 1830, and Telford resubmitted that design, only to see it again rejected. The twelve entries were reduced to four finalists, and Brunel's proposal, modified so that the main span was only 630 feet (192 meters), was placed second. He went to Bristol to meet the committee and convinced them with arguments about the practicalities and the esthetic quality of his tower design. He was appointed as engineer in 1831.

Brunel had an eye for the stunning landscape, with its high wooded cliffs, and his "Egyptian" towers, although not his favorite stylistic alternative, complemented the drama of the place. He had intended to have them inscribed with hieroglyphs and crowned with sphinxes, but the cost was prohibitive. There were delays for other reasons, including the 1831 Bristol riots associated with the Reform Bill, but lack of funds was the main problem. Work did not start until 1836. More financial shortfalls caused an interruption in 1853, and the piers stood untouched for some years, even being threatened with demolition. Reusing chains from another of Brunel's works, the demolished Hungerford Suspension Bridge (1841–1845) in London, the Clifton Suspension Bridge was finally opened in 1864, although the original design was not followed completely. Brunel had died five years earlier.

See also Menai Suspension Bridge; Royal Albert Bridge

Further reading

Body, Geoffrey. 1976. *Clifton Suspension Bridge: An Illustrated History*. Wiltshire, UK: Moonraker Press.
Vaughan, Adrian. 1993. *Isambard Kingdom Brunel, Engineering Knight-Errant*. London: John Murray.

Cluny Abbey Church III
France

The town of Cluny in eastern France's Burgundy region was important because of the Benedictine abbey jointly founded in 910 by Abbot St. Berno of Burgundy and William the Pious, Duke of Aquitaine. The third convent on the site, the great Basilica of St. Peter and St. Paul known as Cluny III (mainly 1088–1130), was the largest church, monastic or otherwise, in the world until St. Peter's, Rome, was completed in the seventeenth century. Cluny III was the high point of Romanesque architecture in France, and, heralding the Gothic, it emphasized the *continuity* of architecture. Its form and detail repudiate the idea of a succession of discrete styles, each somehow frozen in time.

The reformist Benedictine community that originally occupied a Gallo-Roman villa in Cluny eventually developed an innovative system of centralized ecclesiastical government: by the fourteenth century the abbey controlled over 1,450 Cluniac foundations or priories from England to Poland to Palestine, which together could boast a complement of over 10,000 monks. After the pope himself, Cluny's abbots were the most powerful clerics in the Roman Catholic Church and were at the epicenter of religious influence in Europe.

Two earlier abbey churches—the first, dedicated in 927, was succeeded by a larger building in 955–981—were replaced at the end of the eleventh century by Cluny III, which commenced soon after the other monastery buildings had been rebuilt (1077–1085). The new church was over 440 feet (136 meters) long; the narthex and towers added in the late twelfth and thirteenth centuries brought the total length to 600 feet (180 meters). The barrel-vaulted ceiling, especially acoustically suited to the Cluniac uninterrupted sung liturgy, soared 98 feet (30 meters) above the floor. There were double transepts and double aisles to both the nave and choir; the chevet end had five chapels. The ceiling of the crossing under a central tower was 119 feet (36 meters) high. Yet Cluny III was remarkable not just for its size.

Its form, emerging over more than a century, demonstrated the perpetual development of Western

Cluny Abbey Church III, France; architect(s) unknown, ca. 1088–1130. Artist's reconstruction (aerial view).

religious architecture. Since about 1000, the itinerant mason-architects of Europe had addressed their ecclesiastical clients' demands for stone-ceiling churches (perhaps prompted by fear of fire), dealing with the major structural problems that entailed. The need to manage the huge loads and thrusts involved had led (although not all at once) to a number of architectural and engineering innovations. Cluny III, a mature expression of the new form, incorporated them all, masterfully blending liturgical and structural necessities—the two towers at the west end to provide longitudinal stiffening; vaulted aisles to brace the walls of the nave against the thrust of the stone vaults; massive side walls reinforced with even thicker buttresses, employed for a similar reason; small windows, creating the appearance of what someone called "the fortresses of God"; and a complex east end, where apsidal chapels with hemidomes completed the lucidly articulated building, which showed exactly how the vast weight of the superstructure was gently coaxed down to the supporting earth.

At the same time, Cluny III had many features that foreshadowed what would be commonplace just a few decades later: piers disguised as clusters of narrow columns, elegantly tall proportions, pointed arches (a lesson from Islam), and sophisticated vault construction. It also had beautifully carved decorations, giving a glimpse of the reemergence of naturalism. Some sources claim that here were to be found some of the first medieval sculptural allegories (dating from 1095) and the prototype for many carved and painted west portals (dating from 1109 to 1115).

Cluny III influenced a few great buildings (for example, Paray-le-Monial, La Charité-sur-Loire, and Autun Cathedral). But clergymen are notoriously conservative, and the impact of its avant-garde architecture was therefore limited. Indeed, the design was attacked in a Cistercian polemic even before the work was completed. Pope Urban II, who had been a novice and later prior at Cluny, consecrated the high altar of the unfinished church on 25 October 1095.

He announced that its community had reached "so high a stage of honor and religion that without doubt Cluny surpassed all other monasteries, even the most ancient."

The abbey and the town both suffered in the religious wars of the sixteenth century. Early in the French Revolution the abbey was suppressed and then closed in 1790. Most of the basilica was demolished a few years later, and only ruins of the main southern transept and bell tower hint at what was once the greatest church in Christendom.

Further reading

Aubert, Marcel, and Simone Goubet. 1966. *Romanesque Cathedrals and Abbeys of France*. London: Vane.

Conant, Kenneth John. 1959. *Carolingian and Romanesque Architecture*. Harmondsworth, UK: Penguin.

CN (Canadian National) Tower
Toronto, Canada

The CN Tower, next to the city hall on Front Street, Toronto, stands on the shore of Lake Ontario. It transmits television and FM radio for more than twenty broadcasters, as well as serving various other communications purposes. Including the masts, it is the tallest freestanding structure in the world; the top of the transmission antenna is over 1,815 feet (553 meters) high. But at the beginning of the twenty-first century, as technically demanding as it is, height alone does not constitute an architectural feat. The twin Petronas Towers in Kuala Lumpur, Malaysia, currently rank as the world's tallest buildings, at 1,483 feet (454 meters). Others are proposed that will exceed that, including the 1,660-foot (508-meter) Taipei Financial Center on Taiwan, to be completed in August 2002, and the 2,100-foot (642-meter) Russia Tower in Moscow; at 2,755 feet (843 meters), the Millennium Tower in Tokyo will dwarf them all. The CN Tower is remarkable architecture because of its construction technique. For about a year, concrete, mixed and tested on-site to ensure consistent quality, was poured around the clock into a "slip form" that gradually decreased in diameter, to create the elegantly tapered contour of the post tensioned hollow structure.

Slip forming is a rapid construction technique based on extrusion. It employs a self-raising formwork that continually moves upward as the concrete is being placed, at a rate that gives the concrete time to set before being exposed as the formwork rises on a ring of hydraulic jacks, developing enough strength to support the work above. Continuous slip forming obviously speeds up the construction process while enabling excellent quality control, optimizing labor, and reducing the cost of building plant and scaffolding. It also results in monolithic, seamless structures. Developed in North America in the 1920s—The Granary at Logan Square in Philadelphia (1925) was one of the first examples in the United States—it has been widely used to build grain silos, building service cores, and (normally) any tall structures with a consistent cross section.

Early in the 1970s the number of multistory office blocks in downtown Toronto increased significantly, with a consequent interference with television and radio reception in large parts of the city. Toronto needed an antenna taller than any existing office block, indeed, of any that was anticipated, and the CN Tower was proposed to meet that need. The project was initiated in 1972 by the Canadian National Railway, which commissioned John Andrews Architects, working in collaboration with Webb Zerafa Menkes Housden Architects of Toronto. The structural engineering consultant was Roger R. Nicolet of Montreal; the mechanical and electrical engineers were Ellard-Wilson Associates Ltd. of Toronto; and the manager-contractor was Foundation Building Construction.

The original design proposed three concrete towers linked by structural bridges, but that was developed into a single tower with three hollow "legs." As well as serving as electrical and mechanical service ducts, the hollow columns provided the necessary degree of flexibility for such a tall structure. Construction started in February 1973, and in four months a Y-shaped, 22-foot-thick (6.7-meter) reinforced concrete base was founded on the bedrock 50 feet (15 meters) beneath the city. The continuous slip-form process then began. When the tower reached 1,100 feet (336 meters), a seven-story "SkyPod," fabricated on the ground, was raised into position

CN (Canadian National) Tower, Toronto, Canada. John Andrews and Webb Zerafa Menkes Housden Architects; Roger R. Nicolet, structural engineer, 1972–1975. View from Lake Ontario.

and anchored by twelve steel-and-timber brackets that were slowly pushed up the tower by forty-five hydraulic jacks. The concrete-walled SkyPod, reached by four high-speed, glass-fronted elevators, houses a 400-seat revolving restaurant, a nightclub, and indoor and outdoor observation decks. Later, a 2.5-inch-thick (6.4-centimeter) glass floor was installed. Beneath the SkyPod, delicate microwave dishes and other broadcasting equipment are protected by an annular radome. The concrete tower continues to the Space Deck at 1,465 feet (447 meters)—an observation gallery that on a clear day provides a view with 100-mile (160-kilometer) visibility. A Sikorsky Skycrane helicopter lifted the tower's 335-foot (100-meter) communications mast in forty sections, each of about 7 tons (6.4 tonnes), and they were bolted together in place. The mast, erected in three weeks, was covered by fiberglass-reinforced sheathing. The maximum sway experienced at the very top in 120-mph (190-kph) winds with 200-mph (320-kph) gusts is 3.5 feet (1.07 meters).

The CN Tower was completed in June 1975 and officially opened on 1 October. It cost Can$57 million and took about 1,550 workers forty months to construct. It is nearly twice the height of the Eiffel Tower and more than three times as tall as the Washington Monument. Soaring above Toronto, it is struck by lightning about seventy-five times every year.

In 1995 Canada National passed ownership to a public company, the Canada Lands Company. In June 1998, the CN Tower officially opened a 75,000-square-foot (7,100-square-meter) expansion including an entertainment center, shopping facilities, and restaurants.

Further reading

Campi, Mario. 2000. *Skyscrapers: An Architectural Type of Modern Urbanism*. Boston: Birkhäuser.

McDermott, Barb, and Gail McKeown. 1999. *The CN Tower*. Edmonton, Canada: Reidmore Books.

Colosseum (Flavian Amphitheater)
Rome

The Flavian Amphitheater, now in ruins, towers over the southeast end of the Roman Forum, between the Esquiline and Palatine Hills. Its popular name, the Colosseum, was derived from the nearby colossal (120-foot-high, or 37.2-meter) bronze statue of Nero, long since vanished. The most ambitious example of a new building type associated with urbanization, the Colosseum was an architectural feat, even by Roman standards. Its size is awesome, but the logistics of moving crowds to and from their seats was also a major achievement.

The earliest amphitheater on the site was built in timber for the *pontifex maximus* Gaius Scribonius

Curio in 59 B.C.; that was replaced about thirty years later by a stone-and-timber version for Augustus Octavian Caesar, the first emperor. The Colosseum was commissioned in A.D. 69 by Vespasian, whose son Titus dedicated it in A.D. 80. The highest part of that structure was also timber, and not rebuilt in stone until after A.D. 223. It seems that the first three ranges of seats were completed in Vespasian's reign, that Titus added two more ranges, and that Domitian completed the building around 300. Although early sources claim that the Colosseum seated 87,000 spectators, modern scholarship puts the figure closer to 50,000. Other Italian amphitheaters at Capua, Verona, and Tarragona are of similar size. The vast Colosseum, elliptical in plan, measured 620 by 510 feet (189 by 156 meters), covering nearly 6 acres (about 2.4 hectares). Its general height was 160 feet (49 meters).

The structural skeleton of the Colosseum was made of travertine limestone, quarried at Tivoli in the hills near Rome and transported to the site along a specially built road. Travertine blocks, some of them 5 feet high and 10 feet long (1.5 by 3 meters), were fixed together with metal cramps to form concentric elliptical walls. These were linked with radiating tufa walls carrying complex rising vaults of brick-faced concrete, in which volcanic stone such as pumice was used to reduce the weight. The vaults carried the tiers of seats. The Colosseum was built to house extravagant spectacles that took place in an arena measuring 280 by 175 feet (86 by 54 meters). Apart from a number of minor entrances to the arena, there were four principal gates at the ends of the axes, directly joined by passages to the exterior. A 15-foot-high (4.5-meter) wall, probably faced with marble, defined the arena and provided a measure of protection for the spectators. The floor of the arena was made of heavy planks, strewn with sand for the purpose of soaking up the blood of gladiators, prisoners of war, and wild animals that died in their thousands. Such emperors as Caligula and Nero even ordered cinnabar and borax to replace the sand. A labyrinth of chambers beneath the floor possibly housed the participants in the games, and there were complicated machines and hoists to lift men, beasts, and theatrical sets into the arena, adding to the spectacle.

Sometimes the entire floor was removed and the arena flooded by a system of pipes so that galleys could be pitted against each other in mock naval battles.

The terrace on top of the surrounding wall was wide enough to contain two or three rows of movable seats. Undoubtedly the best in the house, they were reserved for senators, magistrates, the vestal virgins, and other important people. The emperor and his immediate retinue occupied an elevated *cubiculum*. Upon entering the Colosseum through numbered arches corresponding to their ticket numbers, other visitors climbed sloping ramps to the *gradus* (bleachers), which were divided into stories and allocated according to gender and social class. The first fourteen rows of marble seats were covered with cushions and set aside for the equestrian order. Above them a horizontal space defined the second range, where a third class of spectators, the *populus*, was seated. Still further up were the wooden benches for the common people. The open gallery at the very top was the only part of the amphitheaters from which women were permitted to watch. There were exceptions, of course. When the games were over, the crowd could quickly disperse through no fewer than sixty-four strategically placed exits, aptly known as *vomitoria*.

The external wall of the Colosseum was divided into four stories, reflecting the circulation corridors within. Its eighty arches, most of which provided access to the interior, were framed by superimposed orders of pilasters (nonstructural columns): Tuscan on the ground floor, Ionic above them, and Corinthian at the top. The fourth story, also embellished with Corinthian pilasters, had stone brackets for the wooden masts from which an awning (velarium) was suspended across the interior to shield spectators from the sun while they watched the slaughter below. Many of the visible parts of the building were enriched with moldings, ornament, facings of marble or polished stone, and statuary. Fountains of scented water were provided for refreshment.

The Flavian Amphitheater was damaged several times by lightning strikes and repaired as often, so that games continued spasmodically until the sixth century, despite the opposition of the church and some Christian emperors. The last recorded slaughter of

Colosseum (Flavian Amphitheater), Rome, Italy; architect(s) unknown, A.D. 69–223. Detail of model in the Museo della Civita, Rome.

wild beasts was in the reign of Theodoric (A.D. 454–526), since when it has been used sometimes as a fortress and (to its detriment) as a quarry. Renaissance palaces in Rome, such as the Cancellaria and the Farnese, and churches including Saint Peter's Basilica, were built with columns plundered from the ancient monument. Various popes made efforts to preserve it, and in 1750 Pope Benedict XIV consecrated it to the martyrs who died there. Surprisingly, and despite popular belief, it was not the main venue for the execution of Christians. In 1996 a U.S.$25 million restoration of the Colosseum was launched. After the cellars were drained, fallen masonry replaced, bushes and weeds cleared from the arena, and the structure repaired and cleaned, the greatest amphitheater was reopened in July 2000 with a season of Greek plays.

See also Circus Maximus

Further reading

Luciani, Roberto. 1990. *The Colosseum: Architecture, History, and Entertainment in the Flavian Amphitheatre.* Novara, Italy: Istituto Geografico De Agostini.

Nardo, Don. 1998. *The Roman Colosseum.* San Diego: Lucent Books.

Pearson, John. 1973. *Arena: The Story of the Colosseum.* London: Thames and Hudson.

Colossus Bridge, Schuylkill River
Pennsylvania

The Upper Ferry bridge built at Fairmount near Philadelphia in 1812 and tragically destroyed by fire in 1838 was the longest single-trussed wooden arch in the United States, spanning over 340 feet (102 meters). It caused a sensation in its day and was inevitably labeled a new "wonder of the world," "the Colossus at Philadelphia," and "the Colossus at Fairmount." This covered bridge, responding to new constraints, took timber engineering to its limits.

At the beginning of the nineteenth century, driven by the need for agricultural growth, the population of the narrow coastal plain of the northeastern United States was spreading beyond the "tidewater" region. Before then, many short streams and estuaries had adequately met communication needs, but the inland farmers demanded roads, fords, and bridges. Water mills, increasing in number as farming increased, were of necessity sited where rivers could not be forded, and they also needed transportation routes. There were good supplies of building lumber in the region and the harsh climate was better suited to wooden construction than to masonry. The earliest bridges were merely logs carried on timber stringers; their spans were limited to the available lengths. As bridge technology developed, longer spans were achieved by joining stringers and employing trusses and arches. Climate was an important factor and the covered bridge soon became not only popular but also necessary. The roof protected the structural timber from alternate wetting and drying, discouraging rot and extending the life of the bridge. There is a story, perhaps apocryphal, of a Virginia builder who observed that bridges were covered "for the same reason that our belles [wear] hoop skirts and crinolines: to protect the structural beauty that is seldom seen, but nevertheless appreciated"—a delightful analogy.

The first covered bridge in the United States replaced a pontoon across the Schuylkill River in Philadelphia and was therefore optimistically called the Permanent Bridge. A stone bridge was originally intended, but when the abutments and piers were completed in 1804, the decision was made to span the river with timber. The New England bridge architect Thomas Palmer designed a structure braced with three arches and multiple king posts, and it was constructed by Owen Biddle, a Philadelphia architect and builder. When it was opened to traffic in 1805 it had no cover, but on Palmer's advice and the prompting of Permanent Bridge Company shareholders, a roof and clapboard siding were soon added. Palmer believed the covering would extend the life of the structure from twelve years to perhaps forty; it was still sound when replaced forty-five years later.

Within five years there was a demand for another bridge across the Schuylkill, to be built at Upper Ferry and connecting the area then known as Fairmont with the western bank. The design was put in the hands of Lewis Wernwag, an immigrant carpenter from Württemburg, Germany, who had already built bridges over Neshaminy and Frankford Creeks.

Wernwag's new bridge, built in 1812, was an elegant single-trussed arch spanning over 340 feet (102 meters)—certainly the longest of its kind in the United States and (according to some sources) the second-longest single-span bridge in the world at the time. The totally enclosed, elegant, low-arch bridge terminated in classical loggias at each end. Ten rectangular windows on each side provided light and ventilation for travelers. Graceful as it was, its achievement does not lie in its appearance but in the genius of its timber engineering. The wooden road deck was supported on five laminated arch beams that rose a little over 3 feet (1.07 meters) at midspan. On each side of the deck the river was spanned by a bow lattice beam, shallower at midpoint than at the ends and stiffened along its length with twenty-eight sets of double diagonal bracing. Iron tension ties anchored the beams to the ground at the masonry abutments, and others complemented the bracing along their entire length.

Wernwag's reputation was established as a builder of long-span wooden truss bridges, and he built several more, including the Hickman Covered Bridge (1838) in central Kentucky. Also known locally as the Wernwag Bridge, it was the longest cantilever wooden bridge in the country. The practice of building wooden covered bridges spread quickly throughout the United States, and literally thousands were built during the nineteenth century. The Covered Bridge Society of America identifies over 1,500 extant covered bridges throughout the world. Over two-thirds of them are in North America. Pennsylvania has 219, over half of which are still in use on public roads.

Further reading

Allen, Richard Sanders. 1983. *Covered Bridges of the Northeast*. New York: Viking Penguin.

McKee, Brian J. 1997. *Historic American Covered Bridges*. New York: Oxford University Press.

Colossus of Rhodes
Greece

One of the seven wonders of the ancient world, the huge statue of the pre-Olympian sun god Helios stood at the entrance to the harbor of Rhodes on the Aegean island of the same name. The work of the celebrated sculptor Chares of Lindos, the giant figure, shown in some representations to be shielding his eyes as he looked out across the sea, towered 110 feet (33 meters) above the entrance to the Mandraki harbor. According to Greek mythology, Helios was the son of the Titans Hyperion and Thea, and brother of Selene, goddess of the moon, and Eos, goddess of the dawn. He was worshiped throughout the Peloponnese, and the people of Rhodes held annual gymnastic games in his honor.

The cast-bronze shell of the Colossus, reinforced and stabilized with an iron-and-stone framework, stood on a white marble base. It has been suggested that, in order to attach the upper parts of the monument, earth ramps and mounds were built. Work commenced around 294 B.C.—although some sources put the date at ten years earlier—and the statue took twelve years to complete. Its size is hard to comprehend, but some idea can be gained from Pliny the Elder, who wrote, "Few people can make their arms meet round the thumb." From medieval times, artists' romanticized impressions have shown the Colossus straddling the entrance to Mandraki harbor, towering over the ships that sailed between his feet. Given its height, the width of the harbor mouth, and the technology available to the builders, that construct is most improbable. The fact is that no one knows exactly what the statue looked like, nor where it stood. Recent scholarship suggests that it stood on the eastern promontory of the Mandraki, or perhaps a little inland.

Rhodes was an important island in the ancient civilization of the Aegean. The Dorians inhabited it in the second millennium B.C., and their city-states of Lindos, Camiros, and Ialysos were vigorous commercial centers with colonies throughout the region. In the fifth century B.C., it belonged to the Delian League, a confederacy of city-states led by Athens, ties they severed in 412 B.C. Just four years later their own confederation was celebrated in the completion of the new city of Rhodes, said to have been designed by Hippodamos of Miletus; it seems more likely that it was laid out according to Hippodamean principles.

In 332 B.C. Rhodes came under the control of Alexander the Great, but following his death nine years later its citizens revolted and expelled the Macedonians. Rhodes's power and wealth reached a zenith in the second and third centuries B.C., and it became a famous cultural center. One badge of that political unity and artistic eminence was the Colossus, built to commemorate the raising of the Antigonid Macedonian Demetrios Poliorcetes' long siege (305–304 B.C.). The metal for the statue was taken from the siege machines abandoned by the invaders when they withdrew. It is said that the dedicatory inscription read, "To you, O Sun, the people of Dorian Rhodes set up this bronze statue reaching to Olympus when they had pacified the waves of war and crowned their city with the spoils taken from the enemy. Not only over the seas but also on land did they kindle the lovely torch of freedom."

A violent earthquake struck Rhodes about 225 B.C. The city was extensively damaged, and the Colossus, broken at the knee, crashed down. Ptolemy III of Egypt offered to meet the restoration costs, but when an oracle warned them against rebuilding, the Rhodians declined. It is ironic that the Colossus was actually lying in ruins when it was accorded a place among the wonders of the world. In A.D. 654 the Arabs invaded Rhodes, and two years later a Muslim dealer—some sources say a Syrian Jew—bought the fragments of the statue as scrap metal and carried them away to be melted down. Tradition has it that they were transported to Syria by a caravan of 900 camels.

In December 1999 the Municipal Council of Rhodes announced an international design competition for a new Colossus. As the island's millennium project, the monument will encompass "modern artistic expression and technical construction that will surpass conventional standards [while borrowing] all the ancient symbolic values of the original." Expected to cost U.S.$2.8 million, it is intended to be finished in time for the Athens Olympic Games in 2004.

Further reading

Clayton, Peter, and Martin Price. 1988. *The Seven Wonders of the Ancient World*. London: Routledge.

Cox, Reg, and Neil Morris. 1996. *The Seven Wonders of the Ancient World*. Parsippany, NJ: Silver Burdett.

Confederation Bridge, Prince Edward Island
Canada

The 8-mile-long (12.9-kilometer) Confederation Bridge, which crosses the Northumberland Strait between Jourimain Island, New Brunswick, and Borden-Carleton on Prince Edward Island, is the longest bridge over ice-covered water in the world. Its daring conception, the quality of its engineering, and the logistics of its realization are among the factors that make it one of the great constructional feats of the twentieth century. The project is also environmentally, politically, and culturally significant.

Prince Edward Island, on Canada's Atlantic coast, is the nation's smallest province, with a population of around 130,000. It lies in the Gulf of St. Lawrence at an average of 15 miles (24 kilometers) across the strait from mainland New Brunswick and Nova Scotia. The strait freezes for up to three months every year, and links with the island historically were expensive, freight and passengers having to be moved by ferry. In 1912 the Canadian government decided to build a railcar ferry to run between Borden-Carleton and Cape Tormentine, New Brunswick, and the *Prince Edward Island* was commissioned in 1917. In the first year she made only 506 round-trips. In 1938, as a response to wider automobile ownership, a car deck was added, and the vessel continued to operate until 1969. The subsequent decades saw improvements to the service, and new ferries now make the seventy-five-minute crossing at hour-and-a-half intervals. Prince Edward Island has become a vacation resort and by the beginning of the 1990s tourism had joined commercial fishing and agriculture as a mainstay of its economy.

Between 1982 and 1986 several consortia approached Public Works Canada (PWC) with proposals for a privately financed permanent link between the island and the mainland. Three were for bridges (the first estimated at Can$640 million), one for a tunnel, and another for a combined causeway-tunnel-bridge link. In December 1986, the central government instructed PWC to commission feasibility studies of fixed-link alternatives. By June 1987 twelve expressions of interest were in hand, and the acceptance of Strait Crossing's proposal was announced in

December 1992. Strait Crossing Development (SCD), a consortium of Janin Atlas, Ballast Nedam Canada, and Strait Crossing, was established to develop, finance, build, and operate the Confederation Bridge.

The proposal, put before the island population in a plebiscite the following January, was generally supported, but lobster fishermen and conservationists raised concerns that led to protracted delays. Their conservation measures won for the contractors the Canadian Construction Association's 1994 Environmental Achievement Award. Working with the Canadian Wildlife Service, SCD provided nesting platforms for endangered osprey in Cape Jourimain National Wildlife Area. The consortium also initiated a Lobster Habitat Enhancement Program, using dredged material to establish new lobster grounds in three formerly nonproductive locations. Construction work commenced in mid-July 1995.

The shore-to-shore Confederation Bridge consists of three parts. The 1,980-foot (0.6-kilometer) east approach from Borden-Carleton and the 4,290-foot (1.3-kilometer) west approach from Jourimain Island, New Brunswick, join the 6.9-mile (11-kilometer) main bridge across the narrowest part of the Northumberland Strait. Its two-lane carriageway rises from 120 feet (40 meters) to 180 feet (60 meters) above the water at the central navigation span. The bridge takes about ten minutes to cross at the design speed of 50 mph (80 kph).

Engineers designed for a 100-year life, taking into account the combined severe effects of wind, waves, and ice. In part, this was achieved by using concrete up to 60 percent stronger than normal in construction. The concrete employed in the 60-foot-diameter (20-meter) ice shields, designed to break up the ice flow at the pier bases, was more than twice normal strength. Because climatic conditions limited on-site construction to six months of the year, the bridge was designed to be assembled in the summers from posttensioned concrete components precast during the winters. The parts of the approach bridges were cast at a staging facility in Bayfield, New Brunswick, transported by land or water to the site, and assembled by a twin launching truss with a traveling gantry crane. Another staging facility was set up in Borden-Carleton to precast the 175 main bridge components. Some weigh as much as 8,000 tons (8,128 tonnes); the main box girders are 570 feet (190 meters) long, yet designed to be joined with tolerances of less than 1 inch (2.54 centimeters).

In August 1995 a purpose-built floating crane, the *Svanen*, began placing the components of the east approach bridge, completing it in November; the west approach was built the following spring. The main bridge followed, and by August 1996 the navigation span was the last to be placed. On 19 November the structure was complete: sixty-five reinforced concrete piers, founded on bedrock, supported the 8-mile (12.9-kilometer) superstructure which curves gracefully across Northumberland Strait. During the next six months, the finishing work—the polymer-modified asphalt cement road surface, traffic signals, emergency call boxes, weather monitoring equipment, closed-circuit television cameras, and toll booths—was carried out, and the bridge was opened on 31 May 1997. The estimated direct construction cost was Can$730 million.

Further reading

Macdonald, Copthorne. 1997. *Bridging the Strait: The Confederation Bridge Project*. Toronto: Dundurn Press.
Thurston, Harry, Wayne Barrett, and Anne MacKay. 1998. *Building the Bridge to P. E. I.* Halifax, Canada: Nimbus.

Coöp Himmelb(l)au

The "maverick Viennese partnership" Coöp Himmelb(l)au (literally, the "Sky Blue Cooperative") was established in May 1968 by Wolf D. Prix and Helmut Swiczinsky. Their architecture has been called expressionistic, spontaneous, irrational—all characteristic of the Deconstructivism that followed them. Why should they be included in an encyclopedia of architectural feats? Because they were the archetypal challengers, not altogether without success, of orthodox architectural thinking at the end of the twentieth century.

Until the late 1970s, when they took a "technological stance," drawing "airy therapeutic machines," their practice focused mainly on interior architecture. Twenty years later they unabashedly aimed to unsettle and create unrest, reacting mostly against the history-plundering aspects of postmodernism. Such contradiction of what were held to be architecture's "eternal truths"—harmony, unity, and clarity—must be seen as neo-Mannerism, playing it for kicks, so to speak. Their interior spaces and elements of their facades, thrust through with girders, giant needles, or spikes, create esthetic emotions of discomfort and disturbance rather than once-prized beauty. Someone has described their work as "an architecture of the chest spiked by the steering column."

Early designs of this kind included the Reiss Bar (1977) in Vienna, whose interior is split by a fissure ostensibly held together by massive turnbuckles. The front door is pierced by two huge spikes. The Red Angel Bar (1980–1981), also in Vienna, uses "tin, steel and glass block [to] embody the form and soul of the hovering angel, the wails of the sinners, and the protests of an antiestablishment youth." To enclose the space, wings spread out from the diagonal spike that forms the structural spine—a frequent motif in their buildings.

This approach climaxed in a number of projects and buildings, including a prizewinning master plan for the new town of Melun-Senart, near Paris, France (1987), a proposed city center for St. Polten, Austria (1989–1990)—urban design schemes in which the excitement of polemic eventually gave place to the pragmatics of city bylaws—and a hilltop studio for Anselm Kiefer in Buchen, Germany (1990). It can be found also in the Funder Factory 3 in St. Veit/Glan, Austria (1988–1989). There, Coöp Himmelb(l)au "dissolved" what might easily have been a boring long-span industrial shed into "an amalgam of more sculptural, functionally differentiated elements," with spectacular results: a main building with a red entrance canopy, a power plant with three 75-foot-high (23-meter) chimneys that lurch drunkenly, and an assembly-line building whose corner is an exploding structure of steel and glass. The same dynamism may also be seen in a strangely compatible penthouse addition (1984–1988) to a neoclassical building in Vienna. It has been described as "biomorphic … an exposed exoskeletal structure" whose boardroom looks like a "dissected ribcage." Outside, it looks very much like a huge beetle with spread wings, scrabbling for a foothold on the roof.

In June 1989 Coöp Himmelb(l)au won (with locally based Morphosis and Burton and Spitz) first prize for a pavilion in a Los Angeles performing arts park. Opening an office in the U.S. city, they secured several commissions in southern California. In 1993 they won a competition for the Jussieu Campus Library of the University of Paris and were commissioned to design the east pavilion of the Groninger Museum, Groningen, the Netherlands, completed in 1995. The following year they represented Austria at the International Architecture Biennale in Venice. They have recently completed the eight-theater UFA Cinema Center in Dresden, Germany (1993–1998), and the SEG Apartment Tower in Vienna, a complex of residential and other towers and a school (1994–1998). Working with a staff of twenty-seven, they undertook a project to convert the shell of a former Vienna gasometer into a multipurpose building; another complex in Hamburg, Germany; an entertainment center in Guadalajara, Mexico; and a building for Expo 2001 in Biel, Switzerland, all between 1995 and 2000.

Further reading

Gruenberg, Oliver, Robert Hahn, and Doris Knecht, eds. 1988. *Coop Himmelb(l)au: Power of the City*. Darmstadt, Germany: G. Büchner.

Peter Noever, ed. 1991. *Architecture in Transition: Between Deconstruction and New Modernism*. Munich: Prestel.

Steinbauer, Jo, and Roswitha Prix, trans. 1983. *Coop Himmelb(l)au, Architecture Is Now: Projects, (Un)buildings, Actions, Statements, Sketches, Commentaries, 1968–1983*. New York: Rizzoli.

Crystal Palace
London, England

The Crystal Palace, a vast demountable building designed by Joseph Paxton for the Great Exhibition of 1851 in Hyde Park, London, was in many ways crucial in the development of architecture: it was the pinnacle of innovative metal structure, it revealed the exciting potential of efficient prefabrication, and it was an early demonstration of the modern doctrine that beauty can exist in the clear expression of materials and function. Altogether, it was one of the most noteworthy buildings of the nineteenth century.

Crystal Palace, (originally in) Hyde Park, London, England; Joseph Paxton, architect, 1850–1851. Contemporary engraving from *The Illustrated London News*.

The idea for a Great Exhibition came from the Society for the Encouragement of Arts, Manufactures, and Commerce, and was given impetus by Henry Cole, then an assistant keeper in the Public Records Office. His wide interests extended to the publication of *The Journal of Design* that encouraged artists to design for industrialized mass production and urged manufacturers to employ them. That, he believed, would raise the quality of everyday articles. Cole was elected to the society's council in 1846, and the following year, with others, he successfully solicited Queen Victoria's consort, Prince Albert of Saxe-Coburg-Gotha, to accept the role of its president. Under Royal Charter, and spurred by the success of French industrial expositions since 1844, the society held Exhibitions of Art Manufactures from 1847 through 1849.

After visiting the exclusively French exhibition in Paris in 1849, Cole realized that an international show would inform British industry of progress (and commercial competition) elsewhere in the world. Prince Albert, convinced that "that great end to which all history points—the realization of the unity of mankind" was imminent, caught the vision. The Royal Commission for the Exhibition of 1851 was established to expedite a self-financing "large [exhibition] embracing foreign productions." It was envisioned as "a new starting-point from which all nations will be able to direct their further exertions," but it was at the same time an expression of British nationalism. Britain had led the world into the Industrial Revolution, and her outlook was smug, to say the least. The Great Exhibition would provide a vehicle to flaunt her industrial, military, and economic superiority and justify her colonialism.

The show was to have a display area of 700,000 square feet (66,000 square meters), much bigger than anything the French had managed. That was too large even for the intended venue in the courtyard of Somerset House, so it was decided to locate it in Hyde Park. An open competition for the design of a building for the "Great Exhibition of the Works of All Nations" attracted 245 entries from 233 architects, including 38 from abroad. The Commissioners' Building Committee liked none of them; besides, it was unlikely that any could have been completed on time. Having prepared its own plan for a large dome standing on a brick drum, the committee called for bids. The result was alarming: building materials alone would have devoured at least half of the available funds of £230,000. Anyway, the design was generally considered ugly, especially by the architects whose proposals had been rejected.

Fox and Henderson and Company, a firm of contractors, engineers, and ironmasters, tendered a price for an alternative, based on a design by the gardener Joseph Paxton. In 1826 Paxton had been appointed head landscape gardener at Chatsworth, the Derbyshire estate of the sixth Duke of Devonshire. He built large conservatories there, including one in 1836–1840 for the giant water lily, *Victoria regia*. Paxton claimed that his design for the Great Exhibition building was inspired by the structure of that lily, whose cross ribs strengthened the main radial ribs.

Learning that the invited architects had been turned down, Paxton had sketched out his proposal on a sheet of blotting paper—romantic tradition says it was during a train journey—and through a lucky meeting with a mutual friend he was able to show it to Cole. The idea was simple: a modular structure of a single cross section, built from prefabricated metal components, could be repeated ad infinitum to produce a building of any size. Paxton promised Cole that he would have detailed designs ready within a fortnight. In fact, they were completed in nine days and passed to Fox and Henderson on 22 June 1850. By then, the provision of a building was becoming urgent. Paxton's proposal had the desirable advantage of rapid construction; moreover, unlike the other schemes, it could later be demounted to leave Hyde Park relatively undisturbed. The commission accepted it; the only modification asked for was a vaulted transept so the building could contain without damage the large elm trees on the site.

The Crystal Palace, as it was soon dubbed, was a single space, 1,851 feet long and 456 wide (554 by 136 meters), rising by 20-foot (6-meter) increments across flanking tiered galleries to a 66-foot-high (20-meter) central nave. It was intersected in the middle by a 108-foot-high (32-meter) vaulted transept. The building covered 19 acres (7.6 hectares) of Hyde Park. A filigree of 330 slender, cast-iron columns and arcades supported its clear glass walls and roofs and the wrought-iron beams that carried the galleries, alternately 24 feet (7.2 meters) and 48 feet wide.

Due largely to Paxton's consummate organizational skills, Fox and Henderson accomplished its construction between September 1850 and January 1851. The Birmingham glassmaking firm of Chance Brothers supplied almost 294,000 panes, which were fixed in a specially designed roof-glazing system based on economical 49-inch-wide (1.25-meter) sheets that determined the module for the entire design. Building work on-site consisted mostly of assembling the 3,920 tons (3,556 tonnes) of cast-iron components that came from ninety different foundries throughout Britain, often cast less than a day before they were fixed. The accuracy obtained through prefabrication and the mechanical fixing dramatically reduced the proportion of nonproductive labor common to traditional construction methods. Cast-iron columns were strength-tested, and on-site milling and machine painting included miles of timber-glazing bars. The building was decorated in red, green, and blue, and the columns were brightened with yellow stripes. The Crystal Palace established internationally a style and a standard for exhibition pavilions, next at Cork (1852), then at Dublin and New York (both in 1853), and Munich (1854).

The Great Exhibition opened on 1 May 1851, with more than 13,000 exhibits from around the world. By the time it closed six months later, over 6.2 million people had visited it. Despite popular insistence that the building should remain, it was scheduled for dismantling. A consortium bought it and it was, under Paxton's supervision, reerected in a modified form in a park designed by him at Sydenham Hill, southeast London. Reopened by Queen Victoria in June 1854, the Crystal Palace became a national center for exhibits of industry, art, architecture, and natural history, all held under the auspices of the Crystal Palace Company. Sporting events took place in the park from about 1857 and for twenty years after 1895 it became the venue for Football Association Cup finals. Motor racing followed in 1936.

In November of that year, the Crystal Palace was destroyed by fire. Only one terrace of the original park now survives, and even that is under threat. The Crystal Palace Partnership, with representatives of five London boroughs and private-sector groups, is undertaking a £150 million regeneration scheme for Crystal Palace Park that includes its "restoration," a concert platform, modernization of the National Sports Centre, and a so-called new Crystal Palace on the surviving 12-acre (4.8-hectare) terrace. The latter, an insensitive proposal for a utilitarian building housing a twenty-screen cinema multiplex with restaurants, bars, and rooftop parking for a thousand cars, provoked local residents to launch the Crystal Palace Campaign in May 1997. A challenge to the scheme is being mounted in the High Court on the grounds that the Crystal Palace Act of 1990 provides that any building on the site should be "in the style and spirit of the former Crystal Palace."

Further reading

Bird, Anthony. 1976. *Paxton's Palace*. London: Cassell.
Elliot, Cecil D. 1992. *Technics and Architecture: The Development of Materials and Systems for Buildings*. Cambridge, MA: MIT Press.
The Great Exhibition: London's Crystal Palace Exposition of 1851. 1995. New York: Gramercy.

Curtain walls

Traditionally, the wall of a building served both structural and environmental purposes. That is, it carried to the ground the weight of the building and its contents and, while admitting air and light through openings, protected the interior from extremes of weather, noise, and other undesirable intrusions. The introduction of structures in which the loads are carried by beams and columns liberated the wall from load bearing, allowing it to function solely as an environmental filter—a relatively thin, light curtain, so to speak. This was first seen in the later medieval cathedrals with their vast stained-glass windows, but it would not be widely developed until the nineteenth century, with the advent of metal-framed architecture and, subsequently, reinforced concrete. The metal-and-glass membrane supported by the building frame, known as the curtain wall, is principally associated with multistory office buildings after about 1880.

Seagram Building, New York City; Ludwig Mies van der Rohe, architect, 1954–1958. Exterior, photographed in 1997.

Although the first skyscrapers, such as the Rookery (1885–1886) and Monadnock Building (1889–1891), both in Chicago and both designed by architects Burnham and Root, had thick conventional load-bearing walls, the twin economic necessities of getting buildings up quickly and optimizing the quantity and quality of interior space soon led to buildings whose outer walls consisted almost entirely of windows supported by perimeter columns and beams. This was a first step toward the development of a true curtain wall, that is, a continuous wall *in front of* the structural frame. The earliest example was Albert Kahn's Packard Motor Car Forge Shop in Detroit (1905). A curtain of glass in steel frames allowed more space and light in the factory, just as it would in an office tower, and Kahn again employed it for the Brown-Lipe-Chapin gear factory (1908) and the T-model Ford assembly plant in Highland Park, Michigan (1908–1909). This rational industrial architecture drew the admiration of Europe and was emulated in Peter Behrens's A. E. G. Turbine Factory (1909–1910) in Berlin and Gropius and Meyer's Fagus Works in Alfeld-an-der-Leine, Germany, of 1911.

It is widely accepted that the first office block with a curtain wall was Willis Jefferson Polk's eight-story Hallidie Building (1917–1918) in San Francisco. Although it was cluttered in places with florid cast-iron ornament, the street facade, suspended 3 feet 3 inches (1 meter) in front of the structure by brackets fixed to cantilevered floor slabs, presented an unbroken skin of glass. Elsewhere, others dreamed of crystal prisms in which the building's whole external membrane was glass: the serried towers of H. Th. Wijdeveld's Amsterdam 2000 (1919–1920) and Le Corbusier's Ville Contemporaine (1922) and—probably best known—the skyscrapers Ludwig Mies van der Rohe projected between 1919 and 1923. But dreams and visions they remained, because the technology was not yet available to turn them to reality. One exception was the A. O. Smith Research Building in Milwaukee (1928–1930) by Holabird and Root, the first multistory structure with a full curtain wall (rather than a single facade) of large sheets of plate glass supported on aluminum frames.

Spin-offs from defense technologies after World War II paved the way for tall curtain wall buildings. Important among them was cost reduction in the production of aluminum, whose corrosion resistance could be improved by a process known as anodizing. This lightweight metal could be extruded into the complicated profiles needed to frame the glass and strengthen the wall against wind loads. Reliable cold-setting synthetic rubber sealants had also become more widely available. These advances were combined with more efficient sheet glass manufacture, especially polished cast glass and, after 1952, the much flatter float glass. Wall elements could be fabricated off-site to exacting tolerances and then transported, assembled, fixed, and glazed with none of the "wet" processes that impede building contracts.

Relevant engineering developments included reverse-cycle air-conditioning—available since 1928—and fluorescent lighting, first demonstrated at the 1938 Chicago World's Fair. All these technologies were exploited in Pietro Belluschi's twelve-story Equitable Building in Portland, Oregon (1944–1948), described by one historian as "an ethereal tower of sea green glass and aluminum." Another writer asserts that it "set styles for hundreds that came after."

The thirty-nine-story United Nations Secretariat Building in New York City followed in 1947–1952. The final design was developed from a proposal by Le Corbusier, and Wallace Harrison acted as executive architect in consultation with him. The curtain walls of the Secretariat Building's east and west facades are all glass, cantilevered 27 inches (80 centimeters) from the line of the perimeter columns; black-painted glass spandrels hide the between-floor spaces. The blue-green tinted windows are of "Thermopane," a special glass that absorbs radiant heat, preventing it from reaching the interior, thus reducing the load on the air-conditioning system. The only breaks in the sheer curtain wall are full-width air-conditioning intake grilles at four levels. Because of its innovation, and no doubt because of its associations, the U.N. Secretariat, together with Mies van der Rohe's Lake Shore Drive Apartments (1951) in Chicago and Skidmore, Owings, and Merrill's Lever House (1952) on Park Avenue, New York, contributed to the universal standard for high-rise buildings.

The latter building, a twenty-four-story, green-tinted glass and stainless steel tower, designed by Gordon Bunshaft, marked a change of direction in American corporate architecture and in the way New Yorkers built. In keeping with the wishes of a client who made household cleaning products, Bunshaft produced an immaculate, clean-lined tower. The architectural critic Lewis Mumford called it "an impeccable achievement." The top three floors are reserved for mechanical services. A mobile gantry carries a window cleaners' platform that serves all faces of the building; such devices became standard for the curtain wall office buildings that followed. Lever House was the first skyscraper to exploit the allowable plot ratios in city planning regulations. By occupying only a quarter of the site, it allowed much more natural light to enter the offices than conventional stepped-back skyscrapers that covered the whole allotment. Lever House is a New York historic landmark, and in November 1999 a $10.7 million contract was let to renovate its curtain walls, designed by Skidmore, Owings, and Merrill under the supervision of the New York City Historical Society.

That leads us to the inherent problems in curtain wall construction, for all of its advantages. In forty-five years, the pristine facades failed in a number of ways—water penetration and consequent damage, corrosion, and broken glass panels. Since their inception, curtain wall systems have been continually revised, most changes geared toward reducing weight while retaining strength. Stiffened sheet aluminum, enameled steel laminated with insulation, and later even thin sheets of stone were used for spandrel panels. The design of joints—problem spots for leaks—was improved and more durable sealants were invented. More recently, the availability of reliable adhesives has allowed architects to indulge in so-called "fish tank" joints between glass panels, doing away with framing bars. Glass technology has also been refined. Double glazing, first manufactured in the 1940s, improves both the sound and thermal insulation of curtain walls. Heat-absorbing glass, already available in the 1950s, evolved in the following decade into reflective glass with thin metallic coatings, also used to reduce heat gain within buildings. In 1984 heat mirror glass was developed; when combined with double glazing, its insulating value approaches that of masonry, but the esthetic effect seems to be a denial of the form of the building: all it does is reflect what's around it.

Given that the two significant advantages of curtain wall construction are the reduction of weight and speed of erection, it might be concluded that it costs less than conventional work. That is not necessarily true, because its behavior as an environmental filter, especially in relation to heat flow, may result in higher air-conditioning costs. Often, the preciousness of the architect's detailing increases costs, as evidenced by Mies van der Rohe's bronze-and-brown-glass Seagram Building (1954–1958) in New York City. It cost $36 million, approximately twice as much as office towers normally did.

The tall glass prism was the major contribution of the United States to the so-called International Style of modern architecture. But its glorious day passed with the rise of postmodernism, and the crystal towers that Frank Lloyd Wright dismissed as "glass boxes on stilts" were replaced with less anonymous designs. Even Philip Johnson, Mies van der Rohe's most ardent disciple, forsook the minimalist forms of curtain-wall architecture in favor of a more congenial architecture.

Further reading

Frampton, Kenneth, and Yukio Futagawa. 1983. *Modern Architecture, 1851–1945*. New York: Rizzoli.

Krinsky, Carol Herselle. 1988. *Gordon Bunshaft of Skidmore, Owings, and Merrill*. Cambridge, MA: MIT Press.

Stubblebine, Jo, ed. 1953. *The Northwest Architecture of Pietro Belluschi*. New York: F. W. Dodge.

Wright, Sylvia Hart. 1989. *Sourcebook of Contemporary North American Architecture: From Postwar to Postmodern*. New York: Van Nostrand Reinhold.

D

De Re Aedificatora

Leon Battista Alberti's theoretical treatise on architecture, titled *De Re Aedificatoria (About Buildings)*, was dedicated in 1452 but not published until 1485. What qualifies it as an architectural feat? It changed the understanding and practice of architecture in much of Europe and continued to influence developments there and in the New World for about 400 years. Although he was gathering the ideas for the book, Alberti (1404–1472) was not an architect but a Catholic priest.

Alberti was born in Genoa, the illegitimate child of Lorenzo, an exiled Florentine from a family of bankers. When he was about ten years old, Battista (he added "Leon" later) entered a boarding school in Padua to receive a basic classical education. Several years of legal studies at the University of Bologna led to a doctorate in church law in 1428, after which he went to Florence. He soon began writing. His first published anthology of poems, *Il cavallo (The Horse)* of 1431, was quickly followed by *Della famiglia (About the Family)*—the first of many philosophical dialogues—and *La tranquillità (Composure)*, a collection of essays, short stories, and plays, both in 1432. By then he was employed as a secretary in the Papal Chancery in Rome and was about to undertake a lives of the saints and martyrs, written, as was fashionable, in classical Latin. Living in Rome opened Alberti's eyes to classicism, although

the city was to remain neglected for another fifteen years. In 1434 he wrote a study about urban design entitled *Descriptio urbis Romae (Description of the City of Rome)*, in which he first explored the classical notion that beauty existed in harmony, achievable through mathematical rules.

Alberti's future lay not in the law but in the church. Taking holy orders, he would eventually become a canon of the Metropolitan Church of Florence in 1447. Other clerical offices and their benefits followed: abbot of San Sovino, Pisa, Gangalandi Priory, Florence, and the rectory of Borgo San Lorenzo in Mugello. In 1436 he completed his first major book, written in classical Latin, that touched upon architecture: *De pictura (About Painting)* was an attempt to bring system to perspective and set down rules for the painter to achieve concord with cosmic harmony. An Italian translation appeared in the same year.

From about 1434 Alberti traveled through northern Italy in the retinue of Pope Eugenius IV, visiting Florence, Bologna, and Ferrara, where, in 1438, under the patronage of Marchese Leonello, he began a more careful study of classical architecture, delving into the ten-part book *De Architectura*, written by one Marcus Vitruvius Pollio around 20 B.C. Alberti returned to Rome six years later and extended that study among the ancient buildings. When Nicholas V succeeded to the papacy in 1447, Alberti was appointed inspector of monuments, an office he held until 1455. *De Re Aedificatoria,* written in classical Latin and structured in ten parts like Vitruvius's *De Architectura,* was completed in 1452. Vitruvius's book was its principal source and model, but Alberti also drew upon Plato, Pythagoras, and the Christian fathers; his own archeological studies; and, importantly, the consensus of contemporary architectural thought. Vitruvius had summarized the architectural practice of his day; Alberti went further to lay down universal rules.

As Italian society and fashions changed, from around 1420 the mason-architect had begun to be displaced, first by the artist-architect and then the courtier-artist-architect. With training in neither building nor art, Alberti wrote a book about the art of building that completed the metamorphosis of the architect into a dilettante-scholar; that made "design distinct from matter," as he put it, and turned the art of architecture into an academic pursuit in which creativity and design skill could be honed to perfection simply by obeying a set of rules. Intuition was replaced with measurable absolutes. It gave architectural design a thoroughly developed theory of harmony and proportion and made it simple—at least in theory. According to some sources, the last Latin edition was a folio version in Bologna, of 1782. Translations and many derivative works found their way through western Europe.

Book I of *De Re Aedificatoria* defined design, set down the criteria for good architecture (convenience, stability, and delight), and discussed the basis of composition and proportion. Book II dealt with matters of professional practice and building materials. Book III addressed practical building construction, Book IV covered many aspects of civic design, and Book V dealt with plans for various building types. The next book explored the esthetic dimension of architecture, defining beauty as "a harmony of all the parts in whatsoever subject it appears, fitted together with such proportion and connection, that nothing could be added, diminished or altered, but for the worse." It also included a section on mechanical and technical details. Alberti's strong attachment to antiquity was revealed in Books VII and VIII, that took up the subjects of ornament in religious buildings and Roman urban design, respectively. In Book IX the axi-

omatic principle underlying Renaissance architecture was restated: that beauty is an innate property of things, achieved by following cosmic rules. Then there was an assortment of chapters about mostly practical issues. Book X descended to the pragmatic: water supply, engineering, repairing cracks, and even how to get rid of fleas.

Alberti applied his theories in only a few buildings, mostly unfinished renovations or extensions. They included the facades of the Church of San Francesco (otherwise known as Tempio Malatestiano) of 1450, in Rimini; the facades of the Palazzo Rucellai (1446–1451) and Santa Maria Novella (1458–1471), both in Florence; and San Sebastiano (1459) and Sant'Andrea (1470–1472), both in Mantua. His biographer Giorgio Vasari wrote in 1550, "His writings possess such force that it is commonly supposed that he surpassed all those who were actually his superiors in art" and added, "He was a person of the most courteous and praiseworthy manners … generous and kind to all."

Further reading

Alberti, Leon Battista. 1988. *On the Art of Building in Ten Books.* Cambridge, MA: MIT Press.

Borsi, Franco. 1989. *Leon Battista Alberti: The Complete Works.* New York: Electra/Rizzoli.

Vasari, Giorgio. 1991. *Selections from the Lives of the Artists.* Oxford, UK: Oxford University Press.

De Stijl

Founded in Leiden, the Netherlands, in 1916, the group known as De Stijl was Europe's most important theoretical movement in art and architecture until the mid-1920s, when leadership passed to Germany.

In 1916 the architect J. J. P. Oud met the critic and painter Theo van Doesburg and soon introduced him to another young architect, Jan Wils. First forming De Sphinx artist's club in Leiden, the three founded, with the railwayman-philosopher Anthony Kok and the painters Piet Mondrian, Bart van der Leck, and expatriate Hungarian Vilmos Huszár, the group known as De Stijl. Others joined them: the fiery Communist Robert van 't Hoff and the Belgian sculptor Georges Vantongerloo (both in 1917); the furniture designer Gerrit Rietveld (1918); the architect Cor van Eesteren (1922); and the painter César Domela (1924). Later arrivals were balanced by departures.

The first manifesto was issued in November 1918, though not all the members signed it. Therefore, De Stijl should never be thought of as a group in the sense that, say, the Pre-Raphaelites or the Impressionists were groups. The members never reached unity of purpose; there were no meetings; and membership seems to have lain in contributing to *De Stijl*, a polemical journal jealously conducted by van Doesburg. He stretched and frayed their fragile ties by personality issues, and the whole fabric unraveled as members withdrew one by one, unable to work with him. Van der Leck lasted only until 1918; Wils and van 't Hoff left in 1919; Oud and Vantongerloo two years later; and Mondrian in 1925. Others briefly established links with van Doesburg, but after 1925 only he was left to continue the magazine, by then published only spasmodically. He died in 1931.

Many De Stijl members were influenced by Theosophical doctrine and, subscribing to a holistic worldview "in which the geometric [was] the essence of the real," they sought unity within the arts and between art and society. Perhaps because its mysticism, religion, and philosophy offered a palliative for the problems of burgeoning capitalism, Theosophy appealed to many in the industrializing world at the fin de siècle. Socialism was an important factor at the time of De Stijl's birth and for some members social issues were all. They so concerned van 't Hoff that, unwilling to work for middle-class clients, he soon forsook architecture altogether. Seeking an appropriate architecture, the others explored Constructivism, temporarily preached Neoplasticism, and generated what Oud called Cubism, but theory seldom extended to architectural realities. The few realized projects were spectacular: van Doesburg's Café Aubette, Strasbourg (1926–1927, with Jean Arp and Sophie Taeuber-Arp), carried "painting into architecture, theory into practice."

Rietveld's Schröder house demonstrated De Stijl ideas and became an icon of European Modernism. In 1921, Rietveld began to collaborate with the interior designer Truus Schröder-Schrader. The tiny house in Utrecht (1924) that he designed for her expresses, more than anything else undertaken by the group, the principles valued by De Stijl. Earlier, Rietveld had collaborated with his De Stijl colleagues on fragments of schemes and unrealized projects. What they had been able to only dream of or explore in scale models, Rietveld built as his first complete architectural work.

The division among Dutch architects on religious and political grounds prevented wider acceptance of De Stijl's ideas within the Netherlands. *De Stijl* became an international journal (or rather, by van Doesburg's duplicity, an illusion of one), and through its pages and his personal preaching he shared with Europe the message of an architectural climax. De Stijl was moribund when van Doesburg died in 1931, but for a moment or two, through it, the Dutch had supplied a lot of theoretical and rather less practical input to modern architecture. Not least, by commenting upon his work to a wide audience, they provided a gateway for Frank Lloyd Wright's "peaceful penetration of Europe." In 1936 Alfred Barr of the New York Museum of Modern Art perceptively remarked that De Stijl had overshadowed German architecture and art in the mid-1920s. Moreover, had van Doesburg's attempted insinuation into the Dessau Bauhaus succeeded, that critically important school of architecture and design would have been turned toward Russian Constructivism.

Further reading

Blotkamp, Carel, ed. 1986. *De Stijl, the Formative Years, 1917–1922*. Cambridge, MA: MIT Press.

Friedman, Mildred, ed. 1982. *De Stijl, 1917–1931: Visions of Utopia*. New York: Abbeville Press.

Overy, Paul. 1991. *De Stijl*. New York: Thames and Hudson.

Deal Castle
Kent, England

Deal Castle, built in 1539–1540 to stand guard over the town of the same name on the Kent coast of southeast England, is a fine example of a new building type, created in response to major changes in politics and the technology of warfare. With others at Walmer and Sandown, it epitomized Henry VIII's new forts by its assured and concentrated use of the design elements common to all. Deal is the largest, most impressive, and most complicated of the so-called Device forts. It probably looks just as was intended: crouching in wait low above the beach, stocky, powerful, and seemingly impregnable.

In the turbulent years that followed Henry VIII's accession in 1509 he twice made war on France, the second time as an ally of the Holy Roman Emperor, Charles V of Spain. When he realized that France's defeat would give Spain too much power, Henry changed sides, joining France and the pope against the empire. England was financially ruined by the campaigns of 1527–1528, and six years later, Henry's divorce from Catherine of Aragon led to a break with the Catholic Church, isolating him from most of Europe. He tried to drive a diplomatic wedge between France and Spain, but in 1538 they signed a truce, arousing Henry's fear of a joint invasion. He urgently launched an ambitious defense program. Using funds plundered from the monasteries by his religious "reforms," in 1539 Henry initiated a chain of about thirty forts and batteries to defend England's major ports and repel the expected invasion fleet. They included ten Device forts: Portland, Pendennis, and St. Mawes in southwest England; Hurst, Calshott, and Sandgate around the Solent; and Camber, Walmer, Sandown, and Deal on the southeast coast.

The nature of warfare was changing, and the sophisticated defense systems of medieval castles had become obsolete. Built to resist mechanical artillery, they now had to withstand missiles shot with gunpowder. The clumsy bombards of the fifteenth century could be fired only a few times an hour. But by the early sixteenth century cast-iron cannonballs had replaced stone; powder quality had improved; and ordnance was generally smaller, reliable, and accurate. In 1386, Bodiam Castle in Sussex was among the first to replace archers' loopholes with cannon and gun ports. The decline of feudalism also had its effect: enemies were more likely to be foreign than envious neighbor barons.

Finished late in 1540, Deal, Walmer, and Sandown completed the metamorphosis from medieval castle to modern artillery emplacement. Each of these squat, powerful-looking "castles in the Downs"—they were still called castles—comprised rounded bastions radiating from a circular keep. Their thick walls were curved to deflect cannonballs, and their many gun ports were widely splayed for easy traverse. There were three tiers of cannon for long-range offense and two tiers of defensive armaments. Built by an army of workmen at a total cost of £27,000—1,000 years' pay for an artillery officer—and joined by earthen bulwarks (since vanished), they formed a defensive cluster along a vulnerable 2-mile (3.2-kilometer) stretch of coast. Sandown has succumbed to coastal erosion, and Walmer has been converted to a residence for the Warden of the Cinque Ports. Only Deal, overlooking the low-lying marshlands, has been conserved.

Henry VIII's sexual notoriety has overshadowed his considerable abilities as a scholar, poet, and statesman. He took an interest in military engineering and personally amended the proposals for his forts, and the "device" (that is, the design) of Deal Castle has been attributed to him. The temptation to compare the concentric plan to the Tudor rose (as many have done), although alluring, must be resisted. Built with stone quarried from a nearby Carmelite priory, the castle's architectural form was primarily constrained by serious military purpose: to pack the maximum firepower into the most compact possible structure.

Six semicircular bastions, with curved parapets and bristling with gun emplacements, radiate in two tiers from a central, cylindrical barracks-keep; the configuration is repeated in the surrounding moat. The upper tier abuts the tower; the lower forms the curtain wall. The concentric layout allowed ordnance to be effectively positioned and fired simultaneously without impeding each other. Almost 200 openings penetrate the massive walls at five levels, including 119 cannon ports and embrasures. The remaining loopholes and casemates, mostly at the lower levels, were for arquebuses and pistols. Gun positions within the bastions were vented to clear the smoke and gases. It is easy to imagine the withering salvo afforded by such purposeful design, but it has been suggested that Henry was unable to find enough cannon to fully equip his fortresses.

Because architects usually build upon what they know, Deal, simply because it had evolved from the medieval castle, also employed traditional defenses. The entrance was at second-floor level and approached by a drawbridge across the moat; attackers then faced a portcullis, beyond which there were heavy, iron-studded oak doors. The gatehouse ceiling was penetrated by five "murder holes" (gun slots for small arms), and a cannon protected an inner door. In the manner of earlier keeps, the central tower was self-sufficient: its basement had supply and ammunition stores and a well. The garrison was quartered at ground level, with a mess hall with fireplace and bake ovens. The upper story housed, rather more comfortably, the captain of the guard.

The anticipated Catholic assault never came. Although Deal was again readied in 1588, this time to repulse the Spanish Armada, once more no invasion eventuated. Late in the English civil war the fortress was held briefly by the Royalists, but they surrendered after a sustained bombardment. In the eighteenth century Deal's parapets were altered (some say disastrously) in unfulfilled expectation of attacks during the French Revolution, and again during the Napoleonic Wars. No shot was fired in anger until the German bombing of 1941. Since 1984 Deal Castle has been in the care of the Department of the Environment (now English Heritage).

See also Dover Castle

Further reading

Morley, B. M. 1976. *Henry VIII and the Development of Coastal Defence*. London: H.M.S.O.

O'Neil, Bryan H. 1966. *Deal Castle, Kent.* London: H.M.S.O.

Saunders, Andrew D. 1982. *Deal and Walmer Castles.* London: H.M.S.O.

Deltaworks
The Netherlands

The Deltaworks comprises a series of audacious engineering projects that effectively shorten the coastline of the southwest Netherlands by about 440 miles (700 kilometers), seal outlets to the sea, and reinforce the country's water defenses. Taking more than forty years to complete, the works involved the construction of huge primary dams totaling 20 miles (30 kilometers) in length, in four sea inlets between the Western Scheldt and the New Waterway, Rotterdam.

The Netherlands is located in the broad deltas of the Rhine, Maas, and Scheldt, and the small country's history and geography have been greatly influenced by a continuous struggle against the rivers and the sea. Through the coincidence of several events in 1953, the southwestern provinces suffered huge floods in which nearly 2,000 people died and thousands of homes were destroyed. The central government quickly reacted, and the Ministry of Transport, Public Works, and Water Management set up the Delta Committee to devise measures to avert a future disaster. The plan informed the Delta Act of 1958, but its implementation, placed in the hands of a complex instrumentality known as Delta Service, took over four decades to complete.

The major elements of the plan were achieved in the following order: the Hollandse IJssel storm flood barrier (1954–1958); the Zandkreekdam (1957–1960); the Veerse Gatdam (1958–1961); the Grevelingendam (1958–1965); the Volkerakdam (1955–1977); the Haringvlietdam (1956–1972); the Brouwersdam (1963–1972); and the Oosterschelde storm flood barrier (1967–1986). The vast scope of the Deltaworks cannot be fully described here, but it may be measured by a brief overview of the largest, most difficult, and most expensive phase: the Oosterschelde (Eastern Scheldt) storm flood barrier, immodestly referred to by its builders as "the eighth world wonder."

It was originally intended to close off the Oosterschelde with a permanent dam, and work started in 1967. By 1973 joining dams between parts of the coast had closed 3 miles (4.8 kilometers)—more than half—of the river mouth, and three sluices had been built. Then, in response to public protests, it was decided to construct a storm flood barrier instead of completely closing the estuary. Huge concrete pylons standing on the river bottom would support gates that could close to resist storm surges; a concrete roadway would cross the structure. The government signed a contract with the consortium De Oosterschelde Stormvloedkering Bouwkombinatie in 1977. A 3,000-yard-long (2.78-kilometer) access bridge was built to the 50-foot-deep (15-

meter) construction docks needed to fabricate the massive pylons. Commenced in April 1979, the first was finished early in 1983. In the meantime, work began on the sliding gates. Fifty-foot-deep foundations were prepared to support the pylons, and a special dredge was designed to secure the estuary floor against uneven scouring. By the end of 1982, the river bottom was secured by vast mats laid by purpose-designed vessels. All was ready for placing the pylons.

The construction docks were flooded and the pylons, each weighing 21,600 tons (18,300 tonnes) and between 100 and 135 feet (30 and 40 meters) high, were floated into position, then sunk to the prepared floor. Sixty-five pylons formed the spine of the barrier: sixteen in the northern opening, seventeen in the central, and thirty-two in the southern. They were connected by prefabricated elements, and the sliding gates, each 150 feet (45 meters) long and weighing 1,440 tons (1,220 tonnes), were then installed, a task that took a little under two years to complete. Then followed the fixing of each of the sixty-two 3,000-ton (2,270-tonne) precast concrete elements that carried the roadway across the barrier. The Stormvloedkering Oosterschelde was officially opened on 4 October 1986. It cost about a sixth of the 11 billion guilder (U.S.$5.5 billion) total of the Deltaworks.

The danger of overflowing rivers in the winter and early spring also threatens large parts of the Netherlands. Several inland engineering works—the Philipsdam (1976–1987); the Oesterdam (1977–1988); the Markiezaatskade (1980–1983); and the Bathse Spuikanaal and Spuisluis (1980–1987)—were adjuncts to the primary dams of the Deltaworks.

Holland's struggle against the water continues. Despite the pleas of regional and local water authorities for river dike reinforcement, the national government concentrated its funding for forty years upon the Deltaworks. Moreover, conservationists oppose any dike improvements that would spoil the landscape. The Boertien Commission was established early in the 1990s to address potential problems, and it produced the Great Rivers Delta Plan, which involved reinforcing nearly 190 miles (300 kilometers) of river dikes and embankments. The first phase was completed by the end of 1996; the second, covering another 280 miles (450 kilometers), was finished by 2001. But that will not solve the problem; if nothing else is done, the next generation of Hollanders will have to raise the dikes again. Climate changes, deforestation, urbanization, and drainage in their upper reaches mean that the river systems will carry increasingly large peak volumes. Cooperative policy and water management must be integrated internationally, from the sources to the deltas.

See also Afsluitdijk; Storm Surge Barrier

Further reading

Boermans, Anne, and Herman Hoeneveld. 1984. English summary of *Tussen land en water: Het wisselende beeld van de Deltawerken*. Amsterdam: Meulenhoff.

Haan, Hilde de, and Ids Haagsma. 1984. English summary of *De Deltawerken: Techniek, politiek, achtergronden*. Delft: Waltman.

Meijer, Henk, ed. 1998. *Het Deltaplan in beeld*. Utrecht: IDG.

Ditherington Flax Mill
Shrewsbury, England

The Industrial Revolution gave rise to a new building type: the factory, where a managed workforce could operate machines that were driven by steam power. The advent of machines also created a demand for iron to be produced on a large scale; in addition to being used to build machines, it soon became apparent that iron could be used to construct industrial buildings. The forerunner was the prefabricated cast-iron bridge at Coalbrookdale, England, of 1775–1779. But the factories, especially textile mills, involved problems other than the structural ones. Because they handled large quantities of cotton, flax, and wool, and because their wooden floors were quickly saturated with the oil used to lubricate the machines, they presented a fire hazard. The earliest textile mills had timber floor and roof framing and solid masonry external walls. Cast iron was noncombustible, and it was believed that it offered, as well as greater strength, a measure of fire resistance. Designed in 1795 and built the following year by the engineer Charles Bage of the milling firm of Bennion, Bage, and Marshall, the Ditherington Flax Mill, in the Shropshire town of Shrewsbury, was the world's first iron-framed building, the predecessor of most modern factories and even office blocks.

Ditherington was the largest flax mill of its day and one of the largest textile mills of any kind in Britain. The five-story building has conventional load-bearing masonry external walls with very large windows. Internally, it is divided into four bays by three rows of slender, cruciform-section, cast-iron columns, extending for eighteen bays on a north-south axis. Each bay measures about 10 feet (3 meters) square, and the average ceiling height is about 11 feet (3.4 meters). The columns support cast-iron beams spanned by the brick vaults that form the floor above.

The nearby warehouse and cross mill, also iron framed, were built soon after. In 1846 Professor Eaton Hodgkinson published *Experimental Researches on the Strength ... of Cast Iron,* a definitive work that established a design methodology for cast-iron structures; together with Sir William Fairbairn he made a major contribution to the theory of nineteenth-century bridge construction. Cast iron is not fireproof; in fact, it fails structurally and rather dramatically at relatively low temperatures. Consequently, the designers of later iron-framed buildings found ways to protect the columns, often by encasing them in non-load-bearing masonry.

The Ditherington Flax Mill survives, reasonably intact. In 1886 the mill ceased operations, and the building was vacant for ten years. For another century, probably because it had large expanses of open floor space, it was converted to maltings for a brewery. It was empty again from 1987, when the brewery closed down, and has been quite badly vandalized since. In the mid-1990s proposals were put in hand for the refurbishment of all the buildings on the site, with the help of a grant from English Heritage. The project included the creation of shops, restaurants, a heritage information center, leisure facilities and offices, an art gallery, and some housing. In March 2000 Advantage West Midlands announced a £2.8 million (U.S.$4.1 million) grant for the restoration of the mill.

Further reading

Briggs, Asa. 1979. *Iron Bridge to Crystal Palace: Impact and Images of the Industrial Revolution.* London: Thames and Hudson.

Jones, Edgar. 1985. *Industrial Architecture in Britain: 1750–1939.* New York: Facts on File.

Mantoux, Paul. 1983. *The Industrial Revolution in the Eighteenth Century: An Outline of the Beginnings of the Modern Factory System in England.* Chicago: University of Chicago Press.

Dome of the Rock (Qubbat As-Sakhrah)
Jerusalem, Israel

Jerusalem is a city holy to Judaism, Christianity, and Islam. At its center, the rocky outcrop known as Mount Moriah was the site of three successive Jewish temples, then a sanctuary of the Roman god Jupiter, before it was capped by the Arabic Dome of the Rock, which was for a short while Islam's most important sacred site. During the Crusades it was commandeered as a Christian shrine before returning to Islamic hands. Today it is at the very core of bitter dispute between Palestinians and Israelis. Although sometimes referred to as the Mosque of Omar, the Dome of the Rock is in fact not a mosque. Nevertheless, as the oldest extant Islamic monument, it served as a model for architecture and other artistic endeavors across three continents for a millennium.

About 1000 B.C. King David of Israel captured the Jebusite town of Urusalim. He renamed it Jerusalem, established his capital there, and chose Mount Moriah—already held sacred as the place where Abraham was prepared to sacrifice his son Isaac—as the site of a future temple. Solomon's Temple was completed in 957 B.C., only to be destroyed by the Babylonians in 586. The Second Temple was completed by 515 and enlarged and refurbished by Herod the Great (reigned 37–34 B.C.). It was leveled by the Roman legions of Titus in A.D. 70 and has never been rebuilt. The Roman emperor Constantine (reigned A.D. 306–337) decriminalized Christianity in 313. Soon afterward his mother Helena visited Jerusalem, where, according to mythology, she identified the locations associated with Christ, generating a tradition of Christian pilgrimages that continued until the invading Persians destroyed all the churches in 614.

Dome of the Rock (Qubbat As-Sakhrah), Jerusalem, Israel; architect(s) unknown, 688–692. Restored 1992–1994.

Twenty-four years later Jerusalem was captured by Caliph Umar Ibn al-Khattab, who renamed it *Al-Quds* (The Holy). Umar cleared the accumulated debris on top of Mount Moriah (Haram al-Sharif) and had a small wooden mosque built on the vast rectangular platform of the demolished Jewish temples.

The Dome of the Rock was built between A.D. 688 and 692 for the tenth caliph, Abd al-Malik ibn Marwan. It is an elaborate canopy encircling the bare rock summit of the mount, the *sakhra* from which Mohammed was miraculously carried through the heavens into the very presence of Allah to receive the tenets of the faith. There is a tradition that, by building the dome, Abd al-Malik was attempting to transfer the Islamic hajj (pilgrimage) to Jerusalem from Mecca, where his rival, Abdullah ibn al-Zubayr, had rebuilt the Kaaba in 684. It is also possible that Abd al-Malik wished to make some tangible statement about Islam's superiority over Judaism and Christianity, a motive suggested by the form of his building. The Dome of the Rock is more Roman or Byzantine than Islamic, and the caliph's Byzantine Christian architects employed architectural language understood by Muslims and Christians alike. Because Islamic architecture had not yet established a tradition, they referred to the best Byzantine models, and the congruence in plan and decoration between the Dome of the Rock and the centrally planned church of San Vitale (525–548) at Ravenna, Italy, is not coincidental.

The 60-foot-diameter (18-meter), timber-framed double dome, covered internally with colored and gilded stucco and originally roofed with lead covered in gold, rises 115 feet (35 meters) over the holy rock. It is carried on a tall drum, originally faced with glass mosaics, that rests in turn upon a circular arcade of twelve Corinthian marble columns, set in threes between four large rectangular piers. At the top of the drum, sixteen colored glass windows light the central space. Surrounding the circle is an octagonal, marble-flagged, 30-foot-high (9-meter) ambulatory of twenty-four piers and columns, reached from outside through four doorways with porticoes facing the cardinal directions. The ambulatory is screened from the sanctuary by half-height walls. The columns and most of the capitals were quarried from older buildings. The marble-faced outer walls of the building also describe an octagon; each side is about 60 feet (18 meters) long. Inside and outside, the Dome of the Rock was enriched with marble columns and facings and floral patterns of mosaic. The total effect must have been awesome: "thousands of lights … supplemented the meagre illumination from the windows, making the mosaics glitter like a diadem crowning a multitude of columns and marble-faced piers around the sombre mass of the black rock surmounted by the soaring void of the dome" (Ettinghausen and Grabar 1994, 30).

The tolerant Arabian caliphs allowed pilgrims of other faiths access to Jerusalem. Not so the Egyptian Fatimid caliphs who gained control of the city in 969, destroying all the synagogues and churches. In 1071 the Seljuk Turks closed the pilgrimage routes, provoking the Crusades and resulting in the European seizure of Jerusalem in 1099. The Dome of the Rock was converted to Templum Domini, a Christian shrine. The Muslims recaptured the city in 1187, and Jerusalem remained under Islamic control until the nineteenth century.

Although the building has survived in much of its original form, changes have occurred over the centuries. Repairs were made under Caliph al-Mamun (reigned 813–833), and the dome was replaced in the twelfth century; before the successive restorations, its curve was probably slightly horseshoe shaped. More recently, its lead roof has been replaced with aluminum. The glass mosaics that covered the drum of the dome and the exterior walls above the sill line were replaced by ceramic tiles in 1554, when the lower windows were also replaced. In modern times, restorations were carried out in 1924 and 1959–1964. The most recent took place between 1992 and 1994;

financed by the late King Hussein of Jordan, it included gilding the dome with 5,000 gold plates and cost U.S.$8 million.

See also Masjed-e-Shah (Royal Mosque); Sultan Ahmet Mosque

Further reading

Bloom, Jonathan, ed. 2000. *Early Islamic Art and Architecture*. Burlington, VT: Ashgate.

Ettinghausen, Richard, and Oleg Grabar. 1994. *The Art and Architecture of Islam, 650–1250*. New Haven, CT: Yale University Press.

Frishman, Martin, and Hasan-Uddin Khan eds. 1994. *Mosque: History, Architectural Development and Regional Diversity*. New York: Thames and Hudson.

Dover Castle
Kent, England

The science of medieval warfare and the design of castle architecture developed side by side until the latter reached its highest degree of sophistication in the almost impregnable concentric castle, exemplified in the royal castle at Dover, known as the "key of England," the first castle of its kind in western Europe. On a clear day the French coast, 21 miles (37 kilometers) across the English Channel, can be seen from the ramparts above the famous white cliffs of Dover, Europe's historical gateway to Britain.

In 55 B.C. Julius Caesar landed his reconnaissance force nearby, and following a full-scale invasion in A.D. 43, the Romans built a walled town, Dubris (from which Dover is derived). They built an 80-foot-high (25-meter) flint *pharos* (lighthouse) on the nearby 375-foot (114-meter) Castle Hill, the site of an Iron Age earthworks that had existed long before. It was inevitable that the commanding position would continue to be used for defense. In the fifth century the Angles and Saxons came in the wake of the Roman withdrawal and founded a fortified town on the hill, employing the ancient defenses. Once Christianized, they built the church of St. Mary-in-Castro (St. Mary in the Fortress) as a chapel for the castle garrison and adapted the Roman lighthouse as part of its bell tower.

William I (the Conqueror) also recognized the strategic value of Dover. He instructed his half brother,

Odo of Bayeux, should the Norman invasion succeed, to land there with building materials for a castle. It took just eight days in 1066 to construct the fortress—probably a motte and bailey—within the Anglo-Saxon earthworks. Nothing of it remains. The motte was an earth mound crowned with a wooden keep and guarded by a wooden palisade; the bailey was a defensible area, also with a palisade and connected to the motte by a bridge. All was surrounded by a ditch. The earliest stone castles were organized in the same way.

Castles multiplied in Britain after the Conquest, responding to the internal tensions created by the feudal system. Dover continued to be strategically important in an international context, a "royal castle" that was not for a feudal baron but for the defense of the realm. Its evolution into a finely tuned concentric castle was a response to changes in medieval military technology and the science of war. Little is known of its earlier defensive works, but extensive rebuilding was undertaken after 1168. Most work was carried out in the 1180s under the supervision of King Henry II's chief architect, a master mason known only as Maurice. Richard I (the Lionhearted) almost completed it in 1189–1190, and his brother John extended the outer curtain wall at the north side so that the outer bailey had been enlarged to include most of the hilltop. The "completed" castle dates from about 1200. Repairs and extensions were necessary after a siege by rebel barons and their French allies in 1216, during which, despite the collapse of the east tower, it was successfully defended by a force of only 140 knights and men-at-arms. By 1256 Dover Castle reached its maximum strength and size, its outer walls then extending to the cliff's edge.

Concentric castles comprised a carefully designed keep that was the last line of defense, surrounded by a curtain wall that enclosed a large bailey. Sometimes there was a second, slightly lower curtain wall

Dover Castle, Kent, England; architect(s) unknown, ca. 1168–1200. View showing concentric curtain walls and keep.

(as at Dover) or even a third. Most functions were served by buildings in the bailey. Dover's daunting keep—the largest in England—was almost 100 feet (30 meters) square and 95 feet (29 meters) high; in places its walls were 21 feet (6.5 meters) thick. It was defended by an inner curtain wall with fourteen projecting "mural towers"—the first in England—which allowed archers to shoot toward any point at the base. The outer curtain wall at Dover was nearly 1 mile (1.6 kilometers) in circumference, with 20 similar towers. Each wall was interrupted only by fortified gatehouses with barbicans. When gunpowder was introduced into the country in the fourteenth century, cannon were developed that could shoot missiles 3 miles (5 kilometers). Given the thickness of its walls, that was of little consequence to Dover Castle. It has been involved in almost every conflict since the Middle Ages. Small wonder it has been called England's greatest castle.

Changes to artillery were not the main reason for the demise of castles; rather, the feudal system gave place to centralized government and the power of the monarch. In Tudor times, the design of castles was to alter dramatically. As a royal castle, with an eye on the Spanish, Dover was heavily fortified with cannon in the reign of Elizabeth I. It continued to function well beyond that: it was "modernized" during the Napoleonic Wars. Caves were excavated to hide troops waiting in ambush should the French invade. The towers were truncated—some say vandalized—to serve as gun platforms. The caves were again used as headquarters of the Dover Patrol in World War I and as bomb shelters and a hospital in World War II. The castle remained in the hands of the British army until 1958; five years later it was put in the custody of the Department of the Environment (now English Heritage) as a national monument. Conservation work continues.

See also Deal Castle; The Krak of the Knights

Further reading

Brown, Reginald Allen. 1974. *Dover Castle, Kent*. London: H.M.S.O.

Coad, Jonathan. 1995. *Book of Dover Castle and the Defences of Dover*. London: Batsford.

Durham Cathedral
England

Durham Cathedral, built principally between 1093 and 1133 to house the relics of the Northumbrian evangelist St. Cuthbert of Lindisfarne and the Venerable Bede, is the finest example of Early Norman architecture in England. Its significance in the development of Western architecture lies in the use of rib-and-panel vaulting, the pointed arch, and flying buttresses in the gallery roofs—all prophetic of the elegant structural system that we now know as the Gothic.

The cathedral stands in a hairpin bend of the River Wear in County Durham. William I (the Conqueror) selected the naturally defensive site, and by 1072 a castle was commenced on the neck of the steep-sided peninsula to defend the northern region of Norman Britain against the Scots. In 1091 an earlier Saxon church was demolished, and two years later work commenced upon the great building dedicated to Christ and the Virgin Mary. It was to form part of the Benedictine monastery that had been started about a decade before, and the whole precinct soon became the seat of the powerful feudal prince-bishops of Durham. Early in the twelfth century the peninsula was encircled by a wall, much of which survives.

Serious attempts to build "in the Roman manner," with semicircular stone arches, vaults, and domes—its architecture has been categorized as Romanesque—date from the second half of the eleventh century. The earliest examples saw barrel (or wagon) vaults used in such churches as Santiago de Compostela, Spain (begun 1078), and St. Sernin, Toulouse (begun 1080). These roofs exerted continuous sideways thrust on the side walls, creating the need to build those walls thicker (to prevent overturning); windows were small, in case they diminished the strength of the walls. Sometimes the walls were braced with arches above their piers. Experiments were also made with the Roman cross or groin vault, in which the church was divided into square bays, each of which was covered with a ceiling made by intersecting two barrel vaults at right angles. Although the groin vault transmitted the loads to the walls at equidistant points (thus allowing for thinner

Durham Cathedral, Durham, England; architect(s) unknown, 1093–1133. Interior of nave, looking east toward choir.

side walls with more and larger openings, braced at intervals with massive piers), most of the stress in the vault itself was at its weakest part: the groin. The system can be seen in parts of Durham and in Speyer Cathedral, Germany (originally 1030–1065).

Instead of groin vaults, the nave and choir (ca. 1104) of Durham Cathedral are covered using a revolutionary technique: the bays are framed by lateral, transverse, and diagonal beams or "ribs"—forerunner of the steel- or concrete-framed buildings of modern times—with panels of stone spanning the much smaller areas between them. The most exciting innovation among several at Durham, these are the first known examples of pointed ribbed vaults. The ribs carry their own weight and that of the stone roof to collection points above the piers, and the complex dynamic nature of the loads is thus cleverly resolved. It seems that the northern Italian clerics behind the development of Norman Christianity knew something of ribbed-vault construction, which the invaders took to England. Some sources believe that Lombard experiments may—and only may—have been as early as 1080, but there are certainly no examples on such a large scale as Durham, which therefore preempts by almost a century the key to the dramatic Gothic constructional system.

The church consists of a western galilee, or Lady Chapel; an aisled nave with two western towers; transepts flanking a taller tower above the crossing; and an aisled chancel (which was reduced in length during the thirteenth century). The eight bays of the nave are divided by piers disguised as clusters of columns, alternating with massive circular columns. The same articulation can be found in the choir and transepts. On the face of each pier is a tall shaft rising from the floor that appears to carry the slightly pointed transverse arches that support the vault, nearly 80 feet (24 meters) above. At the triforium (second level), each arch of the arcade is subdivided into two, and on the clerestory (the highest level), arches are supported by a pair of freestanding columns. The nave vault is laterally braced by quadrant arches—heralding the flying buttress of Gothic architecture—concealed in the triforium galleries. The substructures of the 218-foot-high (65.4-meter) central tower and much of the transepts were begun before 1096. The 155-foot-high (47-meter) vault of the crossing, not completed until the fifteenth century, is carried by four huge arches. The original roof of the choir was replaced by the present vault around 1250.

Like many medieval churches, Durham Cathedral has undergone alterations and additions (and, on occasion, what passed for restoration) through almost nine centuries. None has diminished the first impression of overwhelming power and stability experienced by the modern visitor when entering this "fortress of God" at the frontier of the Normans' domain.

See also Chartres Cathedral (Cathedral of the Assumption of Our Lady); Cluny Abbey Church III, St. Denis Abbey Church

Further reading
Shipley, Debra, and Angelo Hornak. 1990. *Durham Cathedral*. London: Tauris Parke Books.
Stranks, C. J. 1973. *This Sumptuous Church: The Story of Durham Cathedral*. London: S.P.C.K.

E

Eames House
Pacific Palisades, California

The architect Charles Ormand Eames (1907–1978) and his designer wife Ray Kaiser Eames (1912–1988) moved to southern California in 1941 into a new apartment building designed by Richard Neutra. Between 1945 and 1949 they designed and built their family home at Pacific Palisades. Known simply as the Eames House, the unconventional residence can be considered an architectural feat in that it was economically constructed entirely from "off-the-peg" components, most of which were available at any building materials suppliers. In the difficult years immediately after World War II, the designers thus demonstrated to the United States that good design need not be expensive, a mission they continued to fulfill for the rest of their lives.

The house was commissioned as part of the Case Study House Program, sponsored by John Entenza's West Coast journal *Arts and Architecture*. The periodical, setting out to promote good design, was seeking ideas for the creative application of the new technologies and materials developed during the war. Of course, as thousands of GIs were demobilized, one of the objectives of the program was to build "homes fit for heroes." Each house had a hypothetical client, and the Eameses designed one that combined a living space and studio for their own family setup, a working couple with grown children.

The first version was produced through a collaboration between Charles and the Finnish-American architect Eero Saarinen, under whom he had studied at Cranbrook Academy, Michigan. They also worked together on Entenza's Bridge House for the lot next door. Then the Eameses together developed their own house, first proposing a single-story box on stilts, typical of the "International Style" brought to the United States by European émigrés. Receiving permission to build, they had the structural steel delivered to the site. But concerned that the house would look too much like the minimalist houses being provided by Ludwig Mies van der Rohe, they revised the design.

By 1949 they had generated a proposal for two full-height pavilions, separating the living and studio functions by means of an open courtyard. They managed to enclose a much greater volume of space with the need for only one more steel beam. This design was eventually built. The standard 7.5-foot (2.25-meter) bays of framing were assembled mostly from industrially made, black-painted steel window and door modules. They held clear panels of wired or translucent glass and other opaque ones of aluminum, timber, fiberglass, asbestos cement, or stucco, painted white, blue, red, or black. Some were even covered in gold leaf. There was a full-height living room at the south end, whose sliding windows opened to decks made of railroad ties. A spiral ship's stair

Eames House, Pacific Palisades, California; Charles and Ray Eames, architects, 1945–1949. Interior of living room, photographed in 1978.

(ordered from a marine catalog) led to bedrooms on a mezzanine; under the mezzanine was a small alcove with built-in seats and bookcases, and a kitchen. Charles and Ray Eames lived in their "house of parts" for the rest of their lives.

By precept and practice, and by the exploitation of technology, they made an invaluable social contribution by promoting the awareness of good design throughout the United States. Their design partnership encompassed many fields: architectural works from houses to exhibitions; innovative plywood, aluminum, or fiberglass chairs; graphics; toy animals; and even a carousel. They also produced more than 120 films. Recently, an international exhibition of their achievements was mounted at Vitra Design Museum in Weil am Rhein, Germany. Vitra was the European manufacturer of their furniture for more than thirty years. The show, The Work of Charles and Ray Eames: A Legacy of Invention, traveled to Bilbao, Paris, Copenhagen, and London before beginning its American tour at the Library of Congress in May 1999, after which it moved to New York, St. Louis, Los Angeles, and Seattle.

Further reading

Dunster, David. 1990. *Houses, 1945–1989*. Vol. 2 of *Key Buildings of the Twentieth Century*. Boston: Butterworth.

Neuhart, John, Marilyn Neuhart, and Ray Eames. 1989. *Eames Design: The Work of the Office of Charles and Ray Eames*. New York: Abrams.

Steele, James. 1998. *Eames House: Pacific Palisades 1949, Charles and Ray Eames*. London: Phaidon.

Eiffel Tower
Paris, France

The Eiffel Tower was built between 1887 and 1889 as the entrance arch to the International Paris Exhibition, held to celebrate the centenary of the French Revolution. Conceived in 1882 by Gustave Eiffel's chief research engineers Maurice Koechlin and Emile Nouguier, and constructed in collaboration with architect Stephan Suavestre, the tower is a graceful and imaginative puddled iron lattice pylon. It soars to 1,020 feet (312 meters), the first building in almost 5,000 years to surpass the height of the Great Pyramid. Preliminary sketches were made in June 1884 and in September Eiffel, suddenly interested in the project, registered a patent "for a new configuration allowing the construction of metal supports and pylons capable of exceeding a height of 300 meters."

A careful and innovative assembly of over 18,000 small lightweight parts, the Eiffel Tower demonstrated to fullest advantage the structural possibilities of wrought iron. The world's tallest structure until the Chrysler Building was constructed in 1929 in New York City, it became (and still is) a landmark synonymous with Paris. Intended as a temporary exhibit and scheduled for demolition in 1909, it was saved by its tourist potential and its usefulness as a communication antenna. A radio tower added in 1959 increased its height by 56 feet (20 meters).

Eiffel had specialized in metal construction during his studies at the École Centrale des Arts et Manufactures in Paris. Prior to the acceptance of his design for the tower, he had built in iron and steel, notably the Maria-Pia railway bridge over the Douro River in Oporto, Portugal; the Truyere Bridge near Garabit, France; locks on the Panama Canal; and the internal frame for the Statue of Liberty. Whilst the Parisian tower drew on the outcomes of these projects, it was nonetheless a unique scientific and engineering challenge: its great height meant that wind loads had to be calculated in the design as well as the effects of gravity. Eiffel chose open lattice and splayed legs so that the wind would pass through the structure. In gale-force winds the movement of the tower is estimated to be a mere 4.5 inches (11 centimeters). Speedy and safe transportation of workers and materials (and later of visitors) was another challenge. Eiffel installed elevators that ran on inclined tracks within the tower's legs; the guide rails were used as tracks for climbing cranes during construction.

The Eiffel Tower weighs over 13,200 tons (11,180 tonnes), more than 70 percent of which is metal. Its 412-foot-square (126-meter) base is defined by the four huge masonry foundation piers set in bedrock; each supports a leg, and the legs converge to form the shaft. Eiffel employed a team of 50 engineers to prepare 5,300 drawings to his specifications, 100 workers to fabricate the components in the Eiffel factory at Levallois-Perret on the outskirts of Paris, and

between 150 and 300 site laborers. His calculations were so precise that no revisions were required during construction. Work began on 1 July 1887 and the project was finished in a little over twenty-six months. Eiffel was awarded the French Legion of Honor.

On the tower's completion, opposition to its erection was silenced. An earlier protest published in *Le Temps* had been signed by such illustrious Frenchmen as the writers Guy de Maupassant and Alexandre Dumas Jr. and the architect Charles Garnier. Others had described the proposal as a "truly tragic street lamp" and a "carcass waiting to be fleshed out with freestone or brick, a funnel-shaped grill, a hole-riddled suppository." But it was an instantaneous popular success. In the last five months of 1889, over 1.9 million people visited it. Each paid an entrance fee to help defray the cost—a little under Fr 8 million (about U.S.$1.5 million).

Three viewing platforms—at 186, 376, and 900 feet (57, 115, and 276 meters)—were provided for visitors. At the first, where there were restaurants and a theater, arches linked the four legs; applied after the construction of the legs and platform, they were purely ornamental. Visitors were taken to the first and second platforms in double-deck, glass-enclosed hydraulic elevators. Stairs led to the third platform, and an elevator gave access to the top of the tower, where Eiffel originally had his studio and office (now restored). Each level offered a panoramic view of Paris and beyond for about 50 miles (80 kilometers). From the Eiffel Tower, people were afforded, for the first time, the unique opportunity of seeing the earth from far above.

When the Société de la Tour Eiffel's original operating concession expired in 1980, the city of Paris assumed direct control of the tower through a company called Société Nouvelle d'Exploitation de la Tour Eiffel. From 1980 to 1984 it undertook a restoration and renovation program. The tower was reinforced in places, 1,560 tons (1,320 tonnes) of excrescences were removed, and the elevators were replaced. It requires regular maintenance, including painting every seven years. The Eiffel Tower continues to be a prime tourist attraction, with over 6 million visitors annually. Each of the viewing platforms is accessible and Eiffel's office has been opened to tourists. The exclusive Le Jules Vernes restaurant occupies the second level. During the Paris millennium celebrations of 2000, the tower was covered with thousands of small lights that nightly illuminated the gracious "iron lady" of Paris.

See also Maria-Pia Bridge

Further reading

Marrey, Bernard. 1984. *The Extraordinary Life and Work of Monsieur Gustave Eiffel.* Paris: Graphite.
Sechi-Johnson, Patricia. 1997. *100 Greatest Manmade Wonders.* Danbury, CT: Grolier.
Steiner, Frances H. 1984. *French Iron Architecture.* Ann Arbor, MI: UMI Research Press.

Empire State Building
New York City

For forty-one years from 1931, the Empire State Building was the tallest tower in the world. That distinction has since been wrested and rewrested by a series of successors. The 102-story building, covering its 2-acre (0.8-hectare) Park Avenue site and soaring to 1,252 feet (417 meters), was completed in the incredibly short time of 1 year and 45 days; in fact, the time from the decision to build to the letting of office space was only 27 months. Because of the precise planning and exacting project management that achieved such efficiency, this most familiar of all skyscrapers is one of the great architectural feats of the twentieth century.

The Empire State Company was formed in 1929 by John Jacob Raskob (General Motors' chief executive), the industrialist Pierre S. du Pont, the politician Coleman du Pont, Louis G. Kaufman, and Ellis P. Earl. Raskob invited Alfred A. Smith, the New York State governor until 1929, with whom he had political ties, to become president of the corporation. The two men became the prime movers of the project. The 35-year-old Waldorf-Astoria Hotel, on the corner of Fifth Avenue and Thirty-fourth Street, was bought for about $16 million from the Bethlehem Engineering Corporation and demolished to make way for the new building. The architects Richmond H. Shreve, Arthur Loomis Harmon, and William

Lamb (who did much of the design work) were initially commissioned to create a 50-story, 650-foot-high (195-meter) office block. But the scheme would go through more than 15 revisions before emerging as an 86-story, 1,252-foot (375-meter) tower. Last-minute revisions would further increase it to 102 floors and a height, including its mast, of 1,472 feet (450 meters). The structural engineers were H. G. Balcom and Associates.

Shreve, Harmon, and Lamb produced a steel-framed, art deco tower whose marble-clad, five-story base covered the whole site. From a 60-foot (18-meter) setback at the fifth floor, it rose uninterrupted to the 86th floor. The upper levels were faced with silver buff Indiana limestone and granite, and the verticality of the facade was emphasized by continuous mullions of chrome-nickel steel. The office floors were served by seventy-three elevators.

The esthetics of the design were hardly remarkable, and the building was either ignored or criticized by the aficionados of the sterile European Modernism—so-called international architecture—then being touted in North America. For the present purpose, the Empire State's artistic qualities are inconsequential, because its significance lies in the fact that the architects made a design that, in the contractor's words, was "magnificently adapted to speed in construction." And speed *was* of the essence: the clients announced an 80-story building in August 1929 and forecast the completion date: 1 May 1931.

The firm of Starrett Brothers and Eken won the contract, estimated at $50 million. The Waldorf-Astoria Hotel was demolished within a month, and site excavation began on 22 January 1930, digging 55 feet (16.7 meters) below ground to the gray Manhattan bedrock. Construction started just under two months later, and through the meticulous construction scheduling of the chief engineer, Andrew Eken, it proceeded at record pace. Materials suppliers were asked to deliver goods as they were needed, so there was no need for on-site storage in the downtown area. When materials arrived on-site—at the busiest time, that meant almost 500 deliveries daily—they were immediately hoisted to the appropriate floor and transported by railways to their final location for fixing. The steel frame rose an average of four and a half floors a week, on a forest of 210 steel columns. One fourteen-story section was completed in a week! Altogether, 69,600 tons (58,930 tonnes) of structural steel were placed in only six months. By the middle of November 1930 the building's masonry skin was fixed. This unprecedented logistical feat was achieved by an average workforce of 2,500, which at times reached 4,000. Together, they worked 7 million carefully monitored man-hours, including Sundays and public holidays, to meet the deadline. In fact, the building was completed a few days ahead of its rigorous schedule. On 1 May 1931 President Herbert Hoover pressed the switch in Washington, D.C., that turned on the skyscraper's lights.

The Empire State was one of the last gasps of New York's real-estate boom. From late in the 1800s more than 180 tall buildings, none under twenty stories, had been erected in Manhattan. As that phase was drawing to a close about thirty years later, New York City saw what might be described as a three-sided "skyscraper war." The antagonists were the Empire State, the Bank of Manhattan, and the Chrysler Building. The "cold and nondescript" Bank of Manhattan, designed by H. Craig Severance and completed in April 1929, was, at 927 feet (278 meters), the world's tallest building—at least momentarily. The Chrysler Building, then being built for the automobile tycoon Walter P. Chrysler, was originally planned to be crowned with a dome, bringing it to within 2 feet (0.6 meter) of the height of the bank. Its architect William van Alen obtained permission to add the spire that is now recognized as the building's most distinctive feature. Its components were prefabricated inside the upper floors, and it was placed in just one and a half hours in November 1930, bringing the height of the Chrysler Building to 1,048 feet (314 meters). With the advantage of playing a little behind the game, Raskob and Smith had their architects add six stories to the 1000-foot (300-meter) Empire State Building, originally intended to terminate in a flat observation deck. Above it all soared a 200-foot (60-meter) tower, bringing its total height to 1,250 feet (375 meters).

It was mooted that this tower would serve as a mooring mast for airships. The 86th floor would house passenger lounges, airline offices, and baggage rooms, and the vessels would be moored at the 106th level. One attempt to moor a dirigible succeeded for just three minutes, and a near disaster with a U.S. Navy blimp in September 1931 finally led to the abandonment of the scheme—a decision tragically validated by the fiery destruction of the *Hindenberg* at Lakehurst, New Jersey, in 1937. The two observation decks remained just that, and the mast later formed the base of a television tower.

The Empire State Building cost $24.7 million. Optimistically conceived during a real-estate boom, the success of the venture was dashed by the Wall Street crash of 1929. When the building was opened its owners were hard-pressed to find tenants for the 2.1 million square feet (199,000 square meters) of office space, and some witty New Yorker coined the nickname "Empty State Building." Apart from the impact of the Great Depression, the 350 Fifth Avenue address was too far from the central business district. Eighteen months after opening, only a quarter of the space had been rented; six months later, there were still fifty-six vacant floors and the problems continued throughout the 1930s. After World War II the commercial center of gravity of New York was the Rockefeller Center, the last of whose nine towers was completed in 1940. Although the Empire State achieved 85 percent occupancy by 1944, even now it has a vast number of tenants renting small areas. Over 15,000 people work in it, and up to 20,000 clients, shoppers, and tourists visit daily. Every year, over 3.8 million sightseers and tourists visit the observation levels.

In 1955, the American Society of Civil Engineers named the Empire State Building one of the "Seven Modern Wonders of the Western Hemisphere," and on the occasion of its Golden Jubilee in 1981 it was, not without reason, designated an official New York City landmark.

Further reading

Doherty, Craig A., and Katherine M. Doherty. 1998. *The Empire State Building, Featuring the Photographs of Lewis W. Hine*. Woodbridge, CT: Blackbirch.

Goldman, Jonathan, and Sarah Freymann. 1980. *The Empire State Building Book*. New York: St. Martin's.

Tauranac, John. 1995. *The Empire State Building: The Making of a Landmark*. New York: Scribner.

Willis, Carol, and Donald Friedman, eds. 1998. *Building the Empire State*. New York: W. W. Norton.

Engineering Building
Leicester University, England

The Scots architect James Frazer Stirling (1926–1992) formed a partnership with James Gowan (b. 1923) in 1955 after winning a commission for a low-rise housing development in Ham Common, Middlesex (1955–1958). The design started a trend in England for broadly finished brick and exposed concrete. There followed a couple of domestic scale projects, and in July 1959 their more influential work: the Engineering Building at Leicester University (completed 1963), which has been called the "pinnacle of their mutual achievement." The seminal building, which juxtaposes a glazed office tower with red-tile facings on the massive cantilevered lecture theaters and a single-story workshop, was unlike any postwar architecture elsewhere and broke the hold of Le Corbusier upon British architects. The critic Reyner Banham coined the name "New Brutalism" to describe the new style, which exposed concrete, steel, and brick and rejected the polished and elegant finishes and geometric regularity of the International Modern Movement. The character of the Engineering Building was quickly and widely emulated in Britain; its influence persisted even longer in Japan.

Leicester University was founded as a university college in 1921 and granted its Royal Charter in 1957. The administration appointed the Cambridge engineer Edward Parkes to set up a new engineering faculty, to commence with 200 students. The university also commissioned Leslie Martin to produce a master plan for developing the 9-acre (3.6-hectare) campus; Stirling and Gowan's building was its first major postwar facility. By the end of 1959 they had produced two alternative preliminary designs. The final scheme was approved in March 1960, although the two architects disagreed over the glazing of the tower block. In fact, their partnership was dissolved as soon as the building was completed.

The building has two main elements: a complex, multistory main building that houses two lecture theaters, laboratories, and offices, and a lower level housing workshops. Two cantilevered reinforced concrete lecture theaters (attributable to the structural engineers), their sloping seating expressed on the outside of the building, are set at right angles to each other and are joined by a diagonal ramp. Four stories of laboratories rise beside the smaller theater on tall concrete columns; surfaces are faced with deep red Accrington brick and red Dutch tiles. Above the larger theater—also brick and tile clad—is a six-story, fully glazed office tower, its narrow rectangular form modified by cut-off corners, crowned by a water tank. The spiral staircase that serves it penetrates the cantilevered block. The adjacent ground-level heavy-machinery workshops, covering over two-thirds of the site and designed mainly by Gowan, are clad in part with translucent glass and roofed with long, diagonal, north-facing glass trapezoidal prisms. One historian has commented that "a mannerist taste for distortion and paradox" permeates the building, and that the "diversity of forms ... is a pretext for the liveliest interplay of masses." Such a cynical view undervalues the work of one of Britain's—the world's—greatest twentieth-century architects; indeed, a winner of the prestigious Pritzker Architecture Prize (1981) and "a leader of the great transition from the Modern Movement to the architecture of the New."

Further reading
Futagawa, Yukio, and Kiyonori Kikutake. 1974. *Leicester University Engineering Department, Leicester, Great Britain, 1959–63*. Tokyo: A. D. A. Edita.

McKean, John. 1994. *Leicester University Engineering Building: James Stirling and James Gowan*. London: Phaidon.

Walmsley, Dominique. 1988. "Leicester Engineering Building: Its Postmodern Role." *Journal of Architectural Education* 42 (Fall): 10–17.

English landscape gardens

Most of England's apparently natural countryside is in fact contrived, the result of a revolutionary movement in garden design, the discipline first named "landscape architecture" by Humphry Repton (1752–1818). The English eighteenth-century landscape garden, which would be internationally imitated, was possibly Britain's main contribution to European esthetics. Unlike traditional gardens, it was distinguished by asymmetry and informality. It incorporated artificial hills and free-form lakes; redirected rivers; sinuous pathways and drives; strategically placed stands of trees in grassy fields; and, of course, the great house, from which carefully composed and uninterrupted vistas opened to surrounding parkland. The movement was linked with the notorious Enclosure Acts, which allowed the English gentry to resume what formerly had been common land. Large landholdings brought profit as well as social and political power, displayed in the creation of expansive parks surrounding a country seat.

In a 1713 essay the poet Alexander Pope (1688–1744) suggested that formal English gardens be replaced by the "amiable simplicity of unadorned nature." He entreated, "In all, let nature never be forgot.… Consult the genius of the place." The challenge was taken up by three designers: William Kent, Lancelot Brown, and Repton, who within a century had banished geometry from the English countryside.

Kent (1685–1748), called "father of the English landscape garden," was trained as a sign painter and also worked as a coach builder's apprentice. Although never a successful artist, for ten years he studied painting in Rome, earning his living as an art dealer. In Italy in 1715 he met Lord Burlington, who became his patron. Back to London in the 1720s, he worked with Burlington on Chiswick House before engaging in landscape design. The critic Horace Walpole declared that Kent "leaped the fence and saw that all of nature was a garden." Inspired by the works of Lorraine, Poussin, and Rosa, and believing that "all gardening is a landscape painting," Kent regarded his gardens as classical pictures, replete with antique pavilions and composed to maximize the artistic impact of form, light, and color. In 1737 he was invited to Rousham in Oxfordshire to redesign the seventeenth-century house, as well as a garden laid out by the Royal Gardener, Charles Bridgeman (d. 1738), in the 1720s. Kent added wings and a stable block to the house and made interior alterations, but the garden (completed in 1741) is his best surviving work, thought by many to be "the jewel" of the English landscape movement. It marks a transition between the great English Restoration gardens and Viscount Cobham's house, Stowe, in Buckinghamshire.

At the end of the 1600s, Stowe had a modest Italian-inspired parterre garden. From about 1713 the surrounding park, designed by Bridgeman, was dotted with Baroque pavilions and monuments by the architect John Vanbrugh. Then, in the 1730s, Cobham appointed Kent and the Palladian architect James Gibbs to work with Bridgeman. Kent, convinced that "nature abhors a straight line," began to replace the geometrical gardens with winding, shaded paths and created a series of painterly views that unfolded on a walk through the landscape, the beginning of the most important early English landscape gardens. Bridgeman's Octagonal Pond and the Eleven Acre Lake were given a free form. Other changes followed with the arrival of the greatest champion of the "natural" landscape, Lancelot Brown.

Brown (1715–1783) began his horticultural career as apprentice to Sir William Lorraine. After working for Sir Richard Grenville at Wotton, he moved to Buckinghamshire in 1739, and two years later he was an undergardener at Stowe. He remained for seven years as a disciple (and eventually son-in-law) of Kent and became immersed in the new English style of landscape gardening. Kent was still improving the garden, although he undertook other projects. Brown designed the "Grecian Valley," a composition of landform and forest. Kent died in 1748 and Lord Cobham a year later. Brown left Stowe and in 1751 established a landscape practice based in London. He later became head gardener to the Duke of Grafton and in 1761 was appointed Master Gardener at Hampton Court Palace. During his time at Stowe, his employer had allowed him to accept commissions from a number of his friends, and Brown's practice, thus established, grew rapidly. Because he would enthusiastically tell prospective clients of the "great capabilities" of their properties, he earned the nickname "Capability" Brown.

His grand visions, realized in the gardens of about 170 of England's stately rural houses, have been described as idealizations of the English countryside. They accentuated (read "improved") the undulating natural landscape; their asymmetrical compositions were enhanced with winding bands or clumps of trees and vast, rolling lawns, usually focused on a lake. At Blenheim Palace in Oxfordshire, often hailed as his magnum opus, he removed Henry Wise's extensive parterre and brought the lawns right to the house, planting dark trees to frame the landscape beyond. He was later criticized for such wanton destruction of the works of earlier gardeners. A story underlines his enormous impact upon the English countryside: asked by an English lady to make a plan for her Irish estate, Brown is said to have replied, "No, madam, I can't. I haven't finished England yet."

Humphry Repton, probably England's greatest landscape theorist, was a minor landholder who had failed in business and at farming. In 1788 he took up landscape design when a family friend, the Duke of Portland, asked him to alter his garden. A key to his success was his skill as a watercolorist. He could produce attractive renderings of his schemes—an important factor in his profession because clients needed to visualize what might not be realized for years. Repton freely admitted his debt to Brown and continued many of his practices. Because of the Napoleonic Wars, his opportunities were limited. He produced landscapes, seldom as extensive as Brown's, throughout England, among them many for terraces or smaller gardens close to houses. After about 1790 Repton created a transition between houses and their grounds by means of steps, terraces, and balustrades, through a "natural" park to a distant composed view. His ideals—utility, proportion, and unity—were best expressed at Woburn Abbey, where he augmented an existing landscape garden with a private garden, a flower garden, and what he called an "American garden." In some senses, he began the transition from the informal landscape garden to the formality of the Victorian era. In 1795 he published *Sketches and Hints on Landscape Gardening*.

The esthetic mood in England was changing. From about 1770 Brown came under critical attack for, of all things, his "excessive formalism and lack of 'naturalness.'" The romantic picturesque movement called for an exciting wild landscape—what was seen as a *true* return to nature. Nevertheless, those who could afford it had their architects (including Repton) build bogus ruins and enigmatic grottoes in the grounds of their houses; one eccentric even employed a hermit to live in his grotto. Debate about the classical English landscape garden versus the picturesque garden were exhausted by about 1830, and the "grand vision" dulled to be replaced by the Victorian garden with its rose beds, shrubberies, and rockeries. The acceptability of Brown and his followers—there were many—declined until the early twentieth century, and it was not until 1950 that he was finally recognized as the eighteenth century's "most celebrated English landscape architect."

Further reading

Clifford, Joan. 1999. *Capability Brown: An Illustrated Life of Lancelot Brown, 1716–1783*. Princes Risborough, UK: Shire Publications.

Daniels, Stephen. 1999. *Humphry Repton: Landscape Gardening and the Geography of Georgian England*. New Haven, CT: Yale University Press.

Laird, Mark. 1999. *The Flowering of the Landscape Garden: English Pleasure Grounds, 1720–1800*. Philadelphia: University of Pennsylvania Press.

Erechtheion
Athens, Greece

The Erechtheion, built on the site of ancient sanctuaries on the Athenian Acropolis, is so unlike every other Greek temple that some have dismissed it as an aberration. Rather, it is the result of its architect, probably Mnesikles, applying inventive skill to accommodate a complex web of religious relationships. The Erechtheion provides evidence that the craft tradition of architecture, hobbled by convention, was giving place to a new creative approach to design. That was a great step forward.

The city-states of Athens and Sparta and their respective allies fought the Peloponnesian Wars between 431 and 404 B.C., interrupted by the six-year Peace of Nikias, from 421. Although their popular *strategos* (elected general) Perikles had died in an epidemic in 429 B.C., the Athenians took occasion of

The Erechtheion, Athens, Greece; Mnesikles (?), architect, 421–ca. 406 B.C. Detail of caryatid porch, photographed in the 1930s.

the cessation of hostilities to complete his fifty-year plan to restore the glory of the Acropolis. The last phase of the work was the Erechtheion, commenced in 421 B.C. and finished around 406 B.C. The building was a brilliant response to both spiritual and practical problems.

First, the building, unlike other temples, was to be dedicated to more than one deity, and its precise location, because it was connected with several other important spiritual themes, was especially significant to Athenians. The temple's primary purpose was to provide shrines for both Athena Polias and Poseidon. The sanctuary was also to include the graves of Erechtheos (a mythical king of Athens) with the sacred snake, and of Kekrops, another fabled Athenian ancestor. Moreover, there was the place where Poseidon's trident or Zeus's thunderbolt—the mythology was ambiguous—had struck the ground, the

"Erechtheis Sea," and the *thalassa*, a saltwater spring. The altars of Poseidon, Erechtheos, Zeus Hypatos, Hephaistos, the Boutes, Zeus Thyechoos, and a reliquary for an ancient wooden statue of Hermes were to stand within its *temenos* (sacred courtyard). Besides all that, the building had to accommodate a sacred olive tree in the precinct of Pandrosos, which also included the altar of Zeus Herkeios. The difficulty of laying out the Erechtheion was compounded by the irregular terrain. Eventually, the north and west walls would stand about 9 feet (2.7 meters) below the south and east.

The architect satisfied all these conditions, combining great imagination with pious deference to tradition to produce a spatially ingenious temple that must have at once bemused and delighted his clients. The plan of the Erechtheion was very complicated in comparison with the simple rectangular forms of all earlier Greek temples. It consisted of three almost independent sections (the rather traditional main temple, the north porch, and the famous caryatid portico), each with its separate roof. Because of the steep slope across the site, it was built at four different levels. The *naos* was separated—once again, a major departure from convention—into two main parts, the east *cella* devoted to Athena Polias and another, whose roof is divided by a huge beam, to Poseidon-Erechtheos. Underground rooms housed the statue of Hermes and Erechtheos's tomb.

Ionic orders employing three different proportional systems were incorporated, and graceful statues of *korai* (draped female figures) supported the entablature of the caryatid portico. The building was extravagantly decorated. Although it never seems to have had pedimental sculptures, relief carvings filled the frieze of Eleusinian stone. Fragments survive, but the general theme is not known. The ceiling coffers of the caryatid portico are almost entirely intact. Some scholars think all the portico ceilings were paneled and painted dark blue with gold stars, others that they were inset with colored glass panels.

The unique temple was converted into a church during the Middle Ages, and later it was used as a harem for the ruler of Athens during the Turkish occupation. In 1801 the British ambassador, Thomas Bruce, Earl of Elgin, took a caryatid (which he later sold to the British Museum), replacing it with a plaster cast. The Erechtheion was partly rebuilt by the American School of Classical Studies. Now it again suffers depredations, this time from atmospheric pollution and the increasing pressure of tourism.

See also The Acropolis; Parthenon

Further reading

Harris, Diane. 1995. *The Treasures of the Parthenon and Erechtheion*. New York: Oxford University Press.

Jeppesen, Kristian. 1987. *The Theory of the Alternative Erechtheion*. Århus, Sweden: Århus University Press.

Scully, Vincent. 1979. *The Earth, the Temple, and the Gods*. New Haven, CT: Yale University Press.

Erie Canal
New York State

The 363-mile (585-kilometer) Erie Canal between Lake Erie and the Hudson River, New York, was opened in 1825. Compared with earlier U.S. canal projects, common since about 1785 (none was over 30 miles long), it was a colossal enterprise, incontrovertibly the greatest public works project in the young republic. Despite criticism at its inception, when complete it was acclaimed as the world's greatest engineering marvel. But great engineering achievement that it was, the social significance of the Erie Canal outstripped that feat by far.

Before its creation the Allegheny Mountains were the Western frontier. Beyond them, virtually inaccessible to the European settlers, lay the resource-rich Northwest Territories—later to become Illinois, Indiana, Michigan, and Ohio. The canal made westward migration possible, and within fifteen years of its opening it turned New York, formerly the fifth-largest harbor in the nation, into the busiest seaport and greatest commercial center in the United States. In that same period the value of the city's real estate quadrupled, and mercantile activities multiplied five times.

As early as 1724, a surveyor named Cadwallader Colden speculated on the potential value of a direct water link between the Hudson River and Lake Erie. Sixty years (and the War of Independence) later, a bill proposing improvements to the navigability of the Mohawk and Onondaga Rivers, with an eventual link to Lake Erie, was put before the New York State legislature. It was defeated, but the idea was revived in 1791. Legislation was passed, feasibility studies undertaken, and two companies established to "open a navigable waterway from Albany to Lakes Seneca and Ontario" and "improve navigation between the Hudson and Lake Champlain," respectively. Little came of it.

Between October 1807 and April 1808 a miller named Jesse Hawley published several essays in the *Genesee Messenger* that advocated the construction of a 100-foot-wide (30-meter), 10-foot-deep (3-meter) canal from Buffalo, at the southern end of Lake Erie, to Utica, where it would join the Mohawk River to Schenectady, allowing cargoes to be taken over portage to Albany. His idea was widely derided, even labeled "the effusions of a maniac," but New York City's mayor De Witt Clinton publicly agreed with him. The proposed canal was promptly dubbed "Clinton's Folly."

Others sided with Clinton. State legislator Joshua Forman successfully moved in 1808 that the best route be surveyed. The report was made in 1810 and the issue was kept alive until a law supporting the project was passed the following year. Public pressure to start the canal continued until Clinton became state governor in 1817. When President Thomas Jefferson, believing a national waterway to be "little short of madness," vetoed the proposal, the estimated construction cost of $7 million became New York State's responsibility. Clinton persuaded the legislature to authorize the expenditure, to be funded with bond issues. Much of it was raised from the savings of new immigrants; wealthy, more conservative investors took no risks until the first section of the canal was completed.

The builders of the "Great Western Canal" struck out westward from Rome, New York, on 4 July 1817. Untrained gangs of "canawlers," many of them Irish immigrants from New York City, were paid fifty cents a day and worked under the general superintendence of the chief engineer, Benjamin Wright. Existing streams and lakes were not joined by the canal, which followed an independent course. The 80-mile (128-kilometer) middle section was cut in light soil across level terrain—where locks or aqueducts were not needed—between Rome and Utica, built by Wright's assistant engineer, David Stanhope Bates. It opened in October 1819. Separate contracts were let for various parts of the huge undertaking. A month later the 63-mile (100-kilometer) Champlain branch canal was opened, joining Troy with Whitehall, at the southern end of Lake Champlain. In July 1822 the section from the Genesee River to Pittsford was navigable, although several miles of overland connection were necessary until the Irondequoit Valley embankment was completed in October, also under Bates's direction. In the same month a 180-mile (290-kilometer) stretch was opened between Rochester and Little Falls. The eastern section through the Mohawk River valley was finished a year later, allowing uninterrupted navigation from the Genesee River to Albany and Lake Champlain. The last leg, between Brockport and Albany, was completed in April 1824.

On 26 October 1825 Clinton set out down the Erie Canal from Buffalo in the *Seneca Chief* with his wife and a party of "distinguished citizens"; two other canal boats accompanied them, carrying products from the Midwest, and even a bear and two eagles. Traveling at an average of 3 mph (5 kph), they arrived in New York Harbor nine days later to a boisterous welcome. Clinton poured two barrels of Lake Erie water into the sea, ceremonially marking the "marriage of the waters."

The great waterway, disrespectfully rechristened "Clinton's Big Ditch," was 40 feet (12 meters) wide and only 4 feet (1.2 meters) deep and carried vessels of 90 tons (76.2 tonnes) displacement. It had cost $700,000 over budget, but the outlay of almost $8 million was recouped from tolls within ten years. Only the human cost was not recoverable. Neither was it recorded: many died from malaria, others from smallpox; others were maimed by accidents.

Inland shipping now found its way across eighteen aqueducts and through eighty-three locks, falling 570 feet (174 meters) between Tonawanda and Buffalo, on the eastern shore of Lake Erie, and Troy, on the Hudson. There were ninety-three continuance bridges, where draft animals crossed the water. Travel time between the Great Lakes and the East Coast was halved, and freight costs fell from $100 to $10 a ton. Wheat tonnage carried on the canal increased a staggering 140-fold between 1829 and 1837, and by 1841 the figure had doubled again.

Two more important branch canals were later built: the 24-mile (38-kilometer) Oswego (1828), connecting the Erie with Lake Ontario; and the 27-mile (43-kilometer) Cayuga and Seneca (1829), linking the Erie, west of Syracuse, with Cayuga and Seneca Lakes. The entire system was enlarged in 1835 and again in 1862 and 1895 to cater for heavier traffic. Another modification was undertaken in 1904, "canalizing" the Mohawk, Oswego, Seneca, Oneida, and Clyde Rivers and Oneida Lake and abandoning large sections of the original canal. When the 525-mile-long (845-kilometer), 12-foot-deep (3.7-meter) new system, renamed the New York State Barge Canal, was opened in 1918, it comprised the Erie and all the former branch canals. The Barge Canal can carry 2,400-ton (2,032-tonne) vessels.

After about 1850, burgeoning railroads competed with canal transportation, but the Barge Canal was not made redundant until the construction of the St. Lawrence Seaway a century later. Within a decade or so, mercantile traffic had dwindled to the point of insignificance. However, recreational traffic was growing, and in November 1991 the people of New York rallied to save the canals. In 1996 the federal Department of Housing and Urban Development announced grants totaling $131 million for the Canal Corridor Initiative, a program to rehabilitate the Erie Canal and its branches as a "recreationway."

Further reading

Condon, George E. 1974. *Stars in the Water: The Story of the Erie Canal*. Garden City, NY: Doubleday.

Shaw, Ronald E. 1990. *Erie Water West: A History of the Erie Canal, 1792–1854*. Lexington: University of Kentucky Press.

Sheriff, Carol. 1996. *The Artificial River: The Erie Canal and the Paradox of Progress, 1817–1862*. New York: Hill and Wang.

F

Fallingwater
Pennsylvania

The architect Frank Lloyd Wright designed Fallingwater (1934–1937) for the Pittsburgh department store owner Edgar J. Kaufmann and his wife Liliane, on Bear Run, Pennsylvania, in the mountains southeast of Pittsburgh. The spectacular house cantilevers over a 20-foot (6-meter) waterfall amidst a wilderness. Widely admired for over sixty years, Fallingwater has been called "the most famous residence ever built"; in 1991 the American Institute of Architects hailed it as "the best all-time work of American architecture." It is probably the most beautiful house of the twentieth century, some say of any century. Bear Run Nature Reserve, the 5,000-acre (2,000-hectare) area surrounding Fallingwater, and the house itself are now owned, maintained, and protected by the Western Pennsylvania Conservancy.

Throughout the 1920s and well into the next decade, Wright had little work. He publicized himself through writing and a traveling exhibition (he called it "The Show") in the United States and Europe, but despite the efforts of his friends, his poor financial management put him deeply in debt. In 1932 he established the Taliesin Fellowship in Wisconsin, a residential apprentice system in which aspiring artists and architects paid for the privilege of working for him.

Among them was Edgar J. Kaufmann Jr. Impressed with what he saw during a 1934 visit to Wright's Wisconsin home, Taliesin, and by his son's enthusiasm, Kaufmann Sr. commissioned Wright, who was then sixty-five years old, to design a mountain retreat for his family. The waterfall on Bear Run was a favorite spot of the Kaufmanns', and they wanted to build nearby. There is a tradition that Wright made the final design after only one visit to the site: the surprising idea, accepted by his clients, was that the house should sit over rather than face the waterfall.

Fallingwater's four levels are progressively set back to lie low against the forested hillside; their terraces, apparently suspended in space, echo the form of the waterfall. Wright built the house around a core containing a kitchen on the lowest level and bedrooms on the others; it also housed the service ducts. The house's horizontality—Wright called it the line of domesticity—is juxtaposed against a four-story sandstone chimney. The lowest floor, a huge living room, is carried on four stub walls and provides the widest views of the site; one of its cantilevered terraces faces upstream; the other projects over the great boulders framing the waterfall. From the living room, a short flight of stairs leads to a platform just above the creek. It serves little practical purpose, but in summer air from the running water cools the space above. The bedrooms on the second level each have a narrower terrace, and the rooms—another bedroom and a study—on the third level also open to a terrace. Wright used the stairs, terraces, and windows (full

Fallingwater (Edgar J. Kaufmann house) on Bear Run, Pennsylvania; Frank Lloyd Wright, architect, 1934–1937. Photographed ca. 1987.

height on three sides of the living room) to integrate exterior and interior, house and site. Architectural historian Spiro Kostof comments that Wright "sends out free-floating platforms audaciously over a small waterfall and anchors them in the natural rock. Something of the prairie house is here still.... But the house is thoroughly fused with its site and, inside, the rough stone walls and the flagged floors are of an elemental ruggedness" (Kostof 1985, 737).

All of Fallingwater's vertical elements—walls, fireplace, chimney—are built of sandstone quarried on the site and laid in shallow, regular courses by local craftsmen. The cantilevered, horizontal elements—floor slabs and edge beams, roofs and ceilings—are constructed of in situ reinforced concrete and painted with a sand-textured, warm-buff paint. Setting aside the misgivings of some structural engineers that the daring cantilevers would fail, Wright exploited the

reinforced concrete, a material with which he had worked since 1904. When the contractor hesitated to remove the formwork supports of the main floor slab, Wright insisted that they be struck. The floor did not collapse. The slab deflected more than was expected, and the upper terrace followed because they were tied together, but the ensuing cracks did not threaten structural stability.

The interior floors are paved with randomly shaped flags of sandstone, waxed and polished to give a wet appearance, forming yet another link with the site. The beautifully crafted joinery is of sap-grain walnut, and the windows have steel frames painted Cherokee red. There is a separate guest wing farther up the hillside, reached by a path with a stepped, vaulted canopy that circles a large oak tree. The house was completed in 1937 at a cost of almost $155,000—four times the original estimate.

In the face of what Henry-Russell Hitchcock and Philip Johnson dubbed the International Style and the 1930s immigration of many of its greatest proponents to the United States, Wright, when many critics believed his energy was spent, entered a new twenty-five-year phase of his career with Fallingwater. The house employed the same technologies as the "glass boxes on stilts" then fashionable on both sides of the Atlantic, while challenging their bland esthetic.

See also Frederick C. Robie House

Further reading

Kaufmann, Edgar Jr., et al. 1986. *Fallingwater, a Frank Lloyd Wright Country House*. New York: Abbeville Press.

Kostof, Spiro. 1985. *A History of Architecture, Settings, and Rituals*. Oxford, UK: Oxford University Press.

Stoller, Ezra, and Neil Levine. 1999. *Frank Lloyd Wright's Fallingwater*. New York: Princeton Architectural Press.

Fera (Thera)
Santorini, Greece

Fera—a town where no town should be—is an architectural feat for just that reason. It has been achieved largely without architects, and its builders have developed a remarkable symbiosis with their dangerous host, an active and restless volcano. Fera, a comparatively modern town of about 2,000, picturesquely clusters at the edge of a 900-foot (275-meter) cliff above its harbor. It is the capital of Santorini, the 28-square-mile (73-square-kilometer) main island of the southernmost group of the Cyclades.

The other islands in the volcanic group are Therasia, Aspronisi, Paea Kameni, and Nea Kameni. The latter two were created by eruptions since 197 B.C., and the others are fragments of Stronghyle (literally, the round one) after a cataclysm in the middle of the second millennium B.C. It is estimated that this so-called Minoan eruption spewed about 42 billion tons (35.5 billion tonnes) of volcanic material 23 miles (36 kilometers) into the air, blanketing the remaining islands in pumice and ash to a depth of 100 feet (30 meters). The consequent earthquakes and tidal waves destroyed buildings on the south coast of Crete, 100 miles (160 kilometers) away. Santorini has been inhabited since about 3200 B.C. In 1967 the archeologist Spiros Marinatos excavated the city of Akrotiri from the volcanic ash; its culture bore many similarities to the Minoan, and its seafaring people evidently traded throughout the Mediterranean. About 1000 B.C., the island was colonized by migrating Dorians, who built the first Fera and from there founded the colony of Cyrene, Libya, in 634 B.C. Santorini was taken by Athens in 426 B.C. and subsequently by Egypt, Rome, and Byzantium. The island fell in A.D. 1207 to the Venetians, who named it St. Irini, from which Santorini is derived. From 1537 Santorini was occupied by Turks; liberated in 1821, it became part of modern Greece.

During the Minoan eruption, the removal of so much magma caused the volcano to collapse, producing a caldera 32 square miles (83 square kilometers) in area. In places its sheer walls soar to nearly 1,200 feet (350 meters) above the sea and plunge nearly 1,300 feet (400 meters) beneath its surface. Despite a recent history of eruptions—1866–1870, 1925–1928, 1938–1941, 1950, and 1956—the town of Fera has been continually rebuilt on its precarious perch, the very rim of an active volcano. Most of the houses—the world-famous white-and-blue architecture is a medley of Cycladic and Venetian styles standing cheek by jowl—were built in the nineteenth century after the old Venetian capital of Skaros, immediately to the north, was devastated by earthquakes. Much of Fera was destroyed in the 1956 earthquake, but phoenixlike, it rose again, quite literally from the ashes.

Once, the inaccessible location on the caldera's lip offered security from seaward attack. And the ancient volcanic deposits have provided a rich source of building material and a richer opportunity to improvise. Fera's indigenous architecture, in houses and public buildings alike, responds to the multiple constraints of volcano, earthquake, shortage of timber, and the heat and glare of summer. The abundant volcanic material is used in the lightweight vaults and domes so common in the town. The gleaming white walls reflect the summer sun, generations of layered lime wash making the lines of the structures

Fera (Thera), Santorini, Greece. Aerial view of the town and the edge of the caldera.

fluid. Most of the cliff-clutching houses have cool, lofty, vaulted inner rooms carved from Santorini's soft rock mantle; only the *sala* (front room) is built up. This inexpensive way to enlarge a house gave rise to an architecture and an urban design in which one house's courtyard is the roof of the house below it. More importantly, this widespread building technique has created a town of sweet integration: no collection of competing buildings, this, but a place with an overwhelming sense of community.

Further reading
Boleman-Herring, Elizabeth. 1995. *Aegean Islands, Mykonos and Santorini*. Boston: APA.

Bond, Howard. 1991. *White Motif: The Cyclades Islands of Greece*. Ann Arbor, MI: Goodrich Press.

Doumas, Christos. 1983. *Thera: Pompeii of Ancient Aegean*. London: Thames and Hudson.

Firth of Forth Railway Bridge
Scotland

Nine miles west of Edinburgh, Scotland, the mouth of the River Forth is spanned by Europe's first all-steel, long-span bridge. Completed in 1890 it was then the longest bridge in the world. Until 1917 it was also the largest metal cantilever, and at the beginning of the twenty-first century it remains the second largest ever built. It was a major accomplishment of Victorian engineering.

The extension of the railroad along Scotland's east coast, to complete the direct route between Edinburgh and Aberdeen, was hampered for most of the nineteenth century by two broad inlets of the North Sea: the Firth (mouth) of Tay and the Firth of Forth. The River Forth rises near Aberfoyle and widens into its firth about 50 miles (80 kilometers) from the ocean.

Vessels up to about 300 tons (270 tonnes) could navigate as far as Alloa, about 16 miles (26 kilometers) inland; those up to about 100 tons (91 tonnes) could reach Stirling, a little further on.

After earlier aborted proposals—a tunnel in 1806 and a bridge in 1818—for crossing the firth, little more was attempted for fifty years. In 1865 an act of Parliament sanctioned a bridge across the Queensferry Narrows, where the river passes between steep banks at the neck of the firth. Four railroad companies—North British, North Eastern, Midland, and Great Northern—formed a consortium in 1873 and commissioned Thomas Bouch, engineer for North British, to design the bridge. He proposed a suspension structure with twin spans of 1,600 feet (480 meters). The project was delayed for five years because of lack of funds; by spring 1879 only one pier had been started.

When the much-vaunted Tay Railway Bridge, also designed by Bouch and less than two years old, collapsed in a gale on 28 December 1879 with the loss of seventy-five lives, work on the Forth bridge was immediately suspended by another act of Parliament. In January 1881 a British Board of Trade inquiry found that the Tay disaster was caused by inadequate design and poor supervision. Bouch's Firth of Forth scheme was abandoned. Within months the engineer died, a broken man. The engineers of the Forth consortium's member railways, Thomas Harrison, William Barlow, John Fowler, and Benjamin Baker, had to develop a new design. In May 1881 Fowler and Baker submitted a plan for a continuous girder, or balanced cantilever, structure. In July 1882 yet another act authorized construction. The Tay bridge affair had so undermined public confidence in railroads that the legislation insisted that the Forth bridge should "enjoy a reputation of being not only the biggest and strongest, but also the stiffest bridge in the world." There was to be no vibration, even as trains passed over it. Consequently, it was greatly over-engineered.

Before 1877 steel bridges had been banned by the Board of Trade because the Bessemer conversion process produced steel of unpredictable strength. The Siemens-Martin open-hearth process, developed by 1875, had changed that, yielding material of consistent quality. That kind of steel was used in the Forth bridge, heralding the transition from cast and wrought iron in such structures. A smaller steel cantilever bridge had been built in Germany, but the Scottish project was on a larger scale than had been seen before. There is little doubt that its designers owed much to a U.S. model of several years earlier. Between 1869 and 1874 James B. Eads had designed and built the world's first steel bridge, over the Mississippi at St. Louis, Missouri. Its three-arch superstructure, with a center span of 520 feet (156 meters) and side spans of 502 feet (150 meters), supported by four massive limestone piers, carried a railroad and a road for other traffic on two levels. Other pioneering features of Eads's bridge were adopted by the British: the use of pneumatic caissons (large diving bells fed with compressed air) to excavate the foundation, tubular steel structural members, and a balanced cantilever design that allowed construction to proceed without temporary supports that would have obstructed the waterway.

In December 1882 the contract for the Forth bridge was awarded to a consortium led by Tancred Arrol, an experienced and respected company headed by William Arrol, which already had contracts for the Caledonian Railway Bridge over the Clyde and the replacement Tay bridge. At the height of building activity, there would be 4,600 Britons, Italians, Germans, and Austrians working shifts around the clock. The construction of the foundations and piers took until the end of 1885. Each of the bridge's three cantilever towers stands on four 70-foot-diameter (21-meter) granite piers, founded on the bedrock. Eight of the piers are in water, and their foundations were excavated by men working in wrought-iron pneumatic caissons, sunk up to 90 feet (27 meters) below the river surface. The massive cylinders were prefabricated in Glasgow, then dismantled and taken to Queensferry, where they were reassembled. Once excavation was complete, the air shafts and the working spaces were filled with concrete, and the granite piers rose above them.

Work on the superstructure began in 1886 using 64,800 tons (54,860 tonnes) of steel from two steelworks in Scotland and another in Wales, fixed with rivets from a Glasgow foundry. All the structural

Firth of Forth Railway Bridge, Scotland; Thomas Harrison, William Barlow, John Fowler, and Benjamin Baker, engineers, 1881–1890.

members were fabricated in on-site workshops, predrilled, test-assembled—exact dimensions were needed in a riveted structure—and then dismantled to be painted and carried to the site for erection. Each of the 331-foot-high (99.3-meter) cantilevers consists of two inward-sloping trusses fabricated from huge, internally stiffened tubular members up to 12 feet (3.6 meters) in diameter. They support 680-foot-long (204-meter) cantilever arms that are linked midspan by suspended girders of about half that length, making the distances between the towers about 1,700 feet (540 meters). The length of the bridge between the end piers is about 5,300 feet (1,600 meters). Together with the approach viaducts and arches at each end, the bridge carries the double-track railroad for 2,765 yards (2,490 meters), 150 feet (45 meters) above the surface of the Firth of Forth. The central gap was closed on 14 November 1889, and the Prince of Wales ceremonially opened the bridge on 4 March 1890.

Further reading
Hammond, Rolt. 1964. *The Forth Bridge and Its Builders*. London: Eyre and Spottiswoode.
Koerte, Arnold. 1991. *Two Railway Bridges of an Era: Firth of Forth and Firth of Tay*. Boston: Birkhäuser.
Petroski, Henry. 1995. *Engineers of Dreams*. New York: Alfred A. Knopf.

Florence Cathedral dome
Florence, Italy

The dome of the cathedral church of Santa Maria del Fiore (St. Mary of the Flowers) in Florence, Italy, was designed and built by Filippo Brunelleschi (1377–1446) in the beginning of the fifteenth century. Towering over the immediately surrounding buildings and still visible, almost 600 years later, from any part of the city, it is one of Europe's greatest architectural and engineering achievements—a masterpiece of structural ingenuity.

Brunelleschi's dome completed the building, which had been started in September 1296 by the architect Arnolfo di Cambio. Arnolfo's original design, which included a much lower cupola, went through many changes, although it is probable that his general plan was retained. Work ceased when he died in 1310, and did not resume until 1331, when the powerful Wool Merchants Guild assumed responsibility for construction. In 1334, simply because he was "a very great man," the painter Giotto di Bondone was appointed *capomaestro* to Florence Cathedral, and he designed the freestanding 278-foot (85-meter) bell tower near the southwestern corner of the church. Not finished until two years after his death in 1337, the characteristically Florentine building is faced— as is the church itself—with geometric patterns of red Siena, green Prato, and white Carrara marbles. It was later enriched with relief panels by Luca della Robbia and Andrea Pisano.

The cathedral was beleaguered by further delays caused by political intrigue, a capricious economy, and not least, in 1348, an outbreak of plague that halved the city's population. The following year Francesco Talenti was appointed to oversee the work. Apart from completing the bell tower, he continually revised the design, working in conjunction with Giovanni di Lapo. Talenti's "final" scheme evolved by 1366–1367: the nave, flanked by single aisles, was articulated in four square bays leading to an octagonal sanctuary, from which four chapels radiated. Construction work was well in hand by 1370; the nave vaults were finished in 1378 and the aisles a year later. Important in Talenti's design was a huge octagonal dome over the sanctuary, and construction of its drum had been commenced.

In 1417 a committee, the Opera del Duomo, was charged with the monumental task of building the dome. In 1418 Brunelleschi, who had been recently engaged in bridge building in Pisa, was commissioned to act as adviser. But soon after that the project again lapsed. The dome presented a seemingly impossible problem for the builders because not only did it have to span 135 feet (41.5 meters), but it also had to begin nearly 180 feet (55 meters) above the floor. Nothing of the kind *or* size had been built since the Roman Pantheon about 1,300 years earlier. A com-

petition was announced, and at a March 1419 meeting, solutions were offered by invited master masons from Italy, France, Germany, and England. None was satisfactory. It was impossible to construct scaffolding to support traditional centering at such a height. Supporting permanent masonry piers were out of the question because they would clutter the sanctuary, blocking the view to the high altar and defeating the purpose of the cathedral. Someone even suggested that the sanctuary should be filled with a mixture of earth and coins to enable the erection of scaffolding; then, when the dome was complete, the citizens of Florence could remove the soil as they dug for the buried treasure. Brunelleschi asserted that he could build the dome without confronting any of these problems.

At first, the committee was skeptical, and when he excitedly defended his position, he was forcibly removed from their meeting. Given another chance to present his proposal, he was reluctant to reveal details. His biographer Giorgio Vasari recounts a delightful anecdote: producing an egg and a thin marble slab, Brunelleschi challenged anyone there to balance the egg on the slab. No one could, and the items were returned to Filippo. He cracked the egg, and stood it upright. To the protest "We could have done that!" he replied, "That's what you will say if I tell you how I will build the dome!" He was given the commission.

What qualified this sculptor, who trained as a goldsmith, studied science and mathematics, and dabbled in clock making, engineering, and architecture, to confidently undertake such a daunting project? It has been suggested that, because his father was closely connected with the management of the cathedral, Filippo had known of the problem of the dome since 1402, making his first designs as early as 1409. In the intervening years he had studied classical architecture in Rome, developing his own theory of architecture from about 1410. When his model of the dome was accepted in 1418, Brunelleschi returned to Rome to investigate ancient structures, including the Pantheon. Back in Florence a year later, he built a smaller version of the dome in the Ridolfi chapel in San Jacopo sopr'Arno (since destroyed), perhaps to convince his doubters, perhaps himself. He would repeat the process six years later in a chapel built for Bartolommeo Barbardori in the church of Santa Felicita. Also in 1419, Filippo reluctantly accepted the appointment of his old rival, the sculptor Lorenzo Ghiberti, as coarchitect of the dome. Work began in 1420, and the project occupied Filippo for the rest of his life. Ghiberti's incompetence was soon exposed (not without Brunelleschi's connivance), and in 1423 Filippo was given complete charge.

Besides his genius for design, his success was ensured by his management of the construction site. He personally undertook the quality control of materials; he designed the plant needed to efficiently raise those materials—all 27,000 tons (24,500 tonnes)—to the dome; he resolved industrial unrest by convincing the workers of his own capabilities; he communicated with the masons by modeling details in clay, wood, wax, and even carved turnips; he ensured good working relationships; and he increased productivity by providing "vendors of wine and bread and cooks" in the heights of the dome.

Strictly speaking, Filippo's dome is no dome at all, but a double-shell cupola formed by completely self-supporting rings of diminished diameter, built of "herringbone" brickwork (originally stone was intended but was soon abandoned because of the weight), and stiffened by a frame of eight steeply pointed arches built on centering supported from the drum. Between these ribs, Brunelleschi and his eight assistants—the "masters of the trowel"—constructed a double vault, on which a movable light shuttering supported the brickwork during construction. The finished structure was light, strong, and extremely stable. To ensure against spreading, Brunelleschi tied the bottom of the cupola with a massive chain of ironbound oak. Several reasons may be suggested for its hybridized structure. First, Brunelleschi was, quite naturally, unable to divorce himself from medieval

Florence Cathedral (Santa Maria del Fiore) dome, Florence, Italy; Filippo Brunelleschi, architect, 1418–1436. View from southwest, photographed in 1996.

precedent; second, while he showed great inventiveness, he did not fully understand the structural issues involved; and therefore, third, he took measures to ensure the stability of the dome. Fourth, he may have included the ribs simply to convince his clients that it *was* stable. "Shrewd" is how Vasari described him.

The dome was not a Renaissance building. It did not even herald the Renaissance. It drew upon Brunelleschi's study of ancient techniques from both East and West, principally upon the engineering practice of the Middle Ages. In ingenuity it surpassed them all. The cupola was completed in 1434. Two years later the huge lantern was placed, and the cathedral was consecrated by Pope Eugene IV on 25 March 1436. The four hemidomed tribunes were completed in 1438. The decorations to the lantern were finished by 1446, when Brunelleschi was dying, and the great copper sphere crowned it all in 1474.

See also Foundling Hospital

Further reading

Fanelli, Giovanni. 1980. *Brunelleschi*. Florence: Scala.
Klotz, Heinrich. 1990. *Filippo Brunelleschi: The Early Works and the Medieval Tradition*. New York: Rizzoli.
Saalman, Howard. 1980. *Filippo Brunelleschi: The Cupola of Santa Maria del Fiore*. London: Zwemmer.
Vasari, Giorgio. 1991. *Selections from the Lives of the Artists*. Oxford, UK: Oxford University Press.

Foundling Hospital
Florence, Italy

The Foundling Hospital (known in Italian as the Ospedale degli Innocenti) stands in the Piazza SS. Annunziata, Florence. As its name indicates, it was a refuge for abandoned or orphaned children. Around 1419, over a century after the foundation of the institution, the powerful Guild of Silk Merchants and Goldsmiths (Arte della Lana) funded a new building to house refectories, dormitories, infirmaries, and nurseries, all joined with cloisters and porticoes. Designed by the ubiquitous artist Filippo Brunelleschi (1377–1446), the new *ospedale* was a seminal achievement, representing a change not only in how architecture looked but also in the way in which the

building industry was structured. Some scholars hail it as the first Renaissance building and its author as the sole instigator of those changes, the pioneer of a new phase in western European architecture.

The humanism of the Renaissance should never be confused with humanitarianism. Neither should accounts of urbane courtly life be thought of as accurately reflecting the entire social structure. On the contrary, the Renaissance was socially divisive at many levels, even within the family, regardless of class. Children, especially, were victims of a value system that often counted them as chattels whose sole reason for being was to perpetuate a particular dynasty or expand social and political power by strategic marriages. If they could not be put to such use, they were at best ignored; at worst, they were literally abandoned. Although little was done to overturn the attitudes that created this problem, many institutions were set up to care for foundlings. But in the fourteenth and fifteenth centuries, throughout Italy and most of Europe the orphanages could hardly provide for the number of rejected children. The Foundling Hospital was one such refuge that took them in, raised them, and taught them a trade.

Although he was trained as a goldsmith, a series of events in his native Florence early in the fifteenth century caused Brunelleschi to decide to go to Rome, where for about three years he made a detailed archeological study of ancient monuments. After dabbling in clock making and civil engineering, he turned toward the art of architecture. Untrained in the building profession like the contemporary mason-architect, who was the inheritor of medieval traditions and who to some degree physically built what he designed, Brunelleschi was an artist-architect, independent of long-standing trade and craft conventions. He was therefore able to devise, largely through his own intuition, different ways to build. Moreover, producing his oeuvre several decades before the formal architectural theories of the Renaissance had developed, he was also independent of the unbending "correctness" of later philosopher-architects. In the right place at the right time—the fertile intellectual seedbed of quattrocento Florence—he was free to create a beautiful amalgam, a culturally appropriate new architecture, by reinterpreting classical elements within the graceful tradition of Tuscan Romanesque.

Given the Florentines' admiration for anything of classical Rome, it is hardly surprising that the Foundling Hospital is replete with classical motifs. The loggia, doubtless the most familiar aspect of the building, is drawn from the porticoes that surrounded the Roman forum; like most Roman temples, it stands on a platform above the general level of the piazza. Slender columns, with Brunelleschi's version of Corinthian capitals, support a light, cross-vaulted arcade (incidentally, constructed without scaffolding). Classical moldings abound, and there is an entablature of shallow classical profile. The rectangular upper-story windows have triangular pediments, and the facade is crowned with a classical cornice. The elements of the loggia—indeed, most of the building's exterior—are defined with the beautiful gray-green stone known as *pietra serena*. The round-arched loggia was a familiar element in fourteenth-century Florentine buildings, including hospitals, and Tuscan architecture had long been characterized by the emphasis of structure through the use of darker bands of stone: for example, in the Pisa Cathedral group or San Miniato al Monte in Florence itself.

The interior spatial articulation of the Foundling Hospital is based upon a porticoed courtyard. It is Roman-like, its larger apartments and service rooms symmetrically disposed about an axis. The outer loggia unites it all as well as tying the whole building to the piazza. In the spandrels between arches, Andrea della Robbia (1435–1525) later added colored faience medallions portraying babies in swaddling clothes.

Although he used a Roman architectural vocabulary, Brunelleschi's syntax (to continue the analogy) was decidedly un-Roman. The rigor of archeologically correct classical *grammar* would emerge in the so-called High Renaissance, whose architects would never carry (for example) an arch on a column, because that had not been the Roman way. In fact, the delicately proportioned esthetic of the Foundling Hospital owes as much to medieval precedent as to classical models. Brunelleschi may have confused his chronology, because his contemporaries had a skewed view of history. The architecture of ancient Rome was not republican (as the Florentines wanted to believe) but imperial, and Roman*esque* was certainly not Roman.

See also Florence Cathedral dome

Further reading

Gavitt, Philip. 1990. *Charity and Children in Renaissance Florence: The Ospedale degli Innocenti, 1410–1536*. Ann Arbor: University of Michigan Press.
Klotz, Heinrich. 1990. *Filippo Brunelleschi: The Early Works and the Medieval Tradition*. New York: Rizzoli.
Manetti, Antonio di Tuccio. 1970. *The Life of Brunelleschi*. Philadelphia: Pennsylvania State University Press.

Frederick C. Robie House
Chicago, Illinois

When it was completed in June 1910, one neighbor described the Frederick C. Robie House as a "battleship"; another said it was a "disgrace." But in 1957 its architect Frank Lloyd Wright, never known for his modesty, accurately claimed it to be the "cornerstone of modern architecture." Many critics agree, and the building has been recognized by the American Institute of Architects as one of Wright's major architectural contributions to the United States. In 1963 it was designated a National Historic Landmark, and a Chicago Landmark in 1971. The Robie House is the fullest expression of the dwellings known as Prairie houses. It is no exaggeration to say that, as a decisive, even shocking, contrast to traditional contemporary houses, it revolutionized domestic architecture throughout the world.

Wright and others developed the Prairie style—named for Wright's "Home in a Prairie Town" published in the *Ladies Home Journal* in February 1901—mainly in the Chicago area, as "a modern architecture for a democratic American society." The Prairie house was designed to blend in with the flat, expansive midwestern landscape. What characterized it? Wright's "organic architecture" philosophy is difficult to define in few words, but simply, it was this: the house was a single living space, and everything about it grew from a plan that expressed the owner's individuality; that is, the house fits the family, not vice versa. The openness was achieved by

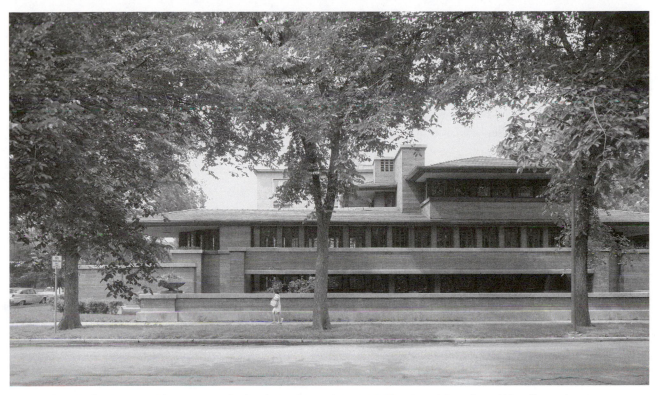

Frederick C. Robie House, Chicago; Frank Lloyd Wright, architect, 1908–1910. View from Woodlawn Avenue.

exploiting technology: central heating defied the harsh prairie winters. Through sensitive use of materials, the spaces became a whole whose external masses, expressing what was within, existed in harmony with each other and the earth itself—the building grew out of the site, so to speak.

In 1908 Frederick Carlton Robie, a bicycle and motorcycle manufacturer, decided to build a house for himself and his wife, Lora. He purchased a city lot at the corner of East 58th Street and Woodlawn Avenue in Chicago's Hyde Park because Lora wanted to live near her alma mater. Robie knew exactly what he wanted: a house with large eaves, broad vistas, and "all the light [he] could get"; avoiding "curvatures and doodads," it would be a house whose rooms were "without interruption." A number of architects advised him to consult Wright, who then had a mostly domestic practice in the suburb of Oak Park. Because of the affinity between client and architect, the design was soon resolved, and construction started in March 1909. The Robies moved in just fifteen months later.

Wright designed the building as two abutting elongated cuboids, separating the living areas from the service areas. The smaller was at the back, and its lower level contained a three-car garage and the main (first-floor) entrance; above it were the servants' quarters, a kitchen, and a guest room. The front part, with stairs ascending through it, had a broad central chimney. A semibasement housed a billiard room and children's playroom, opening to a long, narrow courtyard. Wright "eliminat[ed] the room as a box" by making the living and dining areas into a single space partially divided by the chimney; the living room gave on to a terrace, defined by low walls.

The bedrooms were above, in the center of the house. In a 1918 critique, the Dutch architect J. J. P. Oud praised the functional planning; the plan-generated, three-dimensional form; and the way in which Wright exploited modern materials and technology in the spirit of the age. He pronounced the design a "source of esthetic pleasure for the practised critic."

The Robie House's low-pitched overhanging roof—at the ends it extended 20 feet (6 meters) beyond the walls, supported by steel cantilevers—and the long wall of flat Roman bricks with flush finished joints combined with balconies lined with planter boxes and continuous limestone copings to emphasize horizontality, what Wright called the line of domesticity, the line of repose. The overhang also created a microclimate, in summer shading the glass-walled southern exposure of the house and in winter protecting the windows from rain and snow. Besides that, it provided a sense of shelter and privacy for those within.

Almost 180 patterned lead-light glass windows, screens, and doors add to the fluidity of the inner spaces and serve to coalesce the interior of the house and its surroundings. Wright designed more than a house—he created an environment. He designed rich interiors, glowing with natural oak finishes, patterned glass, furnishing textiles, loose and built-in furniture (also of oak), and carpets. He also carefully integrated the mechanical and electrical services with the overall design.

Sadly, the Robies were soon obliged to sell their wonderful house because of financial difficulties. It remained a residence until 1926, when the Chicago Theological Seminary bought it and used it for a dormitory. When it was slated for demolition in 1957, Wright himself (then ninety years old) led the campaign to save it. The development firm of Webb and Knapp purchased it and six years later donated it to the University of Chicago. Neglect and vandalism (including alterations) took their toll. In 1992 the university began negotiations with the Frank Lloyd Wright Home and Studio Foundation (now renamed the Frank Lloyd Wright Preservation Trust) about jointly undertaking a restoration program. In January 1997 the two institutions entered an agreement with the National Trust for Historic Preservation under which the foundation became responsible for the house. Currently, it is planned to spend more than $7 million on a ten-year conservation project. The house has been nominated for inclusion on UNESCO's World Heritage List.

See also Fallingwater

Further reading

Hoffmann, Donald. 1984. *Frank Lloyd Wright's Robie House: The Illustrated Story of an Architectural Masterpiece*. New York: Dover.

Legler, Dixie, and Christian Korab. 1999. *Prairie Style: House and Gardens by Frank Lloyd Wright and the Prairie School*. New York: Stewart, Tabori, and Chang.

Wright, Frank Lloyd. 1968. *The Robie House*. Palos Park, IL: Prairie School Press.

G

Galerie des Machines (Gallery of Machines)
Paris, France

The Galerie des Machines was designed for the 1889 Paris International Exhibition—L'Exposition Tricolorée—by architect Ferdinand Dutert (1845–1906) in collaboration with engineer Victor Contamin (1840–1893). It was remarkable for its vast exhibition hall, made possible by exploiting a new structural innovation, the three-pin hinged or portal arch. Although used previously in bridge construction, this was the first application of the arch on such a large scale.

The concept of exhibiting to the world a nation's resources and achievements in art, science, and industry has its origins in ancient times. According to the Bible, the fifth-century-B.C. Persian king Xerxes I showed the riches of his kingdom for five months. More recently, fine art exhibitions were mounted, but such shows gradually added inventions. Following the Industrial Revolution and the consequent rise of mechanization, expositions demonstrating industrial progress were held regularly. Before 1900, no fewer than thirteen were organized in the manufacturing centers and capitals of Europe. They were popular events and buildings were purpose-built for them; perhaps the most renowned was Joseph Paxton's revolutionary Crystal Palace, built in London for the Great Exhibition of 1851. In turn, many of those structures became showpieces of structural and technological advances.

Following the celebrated success of the Great Exhibition and Britain's abandonment of such shows after 1862, the French seized the opportunity to take center stage, so to speak. Between 1855 and 1900 five international exhibitions were presented in Paris, boasting of the progress of French industry and the country's rapid transition from a predominantly agrarian to an industrial economy. By 1889 when L'Exposition Tricolorée commemorated the centenary of the French Revolution, the size and variety of machines and other items offered for display were so great that a range of special exhibition spaces was needed. A formal entrance structure was built—the famed Eiffel Tower—and two long galleries were dedicated to the fine and liberal arts and a third to clothing and furniture exhibits. Beyond them and behind the Dome Central that terminated the long axis of the showground rose the vast and impressive Galerie des Machines.

Built principally of iron and glass, the structure employed a three-hinged or portal arch spanning a phenomenal 375 feet (114 meters); the widest previously achieved was 242 feet (74 meters) in the train shed of St. Pancras Station, London, about 25 years earlier. The display hall was 1,270 feet (380 meters) long, and its colossal proportions provided the largest unobstructed floor area of any building in history—an ideal setting in which to show the world the massive engines, transformers, dynamos, and other wonders of the age. The 20 prefabricated wrought-iron trusses of the main span comprised two half-arches, hinged at their meeting point 143 feet (45 meters) above the floor. They curved and tapered to a slender wedgelike base, where their loads were distributed to the ground through a hinged joint. The apparent lightness with which they touched the ground defied the conventional, rational notion that the base was the principal load-bearing component of any structure; here that role was seemingly reversed. The hinges allowed small movements between the foot of the frames and the foundation but made the arches statically determinate. Thus, stresses and reactions at the supports could be calculated beforehand and were only slightly influenced by movements of the supports or thermally induced dimensional variations.

The iron frame of the galerie was exposed at each end in a frank display of its construction system. The walls were generally glazed, in part with colored glass. Paintings, mosaic, and ceramic bricks also formed part of the cladding. Elevated tracks on each side of the long axis carried mobile walkways above the exhibition space, allowing visitors to travel in carriages and to look down on the machines. The interior was lit by electric lights, invented only some seven years earlier. The galerie was more than just a place for displaying machinery; it was in itself, as one historian has observed, an "exhibiting machine." It was enlarged for the 1900 Paris Exposition but demolished in 1910, because (so the reason was given) it spoiled the view of the church of Les Invalides. By then, the three-hinged arch was in wide use.

See also Crystal Palace; Eiffel Tower

Further reading

Mainstone, Rowland. 1975. *Developments in Structural Form*. Cambridge, MA: MIT Press.
Peters, Tom F. 1996. *Building the Nineteenth Century*. Cambridge, MA: MIT Press.
Steiner, Frances H. 1984. *French Iron Architecture*. Ann Arbor, MI: UMI Research Press.

Garden city idea

The garden city idea was conceived in late-nineteenth-century Britain by London-born stenographer and inventor Ebenezer Howard (1850–1928). A garden city movement emerged, inspired by his seminal text *Tomorrow: A Peaceful Path to Real Reform* (1898), revised as *Garden Cities of Tomorrow* (1902), and by the example of on-the-ground models. The movement supported Howard's objectives of improved residential environments and social opportunity. It made an enduring contribution to international planning thought by fostering the growth of planned residential communities and shaping ideas about the form and size of cities and towns.

The process of industrialization placed immense pressure on the physical resources of cities while at the same time depleting the agrarian workforce. Manufacturing processes and products took precedence over workers' needs. Employees toiled long hours and lived in overcrowded, often degraded, accommodations close to their workplace. Parks and gardens—green spaces—were rare, so there was little escape from industrial din and pollution. Social communication waned; crime and immorality increased.

For much of the nineteenth century, the social condition and the issue of land reform occupied reformers, economists, and intellectuals in Britain and elsewhere. In an earnest attempt to find a way forward, societies, organizations, and ameliorative action groups were formed; meetings and debates held; publications released; and theories and schemes advanced. Industrialists made practical efforts to improve employees' working and living conditions. Well-known ventures in England included Lever Brothers' soap factory at Port Sunlight, Liverpool (1888), and the Cadbury chocolate-making enterprise at Bournville, Birmingham (1879). Elsewhere there was Agneta Park near Delft, Holland, and industrial villages outside Noisiel, France, Essen, Germany, and in the United States at Lowell, Massachusetts.

Ebenezer Howard drew from a full larder of antecedents in devising his unique solution to urban disorder and misery. The answer was "Garden City," a town located in a rural setting but presenting all urban functions and services, thus combining the advantages of both town and country life. His scheme affirmed the role of the individual and the home in the urban landscape and the importance of satisfaction with home and place in the building of community. It would provide decent housing, ample opportunity for social interaction, and contact with nature to help keep mind and body healthy.

Garden City was envisioned as a preplanned, self-contained community of about 32,000 people. Its notional circular layout was enveloped and restricted by a greenbelt that offered clean air and space for agricultural and recreational pursuits. The city was divided into six equal segments or wards, separated by boulevards with a public park at the center and designated sites for municipal buildings, dwellings, churches, schools, and playgrounds. Shops were housed in a glass arcade encircling the central park. Facilities were within easy walking distance of all the houses. The industrial sector was placed at the perimeter and segregated from the residential to isolate noise and pollution. The garden city's self-governing community was to own and administer the land on which it was built; revenue would be derived from ground rents and profits returned for the communal benefit. Howard envisaged that once the population reached its limit, a new garden city would be established nearby, eventually creating a cluster of satellite communities—"Social City"—interconnected by a rapid-transit system.

The Garden City Association (1899) and its successor, the Garden Cities and Town Planning Association (1909), supported Howard's idea and promoted the construction of a model. The first was Letchworth Garden City, begun in 1904 in Hertfordshire, England, to a design by British architect-planner Raymond Unwin (1863–1940) and architect Barry Parker (1867–1941). A second garden city was built at Welwyn (1921). As its chief practitioner, Unwin played a vital role in disseminating the goals and principles of garden city planning.

Letchworth demonstrated that Howard's holistic vision was difficult to implement, but it was lauded nonetheless for its exposition of the physical, environmental aspects of the idea (rather than its economic, political, and social themes). It featured low-density development; land-use zoning; separate industrial and residential areas; existing natural features; variation in road width; harmony in building scale, form, colors, and materials; public open space; private and public gardens; and tree-lined streets. These components became popularly accepted as standard for planning "on garden city lines" and informed the design of residential offspring of the garden city—garden suburbs, towns, and villages. The most renowned of these, Unwin's Hampstead Garden Suburb (1907), northwest of London, was the model for subsequent and numerous developments in England, continental Europe, the United States, Australasia, Asia, and Africa. In accord with garden city wisdom, each was adapted to suit local topographical, social, economic, and cultural conditions. However, the theory underpinning the form was the same.

At the metropolitan level, Howard's argument for forward-looking, comprehensive planning and his vision of decentralized satellite cities surrounded by greenbelts offered a new planning approach and paradigm. It demonstrated how the city could be kept in touch with nature and introduced the concept of the master plan for metropolitan and regional development that was taken up as the century progressed. The garden city idea endured and came to the fore in the post–World War II new towns program developed in England to accommodate population overflow in London ("Mark I" towns) and later in provincial cities like Liverpool ("Mark II"). Stevenage (1946) in Hertfordshire was the first of the twenty-eight new towns established in Britain. New towns became an international phenomenon, and present-day planning movements such as New Urbanism acknowledge their debt to the garden city idea.

Letchworth, Welwyn, Hampstead Garden Suburb, and many of their international offspring survive. Some have been designated as heritage conservation areas. These now mature examples of planning "on garden city lines" continue to be attractive and desirable residential environments, proving the soundness of the philosophy that underpinned their plan.

See also Unité d'Habitation

Further reading

Beevers, Robert. 1988. *The Garden City Utopia: A Critical Biography of Ebenezer Howard*. London: Macmillan.

Hall, Peter, and Colin Ward. 1998. *Sociable Cities: The Legacy of Ebenezer Howard*. Chichester, UK: Wiley.

Ward, Stephen, ed. 1992. *The Garden City: Past, Present, and Future*. London: Spon.

Gateway Arch
St. Louis, Missouri

The 630-foot-high (192-meter) stainless-steel Gateway Arch rises from a wooded park within what became the Jefferson National Expansion Memorial Park on the bank of the Mississippi in St. Louis, Missouri. Taller than the Washington Monument in the national capital and twice as high as the Statue of Liberty, the sleek and seamless Gateway Arch (now known as the St. Louis Arch) is a major achievement of twentieth-century architecture and structural engineering.

A decision was taken in 1935 to establish a national monument in St. Louis, Missouri, to commemorate the nineteenth-century westward expansion that pursued Thomas Jefferson's dream of a continental United States. A large tract of riverfront land in the older part of the city was acquired and cleared, but the project was interrupted by the country's involvement in World War II. With the return to peace, in 1947–1948 the Jefferson National Expansion Memorial Association sponsored a design competition for an appropriate monument. The Finnish-American architect Eero Saarinen was awarded first prize.

Work on design development began in 1961, the year of the architect's death, the project being managed by his firm, Eero Saarinen Associates. Fred Severud of the structural engineering practice Severud, Elstad, Krueger and Associates undertook a feasibility study about the buildability of the daring concept, and Dr. Hannskarl Bandel generated exacting calculations for the weighted catenary (an inverted version of the curve of a suspended chain) that forms the basis of the structure. Bruce Detmers and other architects converted the mathematics into working drawings.

The main contractor was MacDonald Construction, and the steel was fabricated and erected by Pittsburgh-Des Moines Steel. The first concrete pour for the massive 60-foot-deep (18-meter) foundations took place late in June 1962 and construction of the arch itself began eight months later. The span of the arch is the same as its height, and the composite structure consists of 142 welded, stainless-steel-faced sections of equilateral triangular cross sections. The length of their side at the base is 54 feet (16.5 meters), and the sections are 12 feet (3.6 meters) high; at the top, they have a side length of 17 feet (5.4 meters) and are 8 feet (2.4 meters) high. The legs have double walls with an inner skin of 0.375-inch-thick (about 10-millimeter) carbon steel and an outer skin of stainless steel, set 3 feet (90 centimeters) apart; at the 400-foot (120-meter) level, the gap between the skins reduces to less than 8 inches (20 centimeters). For the first 300 feet the space between the walls is filled with concrete; above that, to the crown of the arch, the structure is braced with steel stiffeners. It is clear that the engineering design is highly complicated, but all that can be seen from the outside is the sheer skin of polished stainless steel.

The wall components were fabricated and bolted together in Pennsylvania and transported to St. Louis by rail. On-site, the triangular sections were welded by highly skilled tradesmen. In July 1998 their specialized work was recognized by the American Welding Society's Historical Welded Structure Award. The completed 50-ton (45.5-tonne) double-walled sections were transported to the site on a specially designed railroad car and lifted into place. For the first 72 feet (21.6 meters), conventional cranes on the ground were used; above that, purpose-made creeper cranes handled the sections. In effect, each leg of the arch was a vertical cantilever and therefore had no need of scaffolding. But when the 530-foot (162-meter) level was reached, a steel stabilizing truss was lifted into place and fixed to brace the two legs while the remaining twenty-one sections and the central "keystone" were located. The arch was completed on 28 October 1965. As the creeper cranes moved back to the ground, their tracks were dismantled and bolt holes in the stainless-steel surface were made good.

In 1967–1968 passenger trams were constructed in the hollow core of the arch, to carry visitors—there were 4 million in 1999—to a 140-person observation platform at the top, where tiny plate-glass windows give access to views up to 30 miles (50 kilometers) eastward and westward. The total cost of the arch, including $2 million for the internal transportation system, was $13 million. The building received the American Institute of Architects 25 Year Award in 1990.

Further reading

Mehrhoff, W. Arthur. 1992. *The Gateway Arch: Fact and Symbol*. Bowling Green, OH: Bowling Green University Popular Press.

Meyerowitz, Joel, and Vivian Bower. 1988. *The Arch*. Boston: Little, Brown.

Geodesic domes

A geodesic dome is a fractional part of a geodesic sphere, composed of a complex network of triangles. The archetypal geodesic sphere is made up of twenty curved triangles, each corresponding to one facet of the icosahedron, a twenty-faceted solid geometrical figure. The more complex the network (that is, the smaller the triangles), the more closely the form approximates a true sphere. Using triangles of varying size, a sphere can be symmetrically divided by thirty-one great circles (the largest that can be traced on the surface of a sphere). The triangles form a self-bracing framework that develops structural strength with a minimum amount of material. Thus, the geodesic dome combines the sphere (the most efficient container of volume per unit of surface area) with the polyhedron, which has the greatest strength per unit of mass. Developed in the first half of the twentieth century, it provided a completely new way of building light, transportable structures with efficient thermal and wind-resisting properties. For example, the aluminum-and-Teflon geodesic "Pillow Dome" designed by Jay Baldwin is a permanent insulated structure that can resist 135-mph (216-kph) winds and carry tons of snow; it weighs only 0.5 pound per square foot (2.43 kilograms per square meter).

The world's first geodesic dome was assembled on the roof of the Carl Zeiss Optical Works in Jena, Germany, in 1922. The 52-foot-diameter (16-meter) structure, designed by Zeiss's chief designer, Dr. Walter Bauersfeld, was necessary to test what he no doubt regarded as his more important invention, the planetarium projector. He built a complex skeleton of 3,480 light iron rods, accurate in length to 0.002 inch (0.05 millimeter) to form a highly subdivided icosahedron. Twenty-five years later, the American genius Richard Buckminster Fuller (1895–1983) independently derived his geodesic dome (patented 29 June 1954) from general principles, and he is generally credited with the invention of the form.

Fuller was deeply interested in the issues of shelter and housing, and by the end of World War II he had developed industrialized prototypes of the now-famous Dymaxion Houses, which he built for the Beech Aircraft Company in Wichita, Kansas. He then moved his attention to the construction of domes, because he believed that they reflected "nature's coordinate system" and therefore provided the optimally efficient way to enclose space. Through much of the 1940s he worked on small models of spheres or part-spheres made up of intersecting great circles, just as Bauersfeld had done. Fuller coined the name "geodesic dome" because the arcs of great circles are known as geodesics (Greek, "earth dividing"). In 1948 he seized the chance to take part in the summer school of Black Mountain College in North Carolina, taking with him the material needed to build a large-scale geodesic dome. Applying engineering strategy that he dubbed "tensegrity" (a contraction of "tensional integrity")—Fuller loved to invent words, too—he devised a system that depended on a continuous tension element rather than "discontinuous local compression members." He soon built a number of geodesic domes.

In 1953, Fuller built his first commercial dome, for the Ford Motor Company, and it was followed in 1954 by a 42-foot-diameter (12.8-meter) cardboard shelter in his exhibit at the Milan Triennale in Italy; it was awarded the Grand Prize. A few large-scale applications included the Union Tank Car dome (1958). In 1960 Fuller proposed a 2-mile-wide (3.2-kilometer), 200-foot-high (60-meter), temperature-controlled geodesic dome to enclose part of New York's Manhattan Island, claiming that the savings of snow-removal costs would amortize the cost within 10 years. On a more practical level, his domes covered military projects including sensitive radar

Geodesic dome, United States Pavilion at Expo 67, Montreal, Canada; R. Buckminster Fuller and Shoji Sadao, architects, 1967.

installations ("radomes"), emergency shelters, and mobile housing. They were and are also used for weather stations, industrial workshops, and greenhouses. One was even proposed for a cinema, in collaboration with the architect Frank Lloyd Wright.

Fuller's magnum opus is the former United States Pavilion at Expo 67 in Montreal, Canada, designed with Shoji Sadao. The huge, lacy dome, framed with steel pipes enclosing 1,900 molded acrylic panels, has a diameter of 250 feet (76.5 meters) and stands 200 feet (60 meters) high, "weightless against the sky." It has been adapted by Environment Canada and the city of Montreal and is now known as the Biosphere, an environmental water-monitoring center on the St. Lawrence River.

Further reading

Pawley, Martin. 1990. *Buckminster Fuller*. London: Trefoil.
Rosen, Sidney. 1969. *Wizard of the Dome: R. Buckminster Fuller, Designer for the Future*. Boston: Little, Brown.
Sieden, Lloyd Steven. 2000. *Buckminster Fuller's Universe: An Appreciation*. Cambridge, MA: Perseus.

German Pavilion
Barcelona, Spain

The German Pavilion at the Barcelona Universal Exhibition of 1929, designed by Ludwig Mies van der Rohe, is the first built expression of what he called "the architecture of almost nothing." About a decade earlier he had designed projects for multistory tower blocks, crystal prisms whose uninterrupted glass skins enveloped slender steel frames. They were just ideas, but the Barcelona Pavilion, as it is popularly known, set a standard—some would say, generated a fashion—for the austere minimalist architecture that would be dubbed the International

Style at an exhibition in New York's Museum of Modern Art just three years later. Esthetically, it was a major development in modern architecture.

The temporary single-story building, constructed in 1928–1929 and opened in May 1929, exhibited nothing but itself, a pristine, new kind of architectural space. The only purpose it had to serve was brief: to house a reception for the king and queen of Spain when they attended the official opening. For them, Mies designed the now-famous Barcelona chair, handcrafted from stainless steel and covered in white pigskin. The commanding location of the pavilion took advantage of the flow of visitors between the other display halls and the rest of the exhibition. It stood on a low travertine platform that gave a good view of the grounds and beyond to the city. The northern half of the podium was covered by a flat roof, carried on two rows of equally spaced, cruciform steel columns and an asymmetrical series of discontinuous walls of marble, glass, and onyx, parallel or perpendicular to each other. None of the rectilinear spaces thus formed was fully defined—that is, they formed an open plan—and the interior and the exterior of the pavilion were treated in the same way. This was the kind of spatial organization that Mies had observed in the work of Frank Lloyd Wright twenty years earlier. The attention to reductive detail and fine finish was the German's own. His most often quoted axioms were "Less is more" and "God is in the details."

A minimalist approach probably was justifiable for the Barcelona Pavilion because the building had no set functional program. It was in essence architecture as sculpture, an end in itself. But Mies also applied the philosophy to more functionally complex buildings. An almost contemporary example was the Tugendhat House in Brno, Czechoslovakia; commissioned in 1927, it was completed in 1930. Then in 1945 he designed a small weekend house in 9 acres (3.6 hectares) of woodland and fields on the bank of the Fox River south of Plano, Illinois, for his mistress, the Chicago physician Edith Farnsworth—a single room partitioned by a core that includes a kitchen, a fireplace, bathrooms, and a service area. The house is a mechanically perfect cuboid carried on a skeleton frame of sandblasted steel channels and defined by 9-foot (2.7-meter) glass walls and concrete floor and roof slabs. Interior finishes include a travertine floor, natural timber fittings, and a stainless-steel counter in the kitchen. Such obsession with refinement, causing Mies to take his architecture of almost nothing almost to the limit, did little to create a comfortable living space. It may have been admirable architecture; it was hardly congenial. It is emphasized that the issue was unimportant in the case of the German Pavilion at Barcelona, which was built simply to be seen and admired—as someone called it, "a place for contemplative lingering."

When the Barcelona Universal Exhibition closed, the German government tried to sell the pavilion to the municipality, without success. It was taken down in January 1930. It was not until 1983 that the Mies van der Rohe Foundation was established to reconstruct the building in Montjuïc Park, Barcelona, under the superintendence of the architects Cristian Cirici, Fernando Ramos, and Ignasi de Solà-Morales.

Further reading

Futagawa, Yukio, and Ludwig Glaeser. 1974. *Farnsworth House, Plano, Illinois, 1945–50/Ludwig Mies van der Rohe*. Tokyo: A.D.A. Edita.

Mies van der Rohe: European Works. 1986. London: Academy Editions.

Spaeth, David. 1985. *Mies van der Rohe*. New York: Rizzoli.

Golden Gate Bridge
San Francisco, California

When it opened to traffic in May 1937, San Francisco's Golden Gate Bridge boasted the longest single clear span in the world, a claim held true for twenty-seven years. The center span, at 4,200 feet (1,285 meters), was three times longer than the Brooklyn Bridge and 700 feet (214 meters) longer than the recently completed George Washington Bridge in New York. Including the two side spans of 1,125 feet (344 meters) and the 90-foot-wide (27.5-meter) road approaches, its total length was 8,981 feet (2,746 meters). Its towers were the tallest, its main cables the thickest and longest, its submarine foundations the largest ever built. Moreover, the foundation piers of the Golden Gate Bridge were built in the surging currents of the sea and its superstructure was erected across a canyon through which the wind howled at speeds up to 60 mph (96 kph). And all this was achieved without government funding in the midst of a deep economic depression. Against all the odds, the Golden Gate Bridge was a brilliant answer to a whole cluster of "insoluble" problems.

On 5 August 1775 Lieutenant Don Juan Manuel Ayala of the Spanish navy sailed the *San Carlos* from the Pacific Ocean into San Francisco Bay through the 3-mile-long by 1-mile-wide (4.8-by-1.6-kilometer) strait now known as the Golden Gate (one Captain John Fremont of the U.S. Army Topographical Engineers named it some 60 years later after Turkey's Golden Horn).

There is a compelling myth that the San Francisco eccentric Joshua Norton, self-styled "Norton the First, Emperor of the United States and Protector of Mexico," decreed in 1869 that a bridge be built across the Golden Gate. The story may have become confused with his pronouncement of March 1872 ordering a bridge across the Bay between Oakland Point and Goat Island, an idea he probably gleaned from well-publicized current transportation debates. In fact, the possibility of spanning the Golden Gate was first raised in 1872 by the railroad owner Charles Crocker, who naturally wanted to build a railroad bridge. But little more was heard of the matter until July 1916.

James Wilkins, editor of the *San Francisco Call Bulletin*, began a campaign that provoked City Engineer Michael O'Shaughnessy to seek, nationwide, the opinion of engineers on the project. Most said a bridge could not be built; the objections raised included the width of the strait, persistent foggy conditions, high winds and ocean currents, and not least, the high cost: some forecast $100 million. However, the experienced Chicago bridge builder Joseph Baermann Strauss (1870–1938) believed that a bridge was feasible and that it could be built for under $30 million. In June 1921 he proffered a preliminary design for a railroad trestle with a cost estimated at $27 million. Then he energetically tried to convince local politicians that he was right. Although urban growth and traffic congestion led to an urgent need to cross the Golden Gate, all available state and federal finance had then been diverted to other projects.

In 1922 O'Shaughnessy, Strauss, and Edward Rainey, secretary to San Francisco's mayor James Rolph Jr., proposed the formation of a special bridge district comprising the twenty-one affected counties to oversee financing, design, and construction of the bridge. The California State legislature passed the Golden Gate Bridge and Highway District Act in May 1923. In December 1924 the War Department authorized San Francisco and Marin Counties to construct a bridge. Despite opposition from vested interests, the Golden Gate Bridge and Highway District was immediately formed to realize the project.

Eleven engineering firms submitted proposals, and Strauss, assisted by Clifford Paine, was selected as chief engineer in August 1929. Consulting engineers Othmar Amman and Leon Moisseiff, both of New York, and Professor Charles Derleth Jr. of the University of California were appointed. The consulting architects were the husband-wife team of Irving and Gertrude Morrow. Strauss, who had never designed a suspension bridge, first proposed an inelegant cantilever-cum-suspension structure, but Moisseiff, convinced that a simple suspension bridge was possible (although such a span had never been attempted), helped refine the design that was eventually built. The architects did their part, too, designing handrails and light poles, tapering the tower portals, and designing lighting, all to emphasize the bridge's simple beauty. And, setting aside the conventional paint colors used on bridges, they selected the distinctive "international airways orange" for which the Golden Gate Bridge is famous. That, Irving Morrow believed, would look better in the spectacular landscape and would be more visible in the sea mists for which the Bay Area is noted.

In August 1930 the War Department approved a 4,200-foot (1,285-meter) main span, 220 feet (67 meters) above the sea. Although the United States was sunk in the Great Depression, a $35 million bond issue to finance the bridge received overwhelming popular support. Contracts worth $23.8 million were let in November 1932, and construction started the following January. Over the next four years it proceeded in the face of many natural problems—rapid sea currents, frequent fogs, and high winds—and technical ones, especially the construction of earthquake-resistant piers in 100 feet of open water. The latter was solved by building elliptical concrete fenders, 300 feet long and 155 wide (92 by 48 meters), within which the 148,000-ton (134,500-tonne) concrete piers could be poured; rising 15 feet (4.6 meters) above high-water mark, the fenders also protect the piers from the onslaught of the sea. The piers and the approach trestles were completed by December 1934 and the 121-foot-wide (37-meter), 750-foot-high (230-meter) towers were standing a little over

six months later. The steel sections for the towers, fabricated in Bethlehem, Pennsylvania, were sent via East Coast seaports through the Panama Canal to McClintic-Marshall's yards in Alameda. Then they were carried by lighters to the site, lifted by cranes, and erected by teams of riggers.

Catwalks spanned the Golden Gate by July 1935, and John A. Roebling and Sons of New Jersey began spinning the two main cables from the San Francisco and Marin anchorages four months later. Each galvanized steel cable is 36.375 inches (920 millimeters) in diameter, comprising 61 strands of 452 wires. They were completed by March 1936, and the roadway steel was placed from June through November, allowing construction of the flexible in situ concrete road deck, finished by April 1937. The bridge was opened to pedestrian traffic on 27 May 1937 and to vehicles at noon the following day. It had been achieved ahead of schedule and under budget. An estimated 200,000 people walked over it on the first day, and a weeklong Golden Gate Bridge Fiesta celebrated the event with fireworks, parades, and other entertainment.

Further reading

Adams, Charles F. 1987. *Heroes of the Golden Gate*. Palo Alto, CA: Pacific Books.

Zee, John van der. 1986. *The Gate: The True Story of the Design and Construction of the Golden Gate Bridge*. New York: Simon and Schuster.

The Grand Buddha, Leshan, Sichuan Province, China, 713–803. The figure's feet can be seen at the edges of the photograph.

Grand Buddha
Leshan, China

Dàfó (Grand Buddha), the world's largest figure of Buddha, provides tacit testimony to the engineering skills of medieval Chinese civilization. It is carved from the Xiluo Peak of Mount Lingyun, facing the town of Leshan in the Sichuan Province of the People's Republic of China. Work began on the 229-foot-high (71-meter) seated figure in A.D. 713 and took 90 years to complete.

Comparisons may give an idea of the ambitious scope of the project. Seated, the Grand Buddha is about 80 feet (25 meters) taller than the figure of the Statue of Liberty; were he standing, he would be over twice her height. His shoulders are 92 feet (28 meters) wide and his head 48 feet (14.7 meters) high; the Washington head on Mount Rushmore is 60 feet (18.3 meters). The Buddha's big toe is 27 feet (8.3 meters) long, and 100 people can easily stand together on his instep. The practice of creating large statues of the Buddha probably began in India and spread throughout Asia. Standing and seated figures and others in the lotus position can be found in Bilingsi, China; Wat Thatorn, Thailand; and Kamakura, Japan. But the only ones that approached the size of the Leshan Buddha were in the Bamian valley of Afghanistan. Tragically, the two third-century-A.D. sandstone carvings, 182 feet (55 meters) and 125 feet (38 meters) high, respectively, were wantonly destroyed by the fanatical Taliban in March 2001.

The Leshan Buddha is in a serene region known as "Buddhist Paradise" and "Celestial World on Earth," long associated with the religion in China. Nearby, the 10,000-foot (3,060-meter) Mount Emei, one of four sacred Buddhist mountains, rises steeply above the Dadu River. Once there were perhaps 100 pilgrimage sites—temples and monasteries—throughout its abundant forests. Many of them were originally Taoist foundations established during the Eastern Han dynasty under Emperor Ming (A.D. 58–75); others were added during the Ming and Qing dynasties (1368–1911). Not all have survived.

In the eighth century, Leshan, then known as Jiazhou, was a prosperous inland port and trading center. Silk and textiles from Chengdu, 105 miles (168 kilometers) to the northeast, and the agricultural bounty of the Chuanxi Plains were shipped down the Minjiang River to join the Qingyi Jiang and the Dadu, waterways that opened trade routes to much of China. The confluence of these fast-flowing streams created dangerous turbulence above a deep hollow, and boats often capsized. There is a tradition that a monk named Hai Tong, from the nearby Lingyun Monastery, initiated the carving of the Grand Buddha to quieten the waters. Ironically, because of the magnitude of the undertaking, involving a large workforce for almost a century, he did not live to see the figure completed. It may be difficult for the modern mind to grasp the singleness of purpose, on the part of Hai Tong and the builders alike, necessary to sustain such a project for so long.

Romantic tales are attached to the statue and the determined man who conceived it: he is said to have gouged out his eyes in some ruse to keep funding for the statue, and spent the remainder of his life in an abandoned cave tomb. As to the river, there is a tradition that it *was* calmed, perhaps because countless tons of discarded rock were thrown into the pool that caused the problem, perhaps by the watchful presence of Dàfó. Since 803 he has sat majestic, serene, and complete in his gigantic niche, flanked by standing warriors, also carved from the mountain. A path with nine turns winds down the rock face from the top of his head to the riverside platform before his feet; close to his right shoulder is the Great Buddha Temple (Da Fu Si). Until it was destroyed during a war toward the end of the Ming dynasty (1368–1644), a huge building covered the figure, although it seems not to have been to protect Dàfó from the weather. The ancient engineers devised a clever internal drainage scheme to prevent deterioration of the surface by diverting runoff—a system that until recently proved most effective. Now, rainwater seeping into the back of the statue is causing severe erosion. Other problems have arisen because of poor maintenance and repair practices in the past. A conservation plan has been developed, and in July 1999 the World Bank announced a U.S.$2 million loan for China to restore the Grand Buddha by improving internal drainage and reinforcing the base of the statue.

Further reading

Holdsworth, May, and How Man Wong. 1993. *Sichuan*. Lincolnwood, IL: Passport Books.

Snellgrove, David, ed. 1978. *Image of the Buddha*. Paris: UNESCO.

Summerfield, John. 1994. *Fodor's China: The Complete Guide*. London: Fodor.

Grand Coulee Dam
Washington State

Commenced during the Great Depression, Washington State's Grand Coulee Dam, on the Columbia River about 88 miles (142 kilometers) west of Spokane, is a monument to engineering prowess and to the resolve of those people who for 23 years fought for its

Grand Coulee Dam, Washington State, 1933–1951. View from below the dam.

creation. The key to the Columbia Basin Irrigation Project, it provides the region with electric power, irrigation, and flood control and contributes to wildlife conservation. The Grand Coulee Dam is the largest concrete structure ever built in the United States and the nation's largest hydroelectric facility. Its 550-foot-high (168-meter) gravity-type concrete wall, 500 feet (152 meters) thick at the base, spans a little under 1 mile (1,592 meters) and raises the water surface 350 feet above the former riverbed. Nearly 12 million cubic yards (over 9 million cubic meters) of concrete were needed to build it. Franklin Delano Roosevelt Lake (often simply called Roosevelt Lake), created by the dam, has a 600-mile (960-kilometer) shoreline and extends 150 miles (240 kilometers) to the Canadian border.

After several ruinous years of drought in the Northwest early in the twentieth century, the U.S. Reclamation Service Bureau (now the Bureau of Reclamation) considered pumping water from the Columbia River to irrigate agricultural land in eastern Washington, a region then served by artesian wells. In 1917 an Ephrata attorney named William Clapp proposed an alternative: build a high-level dam on the Columbia and raise water to the Grand Coulee, the 50-mile-long (80-kilometer) natural channel of the old riverbed, thereby opening up more than 1 million acres (403,230 hectares) of irrigated farmland. Rufus Woods, editor of the *Wenatchee Daily World*, publicized the notion a few months later. In 1919 the Michigan lawyer James O'Sullivan became interested enough to put it before the Reclamation Service, which directed Washington State's Columbia Basin Survey Commission to include it in a current feasibility study focused on irrigating the basin by gravity canals from the Pend Oreille River. In the teeth of opposition from vested interests connected with the latter scheme, the dam's protagonists managed to enthuse, among others, A. P. Davis, director of the Reclamation Service. At O'Sullivan's prompting, Davis suggested that the state commission an objective report from Seattle engineer Willis Batchelor, who in 1921 recommended a dam on the Columbia, 220 feet (67 meters) above river level.

Several years of argument followed. In 1923 George Goethals of Panama Canal fame—apparently a paid prophet—endorsed the canal system, and two years later the federal Columbia Basin Survey Board of Engineers supported his view. But O'Sullivan, Woods, Clapp, and others unflaggingly kept the dam project alive, and in 1927 the U.S. Senate authorized the Army Corps of Engineers, under Major John Butler, to look for possible sites during a 1929 survey of the upper Columbia River. In June the Columbia River Development League was formed with Woods as president and O'Sullivan as secretary. The *Wenatchee Daily World* became its mouthpiece. Late in 1931 Butler told Congress that a dam was more economical than a gravity canal: besides providing irrigation and flood control, it would raise revenue from electrical energy. O'Sullivan lobbied for authorization, and the Bureau of Reclamation soon recommended development of the project, almost in the form in which it was eventually realized. In 1933 the

Columbia Basin Commission was established, and the state of Washington committed $377,000 to the Grand Coulee Dam. Recently elected President Franklin D. Roosevelt allocated $63 million under the Public Works Administration—a New Deal program. Through the Great Depression men and women from all over the United States would find work at the dam site: averaging 3,000, the labor force peaked at 6,000.

Excavation began in December 1933, and seven months later a $29.34 million contract for the foundation work was awarded to MWAK, a consortium formed by Silas Mason Company of New York City; Walsh Construction Company of Davenport, Iowa; and the Atkinson-Kier Company of San Francisco. Such a large undertaking called for a complex infrastructure: high-tension power lines were set up, the Columbia River was bridged, and over 30 miles (nearly 50 kilometers) of railroad and 60 miles of sealed roads were constructed. A contractor's town, Mason City, and Coulee Dam, a government town, were built at the site. Four years later, a consortium formed by linking MWAK and the Six Companies—Kaiser Construction of Seattle; Morrison Knudsen of Boise, Idaho; Utah Construction; J. F. Shea Pacific Bridge; and McDonald and Kahn (all of San Francisco) and General Construction Company of Seattle—won the $34.4 million contract for the completion of the dam. Their bid was 80 percent of the only other tender.

The proposed height of the dam had been determined by the rather parochial notion that the impounded water should not back up beyond the Canadian border. Then the project's main reason for being was irrigation—there were more droughts in the early 1930s—and flood control, rather than power generation. The Pacific Northwest had plenty of electricity and there was little prospect of industrial expansion. Therefore the original designs included a 350-foot (107-meter) "low" dam about 3,500 feet (1,070 meters) long, which would bring the water surface to only 150 feet (46 meters) above the river level. Should the demand for power increase, it was intended to later raise the wall. That was flawed thinking. Achieving a tight joint between the two parts of the wall would have been difficult, even dangerous; later changes to the turbines would be costly; and it was more expedient to construct the concrete foundation of a high dam at the start of the project. So, with the approval of Congress, the contracts were redrawn in June 1935 to build the high dam to plans by John Lucian, chief designer of BuRec engineers.

The main dam was completed by 1941 and work commenced on the pumping plant and powerhouses. The entry of the United States into World War II meant dramatic changes in priorities for the dam. Power generation was given first place because the region's aluminum industry, a large consumer of electricity, was critical to the defense effort. Six generators were commissioned at the Grand Coulee, and two more were borrowed from the incomplete Shasta Dam project in northern California. Soon after the war, construction resumed on the pumping plant and in 1951 the irrigation system was inaugurated. Six huge pumps lifted water through 280 feet (85.6 meters) from Roosevelt Lake to Banks Lake equalizing reservoir in the Grand Coulee. In 1973 two more reversible pump-generator units were installed, followed by another four late in 1983. Feeding more than 300 miles (480 kilometers) of associated canals, and nearly 5,500 miles (8,800 kilometers) of laterals, siphons, and drains, the pumps can fully supply almost 1.1 million acres (about 440,000 hectares) of formerly dry land. They are not yet being used at their full capacity.

The reversible pump-generators installed could of course be used for power generation, augmenting the already remarkable output of the Grand Coulee Dam, whose power production facilities are by far the largest in North America. Two plants, with a total of eighteen generators, were operational by 1951. A third, coming on line in 1975, increased capacity to about 7,200 megawatts. By 1978 the three were producing over 6,000 megawatts, and subsequently additional generators—the total number is now 33—have achieved an output of over 6,800 megawatts.

In the 1950s the American Society of Civil Engineers included the Grand Coulee Dam and the Columbia Basin Project among the seven civil-engineering wonders of the United States. The project has also been popularly and superlatively dubbed "the Eighth Wonder of the World," "the Greatest

Structure in the World," "the World's Greatest Engineering Wonder," and "the Biggest Thing on Earth."

Further reading

Downs, L. Vaughn. 1993. *The Mightiest of Them All: Memories of Grand Coulee Dam.* New York: ASCE Press.

Ficken, Robert E. 1995. *Rufus Woods, the Columbia River, and the Building of Modern Washington.* Pullman: Washington State University Press.

Pitzer, Paul C. 1994. *Grand Coulee: Harnessing a Dream.* Pullman: Washington State University Press.

La Grande Arche
Paris, France

La Grande Arche is the paramount landmark, the crowning monument of Paris's Place de la Défense. It is the eastern terminus of the monumental Voie Triomphale (Triumphal Way), extending from the Cour Carrée of the Louvre through the Tuileries Gardens and down the Champs-Elysées to the Arc de Triomphe; the axis then continues for almost 4 miles (6 kilometers) along the Avenue de la Grande Armée and through La place de la Concorde to cross the Pont de Neuilly and enter La Défense.

La Défense is dominated by ultramodern geometric office or apartment towers, 30 stories high and more, apparently randomly arranged over a large, paved plane. It also boasts conference centers, an exhibition hall, gardens, and a massive public pedestrian open space, beneath which is Paris's largest shopping complex, restaurants, and a cinema. It was conceived in 1931, when a competition was held to extend the Louvre–Champs Elysées axis. None of the thirty-five classical revival or modernist entries from French architects was realized. The aim had been to continue the French tradition of innovative architecture but for various reasons, no doubt including the 1930s Depression and World War II, little of the kind was built. In 1951, La Défense was zoned for commercial use, and seven years later a specifically appointed agency produced a thirty-year master plan;

La Grande Arche, Place de la Défense, Paris, France; Johann Otto von Spreckelsen, architect, 1982–1989.

revised in 1964, it provided for twenty towers, each of twenty-five stories. Developers and the public disagreed over taller buildings, but the mediocre development—someone has described them as "all of the postmodernist 'could-be-anywhere' style"—emerged as "a forest of towers" of various heights. Finally, a new monument was built at La Défense "as a counterweight for the Arc de Triomphe": La Grande Arche—innovative architecture par excellence and a daring technical achievement.

The construction of La Grande Arche was among the more controversial and certainly the grandest of President François Mitterrand's so-called Grands Projets; initiated at a cost of Fr 15 billion (about U.S.$2.2 billion), the program was intended to build a series of monuments that symbolized France's central place in the world at the end of the twentieth century. A design competition was held in July 1982, and the 424 entries were reduced to a short list of four for a second stage in April 1983. That was won by the Danish architect Johann Otto von Spreckelsen (1929–1987), in collaboration with the civil engineer Erik Reitzel; their scheme was chosen by an international panel of judges for its "purity and strength." Work began in 1985 in the hope that the building would be completed in 1989 to coincide with the celebration of the bicentenary of the French Revolution. The budget for the project was Fr 2.9 billion (U.S.$420 million). Von Spreckelsen, frustrated by French bureaucracy and dissatisfied with his own design, later withdrew from the project. He died before the building was finished.

La Grande Arche, dedicated to the French concept of *Fraternité*, is in fact a *square* arch, a 330,000-ton (300,000-tonne), 352-foot (110-meter), hollowed-out, chamfered cube that houses in its massive legs thirty-five stories of offices, reached by elevators in freestanding transparent shafts. Most offices are occupied by French government ministries, as well as the Fondation des Droits de l'Homme (Human Rights Commission), the World Road Association, and some large private companies. There are also showrooms, a large exhibition hall, and a conference center. The narrower surfaces of the prestressed concrete frame building are faced with Italian Carrara marble and gray granite; glass walls facing

into the hollow provide daylight to the offices. The imposing structure is rotated very slightly off perpendicular to the grand axis, in order to accommodate the placement of foundation piles. Around its base and under a suspended fabric canopy known as "the Cloud" are fountains and sculptures by famous artists, including Joan Miró.

Further reading

Chaslin, François, and Virginie Picon-Lefebvre. 1989. *La Grande Arche de la Défense*. Paris: Electa Moniteur.

La Tête Défense. 1990. *Arkitektur DK* 34 (January/February): the issue.

Vonier, T. 1993. "Critique: Non-Parallel Parking—Two Divergent Approaches to Urban Parks in Paris." *Progressive Architecture* 10: 66–71.

Great Pyramid of Cheops
Giza, Egypt

In the western suburbs of modern Cairo, 130 feet above the Nile, stands a 1-mile (1.6-kilometer) square artificial rocky plateau called Giza (El-Jizah) by the Arabs. It is the site of three Fourth Dynasty pyramid tombs—Cheops', Chephren's, and Mycerinus's—named by the ancients among the seven wonders of the world. The largest of them, built at the command of Cheops, has been called a "unique monument" because of its internal disposition. While it is clearly part of an evolving architectural type, there is little doubt that in terms of engineering and logistics, this so-called Great Pyramid was a superlative achievement.

Cheops, also known as Khufu (Khnum-Khufwy, "Protected by Khnum"), was the second king of the Fourth Dynasty and reigned from 2589 to 2566 B.C. Although little is known of him, he is believed by some scholars to have been a tyrannical and cruel ruler. Whatever the case, clearly he was able to lead and coordinate, because the building of his tomb involved sophisticated social planning to harness an immense team of workers, both on and off the site, together with all the backup resources needed for such a daunting task. The fifth-century-B.C. Greek historian Herodotus calculated that 100,000 slaves would have taken 30 years to build the Great Pyramid. But it was not constructed by slave labor; rather,

Egypt's peasant farmers, displaced from July through November when their fields were inundated by the annual flooding of the Nile, were deployed on the project, as well as on other public works. The cost of their food and shelter (there were workers' villages built nearby) was met from their own surplus production, levied as taxes. Modern scholarship suggests that only 20,000 men could have completed Cheops' tomb in only 20 years.

The base of the Great Pyramid (Akhet Cheops, the Horizon of Cheops), oriented within 0.3 minute of accuracy to the cardinal compass points, is 756 feet (230.5 meters) square, covering 13 acres (5.2 hectares). The extensive base means that the tremendous weight of the tall 479-foot (146-meter) building, amounting to an estimated 6.99 million tons (6.35 million tonnes), does not overload the foundation; it is also very stable because its center of gravity is very low. Although of simple design, such an engineering feat challenges even the modern imagination. The pyramid is estimated to contain 2.5 million limestone blocks, each weighing anything from 3 to 17.7 tons (2.5 to 15 tonnes), rising in 200 steps to the height of a 40-story office block. The joints between the blocks are about 0.02 inch (0.5 millimeter). As originally designed, the pyramid was encased in a 16-foot-thick (5-meter) layer of polished white limestone won from the quarries at Tura, east of the Nile. Most of it was plundered in the sixteenth century and used to build mosques in Cairo. At the pinnacle of the Great Pyramid there was a solid capstone of polished Aswan granite, standing on a 33-foot (10-meter) square platform.

All this, from quarrying to setting the stones, was achieved with copper and stone tools. Barges were used to transport blocks from a quarry on the far side of the Nile. How were they raised as the pyramid progressed? It is thought that ramped causeways, lubricated with water, were used to haul the sleds; these may have been built at different levels on each side of the pyramid, or a single ramp may have wound around the whole structure as it rose. While oxen were used to move stone blocks in the quarry, the accuracy demanded on-site required wooden sleds hauled by men, and fewer than ten were needed to maneuver a block into place using wooden rockers.

For all the looming size of the Great Pyramid, its interior spaces are relatively tiny. An entrance passage—not the original—connects with a narrow, 345-foot-long (105-meter) descending passage that leads to a 46-by-27-foot (14-by-8.3-meter) subterranean room, a little over 11 feet (3.5 meters) high. It has been suggested that this was the first location chosen for Cheops' burial chamber; that was quickly abandoned, probably on theological grounds. From the junction of the two passages, a 129-foot-long (39-meter) ascending passage leads to the outer end of the "great gallery." From that point, a horizontal corridor gives access to the so-called Queen's Chamber, vaulted with inclined blocks; a second alternative burial chamber, it was never completed and never used. The 154-foot-long, 28-foot-high (47-by-8.5-meter) great gallery, with a finely crafted corbel vaulted ceiling, leads upward to the final location of the King's Chamber, built of pink Aswan granite. The chamber still contains the huge red granite sarcophagus that must have been put in place while the pyramid was being built. Above it a series of five relieving chambers distributes the weight of the structure above away from the chamber. There are two shafts sealed at the extremities, through which the king's *ka* (spirit) could come and go from the underworld.

Several ancillary buildings were associated with Cheops' pyramid. Members of the royal family were buried in mastaba tombs, and three small pyramids to the east were probably for his sister-wife, Merites, and perhaps other queens. Nobles and courtiers were interred in the royal cemetery to the west of the Great Pyramid, where there were also funerary temples and processional ramps. All that remains of Cheops' Mortuary Temple is some of the basalt paving.

Since the early 1990s, there have been serious attempts to preserve the fabric of the Great Pyramid. It was restored in 1992. Recurring salt deposits, cracking, spalling of the limestone, and the appearance of black spots, all resulting from increases in humidity and carbon dioxide caused by large numbers of tourists, necessitated further action. Early in 1998 the building was closed to the public while a more efficient mechanical ventilation system was installed. It changes the air every 45 minutes, employing the original ka shafts from the King's Chamber as exhaust ducts and drawing fresh air through the access passage. The number of daily visitors has been severely limited and airlines have been warned of a "no-fly" zone above the site.

Further reading

Pemberton, Delia. 1992. *Ancient Egypt*. San Francisco: Chronicle Books.

Stierlin, Henri, and Anne Stierlin. 1995. *The Pharaohs, Master-Builders*. Paris: Terrail.

Great Wall of China

The largest man-made structure in the world, the Great Wall once stretched more than 4,500 miles (7,300 kilometers) from the Jiayu Pass in Gansu Province in the west to the mouth of the Yalu River in Liaoning Province in the east. The ravages of time and vandalism have reduced it to 1,500 miles (2,400 kilometers). It has been called an "engineering marvel of stone, earth and brick."

From 475 to 221 B.C., there were seven warring states in Chou dynasty, China—Qi, Chu, Han, Wei, Qin, Yan, and Zhao. The borders of the latter three were frequently plundered by the nomadic Xiongnu (Huns) and Donghu tribes, so they built high earth walls as a defense against them. For their part, the remaining states took similar action, fearing attacks from their capricious neighbors. Soon after he had unified China in 221 B.C., the first emperor, the despotic Ch'in Shi Huangdi, set about reinforcing his defenses against the Xiongnu by joining four earlier fragmentary walls and building new sections to extend them to 3,100 miles (5,000 kilometers). In 214 B.C. he sent General Meng Tien, with an army of 300,000 conscripted workers and countless prisoners, to the northern frontiers of his empire to begin the building the Great Wall. Garrisons of soldiers along the wall served a double purpose: they stood guard over the workers and defended the northern boundaries. Much of the Ch'in wall was constructed with dry-laid local stone, but in remote places, where stone was unavailable, the builders used earth, compacted in 4-inch-thick (10-centimeter) layers. Watchtowers were spaced two bow shots apart.

Great Wall of China, 221 B.C.–A.D. 1640. Detail of the Badaling section over mountainous terrain near Beijing. 1977 photograph.

Ch'in Shi Huangdi's policies of heavy taxation and forced labor to pay for foreign wars, the Wall, and other extravagant public works inevitably created social unrest. When he died in 210 B.C., his empire collapsed. Following years of chaos, the Han dynasty (206 B.C.–A.D. 220) was founded. Under Wu-Di (reigned 140–87 B.C.), the Han expanded into southern China, Vietnam, and Korea and opened trade routes through the wilderness of central Asia to India, Persia, and the Western world. Wu-Di controlled the Xiongnu incursions by invading their lands south of the Gobi Desert and colonizing the region with his own people. That strategy, incidentally, forced the Huns to move westward, part of a chain reaction that eventually brought about the demise of the Roman Empire. To protect what he had gained, Wu-Di inaugurated the third major phase of the Great Wall. He restored the Ch'in wall—neglected for years, the earthen parts had begun to collapse—and extended it 300 miles (280 kilometers) across the Gobi Desert. Han builders corrected the problem of the sandy soil by reinforcing the compacted earth with willow reeds. They also built beacon towers at 15- to 30-mile (25- to 50-kilometer) intervals and used smoke signals to warn of attack. All trade routes passed through the Wall.

The final construction phase, which gave the Wall its present form, was undertaken early in the Ming dynasty (1368–1644). Having finally expelled the harassing Xiongnu and their Mongol rulers, the Ming emperors set about securing their empire. They repaired and enlarged the Wall, constructing extensions of tamped earth between kiln-fired brick facings across some of China's most mountainous terrain. The Ming wall averaged 25 feet (7.6 meters) in height; it was 15 to 30 feet (4.5 to 9 meters) thick at the base, sloping to 12 feet (3.7 meters) at the top. The watchtowers were redesigned and cannon, bought from the Portuguese, were strategically deployed.

For all its size and splendor, the Great Wall seems to have been a functional failure, with little military value. Only when China was weakened internally were northern invaders—the Mongols (Yuan dynasty) in 1271 and the Manchurians (Qing dynasty) in 1644—able to seize power without engaging in an attenuated war. Since the seventeenth century parts of the Great Wall have been quarried for their brick or stone; others have simply crumbled, while those in marshy areas have been buried by silt. Two stretches—the Badaling and Mutianyu sections—north of Beijing have been reconstructed and opened as a tourist attraction. In 1979, the Chinese government declared it a National Monument, establishing a commission to oversee its preservation; in 1987 it was inscribed on UNESCO's World Heritage List.

Further reading

Luo, Zewen, et al. 1981. *The Great Wall*. New York: McGraw-Hill.

Schwartz, Daniel. 1990. *The Great Wall of China*. London and New York: Thames and Hudson.

Waldron, Arthur. 1990. *The Great Wall of China: From History to Myth*. Cambridge, UK: Cambridge University Press.

Great Zimbabwe
Republic of Zimbabwe, Africa

The ruins of Great Zimbabwe (Bantu for "stone house") stand about 17 miles (30 kilometers) southeast of the modern provincial capital, Masvingo, and east of the Kalahari Desert between the Zambezi and Limpopo Rivers. They cover about 200 acres (80 hectares). The largest of about 300 such sites in the region, Great Zimbabwe was once the greatest city in sub-Saharan Africa. Misguided and racist Victorians—and others since—thought Africans incapable of such sophistication, and they therefore incorrectly concluded that ancient Phoenicians, Romans, or Hebrews created the amazing structures. The British archeologists David Randall-MacIver (1905) and Gertrude Caton-Thompson (1929) carried out excavations and discovered that the place was indeed indigenous African in origin. Their conclusions were confirmed by further investigations made by R. Summers, K. R. Robinson, and A. Whitty in 1958. The builders were ancestors of the modern Shona people of Zimbabwe. Even in ruin, Great Zimbabwe has been called "remarkable," "majestic," "awe-inspiring," and "timeless"; when intact, it was an architectural masterpiece. Now known as the Great Zimbabwe National Monument, the site was inscribed on UNESCO's World Heritage List in 1986.

The area was first settled by Bantu-speaking farmers, perhaps in the second and third centuries A.D. A second phase of occupation began about A.D. 330. The grasslands in the foothills of the Mashonaland plateau provided excellent pasture, and between 500 and 1000 the cattle-herding Gokomere people overran and absorbed the earlier inhabitants. Rich local gold deposits were later utilized, and it seems that some stone walls were built toward the end of that period. Scholars remain divided on how the Mwenemutapa of Great Zimbabwe attained their high lifestyle and widespread influence. Some believe that it came from cattle wealth and coastal trade in gold (with contacts as far afield as India and China). Others suggest that a powerful politico-religious ideology "gave them a competitive edge over neighbors" so that they could coerce the human resources needed to build their city. But as yet there is no evidence that their success depended upon a single factor.

Great Zimbabwe, Republic of Zimbabwe, Africa, thirteenth to fifteenth centuries A.D. Detail of masonry walls, photographed in 1989.

Anyway, between the thirteenth and fifteenth centuries Great Zimbabwe, capital of the wealthiest society in the region, dominated the area that now encompasses eastern Zimbabwe, Botswana, Mozambique, and South Africa.

There were three main groups of buildings: they are now designated the Hill Complex, the Great Enclosure, and the Valley Complex. The approximately oval Hill Complex (ca. A.D. 1250) was clearly a defensible retreat. Measuring about 330 by 150 feet (100 by 45 meters), it crowned a steep rocky prominence 260 feet (80 meters) above the valley at the north end of the settlement. It comprised several enclosures connected by a network of narrow walled alleys. Drystone walls of dressed rectangular granite blocks linked the large natural outcrops to fortify a number of areas. The large western enclosure, its 20-foot-high (6.1-meter) perimeter wall nearly 17 feet thick (5 meters), is thought to have been a religious precinct. The smaller eastern enclosure was probably residential, perhaps for the royal court or the chief shaman. At least three routes approached the hill from the west.

Most of the buildings in the Great Enclosure at the southern edge of the site date from the early fourteenth century; its elliptical perimeter wall was first built nearly 100 years later and subsequently restored a number of times. The 16-foot-thick (5-meter), 36-foot-high (11-meter) wall, its drystone faces meticulously built from dressed granite (the core is rubble), contains a space 840 feet (255 meters) in circumference and 330 feet (100 meters) across; there are three unfortified doorways. The enclosure embraced a few elite residences, including the royal compound and an enigmatic 33-foot-high (10-meter) conical tower—a solid, granite affair with a 16-foot (about 5-meter) base diameter. There is also a smaller tower.

The rambling Valley Complex, between the other nodes of the city but closer to the Great Enclosure, dates from the early fifteenth century. It once comprised several stone-walled irregular yards—one writer calls it an "archipelago"—around the houses of more important citizens. However, most families probably lived in densely packed *dhaka* (mud) huts with thatched roofs, clustered between these "islands." More small towers, possibly of religious significance, dotted the area. The surviving masonry strongly suggests that the Zimbabweans independently evolved a technology that optimized immediately local resources. Their construction system began as a simple response to necessity and ended with sophisticated work that expressed joy in building, as it employed herringbone, chevron, and other decorative bonding, all made without mortar.

Great Zimbabwe was fully occupied for about 300 years, but by the end of the fifteenth century it lay abandoned. Although grazing land was at first abundant, the poor soil could not have supported crops enough to sustain the city's increasing population—by the late fourteenth century it may have reached 18,000—so some food necessarily would have been imported. By the fifteenth century Great Zimbabwe's power had begun to fade, coincident with the rise of Torwa and Mutapa, the neighboring states. The reasons for its demise are unclear, but a familiar pattern is likely. Urban growth overtaxed the immediate environment, and the pressure put on resources by people and herds alike probably led to decline, resulting in social, economic, and political instability, and finally fragmentation. Many cultures have ended thus, not with a bang but a whimper.

Further reading

Caton-Thompson, Gertrude. 1971. *The Zimbabwe Culture: Ruins and Reactions*. London: Cass.

Garlake, Peter S. 1985. *Great Zimbabwe, Described and Explained by Peter Garlake*. Harare: Zimbabwe Publishing House.

Mufuka, K. Nyamayaro, et al. 1983. *Dzimbahwe: Life and Politics in the Golden Age, 1100–1500 A.D.* Harare: Zimbabwe Publishing House.

Hadrian's Wall
Northumberland, England

The most audacious building project among many initiated by the Roman emperor Publius Aelius Hadrianus (known as Hadrian) was the defensive rampart across the entire width of Britain that marked the northern frontier of the Roman Empire for almost 300 years. Started in A.D. 123 Hadrian's Wall was about 73 miles (118 kilometers) long, stretching from what is now the town of Wallsend (Roman Segedunum) on the River Tyne in the east to modern Bowness (Roman Banna) on the Solway Firth in the west. From there, seaward defenses, somewhat less substantial, turned south along the Cumberland coast for another 40 miles (65 kilometers).

Spurred by his conquest of Gaul, Julius Caesar undertook a reconnaissance of Britain in 55 B.C. A full-scale Roman invasion took place in A.D. 43, when Claudius was emperor, and there followed decades of resistance by various local tribes. But Britain, soon known as "the food basket of Rome," was too rich a prize to surrender. Under the governor Petilius Cerealis, the legions marched north into the territory of the Brigantes and established a base at York (Roman Eboracum) in A.D. 71. About ten years later, they pushed forward into Scotland, creating a temporary frontier between the Rivers Forth and Clyde. They intended to consolidate their new conquests by constructing roads and forts *(caestra)*, but the northern tribes proved too warlike, causing the Romans to strategically withdraw.

Hadrian, the adopted son of Trajan, reigned from A.D. 117 until 138. He loved to build: among his architectural schemes in and near Rome were his own tomb (later known as the Castel Sant'Angelo), the Pantheon, and a luxurious country villa at Tivoli. He was also an inveterate traveler and for over half of his reign he was away from Rome, mostly touring the eastern provinces and North Africa. On a visit to Britain in 122 he appointed a new governor, Aulus Pletorius Nepos, and in order to establish a presence in the far north, he commissioned the construction of the Wall to "separate the Romans from the barbarians." Work started the following year. As planned, the eastern sector between Wallsend and the River Irthing was to be a stone structure, about 10 feet (3 meters) thick and 15 feet high to the rampart (the parapet was 5 feet higher). As it was eventually built, the thickness along the wall varied; faces were of dressed stone and the infill of rubble. From Irthing to Bowness a turf-and-timber wall, about 20 feet (6 meters) thick at the base, was initially built and replaced with stone within a few years.

Immediately south of the wall—except in the craggy terrain across the Pennines—there was a continuous ditch (Roman *vallum*), 10 feet (3 meters) deep and 20 feet wide at the top, with a flat bottom

Hadrian's Wall, Northumberland, England, 123–130. Section of wall looking east toward Housesteads, photographed in 1943.

8 feet (2.4 meters) wide. It was flanked, 30 feet away on each side, by wide earth mounds. These earthworks defined the southern limit of the military zone—in effect, like a customs zone at any modern border.

It was originally intended that the wall would be manned by patrols from small forts called "milecastles" at 600-foot (184-meter, the Roman stadium) intervals. Military and logistical backup would come from established but widely spaced fortresses like Corbridge (Roman Corstopitum), usually at the junction of principal roads. Plans changed during the eight years taken for the building. A total of seventeen forts, some for 1,000 foot soldiers (e.g., Housesteads, Roman Vercovicium) and others for elite, 500-strong cavalry regiments (e.g., Chesters,

Roman Aesica), were built at roughly evenly spaced locations along the wall. Each milecastle, serving as a controlled crossing place to the north, had a gate reached by stone causeways across the vallum. Each housed only about two dozen men. Between them, Hadrian's Wall had two evenly spaced stone observation and signal turrets that were manned by legionaries from the milecastles. Construction was practically complete by A.D. 130, although some work seems to have continued for another eight years. Most of the labor was provided by ordinary soldiers of the three Italian legions then based in Britain, who moved about 1.7 million cubic yards (1.3 million cubic meters) of turf and stone.

The total garrison probably numbered about 12,000, mostly drawn from auxiliary legions raised in different provinces of the empire. It is clear that such a well-manned outpost was not intended merely for defense; it was used to attack the hostile northern tribes. Moreover, Hadrian's Wall identified Rome by creating a highly visible boundary. Because traders had to use the milecastles as crossing points to the unconquered territories beyond, and because there was a concentration of population, markets and other social structures developed in some areas. Hadrian was succeeded by Antoninus Pius, who in A.D. 139 commanded another advance into Scotland, reestablishing the frontier. The 37-mile-long (59-kilometer) Antonine Wall was built around A.D. 142 between what is now Old Kilpatrick on the Clyde River and Carriden on the Forth. The 9-foot-high (2.75-meter) turf-faced soil rampart stood on a stone foundation. There was a 40-foot-wide (12-meter) vallum, 12 feet (3.7 meters) deep on its north side. Small forts were located at about 400-yard (370-meter) intervals.

By about A.D. 155 the Romans again retreated from Scotland, to return only briefly between A.D. 159 and 163. Hadrian's Wall regained its former importance; the vallum, which had been partially filled, was finally reconstructed by about A.D. 208. Breached only three times during the remainder of the occupation—in A.D. 197, 296, and 367—it was retaken on each occasion and rebuilt where necessary to remain the frontier of Roman Britain until the last legions departed in A.D. 410.

Further reading

Bruce, J. Collingwood. 1978. *Handbook to the Roman Wall, with the Cumbrian Coast and Outpost Forts*. Newcastle upon Tyne, UK: H. Hill.

Divine, David. 1969. *Hadrian's Wall: A Study of the North-West Frontier of Rome*. Boston: Gambit.

Shotter, David. 1996. *The Roman Frontier in Britain: Hadrian's Wall, the Antonine Wall, and Roman Policy in the North*. Preston, UK: Carnegie.

Hagia Sofia
Istanbul, Turkey

The great Church of the Holy Wisdom, known as Hagia Sofia or Sancta Sofia, in Istanbul, is the high point of Byzantine ecclesiastical architecture, remarkable for its revolutionary dynamic structural system and the ingenuity of a plan that subordinates liturgy to form. It was dedicated by the Byzantine emperor Justinian in December 537. Like many churches, it was built on the site of former sacred structures, some of which predated Christianity. The earliest church had been replaced in 361 by the timber-roofed basilica Megala Ekklesia. Damaged during religious riots in 404, this second building was restored eleven years later under Emperor Theodosius II, only to be burned down in another uprising in 532. Within weeks Justinian commissioned the great church of Hagia Sofia.

He had been crowned in 527. Despite the fall of the Western Empire to Germanic invaders in the late fifth century, Justinian ensured that his Eastern Empire survived. He and his wife, Theodora, reigned as unofficial joint rulers, together transforming Constantinople into a city that was universally admired and envied. Justinian employed the architects Anthemios of Tralles and Isidor of Miletus to build a church of great size and magnificence, sparing no expense. Materials were transported from all over his domain. Dressed marble was plundered from classical pagan buildings; it is said that eight red porphyry columns were brought from the Artemiseion at Ephesus; new stone came from the finest marble quarries in Phrygia, Egypt, Thessaly, and the Morean Peninsula. The interiors were decorated with mosaics of gold, silver, glass, marble, and granite tesserae. Because of the urgency, tradition has it, 1,000 masons and 10,000 apprentices worked on the building. It was completed in just twenty days under five years. There is a story, perhaps apocryphal, that upon first visiting the completed church Justinian exclaimed, "Oh Solomon! I have excelled you!"

The central dome, framed with forty brick ribs, is slightly elliptical, its base measuring 101 by 104 feet (30.3 by 31.2 meters). It springs from pendentives at 183 feet (54 meters) above the floor and rises to 226 feet (67.8 meters). There is a window between the bases of each pair of ribs, and the resulting ring of light creates the illusion that the dome is poised in the air with little apparent support. The true massiveness of the masonry structure is replaced with a virtual building created from light—not only because

Hagia Sofia, Istanbul, Turkey; Anthemios of Tralles and Isidor of Miletus, architects, 532–537. Exterior view; the minarets were added when the church was converted into a mosque after 1453.

the penetrating rays of the sun constantly change in angle and direction, but also because of the scintillation of the mosaic-covered surfaces as the light skips from facet to facet.

Awesome in size and opulent in finish though it was, it is not for these reasons that Hagia Sofia is an architectural feat. It is because of its structural brilliance and the subtlety with which the spaces are articulated—underlining the difference between the practical directness of Western architecture and the nuances of oriental. Nevertheless, the church remains Roman. The subdivision of its spaces according to their purposes coincides with contemporary Western basilicas: atrium, narthex, nave and aisles, sanctuary, apse, vestries, and altar are all present.

There the similarity ends. Western churches were long and narrow, and their slender parallel walls supported timber-framed roofs. The plan of Hagia Sofia is almost square, approximately 250 by 220 feet (75 by 67 meters), and the four massive piers, each about 25 by 60 feet (7.6 by 18.3 meters), carry a domical roof. Yet when the spatial arrangement of the church is considered, it can be readily seen that by the use of elegant screens to separate aisles and nave, and the placing of the apse, the architects skillfully manipulated a vast single space to meet the liturgical program of the clergy. The interior space is made cruciform by projecting a large hemidome over the apse and smaller ones above the aisles. This daring experimentation with space was made possible through the use of the pendentive, a structural device that allowed Byzantine architects to satisfactorily roof a cubical volume with a dome. That had never been achieved in the West, and never before on such a scale in the East, from whose vernacular architecture it had been drawn.

Hagia Sofia has undergone many changes in its 1,500-year lifetime, with both natural forces and desecration taking their toll. The church was structurally damaged by earthquake only a year after its dedication, and again in 557 and 559. In 562 it was restored and reinforced by Isidoros, nephew of the original architect, who also raised the dome by about 20 feet (6.25 meters). Further earthquake damage in 869 and 889 closed it for five years. The Iconoclasts vandalized the original mosaics in the eighth and ninth centuries, but most were replaced. Hagia Sofia's finest ornaments were plundered by the Fourth Crusade in 1204, and the building was seriously damaged. Large buttresses were added to the north and south facades in 1317, but that did not prevent considerable earthquake damage about thirty years later. Mehmet the Conqueror took Istanbul for Islam in 1453, and Hagia Sofia, although retaining its name, was put to use as a mosque. Large timber medallions with Koranic texts were hung on the walls of the interior and the Christian mosaics whitewashed over. Minarets were added at various times during the Ottoman period. The building became a museum in February 1935.

At the end of the twentieth century Hagia Sofia stood on the United Nations World Heritage Watch List, one of the world's 100 most threatened buildings. "Despite … ongoing support, including a grant [$100,000 in 1997] from American Express, water penetration, tourist control, and uncertain structural conditions remain threats. Areas of the lead roof have cracked, roofing members have weakened, and leaks are damaging frescoes and mosaics." Restoration and repairs of the roof have been effected, but more money is needed to prevent further structural damage and to install a long-term dilapidation monitoring system.

See also Sultan Ahmet Mosque

Further reading

Mainstone, Rowland J. 1988. *Hagia Sofia: Architecture, Structure, and Liturgy of Justinian's Great Church*. New York: Thames and Hudson.

Mark, Robert, and Ahmet S. Çakmak, eds. 1992. *Hagia Sofia from the Age of Justinian to the Present*. Cambridge, UK: Cambridge University Press.

Halles Centrales (Central Markets)
Paris, France

Once described by novelist Emile Zola as the *"ventre de Paris"* (belly of Paris), Les Halles, situated in a square northeast of the Louvre, was the popular and vibrant market quarter. It was alive during the day with merchants and shoppers and at night with vehicles bringing produce from the French provinces and other Mediterranean countries, night butchers preparing meat for the next day's business, and inquisitive patrons from nearby restaurants and bars. Originally the market comprised open-air stalls, but between 1853 and 1866 a series of pavilions was built to create a covered market of grand scale. Known as the Halles Centrales and designed by architect Victor Baltard (1805–1874) with Felix-Emmanuel Callet (1791–1854), the project was commissioned by Emperor Napoléon III as part of the mid-nineteenth-century remodeling of Paris planned by Baron Georges-Eugene Haussmann.

Influenced by his experience of "modern" life in London, Napoléon III was intent upon establishing Paris as an imperial city capable of exploiting new developments in industry, trade, and transport. He aimed to improve housing conditions, remove slums (home to many of the insurgents of the French Revolution of 1789 and nineteenth-century uprisings), establish public parks, and construct grand streets, public buildings, and monuments. The gigantic Halles Centrales was an iron-framed complex that became the prototype for covered market buildings in France and elsewhere, just one of many new structures that emerged during the "Haussmannization" of Paris.

Baltard's first design was for a classical building with masonry walls. However, the emperor requested that he use iron instead, as a demonstration of France's industrial prowess. Pressure from a public wanting a spacious, well-lit, and well-ventilated structure forced the architect to adopt a design not unlike the railroad sheds of the 1830s and 1840s. He planned a series of rectangular pavilions laid out in a grid pattern and connected by broad streets, all but one of which was covered. Initially there were six pavilions, but the number was soon extended to ten; a further two were added in 1936. Based on a

19-foot (6-meter) module, they measured either 137 by 197 feet or 197 by 197 feet (42 by 54 meters or 54 by 54 meters). At one end of the long axis there was a rotunda, near to which were the administration and public services. A vast basement housed food stores and such facilities as a butter-mixing room and poultry abattoir. Externally the building frame comprised hollow cast-iron columns, which acted as downpipes for rainwater; they were connected by arched girders. The interior columns, also of iron, supported clerestory walls that rose above the eaves of the pavilions. All was covered with a glazed roof. The infill walls were usually single-skin brick, with stone dressings at the top and bottom; above them were horizontal bands of timber-framed opening windows and fixed louvers.

Between 1962 and 1969, the food markets were moved to Rungis, south of Paris. The Halles Centrales site was earmarked for renewal, and while debate raged over how it would be utilized, its former pavilions were home to exhibitions and other cultural events. In the early 1970s ten of the graceful buildings were demolished; two others were dismantled and reassembled, one in Nogent-sur-Marne, France, and the other in Yokohama, Japan. Les Halles was replaced by the Forum des Halles, an underground *métro* station with a regional railroad link (1977) and a multistory shopping center (1979). Popular opposition to the demolition of the Halles Centrales led to a wider movement for the conservation of France's nineteenth-century industrial heritage.

Further reading

Peters, Tom F. 1996. *Building the Nineteenth Century*. Cambridge, MA: MIT Press.

Steiner, Frances H. 1984. *French Iron Architecture*. Ann Arbor, MI: UMI Research Press.

Halles Centrales (Central Markets), Paris, France; Victor Baltard and Felix-Emmanuel Callet, architects, 1853–1866. From a French engraving of ca. 1860.

Hanging Gardens of Babylon
Iraq

The ancient city of Babylon stood on the east bank of the Euphrates River about 30 miles (50 kilometers) south of modern Baghdad. Philo of Byzantium, writing in the third century B.C., listed so-called Hanging Gardens among the seven wonders of the world. Tradition has it that the gardens were built by King Nebuchadnezzar II (ruled ca. 605–561 B.C.) for his wife Amytis, because she missed the mountainous landscape of her native Media. They may have been commissioned by the half-legendary Queen Sammu-ramut (known as Semiramis) some 200 years earlier. Contemporary Babylonian clay tablets intriguingly ignore them amid lucid descriptions of Nebuchadnezzar's palace and the city and defenses. Neither the Babylonian priest Berossus nor Philo and other Greek writers—the geographer Strabo and the historian Diodorus Siculus—who centuries later described the gardens ever saw them, and no *certain* traces survive. Some historians suggest they were merely romantic constructs upon accounts of Mesopotamia carried to Greece after the Macedonian conquest in 330 B.C.

The German archeologist Robert Koldewey believed he had found the substructure of Nebuchadnezzar's gardens around 1899 when he uncovered several unusual vaulted foundation chambers, atypically built of stone, and a well in the northeast corner of the palace. From more recent excavations concentrated on the southern palace, archeologists surmise that they were in another building, hundreds of meters from the river. Because Strabo's description had placed them close to the Euphrates, other scholars disagreed. There is another possibility.

More recently, the suggestion has been made that the classical writers were confused, and that the gardens were not in Babylon at all, but in the Assyrian city of Sennacherib, Nineveh, which stood on the Tigris 250 miles (400 kilometers) to the north. Nineveh was about 1,800 acres (700 hectares) in area, enclosed by 10 miles (16 kilometers) of 50-foot-high (15-meter) walls. Within and outside its defenses, Sennacherib created lush parks and gardens, full of exotic plants and watered from a complex system of aqueducts and canals. They are described on a clay prism dating from about 690 B.C. So is the way in which the huge volume of water needed for irrigation was raised to the highest terrace to flow to lower levels through sloping channels. The king had great brass archimedean screws cast (four centuries before Archimedes!) to lift the water from the ample supply. His description matches those of the later writers. For example, Diodorus Siculus portrays a garden (supposedly in Babylon), whose approach "sloped like a hillside" and whose structure rose "tier on tier," adding that "water machines [raised] the water in great abundance from the river, although no one outside could see it." Whether the Hanging Gardens existed or not, or whether they were in Babylon or Nineveh, the descriptions were evocative.

See also Babylon: Nebuchadnezzar's city

Further reading

Clayton, Peter, and Martin Price. 1988. *The Seven Wonders of the Ancient World*. London: Routledge.

Murphy, Edwin, trans. 1989. *The Antiquities of Asia … of Diodorus Siculus*. New Brunswick, NJ: Transaction.

Russell, John Malcolm. 1991. *Sennacherib's Palace without Rival at Nineveh*. Chicago: University of Chicago Press.

Hezekiah's Tunnel
Jerusalem, Israel

Hezekiah's Tunnel, an eighth-century-B.C. subterranean aqueduct in Jerusalem, was a magnificent engineering achievement. Teams of stonecutters, working no more than two abreast and using hand tools, cut the 1,730-foot (576-meter) passageway of bedrock, probably in about seven months. Starting from both ends, between 33 and 150 feet (10 and 45 meters) underground, without sophisticated surveying instruments or contact with the surface, they were able to reach a meeting point.

The Canaanite citadel called Jebus stood on a slope that fell away into a deep valley outside the present-day walls of Jerusalem's Old City. It had a defensible water supply upon which the conquering Israelites were to build, reaching a climax in the reign of

Hezekiah, King of Judah (727–698 B.C.). Jerusalem depended on a single source of water: the Gihon (or Gichon) Spring. Fed from underground streams and hidden in a small cave on the city's eastern slope, it also irrigated surrounding farmland through canals built along the Kidron creek bed. Archeologists have found evidence of Canaanite fortifications designed to protect the spring. Gihon's name describes its erratic nature: the Hebrew word means "eruption" or "gushing." Although reliably producing up to 245,000 gallons (1.1 million liters) a day, the spring would flow profusely for half an hour, then reduce to a trickle for between four and ten hours—longer intervals in summer, shorter in winter.

A response to siege warfare generated the sophisticated water-reticulation systems that culminated in Hezekiah's Tunnel, one of the great engineering achievements of ancient Jerusalem. Possibly as early as 1800 B.C., the Jebusites were able to reach Gihon from within their walls: a diagonal tunnel, like others in the region, followed a natural rock fissure to a point from which pitchers could be lowered to the spring. Some scholars believe that this was the passageway mentioned in the Bible, through which Joab led King David's men into the city, which they then overthrew. It is known as Warren's Shaft, for Colonel Charles Warren, an Englishman who discovered it in 1867. The debate continues over its date and who built it.

The Israelites augmented this basic system in two stages. First, they built the Siloam (Shiloah) Channel, probably during the peaceful reign of King Solomon (970–928 B.C.). From Gihon a part-open, part-tunneled conduit ran south along the Kidron brook to a reservoir in the HaGai (Tyropoeon Valley) at the southwestern corner of Jerusalem, which by then had been extended to what are now known as the Jewish and Armenian Quarters. Sluices along its eastern side had stone gates that could be opened to irrigate the gardens and fields in the valley below.

Ancient Jerusalem's most extraordinary hydraulic engineering project—perhaps better classified as a civil defense undertaking—was Hezekiah's Tunnel, discovered in 1838 by the American scholar Edward Robinson. Under the implacable Sennacherib (705–681 B.C.), the Assyrian Empire extended from the Persian Gulf to the Black Sea, and westward to the Nile valley. His father Sargon had overrun the northern kingdom of Israel, and Sennacherib was concerned with consolidating the family conquests. Hezekiah, the charismatic ruler of the relatively puny kingdom of Judah, reassured by the prophet Isaiah that God would protect Jerusalem, stood against the Assyrian might. He stockpiled weapons and extended the city's defenses by building the 23-foot-thick (7-meter) Broad Wall. And at the first inkling of invasion he had devised a measure that would help his people survive a siege. He planned "to stop the water of the springs that were outside the city [and] closed the upper outlet of the waters of Gihon and directed them down to the west side of the City of David" (2 Chron. 32:30).

Hezekiah's Tunnel (701 B.C.), still a functioning watercourse almost 3,000 years later, connects the Gihon Spring and the Pool of Siloam (or Hezekiah's Pool), specially constructed at the south end of Jerusalem, where the king had extended the outer defenses. Thus the Bible calls the pool "the reservoir between the two walls" (Isa. 22). The direct distance between spring and reservoir is about 1,100 feet (330 meters), but the winding tunnel is 1,730 feet (576 meters) long. On average, it is about 3 feet (900 millimeters) wide and varies between 3 and 9 feet in height; in places, it is 150 feet (45 meters) beneath the surface of the hilly city. The fall from Gihon to Siloam is about 6 feet (1.8 meters), that is, a grade of about 1 in 70. The tunnel was excavated by two groups of workers, starting at each end and cutting toward each other through the rock to eventually connect. The Siloam Tunnel Inscription, engraved on one of the walls and found in 1880, celebrated their meeting:

While there were still three cubits to be cut through, [there was heard] the voice of a man calling to his fellow, for there was an overlap in the rock on the right [and on the left]. And when the tunnel was driven through, the quarrymen hewed … each man toward his fellow, axe against axe; and the water flowed from the spring toward the reservoir for 1,200 cubits, and the height of the rock above the head[s] of the quarrymen was 100 cubits.

Intriguingly, the indirect course of Hezekiah's tunnel penetrated hard rock while missing softer deposits. Several explanations have been offered for its meandering course. Perhaps the diggers followed a sequence of fissures and crevices that pock the limestone under Jerusalem; perhaps they tried to avoid disturbing the tombs of the kings; or perhaps there was a need to continually reorient themselves—evidenced by a number of false starts and the angle at which the two parts meet.

Further reading

Auld, Graeme, and Margreet Steiner. 1996. *Jerusalem.* Macon, GA: Mercer University Press.

Geva, Hillel, and Joseph Shadur, eds. 1994. *Ancient Jerusalem Revealed.* Israel: Exploration Society.

Levin, Gabriel. 1997. *Hezekiah's Tunnel.* South Kent, CT: Ibis.

Hippodamos of Miletus

The fifth-century-B.C. Greek architect Hippodamos of Miletus has long been known as the "father of city planning." Although the claim has been challenged by some historians, his contribution (at least in the West) was the notion of ordered city planning, as opposed to the uncontrolled growth of earlier times. For example, fifth-century-B.C. Athens, the dominant Hellenic city, was an undisciplined accretion of houses lining crooked narrow streets and lanes whose routes were determined by the topography around the great Acropolis. Hippodamos has been credited with the introduction of the orthogonal plan—a "gridiron" with streets at right angles dividing the city into the kinds of blocks we are familiar with. His ideal plan was zoned by land use, with blocks reserved for public buildings and open spaces, integrated with the houses to provide a cohesive social environment.

However, some elements of the Hippodamean city can be found in earlier Greek settlements. For example, the colony of Smyrna, near what is now the Aegean coast of Turkey, was rebuilt in the seventh century B.C. with parallel north-south streets. Therefore, it may have been that Hippodamos simply formalized generally held conventions in his theoretical writings and applied them in the cities he designed.

In *Politics*, Aristotle remarks upon the Miletian's long hair and eccentric dress and notes his wide interest in natural philosophy. It was unusual for an architect to discourse upon the best form of government, but that did not prevent Hippodamos from doing so. His theories of physical planning were linked to social planning; clearly he saw the planner's role not only in terms of functional and esthetic design but also in human organization of religious, civic, and commercial activities. Adopting what today would be called a determinist approach, Hippodamos divided his optimum population of 10,000 into three: artisans, farmers (every Greek city had its agricultural hinterland), and military. Then he divided the city into three parts: one for worshiping the gods, one to support the soldiers, and the third private, the property of the common people. He went further, categorizing laws into three sorts: insult, injury, and homicide. The political scientist Daniel J. Mahoney has commented, "Hippodamos characteristically divided everything—the population, laws, and land—into threes because he wrongly thought that human nature was amenable to mathematical manipulation." Yet Hippodamos is not remembered for his utopian social views, but for the physical form of his cities. Several have been attributed to him, including his birthplace, Miletus.

The prosperous fortified Aegean port stood on a peninsula at the mouth of the Meander River. Established by the Mycenaeans in the middle of the second millennium B.C., it grew to be one of the largest cities in Anatolia, a commercial center with a population said to have reached 100,000. In 499 B.C., with sister Ionian cities, Miletus rebelled against its Persian occupiers. They responded by razing it. Liberated after the Persians' defeat at the naval battle of Mycale (479 B.C.), the Miletians rebuilt their city according to Hippodamos's orthogonal plan: a repeated pattern of identical blocks with wide main streets crossed by minor thoroughfares. The commercial and religious buildings occupied multiple blocks, and all was enclosed by a defensive wall.

Refounded on the site of an ancient city in the mid–fifth century B.C., the smaller port of Priene, north of Miletus, was set out on a Hippodamean grid. Its plan comprised 84 rectangular 120-by-160-foot (37-by-49-meter) blocks, covering 93 acres (37 hectares) and descending toward the sea from the base of a 1,000-foot (306-meter) cliff on Mount Mycale. The north-south streets were steep, even needing to be stepped in places; the east-west streets, approximately following the contours, were easier to negotiate. Provision was made for city growth within the encircling walls. In the event, the population remained at 3,000 and more than half the enclosed area was never developed. Reserves for public spaces were part of Hippodamos's plan, and the agora stood upon a central terrace.

Around 450 B.C., Hippodamos was commissioned by Perikles to redesign parts of Piraeus, the port of Athens. It stood less than 6 miles (9.6 kilometers) southwest of the city on a peninsula surrounded by the Saronic Gulf. He rebuilt the original fortified Themistoclean port, by then about thirty years old, with a well-defined grid of broad streets defining long rectangular blocks. His plan gave better access to the three harbors, dedicated respectively to grain vessels, general cargo ships, and the navy. The parallel Long Walls, about 600 feet (183 meters) apart, were completed in 431 B.C. to protect the supply line between Athens and its port during the Peloponnesian War with Sparta.

There are several other attributions. Hippodamos almost certainly had a hand in the foundation of the colony of Thurii in southern Italy around 444 B.C. Very regular orthogonal extensions to the city of Olynthos, in what is now Macedonia, were laid out soon after 432 B.C. But it may be that Olynthos and the much later city of Rhodes (408 B.C.) on the Aegean island of the same name were laid out by others who implemented the Hippodamean form. That easily surveyed orthogonal form continued to be influential and was perhaps modified by the Romans in any number of their colonial towns. It was revived in the fifteenth century as one of the theoretical bases of Renaissance urban design. Much later, the planners of cities in the New World employed the grid: Savannah, Philadelphia, Chicago, and New York City are all evidence of that. So is San Francisco, where its imposition on a hilly site, even if it provides locations for exciting movie car chases, underlines its suitability for little but the flattest terrain.

See also Timgad, Algeria

Further reading

Castaglioni, Ferdinando. 1971. *Orthogonal Town Planning in Antiquity*. Cambridge, MA: MIT Press.

Mahoney, Daniel J. 1998. The Conservative Critique of Social Engineering. *The American Enterprise* (September): 43–44.

Murray, Oswyn, and Simon Price, eds. 1990. *The Greek City: From Homer to Alexander*. Oxford, UK: Oxford University Press.

Hydraulic boat lifts

When inscribing the Canal du Centre boat lifts in Belgium on its World Heritage List in 1998, UNESCO commented that they "represented the apogee of the application of engineering technology to the construction of canals." That holds true for each example described here. The boat lifts exemplify the seemingly limitless mechanical ingenuity of the Victorian Age. The Industrial Revolution, first in Britain and then in the rest of Europe and North America, saw the necessarily rapid growth of inland transportation networks. Although they were soon augmented (and often replaced) by railroads, canals were the main arteries of industry and commerce. Differences in water levels along their length and at their junctions with rivers were normally overcome by building locks. In order to save time, creative engineers developed a hydraulic mechanism known as a boat lift, which could replace several conventional locks. Among the most ingenious devices of the machine age, the boat lift continued to be refined into the early twentieth century. The principle was simple: a boat or barge entered a watertight trough that was raised or lowered by filling or emptying a counterbalancing trough.

It is likely that the first commercial boat lift was built in 1838 on the Grand Western Canal in the English county of Devon. The canal, first suggested in 1768, was intended to link the Bristol Channel on the west coast and the English Channel on the east. Construction did not begin until 1810 and four years later an 11-mile (17.6-kilometer) stretch was completed. Extensions were built, and by 1838 the canal reached as far as Taunton in Somerset. A decade later the Great Western Railway linked Bristol and Exeter, and work on the canal was discontinued. But the boat lift served vessels carrying limestone from Tiverton in Devon. Consisting of a pair of 30-foot-long (9-meter) wooden troughs joined by chains, it was capable of raising nearly 10 tons (8.14 tonnes) through the 47 feet (14 meters) that separated two sections of the canal.

The most important English model for others in Europe was the Anderton Barge Lift, built near the English salt-producing town of Northwich between 1872 and 1875. It lifted barges over 50 feet (15 meters) between the Weaver Navigation and the Trent and Mersey Canal. Designed by the engineers Edward Williams and Edwin Clarke, the mechanism comprised two sets of connected hydraulic cylinders and pistons, each supporting a 76-by-15-foot (23-by-4.7-meter) boat tank. In order to lift a boat, a little water was released from the lower tank; as the then heavier counterbalancing tank moved downward, the hydraulic system was activated to raise the lower tank, boat and all. The process was augmented by a steam-powered hydraulic pump. The mechanism lasted for about thirty years, but corrosion problems in the hydraulic system led to the construction of a replacement (albeit incorporating several parts of the original structure) between 1906 and 1908. The new lift continued to carry commercial traffic until the mid-1960s and recreational boats until 1982.

Early among the European clones was the lift at Les Fontinettes on the Neuffossée Canal in northern France. Built in 1888 to raise 340-ton (305-tonne) canal boats 43 feet (13 meters) from the River Aa to the canal, it replaced no fewer than five eighteenth-century locks, dramatically reducing the time needed to negotiate the network of inland waterways linking Calais and Dunkerque with the industrial center of Lille. It was replaced by a single modern lock in 1967.

Proposed in 1879, the 17-mile (27-kilometer) Canal du Centre in Belgium's industrial Scheldt-Meuse-Rhine Delta integrates Europe's inland waterways. Because they survive in working condition, four lifts near La Louvière, also based on the Anderton model, are unique among their contemporaries. Each lifts boats through 57 feet (17 meters). The first, with a capacity of 450 tons (407 tonnes), was built around 1889; the remaining 340-ton (305-tonne) lifts followed between 1908 and 1917. In 1999, as part of a long-term program to increase the capacity of Belgium's major waterways, a single hydraulic elevator was completed at Strépy-Thieu on a new section of the Canal du Centre. It is capable of moving barges of 1,500 tons (1,370 tonnes) dead-weight vertically though 243 feet (73 meters)—the highest lift in the world—in tanks that weigh almost 9,000 tons (8,150 tonnes).

Because of growing industrialization in the late nineteenth century, Germany's River Ruhr needed a transport network for raw materials and manufactured goods. In 1899 the Dortmund-Ems Canal was built to connect North Sea harbors to the Ruhr region. The Rhine-Herne Canal, completed in 1914, linked the Rhine with Rotterdam and Amsterdam. The two artificial waterways are joined by the 45-foot (13.5-meter) Henrichenburg boat lift at Waltrop. Constructed between 1894 and 1899 it was replaced in 1958–1962.

Another early hydraulic lift system was built in the New World: the Peterborough Lift Lock on the Trent-Severn Canal, connecting Lake Ontario with the upper Great Lakes and the West. Completed in 1904 it consisted of two ship lifts—each with a mass of 1,900 tons (1,730 tonnes) and rising 49 and 65 feet (14.8 and 19.8 meters), respectively—within the 4-mile (6.5-kilometer) canal, replacing eight conventional locks.

Further reading

Hadfield, Charles. 1969. *British Canals: An Illustrated History*. Newton Abbot: David and Charles.

———. 1986. *World Canals: Inland Navigation Past and Present*. Newton Abbot: David and Charles.

Lead, Peter. 1980. *The Trent and Mersey Canal*. Ashbourne: Moorland Publishing.

Industrialized building

In the second half of the 1920s the modernist architects of Europe, perceiving an urgent need to reform city planning and especially public housing policies, sought to address the social changes resulting from industrialization. At a 1928 meeting at La Sarraz, Switzerland, architects from Austria, Belgium, France, Germany, Holland, Spain, and Switzerland formed the Congrès Internationaux d'Architecture Moderne (CIAM), agreeing that rationalization and standardization were the chief ways to solve the housing problems each country then faced. CIAM reconvened in Frankfurt in 1929 to discuss the pragmatic issue of *existenzenminimum*—low-cost residential units. That "unit" should replace "house" in its lexicon is an indicator of pervasive socialist thinking; indeed, politics could not be excluded from any debate on urbanism and housing policies. In its *Athens Charter*, derived in 1933 and published ten years later, CIAM offered modern technology as the generic solution to the urban problems that would be exacerbated by World War II. That is, they called for a new way of building, and that displacement of conventional thinking with a "problem-solving" approach was an architectural feat in itself. Success is a different matter.

It is one thing to theorize, quite another to find real solutions. Designers on both sides of the Atlantic were investigating industrialized construction techniques as a means of making better, affordable housing. As early as 1910 the German architect Walter Gropius advocated the industrial production of interchangeable housing components, and in 1914 Le Corbusier's Domino house system employed a standardized framework. It was perhaps inevitable that many of the resulting products were mechanistic and austere, emphasizing structure and detail at the expense of esthetic considerations. This new, efficient way of making architecture was grasped as an opportunity to realize the house as "a machine for living in." The first half of the twentieth century is replete with designs for systems and components, too numerous to include here. Suffice it to identify a few key individuals.

The French blacksmith and steel fabricator Jean Prouvé (1901–1984) began experiments with prefabricated construction in 1925, in partnership with Aluminium Française and the car manufacturers Citroën and Renault. Given impetus by electric welding technology, after 1931 he produced building components and entire prefabricated structures. In collaboration with the architect Eugene Beaudouin and engineer Marcel Lods, Prouvé built a transportable structure for the Rolland Garros Aeroclub in Buc, France (1935), a forerunner of their Maison du Peuple in Clichy (1936–1939). By adjusting its movable floors, sliding partitions, and openable roofs, the Maison could be adapted within an hour to become a covered market, a meeting hall, or a cinema. In 1945 the French Ministry of Reconstruction and Planning commissioned Prouvé to produce prototype low-cost, mass-produced housing units. Using a technique he had developed with Pierre Jeanneret and Charlotte Perriand in 1939, he designed his "tree" buildings: a folded-steel central portal (erected without scaffolding) carried aluminum roofing, stiffened by profiling and held in place by external "buffers"; wall panels hung as "curtains" from the buffers. In 1951 the twenty-five dwellings were erected as part of an experimental project in Meudon, proof that industrialized houses were economically feasible. The government shelved the scheme. Although the firm was otherwise successful, serving France and her colonies, Prouvé abandoned it in 1953. His biographer John Winter has remarked, "No one can say of industrialized building that it cannot be done, for Prouvé has done it. He has not adapted building to machine processes; he has worked it all out from scratch as if no one had ever built a building before."

The most influential pioneer in the United States was the architect, engineer, and inventor Richard Buckminster Fuller (1895–1983). Enthralled by the possibilities of technology, Fuller began what he called his Dymaxion investigations around 1927, with a design for an affordable, easily transportable, environmentally efficient, mass-produced house. The structure of the hexagonal house employed aluminum upper floor and roof elements suspended from a central mast. The ground level was open. Proposals for a Dymaxion Bathroom Unit and a Dymaxion Deployment Unit (for Butler Manufacturing Co.) followed in 1937 and 1941. Then toward the end of World War II, Fuller persuaded the Beech Aircraft Corporation of Wichita, Kansas, to underwrite production of the Dymaxion house. Although the materials used in airplanes and the house were similar, the cost of tooling was so high that many thousands of units would have to be made. Fuller resisted suggested design changes, and the partnership was dissolved after only two prototypes of the circular "Wichita" house (1945) had been built.

In 1946 the French government commissioned Le Corbusier to build a prototype of his "vertical city" in Marseilles. The underlying idea was that standardized self-contained housing units could be slid like drawers into a building frame; the result was the Unité d'Habitation (completed 1952), a stack of 340 cramped "superimposed villas" penetrated by internal streets of shops and services. Adored by contemporary students of architecture, the block was and is detested by the people forced to live in it. It should have furnished the designers and providers of public housing with a salutary warning, but it did not and many similar schemes followed.

The inexorable march of industrialization meant that by about 1960 standardized building components had become commonplace, simply through market forces. Some architects accepted the modular dwelling unit as the primary element in larger housing developments. Wherever they were, like their counterparts of thirty years earlier, they linked industrialization, housing, and urbanism.

In London, Archigram briefly and brassily emerged. In 1964 Warren Chalk coined the expression "capsule homes"—it has a similar ring to "units"—prefabricated modular dwellings that could be stacked up to form a tower. Another member of the group, Peter Cook, proposed the Plug-in City (1964–1966), in which self-contained living units could be temporarily plugged into structural towers. The individual house became an anonymous, interchangeable, wedge-shaped pod, with all the ergonomic efficiency and technological sophistication of a space capsule. The main components were to be pressed metal, plastic, or even pressed paper. Thankfully, nothing was ever built.

But Moshe Safdie's Habitat was. Constructed as a permanent model community along the St. Lawrence River, and as part of the Montreal Expo 67, Habitat employed 354 prefabricated reinforced-concrete modular boxes, measuring 17.5 by 38.5 by 10.5 feet (5.35 by 11.8 by 3.2 meters), to generate 158 dwellings (900 were originally proposed) of fifteen plan types. By ingeniously stacking the units and fixing them with steel cables, Safdie produced a highly sculptural building with pedestrian walkways, small gardens, and decks. The success of industrialized building lies in the number of units produced, and as in so many other cases, the construction costs of Habitat were prohibitive.

It is clear that a people's image of "house" is a cherished and entrenched cultural value, so innovation and the visionary ideas of architects are not readily accepted. This is one of the main reasons for the repeated failure of industrialized building. Others have been cited as inaccurate reading of the market, excessive profit expectations, inappropriate use of materials, and professional inertia. Nevertheless, experiments continued to the end of the twentieth century. American architect Wes Jones designed the Technological Cabins in the High Sierras for two Californian academics, using standard steel shipping containers as the module, fitting them out before transporting them to the site, and assembling them to form the house. A similar approach at De Fantasie in the Dutch new town Almere, in the late 1980s, proved disastrous from a climatic point of view. Also in the United States, Andres Duany and Elizabeth Plater-Zyberk have maintained the link between the manufactured house and its environment by applying New Urbanism principles when assembling units in a Rosa Vista, Arizona, home park. Perhaps the answer lies in the approach taken by architect Deborah Berke, who assembles modular units to create the conventional and familiar spaces of North American house types.

See also Archigram; CIAM (International Congresses of Modern Architecture); Unité d'Habitation

Further reading
CIAM. 1944. *International Congresses for Modern Architecture*. Proceedings of first 9 conferences. Cambridge, MA: MIT Press. Facsimile.

Cook, Peter, ed. 1972. *Archigram*. London: Studio Vista.

Winter, John. 1994. "Jean Prouvé." In *Contemporary Architects*, edited by Muriel Emanuel. New York: St. James Press.

Inka road system
Peru

The brief but glorious ascendancy of the Inka lasted for about sixty years from A.D. 1476. At that moment their empire, Tahuantinsuyu (Land of the Four Quarters), was the largest nation on earth. Ruled from the Andean capital, Qosqo, it covered 2,000 miles (3,200 kilometers) north to south and 200 miles (320 kilometers) inland. The empire's northern quarter, Chinchaysuyu, extended beyond what is now Colombia; the southern quarter, Collasuyu, reached as far as central Chile; the eastern quarter, Antisuyu, included the eastern Andean foothills in modern Bolivia and Argentina; and the western quarter, Cuntisuyu, embraced the Pacific coast.

A critical means of sustaining Inka power over subject peoples was a system of primary and secondary roads whose total length has been estimated to be 25,000 miles (40,000 kilometers), comparable to the communication infrastructure of the Roman Empire, and achieved without the advantage of the wheel or large draft animals. Quite apart from the variety of the terrain, the Inkan transportation network was a great engineering feat, and the response to that diversity—mountains and valleys, snow, deserts, and swamps—makes the accomplishment the more remarkable. Near the coast they were dusty tracks, sometimes built on causeways to keep them free of blown sand or sometimes simply pegged out; in swamps they were built on stone viaducts; and in high rain- or snowfall regions they were paved with cobbles or flagstones. Steep slopes were negotiated by means of steps, often cut into the living rock.

The roads sat within a hierarchy, at the apex of which were the two north-south royal, or Inca, roads linking Qosqo with the four quarters of the empire. One crossed the Cordillera from what is now Colombia to Argentina, and the other followed the coastal plains from northern Peru to northern Chile. They were linked by several crossroads. The rest of the primary network consisted of "principal" or "rich" roads and "big" or "broad" roads, covering a conservatively estimated 15,000 miles (25,000 kilometers). A secondary system of people's roads joined villages and districts throughout the Tahuantinsuyu, bringing the total length of roads to some 25,000 miles (40,000 kilometers). Inevitably, in mountainous country, bridges of various construction were necessary. These ranged from simple stone slabs, through small log bridges and "flying foxes," to rope-and-leather suspension bridges, some spanning chasms up to 500 feet (150 meters) wide. There were even floating bridges made of rope and reeds.

A corollary of the Incan road system was the army of young athletes called *chaqsi*, who ran in relays be-

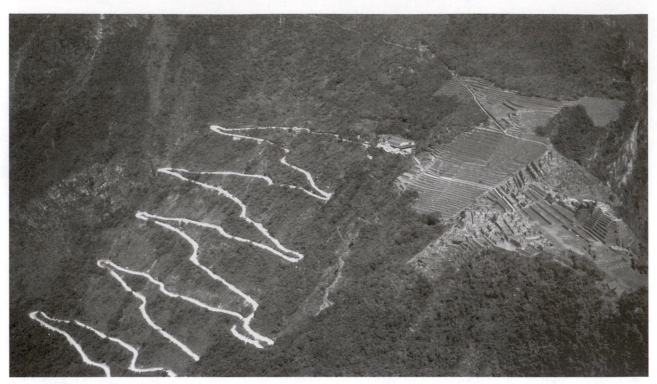

The Inka road to Machu Picchu, Peru, ca. 1400–1532.

tween staging posts *(chasqwasi)* set at 8- to 15-mile (13- to 24-kilometer) intervals. They carried verbal messages and sometimes goods. For example, the royal court at Qosqo enjoyed fresh fish delivered from the coast over 200 miles (320 kilometers) away. The messenger service was continuous, relays of runners covering up to 300 miles (480 kilometers) a day. Armies were deployed along the roads, officials moved between administrative centers, priests traveled to supervise religious services, pilgrims wound their way to shrines, merchants transported their goods by llama or alpaca caravans, and herders coaxed flocks down from the high country. For these more leisurely travelers, services were provided at large villages called *tanpu* along the major routes, strategically located at intervals representing one day's walk, say 25 to 30 miles (40 to 50 kilometers). In the tanpu, lodging, food, and clothing were available for thousands of people at once, because for political or economic reasons, the Inka sometimes would relocate entire populations. These administrative and service centers were as important to the Inkas as the roads themselves; from them, imperial bureaucrats exercised control over the empire. Thus, for example, several centers were established on the royal road at Tambo Colorado

and Huanuco Viejo, each with more than 3,000 buildings to house the civil service, manufacturing and warehouse functions, catering for local food shortages, and so on. Smaller settlements were sometimes built at half-day intervals.

At the beginning of the twenty-first century some 14,000 miles (22,000 kilometers) of Inka roads remain discernible, but much of the continuity has gone, destroyed by modern highways, radio masts, or hydroelectric power stations. Tourism also is taking its toll. Progress is inevitable, but measures are being taken to preserve remnants of the Inka Trail. For example, in the 1990s the Machu Picchu Historical Sanctuary commissioned the British company Mountain Path Repair International to produce a sustainable management plan for the road between Qosqo and the spectacular site and to restore the eroded sections.

See also Qosqo, Peru

Further reading

Burland, Cottie Arthur. 1967. *Peru under the Incas*. New York: Putnam.
Hagen, Victor Wolfgang von. 1976. *The Royal Road of the Inca*. London: Gordon and Cremonesi.

Hyslop, John. 1984. *The Inka Road System*. Orlando, FL: Academic Press.

Inuit snow houses

The Inuit—"the real people"—of Alaska, Arctic Canada, northeastern Siberia, and Greenland sometimes build shelters out of water, or at least water in one of its solid states, snow. The highly sophisticated design and construction of that kind of igloo (the Inuit word for house) is a major architectural achievement, employing a technology that turns a challenging resource to creating a not merely adequate but ideal house form.

The oldest identifiable Inuit date from about 2000 B.C. Some of them followed immense migratory herds of bison, caribou, and musk ox across the Bering Strait into North America. Since two-month summers made agriculture impossible in their harsh, treeless environment, the Inuit relied for their food on hunting and fishing. Although some Inuit have now become westernized and eat supermarket food, fish and sea mammals remain the mainstay of the traditional diet of many, and groups still follow a seasonal nomadic cycle through their lands. In comparison with other hunter-gatherer cultures, the Inuit have highly developed technologies, craftsmanship, and art. The dogsled is used for long-distance transportation of large loads, and the maneuverable kayak (sealskin-covered canoe) has long been a model for Western societies. Inuit weapons are fashioned from ivory, bone, stone, or sometimes copper and often decorated with elaborate carving. Their clothing—parka, trousers, mittens, boots, and snow goggles—is often made of caribou skins.

It should not be thought (as the stereotype has it) that all Inuit live in snow houses. They have *three* traditional dwelling types. A summer house is essentially a caribou-, walrus-, or sealskin tent. A winter house is partially excavated and usually built of stone, with a whalebone or driftwood frame supporting a moss or sod covering. Then there is the circular dome-shaped snow house that some groups use as a winter dwelling. But it is more commonly used by hunters as a temporary shelter while traveling on long journeys.

The igloo is built with carefully shaped blocks of snow about 4 feet long, 2 feet high, and 6 to 8 inches thick (about 1.3 by 0.65 by 0.15 meters), weighing about 45 pounds (20 kilograms). The house can be up to 18 feet (5.5 meters) in diameter, with ample headroom for the occupants. Snow texture and consistency is critical, and the suitable hard-packed snow is usually found on a north-facing slope. Tiny pockets of air trapped between the crystals provide a remarkably effective means of thermal insulation. For maximum structural strength, the first row of blocks is set out in a circle. The blocks are shaped to form a kind of ramp beginning at the front of the igloo, as the base of a self-supporting continuous spiral. As the walls rise to merge into the roof, successive tiers of blocks tilt more and overhang more as they rise, until they converge to form the dome, which is closed with a large fitted cap-block. This method allows the builder to work alone if necessary. The cracks between the blocks are packed with soft snow. Once the first two circuits are completed, it is possible to construct an igloo even during a blizzard, because the structure acts as a windbreak. When intended to be occupied for a long time, the igloo has another low wall of snow blocks placed around it, and the space between the two walls is filled with loose snow, improving thermal insulation.

The entrance is a narrow passage, high enough to admit a crawling person and curved to stop the penetration of cold winds. Additional storage vaults may also form part of the house. The translucent snow provides a little light inside the igloo, and sometimes an ice window is employed. A small ventilation hole is cut in the dome. The floors in larger, long-occupancy igloos are often concave, so that cold air falls into a pool. The remainder of the floor surface is covered with furs, while others hung on pegs trap an air layer against the walls, providing interior warmth without melting the snow. The heat generated by the occupants' bodies and by lamps or camping stoves raises inside temperatures enough to allow the Inuit to move about naked in their houses of snow.

Further reading

Friesen, John W. 1997. *Rediscovering the First Nations of Canada*. Calgary: Detselig.

Steltzer, Ulli. 1985. *Inuit, the North in Transition*. Chicago: University of Chicago Press.

Ironbridge, Coalbrookdale
Shropshire, England

Coalbrookdale is regarded by many as the birthplace of the Industrial Revolution. The town of Ironbridge on the eastern bank of the River Severn is the location of the world's first metal bridge. Designed in 1775, the gracefully arching prefabricated cast-iron structure, appropriately named Ironbridge, was fixed to its masonry abutments in the summer of 1779. Spanning 100 feet (30 meters), the bridge supports itself without a bolt or a rivet in the entire structure! In terms of the creative application of new materials and technology, it remains one of history's great architectural and engineering feats, the product of the fervent inventiveness of optimistic industrialists, opening the way to the modern era of iron- and steel-framed buildings.

Coal and limestone mining and iron smelting made the River Severn, which reaches the sea through the Bristol Channel on England's west coast, one of Europe's busiest waterways. In 1638 one Basil Brooke patented an iron-making process and built a furnace at Coalbrookdale. Seventy years later the operation was acquired and overhauled by the entrepreneurial Bristol Quaker Abraham Darby I, an ironmonger and brass founder. In 1711 he developed a cheaper means of smelting iron by using coked coal as fuel rather than charcoal. The process liberated iron production from fuel restrictions—industrialization initially meant deforestation—as well as making very large castings possible.

Within a couple of years Darby and his partner, Richard Ford, developed what was a minor business producing mainly pots and pans into the world's leading ironworks. After a few decades the Coalbrookdale Company and its subsidiary Lilleshall Company had expanded to own mines, forges, factories, and farms throughout the region. The burgeoning iron-, brick-, and pottery works in the parishes of Madeley and Broseley, facing each other across the Severn Gorge, brought workers flocking to the district. That dramatic population growth and the obvious increase of commercial and industrial traffic meant that the local ferry, precariously approached down steep, slippery banks, soon proved inadequate for local needs.

Abraham Darby II had proposed to bridge the Severn between Madeley Wood and Benthall but the project lapsed when he died in 1763. It was left to his son, Abraham III, to carry out the project. With the eager cooperation of the squire of Broseley, ironmaster John Wilkinson, in 1775 young Darby convened a meeting of potential subscribers to plan a bridge. The group obtained Parliament's approval for a structure of "cast-iron, stone, brick or timber."

The world's first cast-iron bridge was designed by the Shrewsbury architect Thomas Farnolls Pritchard, who two years before had suggested using the new material for such projects. He proposed a single-span bridge, estimated to cost £3,200, because intermediate piers would obstruct traffic on the busy river. Work began in November 1777. When Pritchard died in that year, Darby assumed responsibility for completion. The components were cast in the Upper or Lower Furnace at Coalbrookdale during the winter of 1778–1779, ready for erection the following summer. Some of the castings—there were 453 tons (384 tonnes) in all—were almost 80 feet (25.5 meters) long, and the Coalbrookdale Works had to be altered to accommodate production. Beginning in May 1799, the prefabricated iron structure took only three months to put together. The parts were ingeniously designed to allow assembly simply by fitting projections into slots and tightening the joints with cast-iron wedges—a totally interlocking structure that, as noted, has no riveted or bolted connections. Ironbridge is a semicircular arch of 100 feet, 6 inches (30.5 meters) span, made by joining two half-arches that were each cast as a single piece. It supports a 24-foot-wide (7.3-meter) deck 40 feet (12 meters) above the Severn.

Costing about twice as much as first estimated, it was opened as a toll bridge (to recoup some of the expense) on New Year's Day 1781. Within three years earth movement caused some noncritical cracking in the ironwork. The bridge survived a severe flood in 1795, and in 1802 the masonry abutments were replaced by timber. In turn those were replaced by the cast-iron arches that one sees today. Doubts about

Ironbridge, Coalbrookdale, Shropshire, England; Thomas Farnolls Pritchard, architect, ca. 1776–1781.

the stability of Ironbridge led to suggestions about demolition in 1926 but nothing happened. In 1934 it was closed to vehicles and listed as a historical monument. The Shropshire County Council assumed ownership in 1950, and extensive restoration was undertaken in the 1970s. The bridge now comes under the control of English Heritage. In 1986 UNESCO designated the Ironbridge Gorge a World Heritage Site, noting that "all the elements of progress developed in an eighteenth century industrial region can be found [there]. ... The blast furnace of Coalbrookdale ... is a reminder of the discovery of coke, which, together with the bridge at Ironbridge, the first metallic bridge in the world, had considerable influence on the evolution of technology and architecture."

Further reading

Cossons, Neil, and Barrie Trinder. 1979. *The Iron Bridge: Symbol of the Industrial Revolution.* Bradford-on-Avon, UK: Moonraker Press.

Muter, W. Grant. 1979. *The Buildings of an Industrial Community: Coalbrookdale and Ironbridge.* London: Phillimore.

Trinder, Barrie. 1981. *The Industrial Revolution in Shropshire.* Chichester, UK: Phillimore.

Itaipú Dam
Brazil/Paraguay border, South America

Built between 1975 and 1991, the Itaipú Hydroelectric Power Plant is situated on the Paraná River between the cities of Guaira, Brazil, and Salto del Guaira, Paraguay. Including the reservoir created by its dam, the system extends about 125 miles (200 kilometers) along the river; it is the largest hydroelectric facility in the world. It supplies about a quarter of Brazil's south, southeast, and west-central regional demand, and nearly 80 percent of Paraguay's total electrical energy. In 1995 the American Society of Civil Engineers nominated Itaipú as one of the seven wonders of the modern world, on the basis of its "advances, engineering challenges and long-term significance." Beyond the staggering scale of the engineering project, Itaipú is also important politically (because of the dual ownership of Brazil and Paraguay) and environmentally.

The two nations recognized the energy potential of the Paraná River, the seventh largest in the world, that forms their mutual border. In 1966 they signed the Ata do Iguaçu (Act of Iguaçu), a joint statement agreeing to equally share the energy. In 1970, a consortium comprising the U.S. firm IECO and the Italian company ELC successfully tendered to conduct a thorough evaluation of resources, and in April 1973 the Treaty of Itaipú set out details for the creation of the power plant. A year later the Itaipú Binacional corporation was established to administer the financing, construction, and management of the dam. Construction work began in January 1975 and reached its peak in 1978, when about 40,000 people were engaged on the massive undertaking, described by one source as "a labor of Hercules." The first of eighteen 700-megawatt generating units (nine serve each country) was commissioned in December 1983, and Paraguay's electrical grid went on-line in March 1984. Brazil followed in August, and the whole system, generating 12,600 megawatts, was operational by April 1991. Two more generators will be installed by 2003, bringing Itaipú's capacity to 14,000 megawatts. The dam was projected to cost U.S.$3.4 billion but the final cost reached between $18 billion and $20 billion.

Itaipú's complex series of dams was built after the Paraná was rerouted through a 1.3-mile-long (2.1-kilometer) diversion channel, completed in October 1978, which entailed removing over 50 million tons (45.5 million tonnes) of earth and rock. Together, the great walls stretch 4.8 miles (7.7 kilometers) across the Paraná River, impounding a 125-mile (170-kilometer) reservoir that holds 28.54 billion tons (26.4 billion tonnes) of water. Tourist brochures boast that the dams contain enough concrete to build five Hoover Dams and enough steel for 380 Eiffel Towers. The main hollow gravity-type concrete dam, with a crest height of 640 feet (196 meters), is connected to the spillway (on the right bank) by a concrete buttress-type wing dam, which in turn is linked to a small earth- and rock-fill dike. On the left bank another rock-fill structure links the main dam and an earth-fill dam. The partly submerged, 1,055-yard-long (968-meter) powerhouse sits on the riverbed at the toe of the main dam; fifteen of the generators are in the main powerhouse and the others on the diversion channel.

The resettlement of people—Ava-Guarani Indians and Mestisos—on reservations and the disruption of their lives have had undesirable social effects, both on the displaced people and their new neighbors. Among other consequences, especially in the early stages of dam construction, was the enormous impact on natural vegetation. The binational Forest Management Project, initiated in the late 1970s, aimed to maintain ecological equilibrium and sustainability within the surrounding forests. Reforestation programs and the creation of a number of forest reserves eventually reduced the potential damage by half. It is estimated that 11.5 million plant species were rescued. Fauna rescue and relocation programs saved thousands of animals, birds, and insects, representing over 400 species.

Captive breeding programs will eventually allow the release of rare and endangered creatures into their natural habitat. The Itaipú project has demonstrated that, with careful management, even large-scale socioeconomic development is compatible with environmental conservation.

Further reading

Itaipú Binacional. 1981. *The Itaipú Hydroelectric Project 12,600 MW: Design and Construction Features.* Brasília: Itaipú Binacional.

Kohlhepp, Gerd. 1987. *Itaipú: Basic Geopolitical and Energy Situation.* Braunschweig, Germany: F. Vieweg.

Itsukushima Shinto shrine
Miyajima, Japan

Miyajima is a mountainous island in Hiroshima Bay on Japan's Seto Inland Sea, separated from the mainland by the 550-yard-wide (500-meter) Onoseto Strait. It has long been a sacred site of Shintoism, and renowned for the Itsukushima shrine, built on piles over the water and dedicated to three sea goddesses, Ichikishima-Hime-no-Mikoto, Tagori-Hime, and Tagitsu-Hime. The entire precinct comprises an inner shrine of thirty-seven axially disposed buildings and an outer shrine of nineteen more. The inner sanctuary, the intermediate sanctuary, the hall of worship, the spectacular O-Torii (Grand Gate), several secondary temples, and drama and dance stages are linked by wide covered corridors and galleries known as *Kairo*. All the timber is finished with vermilion lacquer. The Japanese government has named six of the buildings as National Treasures; the rest have been recognized as Important Cultural Assets. The shrine was inscribed on UNESCO's World Heritage List in 1996, and it has been described as one of the great accomplishments of the Shinden-zukuri architectural style of the Heian period (A.D. 794–1184). With a backdrop of mountains and built on tidal land that at high tide gives it the appearance of serenely floating on the sea, the Itsukushima shrine is a magnificent achievement of harmonizing architecture and nature.

Itsukushima is thought to have been first constructed by Saeki Kuramoto in A.D. 593, but the ear-

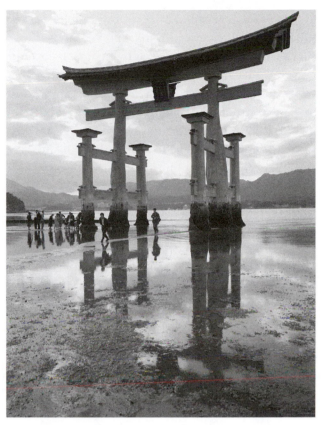

Itsukushima Shinto shrine, Miyajima, Japan; architect(s) unknown, 593–1875. The O-Torii gate (1874–1875) at low tide.

liest historical record dates from 881. It was enlarged in 1168, when Taira-no-Kiyomori was governor of Aki Province, and the Taira clan began to worship there. Fire caused damage early in the thirteenth century, and it is likely that the consecutive restorations included changes to the organization of the buildings. The shrine for the Guest Deity (Sessha Marodo-jinja) was constructed in 1241. The buildings were again restored after being damaged by a typhoon in 1325, since which time the layout has been little changed. By the late twelfth century, the influence of Itsukushima was waning, and by the mid–fourteenth century the buildings had fallen into disrepair. After the warlord Mori Motonari gained control of Hiroshima in 1555, the shrine was restored to its former glory. He commissioned many of the present buildings, including the main sanctuary, in 1571, remaining faithful to the Heian style. Although there are slight stylistic variations in the details—inevitable over so many centuries—the overall architecture of the Itsukushima shrine is remarkably homogeneous.

The approach from the east by boat first encounters the 52-foot-tall (16-meter), vermilion-colored O-Torii, standing in the sea some 220 yards (200 meters) in front of the hall of worship and built on its axis. The eighth since the Heian period, it dates from 1874 to 1875. The great weight of its massive camphor-wood pillars, approximately 44 feet (13.4 meters) tall, together with the 76-foot-long (23.3-meter) hollow cross piece, filled with stones, allows the O-Torii to stand upon the seafloor without being embedded in it.

The main sanctuary *(Honden)*, measuring about 78 by 38 feet (23.8 by 11.6 meters) is crowned with a decorative tile and cypress-bark roof. An offering hall *(heiden)*, a hall of worship *(haiden)*, and a purification hall *(haraiden)* are linked by covered corridors. The main shrine *(Honsha)* has three parts: the inner sanctuary of the goddesses, the sanctuary for the priests, and a space for worshippers. It is faced with turquoise-lacquered folding doors. In front of it is the Broad Stage *(Hirabutai)*, used during the annual midsummer musical festival, *Kangensai*; it has a long, narrow pier extending to the Front Lantern *(Hitasaki)*, used for the departure and arrival of the

sea goddess during those celebrations. The High Stage *(Takabutai)*, standing at the center of the Broad Stage, is used for the performances of sacred shrine music and dancing known as *Bugaku*. The Noh Drama Stage *(Noh Butai)* stands at the end of the structure, and its floor is ingeniously constructed as a sounding board to improve acoustics. Some of the flooring planks are 5 feet wide and 35 feet long (1.5 by 10 meters); they were transported from northern Japan. Their spacing is calculated so that the platforms resist the pressure of high seas. Maintenance of the shrine is continuous because of its exposure to wind and saltwater, and the piles supporting it need to be frequently replaced.

The Itsukushima shrine has graced the island of Miyajima with its elegant presence for 800 years. Its designers and builders, possessors of a grand vision and a deep understanding of the relationship between architecture and nature, remain unknown and unsung.

Further reading

Alex, William. 1968. *Japanese Architecture*. London: Studio Vista.
Suzuki, Kakichi. 1980. *Early Buddhist Architecture in Japan*. New York: Harper and Row.

J

Jahrhunderthalle
Breslau, Germany

The Jahrhunderthalle (Centennial Hall) of 1911–1912 in what was formerly the city of Breslau in Germany (now Wroclaw, Poland) was a major milestone in the development of the enclosure of large public spaces by reinforced concrete structures. It was by far the largest of several pavilions built in Scheitniger Park (now Szczytnicki Park) to house the 1913 centennial of Germany's liberation from Napoleonic rule. The Jahrhunderthalle was intended to serve as an exhibition space, an assembly hall, and a venue for concerts, sporting events, and other entertainment.

Wroclaw in southwestern Poland fell to the Prussian armies of Frederick the Great in 1741, to eventually be renamed Breslau. By the early twentieth century the city had become a major center for the arts, in part because the Expressionist architect Hans Poelzig was director (1903–1916) of the Royal Art and Craft Academy. Breslau's largely German population then exceeded half a million, and the government decided to create what it called a "metropolis of the east." Accordingly, the architect Max Berg, director of Frankfurt am Main's City Building Department, was appointed City Building Commissioner. In Frankfurt, he had been deeply involved with the construction of the city's Festhalle (1907–1909), designed by Friedrich von Thiersch; that ex-

perience was significant for his work in Breslau. He had also designed the development plan for Berlin.

Beginning in the second half of 1910, Berg conceived and developed the structure of the Jahrhunderthalle. Engineering calculations were made by Gunther Trauer of the City Building Department. Trauer described it as an "incredibly clever" design, although he admitted that it was "unusually large and challenging" for him. Nevertheless, he rose to the challenge, and the building is evidence of an admirable symbiosis between architect and engineer. Together they produced two feasibility studies—one that employed a fire-resistant steel structure and another of reinforced concrete—and prepared two sets of contract documents. Because the City Board of Directors was adamant that the exhibition building should be "no-risk [and] fireproof," the former structural system was virtually precluded because of the bulkiness of concrete-cased steel. On the other hand, such a huge reinforced concrete space had never before been built, and conservative members of the board doubted its practicability. However, after six months of deliberations Berg's reinforced concrete proposal was accepted in June 1911 on the condition that the cost be reduced by 10 percent.

The client's insistence on functional flexibility had generated difficulties for Berg. Conventional wisdom pointed to a long space for an exhibition hall and a central plan for the other events. The first design was based upon a longitudinal plan, but that was soon modified to become a central circular space with four semicircular apses that are reached through enormous arches. As built, the hall encloses almost 60,000 square feet (5,600 square meters) of floor space. It provides standing room for 10,000 people; the seating capacity is only 6,000. The 137-foot-high (42-meter) central space is roofed with a 212-foot-diameter (65-meter) dome, formed by 32 half-arches of reinforced concrete—left exposed for acoustic purposes—springing from the massive poetic substructure to a tension ring at the apex. In its day it was the widest monolithic dome in the world. The vast interior is lit by four tiers of curtained clerestory windows, supported by the half-acrches and continuous around the entire structure, which diminish in height as they rise. That gives the dome the appearance of a series of concentric rings. The apses, also structurally formed from reinforced concrete half-acrches, have walls glazed in the same manner, adding to the stunning impact of the space. Although the structural system was revolutionary, the spatial organization (and the overall form that it yielded) had a Renaissance quality, very like the Church of S. Maria della Consolazione (1503) at Todi, Italy, by Donato Bramante and Cola di Caprarola. Berg's inspiration was complex: he drew upon the spirit of Gothic architecture and the esthetic theories of the Frenchman Durand and the Hollanders Lauweriks and Berlage. The monumentality of the huge building evokes the romantic, unbuildable Beaux Arts projects of Boullée and Ledoux; at the same time, Berg avoids ornament for its own sake. The result is that, artistically, the Jahrhunderthalle denies the confident inventiveness of its engineering; at least, that is the impression from *outside* the building.

The contract for the reinforced structure was won by the Dresden firm of Dyckerhoff and Widmann; the Lolat-Eisenbeton Company of Breslau undertook the smaller associated buildings. Work began at the end of August 1911; the foundations were completed two and a half months later, and the building was completed in the amazingly short time of fifteen months, well before the centenary celebrations were due to begin.

The Jahrhunderthalle was a landmark building, and while some architectural writers dismiss Berg's work as "equivocal," others believe that the "structural audacity" he demonstrated in this magnum opus had great influence on the German Expressionist architects (including Poelzig) who flourished between 1910 and 1925. Indeed, Jerzy Ilkosz (1994, 81) has asserted that it "was the first major achievement in the pantheon of Expressionist architecture." Following World War II the Germans were expelled from Breslau and in August 1945 the city, again named Wroclaw, reverted to Poland. The building was renamed Hala Ludowa (the People's Hall).

See also Shell concrete

Further reading

Hasegawa, Akira. 1997. "The Birth and Background of Max Berg's 'Century Hall,' the Inspiration of Artist Max Berg." *Space Design* (December): 83–90.

Ilkosz, Jerzy. 1994. "Expressionist Inspiration." *Architectural Review* 194 (January): 76–81.

James, Kathleen. 1994. *Between Nationalism and Expressionism: Max Berg's Jahrhunderthalle and Bruno Taut's Monument des Eisens*. Berkeley: University of California Press.

Jantar Mantar
Jaipur, India

Jantar Mantar ("instruments and formulae"), the open-air observatory designed by Maharaja Sawai Jai Singh II, India's last great classical astronomer, stands at the entrance to the palace in the old city of Jaipur. Built between 1728 and 1734, the group of large, modern-looking masonry structures is in fact a collection of astronomical instruments. They measure local time to an accuracy of a few seconds; the sun's declination, azimuth, and altitude; the declination of fixed stars and planets; and they predict solar eclipses. It is the largest of the observatories established by Jai Singh II in five principal Hindustan cities; others were in Delhi, Ujjain, Mathura, and Varanasi (Benares). Only two survive: the one at Mathura was quarried for its stone and those at Ujjain and Varanasi are partly in ruins. Jantar Mantar is a remarkable architectural achievement: large build-

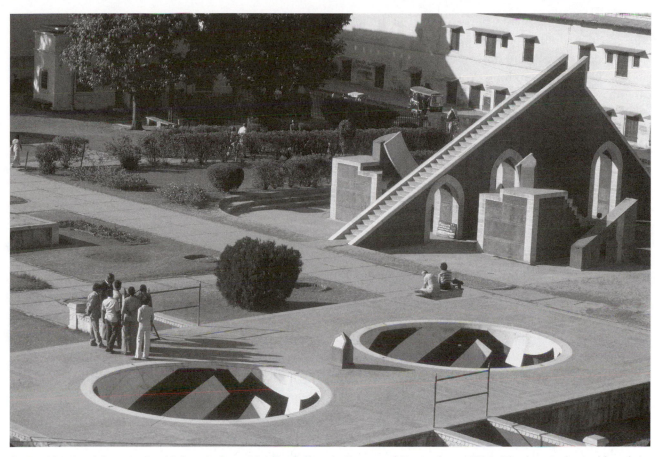

Jantar Mantar, Jaipur, India; Maharaja Sawai Jai Singh II and others, architects, from 1728. The hemispherical bowls in the foreground together form the armillary sphere, and the ramped device at the right is the the equatorial sundial.

ings constructed with such exactness that they can be used as scientific instruments.

Jai Singh II, a member of the Hindu Kachhawaha dynasty, came to power at the age of thirteen. As well as being a capable general, he was so politically and intellectually gifted that the Mogul emperor Aurangzeb conferred on him the title of Sawai (literally, "a man and a quarter"). Mogul power was declining toward the end of the 1720s, but Jai Singh's kingdom was prospering. The water supply in his fortified hillside capital, Amber, was strained by increasing population, so he moved his seat of government to the plains. In 1727 he commissioned the Bengali architect Vidyadhar Bhattacharya to design a new walled city about 125 miles (200 kilometers) southwest of Delhi and named it Jaipur. Unlike the laissez-faire contemporary north Indian cities, Jaipur's plan was based on urban design principles found in the Hindu architectural treatise, the *Shilpa Shastra*. The city was divided by a right-angle grid of wide primary and secondary streets, and further by lanes and alleys, into seven rectangular zones following the caste system, related to occupations and trades. The central rectangle housed the royal complex—the palace, administrative buildings, the women's palaces, and the Jantar Mantar.

Jai Singh II was interested in religion and the arts and sciences and his court became a magnet for savants, artists, and philosophers. He was especially interested in astronomy and acquired a multilingual library on the subject, including the works of Ptolemy and Euclid, Persian and Hindu astronomers, and modern European and Muslim sources. Beginning in 1728, he built the Jantar Mantar in Jaipur. Within high walls on three sides, the observatory covers an area of about 5 acres (2 hectares). It contains fifteen astronomical instruments built of local stone and marble. Six had solar measurement functions, eleven were for observing the night sky, and one was unfinished. These large, architecturally refined devices, capable of achieving much greater accuracy than small brass instruments, were based on Islamic astronomical theories. Most were derived from those commissioned by the fifteenth-century Byzantine ruler Ulugh Begh for the well-equipped observatory built in Samarkand in 1428.

The largest instrument at Jaipur is the equatorial sundial, a 90-foot-long (27.5-meter) straight ramp pointing toward the celestial pole. Graduated masonry quadrants on each side are centered on the nearest edge of the ramp, whose shadow marks local solar time to an accuracy of a few seconds. It was also used to determine the celestial longitude of the sun and to establish the exact time of the equinoxes. The design of another instrument, the Jai armillary sphere, has been attributed to Jai Singh II himself. It comprises two marble hemispherical bowls, each about 13 feet (4 meters) in diameter, set into the ground; their surfaces are inscribed with coordinate lines of celestial latitude and longitude. A small ring was suspended on wires over the exact center of each, and during the day its shadow marked the exact position of the sun. At night an observer could enter a room under the bowls to take sightings on the stars. The two bowls are complementary, and alternating their use within a two-hour changeover allowed continuous observation. There are also several sundials: a vertical one, hemispherical ones, and a smaller equatorial one that can measure time to about 20 seconds' precision. Twelve smaller zodiacal instruments—one for each sign—and similar in design to the equatorial sundial, were used for observing the latitudes and longitudes of the sun and the planets. There are also two sets of tall rectangular columns arranged in circles and calibrated to allow reading of the altitude and azimuth of celestial bodies.

Finally, the astrolabe, a star chart engraved in a metal disc, is about 6.5 feet (2 meters) in diameter— six or seven times the usual size of contemporary examples—and made of a seven-metal alloy that Jai Singh had developed to minimize variations caused by temperature changes. Adjustable rulers allow the calculation of rising and setting points of the stars and planets for the accurate casting of horoscopes. That esoteric function underlines a fact that may become obscured as we marvel at the mathematical sophistication of the Jantar Mantar. It is simply this: that despite Jai Singh II's erudition and urbane universalism, his great observatory and the others like it sprang in part from a religious and not a purely scientific source. In excellent repair after being reconstructed by Chandra Dhar Sharma Guleri in 1901, the Jantar Mantar at Jaipur was declared a national monument in 1948.

See also Taj Mahal

Further reading
Bhatnagar, V. S. 1974. *Life and Times of Sawai Jai Singh, 1688–1743*. Delhi: Impex India.
Kaye, George Rusby. 1973. *The Astronomical Observatories of Maharaja Jai Singh*. Varanasi: Indological Book House.
Sharma, Virendra Nath. 1995. *Sawai Jai Singh and His Astronomy*. Delhi: Motilal Banarsidass.

K

King's College Chapel
Cambridge, England

The architectural historian G. E. Kidder Smith correctly identifies King's College Chapel as "one of the great rooms in architecture." Initiated by King Henry VI in July 1446, it was not completed until 1537. Even then, it was acknowledged by many to be one of Europe's finest late-medieval buildings. It was an architectural achievement in that it epitomized the English High Gothic, its filigreed stone frame, large windows, and exquisite fan vaulting all demonstrating the pinnacle of structural refinement that had taken almost 400 years to achieve.

Henry VI (1421–1471), described as a "a pious and studious recluse" incapable of governing, succeeded his father Henry V as king of England in 1422. Just a month or so after the infant monarch ascended the English throne, he was also proclaimed king of France. Interrupted by the Wars of the Roses in 1461, his reign resumed in 1470, only to be cut short by his murder the following May. When he reached the age of sixteen he was deemed old enough to rule for himself and, despite a reputedly rebellious youth, by the time he was nineteen Henry had grown to be religious. Neglecting matters of government, he turned his attention to the establishment of two educational foundations: Eton College near Windsor (1440–1441) and the Royal College of the Blessed Virgin Mary and St. Nicholas of Canterbury (now known as King's

College) at Cambridge University (1441) provided for seventy scholars drawn from Eton. Henry set out detailed instructions for both colleges and at both his primary concern was for the construction of a chapel. One writer has obsequiously observed that the king's "selfless piety accounts for the form of the chapel at King's, which was conceived ... as a personal testament of faith." Certainly, King's College Chapel had no precedent in other university colleges; its design has more links with the choirs of the great English cathedrals.

Henry VI laid the first stone of King's College on Passion Sunday 1441, and over the next three years he proceeded to compulsorily acquire real estate in the center of Cambridge. Houses, shops, and even a church were demolished as the land was cleared for his grand scheme—a great court of which the chapel was to form the north side. In the event, only the chapel would be built. Henry laid its foundation stone on 25 July 1446; the architect was his master mason, Reginald of Ely. The Wars of the Roses impeded progress, and Henry's chapel was left to be finished by others.

The construction was spasmodic, to say the least. It resumed in 1476 under Edward IV and his architects John Wolrich and Simon Clerk (after 1477) but again ceased when Henry VII succeeded to the throne in 1485. The fellows of the college soon drew his attention to the fact that "the structure magnificently

King's College Chapel, Cambridge, England; Reginald of Ely, John Wolrich, Simon Clerk, and John Wastell, architects, 1446–1537. Interior, showing Wastell's fan-vaulted ceiling. From an illustration of 1815.

begun by royal munificence now stands shamefully abandoned." When the king visited Cambridge in 1506, only five bays had been built, but they had a timber roof, and the open end was boarded and decorated with the arms of the Knights of the Garter painted on paper. Prompted by his mother, and for political reasons, Henry VII decided to finish the chapel. In 1508 work recommenced on a large scale. Although Henry died the following year, his will provided for the work to be completed. By 1512 the stone frame was finished, and his executors found extra money for the magnificent fan vaulting—called by some "the noblest stone ceiling in existence"—designed by the master mason John Wastell. Within three years the structure was complete, and the painted-glass main windows—the most complete set to survive from Tudor times—were finished in 1537. The latter works had been executed under Henry VIII,

and when he died in 1547, King's College Chapel was internationally recognized as an architectural masterpiece.

Much of the unique early design of the chapel can be ascribed to Reginald of Ely, who continued to supervise the work until 1461. The interior is a single vast space 289 feet (88 meters) long and 40 feet (12.2 meters) wide, under a soaring, 80-foot-high (24.4-meter) vault. Some scholars believe that the fan vault was proposed to replace a much simpler lierne vaulting system during master mason Simon Clerk's appointment. King's College Chapel has no side aisles, but ranges of minor chapels and vestries are accommodated between the deep buttresses on the north and south sides. The only subdivision of the entire space, and that just in part, is made by a half-height choir screen, above which the intricate forms of the high vaulting can be seen marching in stately procession toward the altar. The carved oak screen (ca. 1531–1536), in the uninformed mimetic manner of the early English Renaissance, was commissioned by Henry VIII; it is emblazoned with his monogram and Anne Boleyn's. Its clumsy design provides an apt foil for the high refinements of English Gothic architecture seen everywhere else in the building. Almost 70 percent of the walls above the dado (that is, all except the buttresses) are made of painted glass, making the huge interior light and airy and accentuating the stone lacework of the vaults.

Further reading

Heyman, Jacques. 1995. *The Stone Skeleton: Structural Engineering of Masonry Architecture*. Cambridge, UK: Cambridge University Press.

Leedham-Green, Elisabeth. 1996. *A Concise History of the University of Cambridge*. Cambridge, UK: Cambridge University Press.

Leedy, Walter C. Jr. 1980. *Fan Vaulting: A Study of Form, Technology, and Meaning*. Santa Monica, CA: Arts and Architecture Press.

The Krak of the Knights
Syria

Once described as "the key of Christendom," the concentric castle known as the Krak of the Knights stood on the 2,000-foot-high (611-meter) southern spur of the Gebel Alawi, commanding the strategic Homs Gap in the Orontes Valley between Syria's Mediterranean coast and the hinterland. The easternmost in a chain of five castles, it was well placed to control the trade routes between Asia Minor and the Levantine Coast. The formidable fortress represented the height of achievement in medieval military architecture and was described by Lawrence of Arabia as one of the "best preserved and wholly admirable castles in the world."

Medieval warfare was a cycle of conquest and consolidation. Builders were as important as soldiers to an army and throughout the religious wars known as the Crusades (1096–1291) both sides built scores of fortified strongholds, the ruins of which can be found throughout the Middle East. In 1095 Pope Urban II decreed that he would absolve anyone who fought to reclaim the Holy Land for Christendom, a promise that ignited two centuries of conflict. On the face of it, there was a religious reason—pilgrims could not reach Jerusalem—but Urban II's decision was also prompted by a combination of ulterior political motives. The Byzantine Empire was staggering in the face of Turkish expansion; European feudal lords were anxious to profit from their military strength, and some states wanted to exploit their naval might in the Mediterranean. And there was opportunity for the papacy to make the most of rising religious fervor to gain control of the mind of western Europe.

Kings and barons squandered the lives and the wealth of their subjects as they led all social classes against Islam. From time to time the Crusaders controlled parts of Turkey, Syria, Lebanon, and Palestine, capturing Jerusalem from the Seljuk Turks in 1099 and holding it until Salâh al-Dîn regained it in 1187. An earlier Islamic Castle of the Kurds on the site of the Krak was taken, albeit temporarily, by Raymond de St.-Gilles in 1099, and he again laid siege to it, without success, in 1102. Tancred of Antioch occupied it permanently from 1110, and thirty-two years later the Count of Tripoli gave it to the Knights Hospitallers. They invested a great deal of wealth and skill to develop it into "the most distinguished work of military architecture of its time."

The Krak comprised almost 8 acres (3 hectares) and incorporated various buildings in an inner bailey

The Krak of the Knights, Syria, ca. 1110 to early thirteenth century.

with a high curtain wall. That was in turn surrounded by an outer bailey within a second, slightly lower curtain wall—a consistent feature of the concentric castles of the period. The inner bailey was built probably soon after 1110, but it was extensively repaired and refurbished in the late twelfth and early thirteenth centuries after a series of earthquakes after 1157. The later builders thickened the curtain wall and added a glacis and water-filled moat. They also reinforced the towers. The outer bailey, with a complicated, sharply angled (and therefore highly defensible) eastern gateway, was built in stages. The inner curtain, with rectangular towers that hardly projected from the wall, dominated the outer and stood quite close to it. The 16-foot-thick (4.9-meter) outer curtain had eight projecting round towers on the north and west sides, and their loopholes covered every direction. Two towers protected the north barbican, or fortified gateway, from which a narrow ramp led to the inner bailey. All the walls, inner and outer, had castellated galleries, and there were extensive machicolations. The portico of the inner bailey was added around 1250. If necessary, up to 4,000 soldiers and 300 knights with their horses and equipment could be garrisoned in the Krak, and enough provisions could be stored in its warehouses, stables, and cisterns to resist a five-year siege. But early in the thirteenth century only 2,000 men occupied it, with provisions for just one year. Throughout Crusader times, there was a dependent *burgus* (walled suburb) associated with the castle.

The castle was also used as base from which to harass the surrounding country held by Islam. Called by one Muslim writer "this bone in the throat of the Moslems," it easily withstood attacks from Nûr al-Dîn in 1163 and 1164 and by Salâh al-Dîn in 1188. It finally fell in April 1271 after a siege of only a month by the Mameluke sultan, Baybars. He had breached the outer curtain but had no way of overcoming by force the second line of defense with its three massive towers and talus. It was through a forged document commanding the Hospitallers to surrender that the Krak of the Knights was finally taken—by treachery and not by power. In 1291 the Crusaders were expelled from Acre, their last stronghold, and they withdrew to several Mediterranean locations until Turkish expansion in the sixteenth century. The important positive result of the years of war was that cultural exchange with the more scientifically advanced Islamic world contributed much to the enlightenment of Western civilization.

See also Dover Castle

Further reading

Boase, T. S. R. 1971. *Kingdoms and Strongholds of the Crusaders*. London: Thames and Hudson.

Oggins, Robin S. 1995. *Castles and Fortresses*. New York: Metrobooks.

Riley-Smith, Jonathan. 1997. *The Oxford Illustrated History of the Crusades*. Oxford, UK: Oxford University Press.

L

Lal Quila (the Red Fort)
Delhi, India

Lal Quila (the Red Fort) was built between 1638 and 1648 at the command of the Mughal emperor Shah Jahan (who also built the Taj Mahal) as the royal residence in his new capital, Delhi. The fort, representing the highest achievement of Mughal architecture, contained all the accoutrements befitting a center of empire: public and private audience halls, domed marble palaces, luxuriously appointed private apartments, a mosque, and exquisite gardens. Much of the opulence has gone, but in its heyday its magnificence would have been unparalleled, as boasted by an inscription on one of its walls: "If on Earth be an Eden of bliss, it is this, it is this, none but this."

Delhi stands at the western end of the plain of the Ganges. The epic *Mahabharata* speaks of it as a thriving city built about 1400 B.C., although archeological reality suggests it was settled about 1,000 years later. The first city *named* Delhi was founded in the first century B.C. by Raja Dhilu; southwest of the modern location, it had six successors. Its Mughal history is relevant here. In 1526 Babur, the first Mughal ruler, established Delhi as the center of an empire that would unite vast areas of south Asia for the next two centuries. His son Humayun built a new city near Firuzabad but it was leveled when Afghan Sher Shah Suri overthrew him in 1540. He built a new capital, Sher Shahi, as the sixth city of Delhi.

Once more eclipsed when the emperors Akbar and Jahangir moved their courts elsewhere, Delhi reached prominence, even glory, in 1638, when Akbar's grandson Shah Jahan moved his capital from Agra to establish the seventh city of Delhi: Shahjahanabad, now known as Old Delhi. Most of it is still embraced by Shah Jahan's walls, and four of its seventeenth-century gates still stand. He also built Lal Quila as the royal residence within the new city.

Almost immediately, Shah Jahan commissioned the architects Ustad Hamid and Ustad Ahmad to design a fitting royal residence—the Red Fort—at the northeastern corner of Shahjahanabad. It was completed within about ten years. An area of 124 acres (50 hectares) was enclosed within 1.5 miles (2.4 kilometers) of formidable defense walls. It was flanked by the Yamuna River on the eastern side, which fed a moat 76 feet (22.8 meters) wide and 30 feet (9 meters) deep. Thick red sandstone walls (from which the fort derives its name), punctuated by turrets and bastions, rose 60 feet (18 meters) from the river; those on the other side stood up to 112 feet (33.5 meters) above the surrounding terrain. Two of the six main entrances—the Lahori Gate and the Delhi Gate—survive. Now the moat is dry and the Yamuna flows almost a kilometer away, but Lal Quila towers above the modern city of Delhi that spreads out to the west.

The buildings within the walls are all carefully arranged on the long north-south and shorter east-west

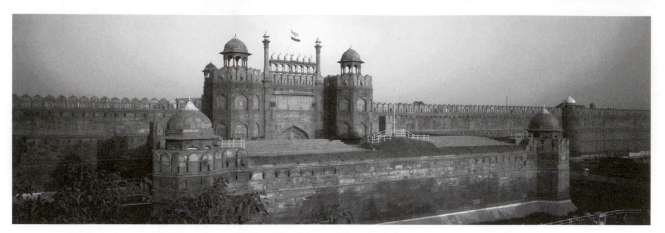

Lal Quila (the Red Fort), Delhi, India; Ustad Hamid and Ustad Ahmad, architects, 1638–1648.

axes of the octagonal plan. Although they reveal the delicate work that can be found in all Mughal architecture, they exemplify the later phase of the style, characterized by the increasing use of marble, elaborate floral decoration of external surfaces, and the proliferation of tall minarets and bulbous domes. Shah Jahan seems to have preferred the flowing plant motifs inspired by the European sixteenth-century herbariums that had been perfected by his father's artists. The walls of carefully cut marble were patterned with precious and semiprecious stones and surfaces were decorated with inlaid flowers of hard stones in many colors.

Immediately inside the fortified Lahori Gate was the Chatta Chowk, a vaulted two-story arcade containing thirty-two shops. East of it, on the same axis, was another gate called Naubat Khana (Drum House), also two stories high, from which musicians played martial music for the emperor five times a day, or announced the arrival of important guests. Further east on the axis and across a courtyard stood the Diwan-i-Am (Public Audience Hall), ornamented with gilded stuccowork and hung with heavy curtains. There the emperor, seated in a canopied, marble-paneled alcove set with precious stones, would hear through his prime minister the complaints and petitions of the commoners. The Diwan-i-Am was also used for state functions. At the eastern terminus of the short axis of the plan stood the Rang Mahal—(Palace of Colors), its roof crowned with gilded turrets. It housed the emperor's wives and mistresses.

The interior was richly decorated with painting. Its ceiling, overlaid with silver and gold, was reflected in a pool in the marble floor. The Nahr-i-Bihist (Stream of Paradise) flowed through its center, feeding small water channels that flowed to cool the other rooms of the Red Fort.

The north-south axis, through the center of a courtyard that separated the Diwan-i-Am and the Rang Mahal, was flanked by sumptuous pavilions. In the Diwan-i-Khas (Hall of Private Audiences), the emperor met with his courtiers and dignified guests. Standing on a plinth and supported by thirty-two pillars, the white marble hall was decorated with floral patterns of precious stones. At its center the fabled Peacock Throne (carried off to Persia in 1739) stood on a white marble dais under a ceiling inlaid with silver and gold. South of that building lay the emperor's private apartments, the Khas Mahal. On their east side was a large sitting room that opened to a cantilevered gallery, where each sunrise the emperor appeared before his subjects. At the northern end of the large square in front of these buildings stood the Hammam (Royal Bath). Built of marble and extravagantly decorated with inlay, glass, and paint, it comprised three apartments that were also used for private meetings. Shah Jahan's son Aurangzeb built the Moti Masjid (Pearl Mosque) within an enclosing wall beside the Hammam in 1659–1660. At the northern end of the long axis stood a three-story octagonal tower, Shah Bhurj—the shah's private working area. At the southern end Shah Jahan built the Mumtaz Mahal, a palace for his favorite daughter Jahanara Begum.

Mughal power waned in the eighteenth century. The British captured Delhi in 1803, and the city was the focus of India's first war of independence—the British still prefer to call it the Indian Mutiny—in 1857. In 1911 the colonials moved their imperial capital from Calcutta to Delhi and began to build the eighth city, New Delhi, officially inaugurated in 1931. India finally expelled the British in 1947, and the nation celebrates its liberty by flying the Indian flag above Lal Quila each 15 August, Independence Day.

Further reading

Blake, Stephen P. 1991. *Shahjahanabad: The Sovereign City in Mughal India, 1639–1739.* Cambridge, UK: Cambridge University Press.
Nicholson, Louise, and Francesco Venturi. 1989. *Red Fort, Delhi.* London: Tauris Parke Books.
Spear, Thomas George Percival. 1994. *Delhi, Its Monuments and History.* Oxford, UK: Oxford University Press.

Lalibela rock-hewn churches
Ethiopia

Lalibela is a village in the mountainous Welo region of northern Ethiopia, about 440 miles (700 kilometers) north of Addis Ababa; in the Middle Ages it was known as Roha and was the capital of the Zagwe dynasty. Standing on a rock terrace at an elevation of 8,500 feet (2,600 meters), it is the site of eleven large rock-hewn monastic churches that date from the late twelfth and early thirteenth centuries. Each is architecturally distinctive and all are finely carved inside and out. Declared a UNESCO World Heritage Site in 1978, they are not the earliest such churches in Ethiopia (others predate them by at least 500 years), but they are widely recognized as the most beautiful. Francisco Alvarez, a Portuguese Jesuit missionary, visited Lalibela in the 1520s, the first European to see the churches. He was reluctant to report to his superiors, fearing that they would not believe his account of buildings "unlike any to be seen elsewhere in the world." Nevertheless, he described them, "hewn entirely out of the living rock, which is sculpted with great ingenuity." The cultur-

ally unique churches are remarkable for that reason: each has been cut from the purple-red volcanic tufa, in some cases 90 feet (27 meters) into the ground. Some of them are connected by tunnels or passageways open to the sky. Even to the modern mind, they are an architectural marvel.

The history of the churches is swathed in mythology. It is probable that King Lalibela (1181–1221) commissioned them. According to legend, angels carried him to heaven when he was affected by a poison that his envious brother had administered; God sent him back to earth with instructions to build the churches and later dispatched angels to continue the work at night. Another account says that the king recruited Indian, Arab, and Egyptian builders, or even "white men" from Jerusalem, a link that is strengthened by the naming of the local river, Jordan. It has been suggested that, upon learning that the Holy City had fallen to Islam, Lalibela wanted to create a "new Jerusalem" in his secure mountain fastness. Tradition has it that the eleven buildings were completed in twenty-four years—archeologists calculate that would have needed 40,000 workers—but the time frame seems too short. Maskal Kabra, Lalibela's queen, is said to have built one of them to his memory.

The churches stand in two groups flanking the Jordan. Four of them—Bet Medhane Alem, Bet Maryam, Bet Amanuel, and the cruciform Bet Ghiorghis, dedicated to Ethiopia's patron saint—are in effect huge blocks of sculptured stone standing in deep excavated courtyards and attached to the rock only by their bases. *Bet* signifies "the house of." They look like normal buildings, but each one is a single piece. The others must be accurately described as *semi*monolithic, because they remain attached to the rock by at least one face, whether the roof or walls. For example, although the twin churches of Bet Golgotha and Bet Qedus Mikael share a roof, they have, respectively, one and three facades exposed. Bet Abba Libanos is isolated from the mother rock except for its roof, which is integrated with the overhanging cliff; in front of it stands a large forecourt, cut from the tufa. The other churches are named Bet Danaghel, Bet Debre Sinai, Bet Gabriel-Rufa'el, Bet Merkorios, and Bet Meskel.

The church of Bet Ghiorghis, Lalibela, Ethiopia, ca. 1181–1221. View of the roof.

The eclectically blended artistic influences are varied—Greek, Egyptian, and even Islamic—and the nature and the extent of the carefully carved exterior and interior walls, ceilings, moldings, and window tracery are just as diverse. Bet Qedus Mikael has smooth exterior wall surfaces, and its interior is austere, decorated with Greek crosses; on the other hand, Bet Golgotha is more ornate, perhaps because it houses the tomb of King Lalibela, and it contains bas-reliefs of saints, the only sculptures in any Ethiopian church. Other churches have painted decoration, mostly with a teaching function, in various states of preservation.

The churches of Lalibela are home to hundreds of monks, clerics, and students, who celebrate liturgies that are the same as they were eight centuries ago. It is the most important pilgrimage site in Ethiopia, a country that includes an island of Christianity in a sea of Islam, and during the major holiday seasons it may be visited by as many as 50,000 devotees. More recently, Lalibela has become a tourist attraction, precisely because of its spectacular churches, and draws over 10,000 secular visitors a year. Inevitably, there is a tension between conservation and development. But because tourism is the village's only real source of wealth and is encouraged by the central government, a compromise must be reached. In 1996 the European Community earmarked EUR4.7 million for shelters to replace the corrugated-steel roofs that covered Bet Medhane Alem, Bet Maryam, Bet Meskel, Bet Amanuel, and Bet Abba Libanos from damage caused by torrential rains, and an international architectural competition was held. Structures designed by the first-prize winners, Teprin Associati of Italy, were completed by December 2000. UNESCO and the Ethiopian Department of Preservation of Cultural Heritage are urging restoration of the deteriorating fabric of the churches.

Further reading
Bidder, Irmgard, and Elfriede Fulda. 1960. *Lalibela: The Monolithic Churches of Ethiopia*. New York: Praeger.

Larkin Administration Building
Buffalo, New York

The Larkin Administration Building (1902–1906) by Frank Lloyd Wright (1869–1959) was his first major public work, built, as he said, "to house the commercial engine of the Larkin Company in light, wholesome, well-ventilated quarters." It was a milestone in the history of commercial architecture, in terms of both its spatial organization and the exploitation of modern technology. Indeed, some historians identify it as the twentieth-century structure that, more than any other, changed the face of architecture; within a few years it was hailed in Europe. Peter Blake has claimed that it was the "first consciously architectural expression of the kind of American structure which Europeans were beginning to discover to their delight: the great clusters of grain silos and similar industrial monuments that [they] found so exciting in the early 1920s." (Blake 1964, 55–56). The Larkin Company's soap-manufacturing and mail-order operations occupied a large urban industrial site between Swan, Exchange, Van Renssalear, and Hamburg streets of Buffalo, in western New York State. Wright's innovative building on Seneca Street, near the corner of Seymour and Swan, housed the firm's administrative functions.

Around 1902 Wright realized that different building types called for different esthetic systems. Thereafter, he developed two patently distinct architectures. In his houses he pursued what might be called prairie horizontality—the line of repose that reached its best expression in the Frederick C. Robie House, Chicago (1908–1910). For nondomestic buildings, such as the Larkin Building; Unity Temple, Oak Park, Illinois (1905–1909); and Midway Gardens, Chicago (1913–1917), he adopted "Cubic Purism," often squat and squarish with symmetrical plans and elevations. The rather severe exterior of the Larkin Building was relieved with sculpture by Richard Bock, who produced a globe of the world, supported by celestial beings and emblazoned with the company name.

The great six-story space in the center of the building—today we think of it as an atrium—was lit by a large skylight. It was surrounded by balconies; lit by high-level windows around the perimeter of the building, they contained the general office spaces, set out (years before their time) on an open plan. In keeping with Wright's views about the nature of work, and no doubt with those of his client John D. Larkin, the interior espoused nonhierarchical, democratic office planning. There was even an employees' lounge with a piano, where the company provided a weekly lunchtime concert for the workers; an organ stood at one end of the third story of the atrium. Many of the 1,800 employees worked at long desks running between the outer walls and the atrium. The lighting was an important part of the design; the desks received daylight from two sides: the exterior windows and the atrium. Electric lamps were mounted at the ends of the tables in the ground floor of the central court so that every office worker had well-balanced, shadow-free light. Wright believed in making total architecture and designed the lighting system himself, as well as the steel office furniture. The employees were protected from industrial pollution and the noise of the nearby rail yards by heavy red brick walls, and from undue interior noise by sound-absorbent surfaces. The revolutionary working environment was also air-conditioned, one of the first in the United States.

Just as he separated service rooms from living rooms in his contemporary houses, Wright gathered the services—electrical and plumbing ducts, stairways, toilets (he introduced wall-hung water closets to make cleaning easier), and heating systems—at the outer corners of the main building. "Beating the box" (as he put it), he expressed the service functions as square towers, "freestanding, individual features."

Responding to criticisms by Russell Sturgess of *The Architectural Record*, who called it "an extremely ugly building" and "a monster of awkwardness," Wright said in 1908, "It may be ugly … but it is noble. It may lack playful light and shade, but it has strength and dignity and power." He went on to claim: "It is a bold buccaneer, swaggering somewhat … yet acknowledging a native god in a native land with an ideal seemingly lost to modern life—conscious of the fact that because beauty is in itself the highest and finest kind of morality so in its essence must it be true." His opinions, even if a little arrogant, were confirmed by the great Dutch architect H. P. Berlage, who spoke of the Larkin Building to attentive European audiences. To call it Wright's magnum opus (exclaimed Berlage) "was not to say enough." It was a building without equal in Europe, and there was "no office building [there] with the same monumental power."

The Larkin Company went into decline in the 1930s, and within a decade its world-famous Administration Building was being used as a showroom. In 1949–1950, for "mysterious and untraceable reasons," it was pointlessly demolished, brick by brick. Today, only a single pier remains—the site was never redeveloped but used as a parking lot—and in 1997 the outline of the building's footprint was painted where once stood one of the most important achievements of twentieth-century architecture.

See also Frederick C. Robie House

Further reading

Blake, Peter. 1964. *Frank Lloyd Wright: Architecture and Space*. Baltimore: Penguin.
Quinan, Jack. 1987. *Frank Lloyd Wright's Larkin Building, Myth and Fact*. New York: Architectural History Foundation; Cambridge, MA: MIT Press.
Wright, Frank Lloyd. 1994. *Frank Lloyd Wright: The Early Work of the Great Architect*. New York: Gramercy Books.

London Underground
England

London's underground railroad system, popularly known as "the Tube," is the oldest in the world. As early as the 1830s Charles Pearson, the city of London's solicitor, suggested that the mainline stations could be linked by an underground railroad with as many as eight tracks. Despite the potential economic and social advantages of the scheme, it could find no financial backing, and Parliament refused to approve it. The city's first above-ground passenger service was the London and Greenwich line, opened in February 1836. Within four years it was carrying nearly 6 million passengers annually between the major mainline train stations on the borders of the metropolis and the edge of the central business district. With an area of 60 square miles (154 square kilometers) and a population of 2.5 million, Greater London was then the world's largest city, and the most crowded, plagued by street congestion.

To find a solution to a worsening problem, the City Terminus Company (CTC) revived the underground railroad idea in 1852 and placed it before Parliament, only to again fail. The following year the Bayswater, Paddington, and Holborn Bridge Railway Company submitted a plan for a different line, ostensibly at half the cost. Parliament endorsed the North Metropolitan line in 1853, and the company promptly had the CTC line approved as part of its own. The Great Western Railway Company agreed to finance construction of the underground in return for direct access to the city. In 1854 an act of Parliament was obtained to begin the Metropolitan Railroad. A sum of £1 million was raised by December 1859, and the following February the first shafts were sunk. The earliest tunnels were made by the "cut and cover" method: a deep trench would be excavated, side walls and roof built, and the ground surface backfilled. The process was expensive and slow, and it created chaos along the route of the railroad, not least of which was the dispossession of citizens and the demolition of buildings, often the homes of the poor. The first trial run was on 24 May 1862, and on 10 January 1863 the Metropolitan Railway opened, the world's first underground line, between Bishop's Road, Paddington, and Farringdon Street. There were 38,000 passengers on that first day, and from that moment the London Underground began to grow. In 1868 the first section of the Metropolitan District Railroad from South Kensington to Westminster was opened.

It was soon realized that a citywide underground network must eventually pass beneath the River Thames. "Cut and cover" methods would not be appropriate to build such lines, but an "old" technology was already in place. Completed in 1843, Marc and Isambard Brunel's Thames Tunnel had been dug using the former's tunneling shield, patented in 1818. The machine had been improved in fifty years, and the engineer James Henry Greathead finally built a lighter and (more importantly) circular version. In 1870, with one Peter Barlow, he drilled the 6-foot-diameter (1.83-meter) Tower Subway Tunnel from Tower Hill to Vine Lane. Its system of elevators and a twelve-seat car, all wound by steam-operated wire cable, was unreliable, and within months it was reduced to a pedestrian passage. Although extremely short-lived, it was the first tube railroad, and the construction method obviated all the disadvantages of "cut and cover." Greathead's tunneling machine had a diaphragm within which segments of the cylindrical, cast-iron tunnel lining were bolted together as the excavator was advanced hydraulically; the gap between the excavation and the lining was filled with cement grout. Because it was circular in cross section, the tube was structurally stronger.

The next route to be completed was the Circle line in 1884. At that time all trains were drawn by steam locomotives, filling the tunnels with smoke and fumes. Steam trains could not operate in the deeper tunnels, and after considering cable-hauled cars, the decision was made to employ electric traction. Most of the transition took place in the first decade of the twentieth century, although the world's first successful electric tube route, the City and South London Railway, was opened in December 1890. In 1902 an American, Charles Tyson Yerkes, financed the expansion of the network and by 1907 five new lines—Central, Northern City, Bakerloo, Piccadilly, and Charing Cross Euston and Hampstead—were opened, and electrification proceeded. Yerkes formed the Underground Electric Railway Company of London (known as the Underground Group). Between 1902 and 1905, they built the world's largest power station, at Chelsea, to electrify the District line. Powering the Tube for almost a century, it was closed in 2000 when the Underground moved to the national grid. By 1913, mergers had brought all lines except the Metropolitan into the group.

Underground services expanded from 1907 through the 1930s. In 1933 the Underground Group and the Metropolitan Railway were subsumed by the London Passenger Transport Board, which managed all public transport systems in the London area. Following World War II (when no fewer than eighty Underground stations served as air-raid shelters for Londoners), the Passenger Transport Board was nationalized and renamed the London Transport Executive, which in turn became the London Transport Board. More administrative changes began in May 2000, with the establishment of Transport for London, an executive body of the Greater London Authority.

In September 1968 the first section of the Victoria line was opened, and extensions were completed by 1971. In May 1979 the Jubilee line opened, bringing the total number of routes beneath London to eleven: Bakerloo, Central, Circle, District, East London, Jubilee, Metropolitan, Northern, Piccadilly, Victoria, and Waterloo and City. Upgrades and improvements continue. Recently, computer signaling was introduced; the Central line was modernized and the Victoria line converted to automatic operation. The most significant addition to the complex system, begun in 1993, was the construction of the £1.9 billion (U.S.$2.8 billion) Jubilee line extension, the largest engineering project undertaken in Europe since the Channel Tunnel. Completed in May 1999, the new route from Westminster Station to Stratford via North Greenwich (to serve the Millennium Dome) involved negotiating the already crowded undercity with its myriad railroad tunnels, cables, drains, and service ducts, as well as overcoming subsidence problems.

Nearly 80 percent of Londoners working in central London travel to work on public transport, most of them on the Tube. Trains traveling at an average speed (including stops) of 20.6 mph (33 kph) move a total of almost a billion passengers annually over a multilevel underground network—some tubes reach 221 feet (67.4 meters) deep—that extends 45 miles (72 kilometers) east to west and 28 miles (45 kilometers) north to south. The first underground railroad in the world, which began with a track a mere 3.57 miles (5.7 kilometers) long, now covers 250 miles (392 kilometers); 42 percent of that is in tunnels.

Further reading

Day, John Robert. 1979. *The Story of London's Underground*. London: London Transport Executive.

Green, Oliver. 1987. *The London Underground: An Illustrated History*. Shepperton, UK: Runnymede.

Lawrence, David. 1994. *Underground Architecture*. Harrow, UK: Capital Transport.

M

Ma'dan reed houses
Iraq

The reed houses that form part of the distinctive culture of the Ma'dan, or Marsh Arabs, of southeastern Iraq are an architectural achievement because they result from pushing available resources to their limits. Descended partly from the ancient Sumerians and Babylonians, this seminomadic people, now numbering perhaps 200,000, have for millennia inhabited Lake Hammar and the surrounding marshlands in the Tigris-Euphrates Delta, about 200 miles (320 kilometers) south of Baghdad. Not only have they developed a sophisticated house form using a single building material—the stalks of the prolific giant reed *(Fragmites communis)*—but they have also created the very land upon which their houses and farmsteads stand.

The Ma'dan villages are irregular clusters of small islands constructed by alternating layers of reed mats and layers of mud dredged from the marsh bottom. Thus, paradoxically, much of the fertile land is actually floating on the water. Each island has its house and buffalo paddock, and communication between them is by means of narrow canoes *(mashuf)* of bitumen-coated wood, propelled through the shallow water with long poles. The Ma'dan fish, hunt waterfowl and pigs, breed water buffalo, and raise crops of paddy rice and great millet. Many domestic necessities—beds, cots, baskets, and canoe poles—are woven from reeds. In short, until recently the Ma'dan have lived in harmony with the ecosystem of their harsh but bountiful environment.

The reed house *(mudhif)* is constructed around a framework made by tying the giant reeds—they can grow to 20 feet (6 meters) long—to make bundles that taper from about 1.5 feet to 6 inches (45 to 15 centimeters). The thick ends are stuck into the mud floor of the island in opposing pairs and then bent and lashed together, with a substantial overlap at the top, to form a row of parallel parabolic arches, at about 6-foot (2-meter) centers. The builders even use a tripod of bundled reeds as scaffolding for this part of the work. The primary frames are stabilized with closely spaced, much thinner reed bundles (like purlins) around the perimeter of the house. The completed framework is covered with intricately woven split-reed mats to form the integrated walls and roof. The upper parts of the end walls are enclosed with a curtain of the same material, and four or five reed "columns" are erected to support a framework to which a decorative lattice is fixed, always to beautiful effect. Depending on the length of the reeds used for the arches, the house can be 12 feet (3.7 meters) wide; the length is indeterminate, and buildings up to 100 feet (30 meters) have been recorded. Furnishings are sparse: the reed floors are covered with carpets, and there is a clay hearth for making coffee. The distinctive house form has a long pedigree, being illustrated on a clay plaque dating from the fourth millennium B.C. found in excavations of Sumerian Uruk. That fact, and the appearance of vegetable forms in stone, such as Egyptian papyrus and lotus columns, has given rise to the speculation that all columnar architecture in the protohistoric civilizations (and perhaps beyond) springs from such construction.

The unique culture of the Marsh Arabs is in danger; indeed, it may already be beyond help. Largely as a result of their isolation, they have maintained their traditions and were untouched even by Turkish and British colonialism. Because of high evaporation, the marshes have long been regarded as wasteful of water that could be used for irrigation; a major drainage scheme was proposed in a 1951 report drafted by British engineers commissioned by the Iraqi government. In the 1970s Turkey dammed the Euphrates. But the Ma'dan's problems started in earnest after 1980, during the Iran-Iraq War. Within two years Iran regained the territory, including the marshlands, taken earlier by Iraq. The marsh dwellers fled as the Iraqi army sent enormous electrical currents through the water to electrocute invading Iranian soldiers. Saddam Hussein's unrelenting destruction continued after the war.

Following Saddam's defeat in the Gulf War in 1991, southern Iraqi Shi'ite Muslims launched a guerrilla offensive against his Sunni Muslim government. The uprising was crushed, and many rebels sought refuge in the marshes, supported by the Ma'dan, who are also Shi'ite. To flush them out, in 1992 Saddam began to drain the region systematically, using the 1951 British report. Within a year a network of 20-foot-high (6-meter) dikes was preventing two-thirds of the normal water flow from reaching the marshlands, thus turning much of it into expanses of dried mud. Between the Tigris and the Euphrates Rivers, the man-made Saddam River carried floodwaters directly to the Persian Gulf. A third of Lake Hammar dried up, and thousands of Marsh Arabs moved deeper into the surviving wetlands or fled to Iran and elsewhere. Some sources estimate that fewer than 10,000 remain in Iraq, recognized as a "persecuted minority" by the European Parliament, to pursue their traditional lifestyle. To compound the offense of ethnocide, Saddam's actions have caused probably irreversible environmental damage. International organizations such as the UN Human Rights Commission, the Supreme Council of the Islamic Revolution in Iraq, and the International Wildfowl and Wetlands Research Bureau have been watching in alarm, but have been powerless to act.

Further reading

Maxwell, Gavin. 1990. *A Reed Shaken by the Wind*. London: Isis.

Salim, S. M. 1962. *Marsh Dwellers of the Euphrates Delta*. London: Athlone.

Young, Gavin. 1977. *Return to the Marshes: Life with the Marsh Arabs of Iraq*. London: Collins.

Maiden Castle
Dorset, England

The ancient British hill fort now known as Maiden Castle (from *mai-dun*, Celtic for "great hill"), about 3 miles (4.8 kilometers) southwest of modern Dorchester, grew from a neolithic village to become the largest pre-Roman fortress among nearly 1,400 in England. Indeed, it was one of the most extensive in western Europe. Still visible 2,000 years after its massive ramparts were completed, the fort crowns a low saddleback chalk hill south of the Frome Valley. Its strength did not lie (as in the case of others) in its siting, but rather in the sheer size and scale of its fortifications. By the middle of the first century B.C., four rings of ditches and steeply sloping earthen walls, in places as much as 90 feet (28 meters) high and reinforced by timber palisades or drystone structures, occupied an area of 100 acres (40 hectares). Within the defenses, the long axis of the fort is over 0.5 mile (0.8 kilometer) and its inner circumference about 1.5 miles. It was a remarkable engineering achievement, not only in terms of its monumentality, but also because of its organic nature, by which it grew over twenty centuries.

Maiden Castle has a long prehistory, revealed by archeological studies first undertaken by Mortimer Wheeler in 1934–1938; further excavation took place in 1985–1986 under the direction of Niall Sharples. The first earthwork was a neolithic causewayed camp (ca. 4000 B.C.) consisting of a single ditch and bank defending an area of about 12 acres (4.8 hectares). It was followed after half a century by a 1,750-foot-long (537-meter) bank barrow, crossing the center of the fort from east to west. About 1,000 years later settlers built burial mounds on the site, after which it seems to have been abandoned for some time.

After about 700 B.C. various tribes settled Britain, and most of the southwestern region now known as Somerset and Dorset was occupied by the Durotriges. They secured their lands against rival tribes with hill forts: such places as Hambledon Hill, Hod Hill, South Cadbury, Spettisbury Rings, and of course Maiden Hill, which some scholars suggest was their capital. Around 600 B.C. these Iron Culture settlers incorporated the existing earthworks into their own defenses—an earth rampart augmented by a timber palisade—enclosing about 15 acres (6 hectares) at the east end of the saddleback. There was continual growth: limestone walls were added to parts of the ramparts, and it seems that around 450 B.C. a westward extension was constructed. Sometime before the third century B.C., the encircling fortifications were enlarged, and entrances with double gates were constructed at the east and west ends; the entire hilltop—some 45 acres (18.2 hectares)—was secured. The height of the earth walls was increased, perhaps late in the second century B.C., and yet another rampart and ditch were built around the perimeter. Further enlargement took place a century later. Although it may be that not all Dorset hill forts were continuously occupied, and that some were simply used as havens in times of danger, evidence suggests that Maiden Hill was a permanent settlement, and at the middle of the first century A.D. perhaps 5,000 people were living within what they believed to be the safety of its walls. There were made streets, and archeologists have discovered graves, storage pits, and other pits for refuse—it might be said, sanitary landfill.

The Romans launched a full-scale invasion of Britain in A.D. 43, moving westward across the country. The Roman historian Suetonius claims that twenty of the southwest hill forts fell quickly to the II Augusta Legion, come from Strasbourg under the general Titus Flavius Vespasianus (later to become Emperor Vespasian). They reached Maiden Castle within the year. The Durotriges were renowned warriors, accustomed to hand-to-hand combat. At longer range, they used slings and were prepared to defend their town with them: ammunition dumps within the ramparts held a reserve of 40,000 large pebbles brought from Chesil Beach. The Romans chose to turn their war machines against the well-defended east gate, defended by slingers on its four ramparts. Overwhelmed by the weight of numbers and the superior tactics and weapons technology of the invaders—especially the catapults that launched missiles from beyond the slingers' range—Maiden Castle surrendered, although not before offering savage resistance.

After three millennia the huge, spectacular hill fort had become obsolete, and it was abandoned within about thirty-five years. Many of the former inhabitants moved to the new Roman town of Durnovaria (Dorchester), others to the century-old Celtic village in the shadow of Maiden Castle. In about A.D. 370 the Romans built a temple in the precincts of the fort, but it too was abandoned when they withdrew from Britain only 100 years later. The site is now maintained and managed by English Heritage.

Further reading

Peddie, John. 1987. *Invasion: The Roman Conquest of Britain*. New York: Saint Martin's.

Sharples, Niall M. 1991. *Maiden Castle*. London: Batsford/English Heritage.

Wheeler, R. E. M. 1943. *Maiden Castle, Dorset*. Oxford, UK: Oxford University Press.

Maillart's bridges

The Swiss engineer, architect, and artist Robert Maillart (1872–1940) exploited the structural strength and expressive potential of reinforced concrete to generate a modern form for his bridges. By using simple construction concepts he developed graceful structures based on flat or curved reinforced concrete slabs. Amongst his radically innovative ideas were the mushroom slab, the deck-stiffened arch, the open three-hinged arch, and the hollow-box arch. Maillart's biographer David Billington (1997, 2) asserts that the engineer's "elegance arose from structure itself and not from an extraneous idea of beauty."

Taken singly or together, Maillart's bridges are engineering and architectural feats that elegantly demonstrated, as Le Corbusier claimed in *Vers une Architecture* (1923), that engineers recognized (long before architects) that beauty could be achieved through thoroughly defining and solving problems. That new approach to design lay at the foundation of modern architecture.

Maillart studied civil engineering at Switzerland's Federal Technological Institute in Zürich under Wilhelm Ritter, an expert on reinforced concrete. Graduating in 1894, he worked in private and government engineering offices, mostly on railroad, road, and bridge projects.

Unreinforced concrete was first used in 1865 for a multiple-arch bridge on the Grand Maître Aqueduct between the River Vanne and Paris. The invention of *reinforced* concrete is credited to Joseph Monier, a French gardener who in 1867 patented molded planters made of cement mortar reinforced with iron-wire mesh. Over the next decade there followed several bridge patents. Because French law prevented him from building bridges, Monier sold the patents to contractors Wayss, Freitag, and Schuster, who built Europe's earliest reinforced concrete bridges in Germany and Switzerland.

Around the turn of the century the French engineer François Hennebique built reinforced concrete bridges at Millesimo, Italy (1898), and Châtellérault, France (1900).Therefore, while his designs were probably the most elegant, Maillart was not the pioneer of reinforced concrete bridges. In 1902 he established his own firm, specializing in reinforced concrete design and construction. By then he had already built a 100-foot-span (30-meter) bridge over the River Inn at Zuoz; its innovative slenderness and flatness created a stir in professional circles. There followed the 115-foot (35-meter) Thurbridge near Billwil (1903) and another single 167-foot (51-meter) arch across the Rhine near Tavanasa (1905, since demolished), identified by some scholars as marking the birth of a modern architecture that reintegrated art and technology.

Maillart moved his practice to Russia in 1912 and produced a number of factories, warehouses, and office buildings in Riga, Charkov, and Kiev. Following the October 1917 Revolution, he returned to Switzerland and in 1919 set up a consulting design practice in Geneva, later opening branch offices in Bern and Zürich.

From 1925 he built several remarkable bridges in Switzerland of two principal structural types: stiffened-slab arches, such as the 140-foot (43-meter) span over the Val-Tschiel near Donath, of 1925; and three-hinged arches in which the arch, roadway, and stiffening girder were integrated into a monolithic structure, exemplified by, among others, the Schwandbach Bridge (1933) in Bern Canton and the Salginatobel Bridge (1930) over the Salgina Valley in Graubunden Canton. The latter is a hollow, box-concrete arch bridge with a span of 295 feet (90 meters) and a rise of 43 feet (13 meters). The slender arch rib deepens from the supports to the quarter-span points, at which it becomes integral with the concrete road deck and tapers again to the midspan hinge. Salginatobel Bridge is widely regarded as an exceptional monument of modern architecture, a piece of structural art. In 1991 the American Association of Architects and Engineers designated it an International Historic Civil Engineering Landmark. It has also been called "the most spectacular and classic example of [its] type in the world."

Ritter had urged his students to think of shapes and forms that could not be analyzed easily by mathematical calculation. Clearly, Maillart had learned that lesson, and his visual imagination and intuition were a major part of his approach to engineering; just as clearly, his subtly beautiful bridges evidence a thorough comprehension of the nature of forces in a structure. His deep appreciation of the properties and behavior of reinforced concrete permitted him to develop innovative, light sculptural forms. Through his bridges, Maillart, virtually unknown and unacknowledged before about 1930, became internationally famous as a designer of sophisticated concrete structures.

See also Reinforced concrete

Further reading
Bill, Max. 1969. *Robert Maillart: Bridges and Constructions*. New York: Praeger.

Billington, David P. 1990. *Robert Maillart and the Art of Reinforced Concrete.* Cambridge, MA: MIT Press.
———. 1997. *Robert Maillart: Builder, Designer, and Artist.* Cambridge, UK: Cambridge University Press.

Maria-Pia Bridge
Oporto, Portugal

Located at the mouth of the Douro River, Oporto is the capital of northern Portugal and the second-largest city in the country, rising steeply from the deep river valley. In 1875 the railway between Lisbon and Oporto was almost complete, and the final problem facing its builders was crossing the Douro. An international competition attracted only four entries, three from France and one from England. Gustave Alexandre Eiffel's winning proposal for the "transparent" Maria-Pia Bridge was not only the least expensive—two-thirds that of the next tender and only one-third of the highest price—but it also involved revolutionary structural design.

Although Eiffel is best remembered for the Eiffel Tower in Paris, much of his professional life was given to building bridges. Upon his graduation from the École Centrale des Arts et Manufactures in 1855, he was employed by a firm in southwestern France that produced steam engines and railroad equipment. In 1858 it won a contract to erect a railway bridge over the Garonne River near Bordeaux; Eiffel oversaw the construction, which was completed in 1865. The following year he set up business as a "constructor," designing and fabricating metal structural work, especially in wrought iron. After 1872 foreign contracts came his way, and three years later he designed the Maria-Pia railway bridge in Oporto.

Eiffel supported the railroad deck 190 feet (57 meters) above the river with a graceful, filigreed wrought-iron arch spanning 525 feet (160 meters); the approaches to the center span were borne on lacy framed pylons of varying heights to accommodate the sloping banks. Construction started in 1877, and

Maria-Pia Bridge, Oporto, Portugal; Gustave Alexandre Eiffel, engineer, 1877–1879.

the bridge was built in just a year and ten months, without the need for temporary scaffolding directly supported on the ground—a masterly piece of design. The structural system involved several other technological innovations, not least the design analysis methods. Civil engineers already knew how to calculate for statically indeterminate beams, but the force method needed to predict the behavior of this kind of structure, although propounded a decade earlier, had been taken seriously only a year before Eiffel designed the bridge. It has been asserted that this was the first application of the analysis of a statically indeterminate structure other than a beam, and that Eiffel discovered the method by himself.

The pioneer technique was to be used in many large arches, including two in Oporto. The first came soon after: the wrought-iron Dom Luís I Bridge for pedestrian and vehicular traffic (1886), designed by the French engineer Téophile Seyrig. It is noteworthy that it weighed almost twice as much as the 1,800-ton (1,630-tonne) Maria-Pia. The second arch in Oporto was built almost eighty years later: the 900-foot-span (270-meter) reinforced concrete Arrábida Bridge (1963) was designed by the Portuguese Edgar Cardoso. And Eiffel himself reused the design in France: in 1880 Leon Boyer of the Ponts et Chaussées (Bridges and Highways Department), who was aware of the success of the Maria-Pia Bridge, invited him to build a bridge across the La Truyère River near Garabit on the railroad between Marvejols and Neussargues. The 550-foot (165-meter) span Garabit Viaduct, completed in 1884, incorporated all the innovations of the revolutionary Portuguese structure: it comprised a 1,500-foot-long (450-meter) wrought-iron truss girder, carried to the arch on variable height piers and extended by brick approach viaducts to a total length of 1,880 feet (564 meters).

In 1996 UNESCO designated Oporto a World Heritage City. The Maria-Pia Bridge, threatened with demolition after it was replaced by a new rail crossing in 1991, is now safe and awaiting a new use appropriate to its significant place in the history of engineering.

See also Eiffel Tower

Further reading

Loyrette, Henri. 1985. *Gustave Eiffel*. New York: Rizzoli.

Marrey, Bernard. 1984. *Gustave Eiffel, the Engineer Who Built the Statue of Liberty, the Garabit Viaduct*. Paris: Graphite.

Marib Dam
Yemen

The Republic of Yemen is located on the southwestern coast of the Arabian peninsula, the region once possessed by the ancient southern Arabian kingdoms that occupied the mouths of large *wadis* (valleys) between mountains and desert. The first-millennium-B.C. kingdom of Saba sprang up in the dry delta of the Wadi Dhana that divides the Balak Hills. In the eighth century B.C., at the height of their prosperity, the Sabaeans had established colonies along both sea and land trade routes to Israel, and they dominated the region. Their capital, Marib, among the wealthiest cities of ancient Arabia, stood 107 miles (172 kilometers) east of Sana'a, the capital of modern Yemen. It is generally agreed that artificial irrigation was practiced near ancient Marib as early as the middle of the third millennium B.C. About 2,000 years later a dam was built to harness the biannual floods and systematic irrigation was introduced. Some scholars believe that the Marib Dam was the "greatest technical structure of antiquity."

Around 685 B.C., under King Karib'il Watar, Saba enlarged its borders. Territories were conquered in the southwest of the peninsula; Ausan in the south was defeated and Sabaean rule extended northwest as far as Nagran. In the second half of the sixth century B.C., two kings successively built the Marib Dam near the mouth of the Wadi Dhana, the largest water course from the Yemeni uplands. By impounding water during the two rainy seasons, the dam provided irrigation for some 25 square miles (65 square kilometers) of fields and gardens. Replenished and enriched by sedimentary deposits, this agricultural land supported a population estimated to be about 30,000.

The first dam was a simple earth structure, 1,900 feet (580 meters) long and probably only 13 feet (4 meters) high, built between rocks on the south side of the wadi and a rock shelf on the north. Its location a little downstream of the wadi's narrowest point permitted space for a natural spillway and sluices. Around 500 B.C. a second 23-foot-high (7-meter) earth dam was built. It was triangular in section; both faces sloped at 45 degrees and the upstream side was faced with stone set in mortar. The final form of the Marib Dam was not built by the Sabaeans.

Late in the second century B.C. the Himyarites, a tribe from the extreme southwest of Arabia, established their capital at Dhafar and gradually absorbed the Sabaean kingdom, gaining control of South Arabia. They undertook the next major reconstruction of the Marib Dam, building a new 46-foot-high (14-meter), 2,350-foot-long (720-meter), stone-faced earthen wall, incorporating sophisticated hydraulic systems. It was nearly 200 feet (60 meters) thick at the base, built on a stone foundation, and created a lake that was probably 1.5 square miles (4 square kilometers) in area. At each end of the wall there were sluices, constructed with what has been described as the "finest ancient masonry…in Arabia," through which water was channeled to extensive irrigation networks on both sides of the valley floor. The southern sluice system had a 10-foot-wide (3.5-meter) spillway about 23 feet (7 meters) below the top of the dam. The northern system included a spillway and a massive channel outlet between the spillway and the earth wall. It carried water via a 3,300-foot-long (1-kilometer), 40-foot-wide (12-meter) stone-lined earthen conduit, rectangular in cross section, to a distribution point that fed 12 irrigation canals. The discharge flowing into the conduit was controlled by a pair of gates; it also passed through a large settling basin.

When the Romans began to trade with India directly via the sea routes, the South Arabian economic monopoly was broken. The overland route declined, and social structure began to disintegrate. The Himyarite dynasty was toppled by an Ethiopian invasion in A.D. 335, reestablished toward the end of the fourth century, and again overthrown by the Ethiopians in 525. The Himyarites were absorbed into the wider South Arabian population.

The Marib Dam was regularly breached, usually by overtopping, during the extreme floods that oc-curred about once in fifty years. Just as regularly—for example, in A.D. 450 and 542—substantial repair work was undertaken. But when it was overtopped in 575, it was not repaired. Its final destruction was later recorded in the Koran (632–650), attributed to the judgment of Allah: "But they turned aside, so We sent upon them a torrent of which the rush could not be withstood, and in place of their two gardens We gave to them two gardens yielding bitter fruit. …" There is also a Yemeni proverb, "The Marib dam was destroyed by a mouse." Archeologists and engineers attribute its collapse to lack of adequate, regular maintenance or to the gradual failure of the foundation. Whatever the case, deprived of their water supply, the lifeblood of their crops and gardens, thousands of people from Marib returned to the nomadic life or migrated northward. The collapse of the dam expedited Bedouin insurgence from the Najd, and Islam was introduced around 630.

In December 1986 a new 125-foot-high (38-meter) earth dam was officially inaugurated. It closes off the Wadi Dhana a little under 2 miles (3 kilometers) upstream of the old dam site. Like its ancient predecessor, it was designed to impound water for irrigating the Marib plains; a 12-square-mile (30-square-kilometer) lake with almost a capacity of 437 million cubic yards (400 million cubic meters) has transformed 45,000 acres (18,000 hectares) of desert into productive farmland.

Further reading

Grolier, Maurice J., et al. 1996. *Environmental Research in Support of Archaeological Investigations in the Yemen Arab Republic, 1982–1987*. Washington, DC: American Foundation for the Study of Man.

Knutsson, Bengt, et al., eds. 1994. *Yemen, Present and Past*. Lund, Sweden: University Press.

Schnitter, N. J. 1994. *A History of Dams: The Useful Pyramids*. Rotterdam: Balkema.

Masjed-e-Shah (Royal Mosque)
Isfahan, Iran

The Royal Mosque, or Masjed-e-Shah (now known as the Masjed-e-Imam), was the major legacy of the Safavid Shah Abbas I (1587–1628), sometimes called Abbas the Great, who established Persia as a unified

Masjed-e-Shah (Royal Mosque), Isfahan, Iran; Ostad Abu'l-Qasim, architect, ca. 1590–1630.

state. The beautiful building, said to "stagger the visitor with its opulence and inventiveness," represents the epitome of Iranian architecture. It merits a place among the world's architectural feats because of the resplendent tile work that covers it both inside and out.

Helped by the British mercenaries Sir Anthony and Sir Robert Sherley, Abbas I defeated the Turks and expelled the Portuguese from the strategically critical island of Hormuz. He unified Persia by enforcing adherence to Shi'ism and establishing Farsi as the official language. His domestic policy focused on providing an economic infrastructure by building roads and bridges, and looking abroad, he also employed Armenian merchants to improve the silk trade with India. But nothing in the entire Safavid period (ca.

1320–1772) is better remembered than the vast amounts he spent developing Isfahan, where he established his capital in 1598.

Beginning in 1602, Abbas I completely rebuilt the city center in the form that survives. He commissioned the grand avenue of Chahar Bagh, the 1,700-by-500-foot (500-by-150-meter) Meidan-e-Shah (central Royal Plaza) and the buildings that surround it: the Bazaar (1619), still the largest in Iran; the Royal Palace of Ali Qapu (1602) facing the Sheikh Lotfallah Mosque (1602); and of course the Masjed-e-Shah. The Bridge of Thirty-Three Arches over the Zilldeh Rudh was built for him, as well as the Jubi Aqueduct to water the gardens with which he bedecked the capital. He also patronized a flourishing school of painting, and rugs for the royal palace and other buildings were woven on the court looms. It was said of Abbas I's unparalleled achievements in art and architecture, "Isfahan is half the world."

The commencement date of the Royal Mosque is uncertain. Some sources give 1590, a little early in the context of other urban development, and others claim that Abbas I laid the first stone in spring 1611. Ali Reza, the calligrapher responsible for the inscriptions in the building, dated the main entrance in 1616. Although Abbas put great pressure on his architect Ostad Abu'l-Qasim and his team of workmen, the mosque was incomplete when the Shah died in 1628 at the age of seventy. It is probable that work was still going on two years after that. The beautiful building certainly set a precedent, for elements of some later mosques are derivative: for example, the dome of the nearby Madrasa Mader-e-Shah (Royal Theological College) of 1714.

There are an estimated 18 million bricks in the Royal Mosque and the exterior reveals of its openings are claimed to be faced with 472,500 tiles. Indeed, the building should be included among the world's architectural feats because of the resplendent tile work on its main facade, its beautiful turquoise dome, and the interior. Tiles were a critical element of Persian architecture for two reasons: first, there was a practical need to weatherproof the clay bricks normally used in construction; and second, artistically, they were used to ornament the building. This was not merely for decoration but to define and articulate the underlying architectural form: tile work emphasized selected motifs and marked transitional points in the design, either by providing a patterned boundary or by the use of calligraphy. The Royal Mosque is widely celebrated for its exquisite *haft rang* (seven-color) tile work—colors were white, blue, yellow, turquoise, pink, aubergine, and green—which was developed extensively during the seventeenth century as the quality of glazes improved. It differed from conventional mosaic in that the full range of colors was used to create sinuous or calligraphic patterns on individual tiles, so that when they were placed, the overall design could be seen.

See also Dome of the Rock; Sultan Ahmet Mosque (Qubbat As-Sakhrah)

Further reading

Ferrier, R. W., ed. 1989. *The Arts of Persia*. New Haven, CT: Yale University Press.

Golombek, Lisa, and Donald Wilber. 1988. *The Timurid Architecture of Iran and Turan*. Princeton, NJ: Princeton University Press.

Honarfar, Lutfullah. 1978. *Historical monuments of Isfahan*. Tehran: Ziba Press.

Maunsell sea forts
England

The coasts of Kent and Essex Counties, England, overlook the Thames Estuary, the only sea route to London. Throughout World War II it was constantly endangered by German minelayers, U-boats, and the Luftwaffe. From 1939 until 1942 the British navy patrolled the area; then a series of seven sea forts was built to permanently guard the river mouth. They were an innovative architectural and engineering achievement. The reinforced concrete and steel structures were entirely prefabricated in a Gravesend dry dock, floated to their locations, sunk, and anchored on the bottom of the sea, up to 9 miles (14 kilometers) off the coast. Although not as large as the now almost commonplace offshore oil and gas platforms around the world, the sea forts predated them by about five years, and the six so-called "Texas Towers" that form part of the U.S. lighthouse system by almost twenty.

Two kinds of forts, one for the navy and another for the army, were designed by the civil engineer Guy Maunsell. Even when war was little more than a threat, he submitted several proposals for seaward defenses, but it was not until October 1940—over a year after the outbreak of war—that the Admiralty commissioned him to design a prototype sea fortress. His initial costly proposal, for a 2,900-ton (2,640-tonne) pontoon supporting a gun battery, was shelved by the government. But when France fell, the Admiralty was moved to action and asked Maunsell to produce five sea forts for the Royal Navy.

The naval sea forts were essentially steel gun platforms with two 6-inch (150-millimeter) cannon and a Bofors antiaircraft gun. The huge structures were assembled by Holloway Brothers at the Red Lion Wharf, Gravesend, towed downriver by three tugs, and sunk by flooding their hollow pontoon base. Two were positioned in the estuary off the Essex coast and two off the Kent coast. Each fortress had a crew of about 100, who lived, provisioned for more than a month, in the two 26-foot-diameter (8-meter), 7-story concrete "legs" that supported the main platform, with its guns, radar, and control tower. The first was sited at The Roughs in February 1942. Sunk Head followed on 1 June, and Tongue Sands was completed about a fortnight later. Knock John was ready for action on 1 August. The fifth was never built.

The army sea forts, also designed by Maunsell, were England's response to German air attacks on the strategic Liverpool docks via the undefended Mersey Estuary. It was decided to build five in the Mersey mouth and seven in the Thames Estuary. Each self-contained fort had living quarters for twenty-four men and comprised seven steel platforms supported on four 160-foot (49-meter) concrete legs. Four were gun towers with 3.7-inch (95-millimeter) cannon; a fifth was armed with a Bofors gun; the sixth was a searchlight tower; and the last was for radar. They were linked high above the sea by tubular steel catwalks that also carried power and fuel lines between the platforms. Their disposition was based upon the proven layout of shore gun batteries. In the event, only three were built on each side of England. Those in the Thames Estuary, constructed by the engineers who built the navy forts, were towed downriver in pairs and lowered by winches at strategic sites: The Great Nore, Shivering Sands, and Red Sand—all rather closer inshore than the navy counterparts. The pontoon bases used in the earlier structures would have been unsuitable in shallower water, where tidal currents constantly shifted the seabed; instead, Maunsell designed a self-burying footing that firmly anchored each tower in place. Construction began in August 1942, and the last tower was completed sixteen months later. At each site, the Bofors platform was erected first to defend the construction crews as they assembled the rest of the fort.

There is now no way to measure the passive deterrent effect of the Maunsell forts, but during their short active life they accounted for the destruction of twenty-two enemy aircraft and about thirty flying bombs. Because the Ministry of Defence believed that a combination of bad weather and tidal action would quickly destroy them after the war, no thought was taken for their disposal. For a few years after 1945 the naval forts were serviced by the Thames Estuary Special Defence unit, and two were temporarily adapted as lightships. Difficulty of access in storms led to that being discontinued; in fact, Tongue Sands was wrecked in bad weather in 1966. Only Knock John and The Roughs survive. After May 1964 the former, together with Red Sand and Shivering Sands army forts, was occupied at various times and for various periods by pirate radio stations, until the last was shut down under the Offshore Broadcasting Act in July 1967. The Roughs continues to have an eccentric postwar history.

It lies slightly north of the Thames Estuary off Harwich, and in September 1967, when it was still outside British territorial waters, a former British army officer named Paddy Roy Bates formally (and it must be said legally) annexed it as the Principality of Sealand, going aboard as the "prince" with his family. In the late 1990s a consortium of U.S. Internet entrepreneurs set up the world's first offshore data haven there, offering prospective clients security for their computer operations, free from the interference of legislation.

The army forts also went into decline. For a short while, under the control of the Anti-Aircraft Fort Maintenance Detachment, they were furnished with improved searchlights and radar installations. Any perceived crisis past, the army stripped all guns and equipment from them in 1956. The Red Sand fort, off the Isle of Sheppey, was abandoned that same year. The Great Nore fort was dismantled in 1958 after being struck by a ship and officially declared a hazard to shipping. In 1959 another vessel collided with the Shivering Sands fort, bringing down one of the towers. Despite their short-lived roles as radio stations, the survivors are now derelict. Their robustness means that their skeletons will stand in the North Sea for some years to come, gaunt confirmation of the proverb "Necessity is the mother of invention." Guy Maunsell did not survive his great sea forts, dying in 1961 after establishing an international civil-engineering partnership, which continues.

Further reading

Turner, Frank R. 1995. *The Maunsell Sea Forts.* Gravesend, UK: F. R. Turner.

Mausoleum at Halicarnassos
Anatolia, Turkey

The tomb of King Mausolos, known as the Mausoleum, was a structure impressive enough to merit inclusion among the seven wonders of the ancient world, and its name has passed into many European languages to describe any imposing funereal structure. It was designed by the Greek architect Pythios (some sources credit Satyros also) and decorated with works by the sculptors Scopas, Bryaxis, Timotheus, and Leochares. Because it survived for sixteen centuries, descriptions abound; combined with archeological evidence, they provide a good idea of the monument's appearance.

Mausolos (reigned ca. 377–353 B.C.) was a Persian *satrap* (governor) of Caria in southwestern Anatolia—a region so remote from the Persian capital that it was virtually independent. With a view to extending his power, Mausolos moved his capital from Mylasa in the interior to the coastal site of Halicarnassos, with its key position on the sea routes and large safe harbor, on the Gulf of Cerameicus. In 362 B.C. he joined the ill-starred rebellion of the Anatolian satraps against Artaxerxes II, but anticipating defeat, withdrew from the alliance in time. From then on he became the almost autonomous king of a large domain including Lycia and several Ionian cities northwest of Caria, later forming coalitions with the island city-states of Rhodes and Cos.

Mausolos undertook major urban design projects in Halicarnassos: a defense system, civic buildings, and a secret dockyard and canal. But the most interesting of all his public works was the planning of his great tomb. Conceived during his lifetime, it was initiated probably after his death by Artemisia II, who was at once his sister and his widow, and who for three years was sole ruler of Caria. She died in 350 B.C. and was buried with Mausolos in the uncompleted tomb. According to Pliny the Elder (A.D. 23–79), the craftsmen, realizing that the tomb was a monument to their own creativity, elected to finish the work after their patroness died.

Sited on a hill above Halicarnassos, the tomb rose 140 feet (43 meters) into the air from the center of a stone podium in an enclosed courtyard. A stair flanked by lions led to the top of this platform, whose outer walls were arrayed with statues, including an equestrian warrior at each corner. Its rectangular, tapered pedestal of white marble, with base dimensions of about 120 by 100 feet (37 by 30 meters), was 60 feet (18.3 meters) high. Its faces were carved with reliefs of Greek legends, including battles between centaurs and Lapiths, and Greeks and Amazons. The pedestal supported a colonnade of thirty-six 38-foot-high (11.6-meter) Ionic columns that housed a sarcophagus of white alabaster decorated with gold in a burial chamber. The tomb was roofed with a 22-foot-high (6-meter) stone pyramid of 24 steep steps, crowned with a 20-foot (6-meter) marble chariot bearing statues of Mausolos and Artemisia. Sculptured friezes of people, lions, horses, and other animals adorned every level of the Mausoleum; tradition has it that each of the famous sculptors was responsible for a side.

Under Memnon of Rhodes, Halicarnassos resisted Alexander the Great in 334 B.C. But it successively fell to Antigonus I (311 B.C.), Lysimachus (after 301 B.C.), and the Ptolemies (281–197 B.C.), after that retaining its independence until the Roman conquest in 129 B.C. Throughout all this conflict and for 1,600 years, the Mausoleum remained intact until a series of earthquakes shattered the columns and damaged the roof, bringing down the stone chariot. By the fifteenth century A.D. only the base remained. When the Crusader Knights of St. John of Malta invaded the region, they built a castle on the site and in 1494 used the stones of the Mausoleum to fortify it against an expected Turkish invasion.

Within twenty-five years almost every block of stone had been placed in the walls of their Castle of St. Peter the Liberator. Before grinding much of the Mausoleum's surviving sculpture into lime for plaster, the knights selected many of the pieces to adorn their castle. They renamed the city Mesy; today the ancient site is occupied by the town of Bodrum.

In 1846 Charles Newton of the British Museum began a search for vestiges of the Mausoleum. By 1857 he had uncovered sections of the reliefs and pieces of the roof. He also found a broken wheel from the stone chariot and, finally, the statues of Mausolos and Artemisia that had ridden in it for twenty-one centuries. All that remains of this wonder of the ancient world can now be found in the Mausoleum Room of the British Museum.

Further reading

Cox, Reg, and Neil Morris. *1996. The Seven Wonders of the Ancient World*. Parsippany, NJ: Silver Burdett.

Hornblower, Simon. 1982. *Mausolus*. New York: Clarendon Press.

Megalithic temples
Malta

The oldest monumental architecture in the world is found on the tiny islands of Malta and Gozo, south of Sicily in the western Mediterranean. There, for perhaps 1,500 years from 3800 B.C., communities of neolithic farmers built about thirty massive post-and-beam temples. None of these megalithic structures has survived intact, but no doubt they were architectural masterworks, the earliest of them a thousand years older than the pyramids at Giza, Egypt. The most striking examples are at Ggantija (the word means "giant") on Gozo and at Hagar Kim, Mnajdra, Tarxien, Ta' Hagrat, and Skorba on Malta.

The islands were first settled, quite separately, by people from southeast Sicily sometime between 5000 and 4000 B.C. These simple agrarian immigrants bred cattle, sheep, and pigs and grew lentils and barley. There probably followed a second wave of colonists from Sicily who absorbed or displaced the original group, and who left evidence of a culture expressed in communal underground tombs, for example, those at Zebbug on Malta and at Xaghra on Gozo. These graves foreshadowed the spectacular subterranean building known as the Hypogeum at Hal-Saflieni, described below. It has been suggested that later temple forms were also derived from these earlier burial places, because both building types consist of irregular compartments joined by short corridors.

Architecture, especially religious architecture, on such a scale indicates that the society produced an agricultural surplus to fund the work, that their organization permitted collaborative effort, and that their religious beliefs were strong enough to inspire and maintain that effort. The temples demonstrate a developing form. The earliest were constructed by piling massive limestone rocks that were neither dressed nor carved. Later temples, like those at Ggantija, Hagar Kim, Mnajdra, and Tarxien, were also built of huge slabs transported from neighboring quarries, but the blocks were set out to a clearly predetermined plan, carefully dressed and fitted and carved with finely detailed ornament. This later phase is lucid evidence of an ingenious people with a well-developed technology. They could transport immense blocks of stone, up to 20 feet (6 meters) high and weighing many tons, and accurately shape them using only flint or obsidian tools. The quality of the decorative work that embellished the structures—spiral carvings, intaglio patterns, and figures—demonstrates creative and artistic skills of a similar order.

The Hagar Kim and Mnajdra Temples stand on rocky ground a few hundred meters apart near the village of Qrendi on Malta's southeast coast. Their layout is difficult to describe. For example, the entrance to the approximately oval compound at Hagar Kim is set in a wall of carefully shaped and fitted rectangular limestone blocks. The doorway itself is a trilithon (three stones). This device, consisting of two uprights supporting a lintel, would remain the essential architectural and structural element of European architecture for the next 3,000 years. But beyond the gate there is a confused assemblage of amorphous rooms and courtyards linked by corridors, whose elaborate arrangement must be seen to be understood.

The three temples and the small enclosure of the Mnajdra complex are built of hard and soft limestones and are rather better defined. Two large elongated elliptical spaces forming a figure eight make up the largest building; they are entered through trilithonic doorways flanked by small square apses. The enclosing walls are built in two layers; internally they present as tall, massive slabs, while the outer face is constructed of masonry blocks. Although

Ruined temple at Hagar Kim, Malta, ca. fourth millennium B.C. Aerial view.

smaller, each weighs several hundred kilograms. It seems that some of the temples once had domed roofs. The complex of three linked temples in the town of Tarxien probably was built sometime later than the others, although the basic form is the same.

Perhaps the most striking prehistoric site on Malta, dating from around 3000 to 2500 B.C., is the Hypogeum at Hal-Saflieni, near the town of Paola. The three-story, 1,600-square-foot (150-square-meter) subterranean curvilinear building was excavated from the soft coralline limestone. Its upper level is a series of irregular, roughly finished burial chambers, very like the earlier rock-hewn tombs found elsewhere on the island and on Gozo. The middle level has twenty larger, more regularly shaped rooms joined by corridors. One is carved from the rock in close imitation of the contemporary aboveground temples, complete with trilithonic forms, roof beams, and other structural devices, none of which (of course) are structural. Some walls are almost covered with painted animals and curvilinear geometric designs. An ante-

chamber known as the Holy of Holies has a stairway leading to the lowest level, 36 feet (11 meters) beneath the surface, which has a maze of chambers and more rock tombs. That section seems to have had little use, but the remains of some 7,000 people have been found in the whole Hypogeum. Perhaps the building was designed as a temple for the dead, since archeological discoveries suggest the spaces were used for rituals other than burials.

The theme of an earth-mother goddess was common throughout the ancient Mediterranean region, but the intricate art of Malta may have been associated with a more complicated cult than fertility worship. Whoever or whatever they worshiped, it seems that this mysterious culture was suddenly terminated around 2000 B.C., when it was at its height. The directors of a joint archeological project between the Universities of Cambridge, Bristol, and Malta have theorized about the reasons for this sudden collapse, attributing it to a combination of several factors: the transition from an egalitarian to a hierarchical social structure, the pressures of increasing population, obsession with temple building that detracted from agricultural efforts, the effects of erosion on productivity, and diminishing trade links with Sicily.

The architecture they left behind was undervalued for centuries by the Maltese authorities, and through exposure to the severe marine climate and more recently shocks from nearby quarries, it inevitably decayed. In 1980 the Temple of Ggantija was inscribed on UNESCO's World Heritage List, and in 1992 the listing was extended to include five more complexes on Malta and Gozo under the title "the Megalithic Temples of Malta." In that year a carefully designed conservation project was launched by a multinational team of experts to save the Hypogeum, whose ochre rock paintings were being badly affected by seepage and eighty years of tourism. Each site presents its individual challenge and further conservation measures are planned for the oldest monumental architecture in the world.

Further reading

Biaggi, Cristina. 1994. *Habitations of the Great Goddess.* Manchester, CT: Knowledge, Ideas and Trends.

Bonanno, Anthony. 1990. *Malta: An Archeological Paradise.* Valletta, Malta: M. J. Publications.

Zammit, Themistocles, and Karl Mayrhofer. 1995. *The Prehistoric Temples of Malta and Gozo: A Description.* Malta.

Menai Suspension Bridge
near Bangor, Wales

The many achievements of the Scots engineer Thomas Telford (1757–1834) include bridges over the River Severn at Montford, Buildwas, and Bewdley, all built in the 1780s. In the following decade, as engineer for the Ellesmere Canal Company, he designed and constructed aqueducts over the Ceiriog and Dee Valleys in North Wales. Temporarily returning to Scotland, with William Jessop he built the Caledonian Canal, more than 900 miles (1,440 kilometers) of highland roads, and harbor works at Dundee, Aberdeen, and elsewhere. From 1810 he was engaged as principal engineer—William Alexander Provis was the resident engineer—to construct a highway between the Shropshire county town of

Shrewsbury and Holyhead in northwest Wales. It is widely agreed that his masterpiece is the Menai Suspension Bridge (1819–1826), which carries that highway across the Menai Strait, linking Bangor in mainland Wales with the island of Anglesey. It was the first large-scale chain-link suspension bridge and at that time the longest span bridge ever erected.

In 1782 a meeting on Anglesey examined complaints concerning the operation of the ferries at Porthaethwy, Llanfaes, Llanidan, and Abermenai that for centuries had been the only means of crossing the Menai Strait to the Welsh mainland. Increasing traffic across had led to delays and overcharging, and many of the boats were neglected and in dangerously poor condition. Alternatives to the ferries were canvassed, including an embankment and stone or timber bridges. With 4,000 vessels passing through the strait each year, those proposals were met with reasonable objections, and nothing was done. In October 1785 the Irish Mail Coach service was inaugurated between London and Holyhead on Anglesey, where travelers took ship for Ireland. The situation was further exacerbated in 1801, when the Act of Union demanded that Irish members of Parliament travel between Dublin and London, partly via the primitive Holyhead-Shrewsbury road and of course the ferry. Nevertheless, it was not until 1810 that Parliament commissioned Telford to recommend the line for a link across North Wales and Anglesey, including a bridge across the Menai Strait.

Attempts to improve only parts of the existing road were disastrous, so in 1816 Telford was appointed its resident engineer. His 69-mile (110-kilometer) stretch of the 93-mile (150-kilometer) toll highway (now the A5 national road) was probably the best road in Britain. It was up to 40 feet (12 meters) wide, with easy gradients and excellent bridges; moreover, its well-designed construction meant that it could accommodate heavy wagons.

Telford offered three alternative designs for the Menai Strait bridge, and that for a suspension structure was accepted. Finally, after forty years of debate and quibbling, the first stone was laid on 10 August 1819, and in the face of opposition from ferry proprietors and businesspeople in the ferry ports, construction work commenced. Including the ap-

Menai Suspension Bridge, near Bangor, Wales; Thomas Telford, engineer, 1819–1826. General view, from an engraving of January 1826.

proaches the bridge is 1,500 feet (459 meters) long. The approaches, completed in the fall of 1824, were carried on seven stone piers—three on the mainland side and four on the Anglesey side—supporting arches. The 579-foot (177-meter) main span, with its 24-foot (7.4-meter) dual carriageway, was suspended 100 feet (30 meters) above the water by sixteen chain cables hung from 153-foot-high (47-meter) massive battered towers—they were called "pyramids"—at each end, built of limestone from Penmon Quarries at the north end of the strait. Telford designed the piers to stand above the low-water mark, to facilitate inspection of the masonry.

The suspension chains were fabricated in wrought iron from Hazeldean's foundry near Shrewsbury. Each consisted of 935, 9-foot-long (2.75-meter) eyebar links, about 3.5 inches (83 millimeters) square in cross section, pinned together. To prevent rusting between fabrication and placement, they were immersed in warm linseed oil. Tunnels were excavated in rock to provide anchorage, and the first section of chain was secured at the mainland end, draped over

the top of the eastern pyramid and left hanging to water level. The procedure was repeated on the Anglesey side. The central section, weighing nearly 28 tons (25.4 tonnes), was maneuvered into position between the towers on a barge and connected to the end sections before being raised to the top of the tower by block and tackle and the strength of 150 men, thus completing the span. The chains were all placed in ten weeks, by July 1825. Iron rods suspended from them were bolted to iron joists that carried a timber deck. The Menai Strait bridge was opened to the public on 30 January 1826. Its completion and Telford's Shrewsbury-Holyhead road reduced the travel time between London and the Irish Sea port by a quarter.

Without stiffening lateral trusses, Telford's bridge soon proved unstable in the winds that swept through the strait, causing the road deck to oscillate. In 1826 a gale caused 16-foot (4.9-meter) deformations in the deck before it failed; although severely damaged, the bridge survived and was strengthened. A more rigid timber deck was incorporated in 1840, and that was replaced by a steel structure in 1893. Further changes were made in a major renovation of 1938–1941, ostensibly to cater for modern automobile traffic (the previous load limit per vehicle was 5 tons [4.6 tonnes], although it might also have been defense related). The arched openings in the towers were widened to allow easier passage of larger vehicles, the carriageway was strengthened, and the chains were replaced with steel cables and realigned. The bridge remains in use.

Further reading

Maré, Eric de. 1975. *Bridges of Britain*. London: Batsford.

Pearce, Rhoda M. 1978. *Thomas Telford: An Illustrated Life of Thomas Telford, 1757–1834*. Aylesbury, UK: Shire Publications.

Richards, James Maude. 1984. *The National Trust Book of Bridges*. London: Jonathon Cape.

Menier chocolate mill
Noisiel, France

The Menier chocolate mill at Noisiel, Marne-la-Vallée, was at the heart of a factory complex of industrial structures associated with Menier's chocolate-manufacturing business. The multistory mill, built between 1872 and 1874, demonstrated an innovative design approach that frankly exposed its structure and materials, using the latter for decorative effect. It is widely regarded as the first building in continental Europe to have been constructed with an iron frame and non-load-bearing masonry walls and has been described as "one of the iconic buildings of the Industrial Revolution."

In 1816 the pharmacist Jean-Antoine-Brutus Menier opened premises in Paris to sell his medicinal powders to chemists and hardware shops. Later he expanded his business to include chocolate-coated medicines and chocolate confectionery. Having outgrown his Paris base, in 1825 he transferred to Noisiel on the River Marne, where he purchased a mill to grind powders. Following his death in 1853, his son Emile-Justin took over the business, transferred its pharmaceutical arm to St. Denis and diversified into rubber production in a factory on the outskirts of Paris. The Noisiel plant was given over entirely to chocolate production.

Between 1860 and 1867 Emile Menier commissioned the architect Jules Saulnier (1817–1881) to redevelop the plant, constructing new buildings and improving the existing premises to better support the chocolate-making process. The factory would earn the nickname "the cathedral" because of its architecture. In 1869 Saulnier, working with the engineers Logre and Girard, prepared designs for replacing the timber-framed water mill that spanned the river, in order to house three new turbine wheels; he first chose stone as the principal material. Interrupted by the Franco-Prussian War of 1870, construction did not commence until 1872. By then Saulnier had revised the design and the outcome has been described as his masterpiece. The structural frame of the six-story chocolate mill was of puddled iron, diagonally braced to achieve a distinctive effect across the upper three levels of the facade; Saulnier likened the resulting pattern to the girders of a lattice bridge. The non-load-bearing, 7-inch (18-centimeter) yellow brick infill walls were decorated with diaper work and ceramic tile inlays with flower and cocoa-bean motifs, mainly in reds, dark yellow, and black. The frame was supported by a skeleton iron structure resting on the substantial stone piers that had carried the earlier timber-framed building, and floors were constructed of shallow brick arches between I beams, which were in turn carried by the main frame. The water-driven turbines were located between the piers. The interior was disposed to house the cocoa-bean milling process, and to free the third level of columns, its floor was suspended from the roof trusses. The spacing of columns and windows varied slightly, and deliberately, contributing to the artful composition of the facade.

Under Emile Menier's entrepreneurial leadership the business continued to expand. In the 1880s it established a factory in London and acquired cocoa plantations in Nicaragua, as well as a sugar refinery and a merchant fleet. It even established a railroad company to move materials and products. More buildings were constructed at Noisiel, utilizing the most advanced constructional methods and materials. A self-contained village was founded in which most of the factory's 2,000 employees lived in detached two-family houses or, if single, in hostels. The complex was set in extensive landscaped grounds. At the turn of the century the Menier chocolate business was the world's largest, and it reached its heyday before World War I. Decline was probably inevitable.

Between 1971 and 1978 the British confectionery company Rowntree-Mackintosh progressively purchased the Menier company, including its Noisiel factory, where chocolate continued to be made until 1993. The multinational Nestlé, which has owned Rowntree-Mackintosh and its subsidiaries since 1988, has taken over the Menier factory as Nestlé-France's headquarters, conserving the original buildings at a reported cost of Fr 800 million (U.S.$107 million). The chocolate mill has been made the focal building in the redevelopment by architects Robert and Reichen, and is used for a boardroom, reception rooms, and directors' offices. The French government has registered it as a Monument Historique.

Further reading

Cartier, Claudine, and Hélène Jantzen. 1994. *Noisiel, la chocolaterie Menier*. Paris: Association pour le patrimoine d'Île-de-France.

Steiner, Frances H. 1984. *French Iron Architecture*. Ann Arbor, MI: UMI Research Press.

Mesa Verde Cliff Palace
Colorado

Mesa Verde National Park is spread over more than 52,000 acres (21,000 hectares) of a well-wooded mesa between Cortez and Durango, Colorado, at a general elevation of 7,000 feet (2,100 meters). Within its boundaries are the ruins of almost 4,000 Amerindian settlements, some up to 1,300 years old. The largest and most remarkable is the so-called Cliff Palace, a multistory building like a modern apartment block built under overhanging cliffs. It accommodated probably 100–150 people in its 151 rooms and 23 kivas, and its size and complexity make it a preeminent feat of "architecture without architects."

Who were these exceptional builders? They are generally known as the Anasazi (Navajo for "ancient ones"), and their civilization was centered around the region where the states of Arizona, New Mexico, Colorado, and Utah now join. Some scholars identify the Anasazi as the ancestors of the Hopi and other indigenous Pueblo groups of the southwest United States, and modern Pueblo Indians prefer to call them "ancestral Puebloans." The precise origins of the Cliff Palace dwellers are unknown: certainly there were permanent settlers in the region before A.D. 500, farming and using caves or adobe structures for shelter and digging covered storage pits. By about 700 villages were being built: those in caves consisted of half-buried pit houses, while those on open ground had straight or crescent-shaped row houses with rooms both above and below ground. For the next three centuries the same house types—though somewhat larger—persisted, and stone masonry began to replace earlier pole-and-mud construction. The pit houses are the predecessors of the kivas, underground chambers common in the next phase of building.

Known as the Classic Pueblo period (A.D. 1050–1300), this was the era of the Cliff Palace and other villages built in similar sheltered depressions, as well as large freestanding apartment-like structures along the walls of canyons or mesas. Most consisted of two to four stories, differing little in construction from the earlier masonry and adobe houses, and often stepped back so that lower roofs formed a sort of patio reached from the floor above. All were built in places difficult to reach, some accessible only via almost vertical cliff faces, hundreds of feet above the canyon floor. The population of the region became more concentrated, perhaps acting upon the conviction that there is safety in numbers.

The Cliff Palace clearly was located with defense rather than esthetic appeal primarily in mind. The only access to it was by hand- and footholds—large enough for only fingertips and toes—carved in the rock. Afraid of something or someone (there is now no indication of what or whom), the Anasazi built fortresses unique among indigenous Americans. Their main building material was sandstone, laid in a mortar made from mud reinforced with tiny stone chips; the masonry was covered with a thin coat of plaster. Deliberately small doorways, set a foot or two above the floor, were probably intended to keep out winter drafts; they could be covered with rectangular sandstone slabs about 1 inch (2.5 centimeters) thick.

Mesa Verde Cliff Palace, Colorado, exact date unknown (A.D. 1050–1300).

The Cliff Palace was first excavated and stabilized by Jesse Walter Fewkes of the Smithsonian Institution, in 1909, more than twenty years after it was first seen by European Americans. Archeological work did not resume until about eighty years later, when evidence was discovered of a hierarchical society: a wall divides the Cliff Palace into two parts. It has also been suggested that the site was not continuously occupied except by a small caretaker population of perhaps 100 people. Then, its twenty-three large kivas would have accommodated larger numbers who gathered there only on special occasions, perhaps for the distribution of surplus food. The kiva, traditionally described as a ceremonial room, was a sunken, usually circular chamber entered through an opening from the "plaza" above. It had a ventilated hearth, and ledges and recesses surrounded the central space. The Anasazi may have used the Cliff Palace as living quarters during the winter lull in the agricultural year. The investigation of the site continues.

Mesa Verde was abandoned quite suddenly, around A.D. 1300. The Anasazi left so much behind that it has been suggested that their departure was hasty.

But that is speculation, and other sources suggest that they depleted the resources of the region, leading, through a tragic path of famine and internal wars, to the demise of their culture. Others cite the migration of Navajos and Apaches from the north, and yet others a fifteen-year drought at the end of the thirteenth century. For whatever reason, the Anasazi departed, leaving behind them the amazing and mysterious ruins of an architecture that is one of North America's greatest archeological treasures.

Further reading

Roberts, David. 1997. *In Search of the Old Ones: Exploring the Anasazi World of the Southwest*. New York: Simon and Schuster.

Smith, Duane A. 1988. *Mesa Verde National Park: Shadows of the Centuries*. Lawrence: University Press of Kansas.

Meteora, Greece

The almost flat valley of the Pineios River, north of the town of Kalambaka in Thessaly, is punctuated by spectacular formations of iron gray conglomerate rock, huge, sheer-sided columns abruptly projecting

Great Meteoron (the Monastery of the Transfiguration), Meteora near Kalambaka, Greece, architect(s) unknown, ca. 1360–1550.

up to 2,000 feet (600 meters) above the plains. On the seemingly inaccessible pinnacles of many of these weathered outcrops there stand, as though growing out of the rock, the monasteries of Meteora. Were they architectural feats? We believe so. Although most conventual buildings by definition demonstrate some degree of preoccupation with solitude, those at Meteora are unique, built where it appears virtually impossible to build. Not only were there no materials in situ, the task of delivering the imported materials to the builders—indeed, of getting the builders themselves to the precarious sites—could hardly have been more difficult. The logistical problems were subordinated to the need for isolation.

Christian monasticism originated in Egypt and spread throughout the Byzantine Empire between the fourth and seventh centuries. For a hundred years after the accession of Emperor Leo III in A.D. 717, the Iconoclasts attacked the eastern monasteries, seizing their treasured relics, thus greatly diminishing their wealth and power. As the rabid movement waned, Christian ascetics, perhaps moved with fear of a recurrence or perhaps with an eye on the restless power of Islam, sought secure places in which to follow their religious exercises. Throughout the ninth century hermits settled in rock crevices and caves in the great brooding pillars of the Pineian valley, long known as a retreat by mystics of pre-Christian religions.

As their numbers increased, the Thebaid of Stagoi Monastery was created at Doupiani, and its community grew during the eleventh century. Meteora became a sanctuary, especially after about 1300, when it provided asylum for secular as well as religious refugees under Ottoman rule. Around 1356 St. Athanassios Meteoritis founded Great Meteoron (from which the region derives its name), and about eighty years later the Serbian Orthodox prince John Uresis joined the community, endowing it with such wealth and privilege that it soon became the region's dominant monastic house. The growth of other foundations—Varlaam, commenced in 1350 and rebuilt in 1518; Holy Trinity of around 1470; and Roussanou, established in 1288 and rebuilt sometime before 1545—led to a golden age of monastic life and produced an environment in which scholarship and Byzantine ecclesiastical art flourished. At its peak the whole community numbered thirteen coenobite monasteries and about twenty smaller foundations. The patriarch Jeremias I (ruled 1522–1545) raised several of them to the rank of imperial *stavropegion*.

The monks set out to create places of inaccessible isolation. In the completed buildings entry could be gained only by a series of vertical wooden ladders of dizzying length (65–130 feet, 20–40 meters), which could be drawn up at night or when intrusion was imminent, or by nets hauled up by windlasses housed in cantilevered towers. Great Meteoron, or the Monastery of the Transfiguration, largest and highest of the houses, stands on the Platylithos (Broad Rock) 1,780 feet (534 meters) above the valley. Varlaam was originally reached by using scaffolding dug into the rock, and its windlass and rope in the tower (built 1536) were used for materials and supplies until 1963. Roussanou is built on a site only just large enough for it, and its walls stand right at the edge of the precipice. Whatever the reason for such a defense against the world—whether to protect the souls and minds of the monks or the wealth of the monasteries—the construction of these buildings in the sky, some of which are large and complex, represents a formidable challenge to the resolve and skill of the builders. It has been well met.

The monasteries generally declined in the seventeenth century (although some had failed long before), and by about 1800 they were little more than a "decaying curiosity," a unique sight for tourists. They surrendered their independence to the Bishop of Trikkala in 1899. At the beginning of the twenty-first century only five are still occupied: the monasteries of Great Meteoron, Ayia Triadha, Varlaam, and the convents of Agios Stefanos and Roussanou.

Further reading

Hellier, Chris. 1996. *Monasteries of Greece*. London: Tauris Parke Books.

Nicol, Donald M. 1975. *Meteora: The Rock Monasteries of Thessaly*. London: Variorum.

Mir space station

Mir (Russian for "peace") was conceived in 1976 as the climax of the (then) Soviet program to achieve the long-duration presence of a man in space. Its first component was launched into orbit ten years later. The first modular station assembled in space, it is *the* pioneer work of extraterrestrial building; constructed in a virtually gravity-free environment, it is unique among architectural and engineering works. Earlier space stations had been integral units, completed before launching.

Mir circled the earth for over fifteen years. As first proposed, it was 43 feet (13.1 meters) long and 13.6 feet (4.2 meters) in diameter; its mass was 46,200 pounds (20,900 kilograms). By 1985 the Russian Space Agency had decided that four to six additional modules, each with a mass of 46,000 pounds (20,800 kilograms), would be moored at docking ports on the station. By the time the final module was in place, the total mass was about 221,000 pounds (100,000 kilograms). *Mir*, humanity's first landmark—if that is the correct word—in space, orbited the earth at an altitude of 225 miles (390 kilometers) and an inclination of 51.6 degrees.

The primary function of the station was as a location for scientific experiments, especially in the areas of astrophysics, biology, biotechnology, medicine, and space technology. At various times, *Mir* was "leased" as a laboratory. Cosmonauts, astronauts, and scientists of many nationalities—Russian, American, Afghan, British, Canadian, German, Japanese, and Syrian among them—conducted over 20,000 experimental programs on board. However, space-watcher David Harland observed that *Mir* was the first station to be permanently manned, extending the time spent in space for periods between one month and six; "learning how the technology degrades, and how to repair it, and do so in space" showed its real mission as a technology demonstrator.

The *Mir* module, the core of the station, was launched on 20 February 1986. Most of it was occupied by the main habitable section—crews' quarters, a galley, a "bathroom" with shower, hand basin, and toilet—and the operational section, forward of which were the primary docking module and air lock. The galley was furnished with a folding table with built-in food heaters and refuse storage. For privacy, each crew member had a separate cubicle containing a folding chair, sleeping bag, mirror, and porthole. To provide a familiar environment in microgravity, the living quarters had identifiable surfaces: the floor, above several storage compartments, was carpeted in dark green; the light green walls had handrails and devices for securing articles; and the white ceiling had fluorescent lights. The other part of the core module was the station's control area, set up for flight control, as well as systems and medical monitoring. There were six docking ports on the core's transfer compartment for secondary modules or the *Soyuz* and *Progress-M* transport vehicles: one on the long axis, four along the radius, and another aft, connected to the working module by a 6-foot-diameter (1.8-meter) pressurized tunnel. The engine and fuel tanks were in the assembly compartment.

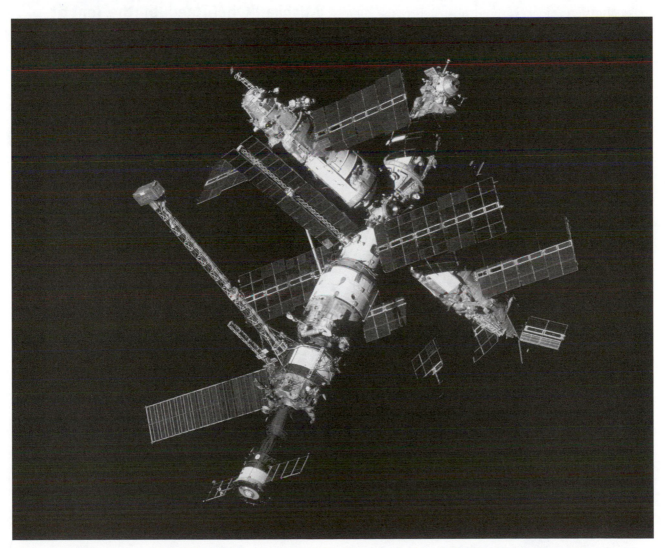

Mir space station, constructed 1986–1996; destroyed 2001. Photographed in 1995 from the space shuttle *Atlantis*.

Five more modules, added between 1987 and 1996, completed the space station. The first, located on the aft docking port, was the astrophysics module known as Kvant–1. Nineteen feet (5.8 meters) long and 14 feet (4.3 meters) in diameter, it contained a pressurized laboratory compartment and a store. Kvant–2, about twice as long as Kvant–1, was the scientific and air-lock module added in 1989 that allowed cosmonauts to work outside the station. It also included a life-support system and water supply. Kristall, a 39-foot-long (12-meter) technological module, was attached to the station in 1990; it carried two solar arrays as well as electrical energy supply, environmental control, motion control, and thermal control systems. In 1995 U.S. astronauts installed a special docking port that allowed the U.S. space shuttle to dock without obstructing the solar arrays. Also in 1995, the Spektr remote-sensing payload arrived at *Mir* with equipment for surface studies and atmospheric research and four more solar arrays. *Mir* was completed when the Priroda remote-sensing module arrived on 26 April 1996.

The station could not remain in orbit indefinitely, and two options for closure were available. *Mir* could be fitted with booster rockets and moved to a higher orbit or simply abandoned and allowed to crash into the ocean. *Mir* fell into an uninhabited part of the South Pacific late in March 2001. That course of action was chosen so that efforts could be refocused on the construction of the International Space Station (ISS). The decision fits in with the claim of NASA (the National Aeronautics and Space Administration) that the nine U.S. collaborations with *Mir* since 1994 formed Phase One of the joint construction and operation of the ISS.

The ISS is a joint venture of the United States, Russia, Belgium, Britain, Canada, Denmark, France, Germany, Italy, Japan, the Netherlands, Norway, Spain, Sweden, Switzerland, and Brazil. The first components of the station, the Zarya and Unity modules, were put into Earth orbit in November and December 1998, respectively. Scheduled for completion in 2004 after a total of 44 launches deliver over 100 components, the ISS will have a mass of 1 million pounds (454,500 kilograms) and measure 356 by 290 by 143 feet (109 by 88 by 44 meters). It will orbit Earth at about the same altitude and inclination as its predecessor. A crew of up to seven will have pressurized living and working space about twice as big as the passenger cabin of a jumbo jet. *Mir* was there first.

Further reading

Brown, Irene K. 1998. *New Millennium NASA: International Space Station and 21st Century Space Exploration*. Houston: Pioneer Publications.

Harland, David M. 1997. *The Mir Space Station: A Precursor to Space Colonization*. New York: Wiley.

Logsdon, John M. 1998. *Together in Orbit: The Origins of International Participation in the Space Station*. Washington, DC: NASA History Division.

Mishkan Ohel Haeduth (the Tent of Witness)

The Mishkan, or sacred tent, was a unique portable temple constructed under the direction of Moses as a place of worship for the Hebrew tribes. It was used during the forty-year period of wandering between their liberation from slavery in Egypt and their arrival in the Promised Land (ca. 1290–1250 B.C.). According to chapters 25 and 26 of Exodus, the warrant and exact specifications for its construction were given by God. The tent seems to have been still in use in the first half of the eleventh century B.C., but it no longer served a religious purpose after Solomon built a permanent temple in Jerusalem in 950 B.C.

Portable shrines existed in Egypt as early as the Old Kingdom (2800–2250 B.C.), and fine examples were discovered in the tomb of Tutankhamen (ca. 1350 B.C.). But they are small in comparison with the Tent of Witness, which differed from all contemporary religious buildings in several remarkable ways. First, it was the *only* temple constructed by the monotheistic Israelites, in contrast to the many—often several dedicated to the same deity—built by their polytheistic neighbors. Second, it was never associated with one particular sacred geographical location, peculiar to the deity; rather, it was set up wherever Yahweh, the God of Israel, indicated, in the belief that his presence made every location sacred. Third, it was small and outwardly unimposing, and although constructed of the choicest durable materials, it did not have (indeed, could not have) the appearance of weighty permanence common to contemporary religious buildings. Fourth, the materials used imparted a brightness that contrasted with the dark tents of the tribespeople who camped around it and that marked it out against the somberness of other shrines. Finally, its construction was not financed by temple taxes but by the voluntary offerings of the Israelites: according to Exodus, they gave 2,800 pounds (1,270 kilograms) of gold, 9,600 pounds (4,360 kilograms) of silver, and 6,700 pounds (3,050 kilograms) of bronze besides the necessary yarn and textiles. Its architectural character was inextricably linked to the Hebrews' nomadic life for the first forty years of its existence. The Law of Moses provided instructions for the Levite families of Gershon, Kohath, and Merari responsible for assembling, demounting, and carrying the Mishkan and its court.

The complex invariably stood at the very center of the Israelite camp. It comprised a large courtyard around a comparatively small building that may be regarded as the Tent of Witness proper. The outer court was enclosed by a white linen wall, 150 feet (46 meters) long by 75 feet (23 meters) wide and 7.5 feet (2.3 meters) high, hung on 60 pillars of the brownish orange wood from the durable desert acacia. The pillars, each crowned with a silver capital, stood on bronze sockets, and their guy ropes were fastened with bronze pins. Access to the court was through a "gate" at the eastern end, also of white linen but distinguished from the general walls by an embroidered pattern in blue, purple, and scarlet and fastened to its pillars with gold hooks. Immediately inside the gate was an altar made of bronze-sheathed acacia wood. It is a comment upon the portability of the sanctuary that, at only 7.5 feet (2.3 meters) square and 4.5 feet (1.35 meters) high, this was the largest of the furnishings, designed to be carried on poles, rather like a sedan chair. Nearby stood a bronze basin holding water used for the priests' ritual ablution.

The Tent of Witness itself stood at the western end of the court. An oblong enclosure, about 45 feet long by 15 feet wide (13.5 by 4.5 meters) and 15 feet high, was framed by walls assembled from 48 gold-sheathed acacia boards, each 27 inches (about 70 centimeters) wide. Standing on foundation blocks of solid silver, the boards were locked together by a system of bars passed through brackets on their outer faces and through their centers.

The plain exterior gave no clue to the richness and brilliant color of the rooms it contained. The ceiling was a draped curtain of the same textile as the courtyard gate, covered with another of goat's hair, then red-dyed rams' skins; an outer layer of porpoise skins provided durable protection. The interior was reached through a door of the same embroidered fabric hung on gold-sheathed pillars. By absolute contrast, the floor was simply the earth of the desert. The first compartment, 30 by 15 feet (9 by 4.5 meters), was called the Holy Place. It was furnished with a gold-sheathed table; a small altar for burning incense, also covered in gold; and a seven-branched menorah (lamp stand) hammered from solid gold. Beyond an inner curtain emblazoned with embroidered cherubim (angelic beings) was the Holy of Holies. The only furniture in that inner sanctum was the Ark of the Covenant, a gold-sheathed wooden box containing the stone tablets of the Ten Commandments. This was the dwelling place of the God of Israel, who sat invisibly enthroned above the gold "seat of atonement" that rested on the Ark. Access was denied to all except the High Priest, and then only on Yom Kippur (the Day of Atonement). Because of the uniqueness of the spiritual beliefs that the Tent of Witness expressed, it was never a prototype for anything else. When Solomon built the great temple in Jerusalem, the architectural emphases were quite different.

See also Solomon's Temple

Moai monoliths
Rapa Nui (Easter Island)

The small Pacific island of Rapa Nui (Easter Island), 2,300 miles (3,680 kilometers) west of Chile, is the most remote inhabited island in the world, with Pitcairn, its nearest neighbor, 1,400 miles (2,240 kilometers) away. The staggering architectural achievement of the people of Rapa Nui was the creation, but especially the transportation and erection, of hundreds of monolithic *moai*—stylized giant human heads-on-torsos—carved in hardened volcanic tufa. On average, the statues are 13 feet (4 meters) high and weigh 14 tons (14.22 tonnes). But the largest ever raised once stood at the prominence known as Ahu Te Pito Kura; nearly 33 feet (9.80 meters) high, it weighed about 91 tons (83 tonnes). Even it would have been dwarfed by another found incomplete in a quarry: measuring almost 72 feet (21.6 meters), its weight was perhaps 185 tons (168 tonnes).

Since the Dutch seafarer Jacob Roggeveen made Rapa Nui known to Europe in the 1720s, scholars have debated the origins of its culture. Local legend has it that the canoes of Hotu Matu'a (the Great Father) arrived from Polynesia around A.D. 400. Some scholars, citing archeological evidence, assert that they came between 300 and 400 years later. Whatever the case, among the lush palm forests the newcomers planted their gardens of bananas, taros, and sweet potatoes. The South American origin of the latter led the adventurer Thor Heyerdahl to conjecture that Polynesia had been colonized by pre-Inka people, a view refuted by later scholars, who cite biological, linguistic, and archeological evidence to support Southeast Asian origins. As compelling as it is, the question is not our present concern.

Unique in Polynesia, the mysterious moai are thought to have been carved between A.D. 1400 and 1600 by specialist master craftsmen using tools made from obsidian found at Orito. The figures, always male, are believed to be iconographic representations of powerful beings—ancestors, chiefs, or others of high rank—rather than portraits. The red volcanic stone for their headdresses *(pukaos)* came from the Puna Punau volcanic crater; their eyes were made of shell and coral. They were the product of a spiritual and cultural imperative that seems to have become an obsession.

The archeologist Jo Anne van Tilburg of the University of California at Los Angeles suggests that the statues acted as ceremonial mediators "between sky and earth, people and chiefs, and chiefs and gods."

The statues were transported, probably by conscripted labor, from where they were quarried and set up on the perimeter of the island, mostly on the southeast coast. Some were moved up to 14 miles (22.4 kilometers) and placed facing inland upon flat mounds or stone pedestals *(ahu)* about 4 feet (1.2 meters) above the surrounding ground. The word *ahu* also conveys a sacred site, and some, comprising massive masonry blocks and tons of fill, supported a whole group of moai.

For fifteen years van Tilburg carried out a "census" of the moai, finding a total of 887 statues. Fewer than one-third (288) had been transported to their coastal locations. She recorded another ninety-two as "in transport," that is, on their way to their intended locations. The remainder were still in the quarries at what van Tilburg calls the central production center, in the volcanic caldera known as Rano Raraku near the eastern end of Rapa Nui. Perhaps they were abandoned because flaws were found in the stone, perhaps they were too large to move, or perhaps deteriorating social conditions forced the work to end.

How were they moved to their solemn stations around the coast of Rapa Nui? Several possibilities have been suggested. There is a local tradition that the moai "walked" to their sites, which led Heyerdahl to conclude that they were stood upright and rocked from side to side, thus "walked" along. A poorly rendered Dutch illustration of 1728, showing a statue standing upright on a base at which people are working, has been interpreted as moving the moai on rollers. Both systems have been tested using pseudo-moai and both worked. Others have suggested that the gigantic figures were laid prone, just as they had been carved, and dragged on sleds. Working from computer models that took account of many variables, including the food needed for the workers, van Tilburg proposed a plausible alternative, tested by experiment: the massive figures were moved in the prone position, supported on long logs that were rolled on smaller ones. In fact, no one knows with certainty how such loads were moved over the difficult terrain of the island.

Around A.D. 1550, Rapa Nui's population reached a peak of about 10,000, placing an untenable load on the tiny island's resources just when moai carving and, more significantly, transportation reached a climax. Over the next century or so, radical change occurred, heralding the collapse of the society. Some scholars lay most of the blame for decline on the compulsion to construct the colossal figures. The once abundant palm forests were cleared for housing and crop production and to provide tools and pathways for moving the moai. Deforestation allowed the erosion of topsoil, and crops failed. Soon, driven by territorial imperatives, the island clans descended into civil war and even cannibalism. All the coastal moai had their eyes smashed out and the statues were toppled and decapitated by the islanders themselves.

Contacts with the West from the beginning of the eighteenth century served only to make matters worse, and in 1862 Peruvian slavers and exotic diseases together ravaged the population, reducing it to little more than a hundred. The process was reversed after Rapa Nui was annexed by Chile in 1888, and in 1965 it received the same privileges as other Chilean provinces. The economy now depends on sheep ranching and tourism. The main attraction for tourists is the mysterious moai, whose uniqueness led to the island being inscribed on UNESCO's World Heritage List in 1995, with the following description:

> Rapa Nui … bears witness to a unique cultural phenomenon. A society of Polynesian origin … established a powerful, imaginative and original tradition of monumental sculpture and architecture, free from any external influence. From the tenth to the sixteenth centuries this society built shrines and erected enormous stone figures, *moai*, which created an unrivalled cultural landscape and which today continue to fascinate the entire world.

Further reading

Heyerdahl, Thor. 1975. *The Art of Easter Island*. Baarn, the Netherlands: De Boekerij.

McCall, Grant. 1994. *Rapanui: Tradition and Survival on Easter Island*. Honolulu: University of Hawaii Press.

Orliac, Catherine, and Michel Orliac. 1995. *Easter Island: Mystery of the Stone Giants*. New York: Abrams.

Tilburg, Jo Anne van, et al. 1995. *Easter Island: Archaeology, Ecology, and Culture*. Washington, DC: Smithsonian Institution Press.

Mohenjo-Daro
Pakistan

The city of Mohenjo-Daro ("hill of the dead") was the largest settlement of a culture that for more than 600 years from 2500 B.C. extended over 600,000 square miles (1.5 million square kilometers) of India and Pakistan—larger than western Europe. The city's ruins, on the west bank of the Indus River about 200 miles (320 kilometers) north of Karachi, evidence careful urban design combined with a sophisticated infrastructure that was undreamed of in the contemporary river-valley civilizations of Egypt and Mesopotamia. Although presented with undeniably nationalistic and political bias, recent archeological evidence from the subcontinent suggests that there, and not in Mesopotamia, was the cradle of civilization. Mohenjo-Daro has been chosen here as simply representative of a great achievement, the invention of city planning.

The first traces of the ancient cities were accidentally discovered on the Indus River floodplain in 1856. The occupying British, building the East Indian Railway between Lahore and Karachi, plundered hundreds of thousands of bricks from the site of Harappâ, a metropolis on the Ravi River, 400 miles (640 kilometers) northeast of Mohenjo-Daro. In 1920 Sir John Marshall, director general of archeology in India, initiated investigation of these "twin capitals," and discoveries were made by Daya Ram Sahni (at Harappâ) and R. D. Banerji. Nani Gopal Majumdar worked in the Sindh region (now in southern Pakistan) from 1927 to 1931. About a decade later, Ernest Mackay discovered Chanu Daro, and Sir Aurel Stein found more sites in Baluchistan and Rajputana. From 1946 into the early 1950s, Sir Mortimer Wheeler continued excavations at Mohenjo-Daro and Harappâ.

The terms "Indus valley civilization" and "Harappân culture" spring from the collective work of all these men, but Hindu scholars in India and Pakistan recently have challenged that nomenclature. Some insist that the civilization was created by Vedic people. Hard evidence is displacing earlier speculative myths about origins. Since the early 1990s archeologists have uncovered several cities east of the Indus. The settlement known as the Dholavira excavation, about 150 miles (250 kilometers) from modern Bhujin in India; another under modern Rakhigarhi in the Haryana district; and Kunal, also in Haryana and close to the dry Ghaggar-Hakra River (thought to be Mother Saraswati of the *Rig Veda*), are evidence of how widespread Harappân culture was. It extended from the mountains of northern Afghanistan south to the Arabian Sea, and from the Baluchistan highlands in the west eastward to the Thar Desert, once a fertile plain watered by the Saraswati. Of about 1,400 sites now uncovered, about 900 are in India, 1 is in Afghanistan, and the remainder are in Pakistan. Because there are more Harappân settlements along the Saraswati than along the Indus, it has been suggested by some Indian archeologists that the name "Indus-Saraswati" civilization be adopted; their peers in Pakistan prefer "Hakra" civilization.

However modern scholars choose to classify them, well-structured barley- and wheat-growing communities existed in the region around 4300 B.C., and by 3200 B.C. large villages stood along the great rivers. The historian Arnold Toynbee has suggested that such locations were chosen not because they offered an easy life but because of the challenges presented by annual flooding. Whatever the case, the Harappâns made a dizzying leap from villages to cities between 2600 and 2500 B.C. As elsewhere, there was no intermediate, evolutionary step. The U.S. archeologist Gregory Possehl describes this as a century of cathartic changes, and Gordon Childe coined the expression "the urban implosion." Quite suddenly, there existed commercial centers with high levels of municipal control; complex social organization; specialized occupational structures; administrative expertise; and tools, such as a system of writing (that of the Indus-Saraswati culture is as yet undeciphered), mathematics and especially geometry, survey instruments, and standard weights and measures.

Although there were other locations of comparable size and importance, Mohenjo-Daro was certainly a principal—and typical—city. This largest Indus valley settlement covered more than 200 acres (80 hectares) and was over 3 miles (5 kilometers) around. Like most Indus-Saraswati cities, it had two principal and functionally disparate districts, each built on a huge mud-brick platform, which raised it above the annual floods. To the west stood the 45-foot-high (14-meter) "citadel," measuring 1,400 feet (430 meters) by 450 feet (140 meters). Fortifications have survived at its southeast corner and the entire platform may have been enclosed by a wall. Several public buildings stood on it. Archeologists once imagined these to include a large granary (with a wooden superstructure), an assembly hall, a college, and a ritual tank, but more recent scholarship has found no evidence for those conclusions. Whatever their purpose, the layout and juxtaposition of the structures demonstrates careful urban design. Noteworthy among the buildings was the so-called Great Bath, about 29 by 23 feet (8.8 by 7 meters) and 8 feet (2.4 meters) deep, entered by steps at each end. It was built of fired bricks laid in gypsum mortar and sealed with bitumen.

Across an area of probably unused land, the residential "lower" city of Mohenjo-Daro occupied a more extensive platform to the east. It also was set out on a north-south, east-west grid of properly drained, brick-paved streets, 30 feet (9 meters) wide, forming rectangular urban blocks measuring about 400 by 270 yards (360 by 240 meters). Tightly packed courtyard houses and places of business, all built of fired brick, faced the streets. It is likely that internal walls were plastered, and most houses had stairs leading to either a second story or a flat roof. The houses normally had a private bathing area supplied with water from their own wells, and a properly drained toilet. A secondary grid of narrow service lanes subdivided the main blocks, and chutes from most residences were connected to a system of covered sewers—more evidence of well-developed municipal controls that still cannot be found in many Asian cities. Because of the high water table beneath Mohenjo-Daro, it is impossible to extend archeological investigation to its foundation level. Exploration continues, and in the early 1980s German scholars discovered a "suburb" about 1 mile (1.6 kilometers) from the "downtown area."

Harappâ and many other Indus-Saraswati sites are almost identical to Mohenjo-Daro in layout and organization, indicating that, at the peak of the civilization, regional centers may have been built to a standard city plan. There are exceptions in detail, if not in overall form.

This remarkable civilization remained unified for nearly 700 years. Then, partly because of overexpansion of its trade networks, after about 1900 B.C. it gradually disintegrated into a regionalized pattern of cultures, referred to as late or post-Harappân. Within 150 years Mohenjo-Daro's efficient urban government had deteriorated. Administrative breakdown was augmented by ecological factors. Recent research has established that most protohistoric cultures suffered three centuries of persistent drought from about 2200 B.C., perhaps activated by a sudden global climate change. The passing of the Indus-Saraswati cities can be attributed in part to changing river patterns, upsetting a river-based agricultural and trade economy that had already outgrown its strength. The Saraswati dried up, the Hakra-Nara's tributaries were diverted eastward to the Jamuna River and westward to the Indus, and the course of the Indus itself began to change, resulting in frequent violent flooding of its southern reaches.

Further reading

Gupta, Swarajya Prakash. 1996. *The Indus-Saraswati Civilization: Origins, Problems, and Issues*. New Delhi: Pratibha Prakashan.

Possehl, Gregory L., ed. 1993. *Harappan Civilization: A Recent Perspective*. New Delhi: American Institute of Indian Studies.

Rao, Shikaripur Ranganatha. 1991. *Dawn and Devolution of the Indus Civilization*. New Delhi: Aditya Prakashan.

Mont-Saint-Michel
Normandy, France

Mont-Saint-Michel is a craggy, conical island, about half a mile (0.8 kilometer) across and standing half a mile from shore in the Gulf of Saint-Malo, near the border of Brittany and Normandy on France's northern coast. The north side of the island is wooded and the west presents a barren face to the sea. A fortified village of fewer than 100 inhabitants huddles on the lower southern and eastern slopes and the great Benedictine abbey, dating from the thirteenth century, crowns the entire mount, towering about 240 feet (73 meters) above. The integration of monastery with village and both with the rock was noted by UNESCO as "an unequalled ensemble" when the site was inscribed on the World Heritage List in 1979. Mont-Saint-Michel is an architectural feat for that reason and others: the audacity displayed by the builders on so difficult a site and the harmony achieved between its parts, which were built in many architectural styles over five centuries.

The place known as Mont Tombe, which became Mont-Saint-Michel, has a spiritual history dating from pre-Christian times. There the Gauls had worshiped Belenus, the god of light, and there the Romans consecrated a shrine to Jupiter. By the fifth century A.D. the secluded crag and the Scissy Forest around it had become a retreat for hermits. There is a tradition that in 708 St. Michael appeared to Aubert, twelfth bishop of Avranches, directing him to build a sanctuary to the archangel on the mount. In October of the following year, Aubert consecrated a simple circular oratory, to accommodate about 100 people, and built cells to replace the earlier huts, but not before an abnormal tide—some sources say a tidal wave—had gouged a channel between rock and shore, creating the islet. At low tide a land bridge connects to the mainland across beaches of gray silt; at high tide it is covered by about 40 feet (14 meters) of water.

Under the sponsorship of Richard the Fearless, Duke of Normandy, Abbot Mainard occupied the island in 966 with twelve Benedictine monks from Monte Cassino. He built a rectangular chapel with 6.5-foot-thick (2-meter) stone walls on the ruins of the oratory. By that time, the Benedictines had enjoyed four centuries of prominence in western Europe and monasticism had reached a zenith. In France, the abbeys—there were about 120 of them—exercised great influence in many spheres: spiritual, artistic, intellectual, economic, and political. Besides the Benedictines, whose other Normandy houses were at Fecamp, Lessay, and Lonlay, the Premonstratensian Canons had established themselves at Ardenne and La Lucerne. Eventually, Mont-Saint-Michel would become a magnet for thousands of the faithful from all over Europe.

The next building phase was initiated by Abbot Hildebert II in 1017. An extensive masonry foundation

Mont-Saint-Michel, Normandy, France; Raoul de Villedieu and others, architects, 1017–1521. View from the southeast, photographed ca. 1950.

leveled the entire top of the island and an abbey church was built on the summit. Mainard's sturdy chapel formed its crypt and was later named Notre-Dame-sous-Terre (Our Lady Underground). The rest of the new cruciform church—with its seven bays, the nave was nearly 230 feet (70 meters) long—was supported on masonry walls and piers. The project, designed in the latest style (now known as Romanesque), was completed by 1135. That was not the end of the architectural development, and about

thirty-five years later Abbot Robert de Torigny commissioned a new west front with twin towers.

In 1203 the French king Philip II Augustus sent an expeditionary force against the abbey, and some of its dependencies were destroyed by fire. To compensate for the damage, a generous endowment allowed Abbot Jordan to immediately commence the granite conventual building known as La Merveille (the Marvel), flanking the church on the seaward side of the rock. Remarkably, the extensive, logistically difficult works were completed by 1228. The Marvel began at 160 feet (49 meters) above the sea and consisted of three terraced levels. The lowest housed the almonry and cellar. The second was taken up by the kitchens; a huge refectory with timber barrel vaults; a guest hall, adorned with tapestries, stained glass, and glazed tiles; and a scriptorium (now called the "hall of the knights"). At the top was the monks' dormitory and a beautiful arcaded, vaulted cloister attributed to Raoul de Villedieu.

In contrast to that tranquil security, the Marvel has been described as "half military, half monastic." Louis IX visited the Mont in 1254 and later helped to pay for its fortification. Strategically located, it acquired a defensive role and housed a garrison jointly paid by king and abbot. Through the fourteenth and fifteenth centuries, both the abbey and the town were enclosed by walls on the land side, adding another texture to the varied architecture of the rock. Frequently attacked, it would never be captured, even remaining unconquered when English armies took most of the fortresses of Normandy early in the fifteenth century.

There was a series of structural failures in the abbey church. In 1300, one of de Torigny's west towers fell down. More serious was the collapse in 1421 of Hildebert's Romanesque choir. France was still at war with England, and all thought of reconstruction was deferred until 1446, when a massive base known as the Crypt of the Large Pillars was built as foundation for a replacement building. Work on the new choir began in 1450 and it was completed in 1521. Apsidal in plan, with radiating chevet chapels, it was naturally built in the contemporary, highly ornate French style, appropriately named *flamboyant* because of the flamelike patterns of its window tracery. Other architectural failures followed: in 1618 the de Toringy west facade started to collapse, and eventually it was pulled down in 1776, together with the three western bays of the nave.

The monastic foundation seemed to decline with the buildings. Although by the twelfth century under de Toringy, the Benedictine abbey of Mont-Saint-Michel had acquired fame for its intellectual life, drawing pilgrims from across Europe, about a century later its power had begun to slowly wane. As the balance of its role tipped from devotion to defense, the size of the community decreased. In 1523 it was granted *in commendam* to Cardinal Le Veneur, the series of commendatory abbots continuing until 1622—by then hardly any monks remained—when control passed to the reformed congregation of St. Maur. In turn, the Maurist monks were dispossessed during the French Revolution. From 1790 the abbey, its name ironically changed to Mont Libre (Mount Freedom), was used to incarcerate criminals and political prisoners. Napoléon III abolished the prison in 1863. Having gone full circle, the buildings were leased to the Bishop of Avranches until 1874, when the Commission des Monuments Historiques appointed the architect E. E. Viollet-le-Duc to restore it. In 1966, in recognition of the monastery's millennium, the French government allowed the resumption of monastic life on Mont-Saint-Michel; since then a community of monks, nuns, and lay oblates lives in a part of the abbey, reviving the ministry to pilgrims.

This has been a complicated story, whose point is just this: the architectural feat of Mont-Saint-Michel was not achieved in a day, a month, or a year. The harmony and the unity of its parts, diverse in date, style, and function, took 500 years to realize.

Further reading

Bony, Jean. 1983. *French Gothic Architecture of the Twelfth and Thirteenth Centuries*. Berkeley: University of California Press.

Braunfels, Wolfgang. 1993. *Monasteries of Western Europe: The Architecture of the Orders*. New York: Thames and Hudson.

Froidevaux, Yves-Marie, and Jacques Boulas. 1965. *Le Mont Saint-Michel: Photographies de Jacques Boulas*. Paris: Librarie Hachette.

Mount Rushmore
South Dakota

The broad granite southeast face of 5,725-foot (1,750-meter) Mount Rushmore, neat Rapid City, South Dakota, is carved with the massive portrait heads of four U.S. presidents—George Washington, Thomas Jefferson, Abraham Lincoln, and Theodore Roosevelt. For its sheer engineering ingenuity and ambitiousness of scale—Washington's head is 60 feet (18 meters) high—the ensemble may be regarded as an architectural feat.

In 1923 South Dakota's state historian Doane Robinson suggested carving giant statues in the Black Hills. Perhaps he was prompted by the knowledge that a colossal Confederate memorial had been commissioned a few years earlier for Stone Mountain, Georgia, but it is more likely that the idea was first conceived as a tourist attraction. Initially, Robinson wanted to have a cluster of tall granite outcroppings known as the Needles carved to form a procession of the Amerindian leaders and European explorers who shaped the Western frontier. Conservationists resisted the idea, and there was no public support. Nevertheless, in 1925 the financial backers of the proposed memorial approached the sculptor Gutzon Borglum, who was known to specialize in large-scale sculpture and was then rather unhappily employed on Stone Mountain.

Borglum suggested that the southeast face of Mount Rushmore would make an ideal site for a monument. He proposed to carve the heads of the four presidents beside a table inscribed with a history of the United States. Such a composition would have more than regional significance; it would commemorate "the foundation, preservation and continental expansion of the United States" and be a shrine to democracy. And behind the figures a hall of records would preserve national documents and artifacts.

President Calvin Coolidge dedicated the memorial in 1927, and Borglum began drilling. But although less than half the time was spent on actual carving, the work would take fourteen years to complete. Most of the delay was due to money shortages during the Great Depression. Borglum lobbied at every political level, playing on nationalistic feelings and stressing that public works created jobs and won votes. As a result of his persistence, nearly 85 percent of the monument's $1 million cost came from federal coffers. The Washington head, 500 feet (150 meters) up the mountain, was formally dedicated in 1930, when the name "Shrine of Democracy" was officially adopted; Jefferson followed in 1936, Lincoln in 1937, and Roosevelt in 1939. Borglum died in March 1941 and his son Lincoln supervised the completion of the sculpture.

Borglum's plaster maquettes were based on life masks, images, and descriptions, but the differences between them and the finished heads demonstrate that the sculptor did not simply transpose from plaster to stone. Once the dimensions were scaled up to the finished size and marked out on the mountain, the team of carvers was faced with the problem of removing the unwanted granite. Despite Borglum's first inclination against its use, dynamite was the only practical way to do that. Once an oval-shaped mass of rock was formed for each head, explosive experts blasted its surfaces to the approximate final measurements. Carvers suspended in bosuns' chairs shaped the features. They used pneumatic drills to cut closely spaced holes that nearly defined the final surface, and the honeycombed granite was ultimately chiseled away to expose the smooth surfaces of the presidents' faces. Viewed from a distance, stone miraculously became flesh; as the architect Frank Lloyd Wright observed, "The noble countenances emerge from Rushmore as though the spirit of the mountain heard a human plan and itself became a human countenance."

A similar feat, already mentioned, deserves a little more detail. The north face of Stone Mountain, 16 miles (26 kilometers) east of Atlanta, Georgia, is carved with a 138-foot (42-meter) equestrian bas-relief of the Confederate heroes Robert E. Lee, Stonewall Jackson, and Jefferson B. Davis. What began in 1915 as a commission for Borglum to produce a 70-foot (21-meter) statue of Lee developed into a proposal for the group portrait. Preliminary work started soon after World War I and carving began in June 1923. Irreconcilable differences with the client caused Borglum to quit in March 1925—just as he received the Black Hills commission—when little more than Lee's head had been finished. Augustus Lukeman replaced Borglum, dynamited most of the earlier work, and started again. Disputes over property ownership halted the project in 1928, and it was not revived until 1960, when an international competition led to the appointment of Walker Hancock as chief carver. He started work in 1964, making only slight modifications to the Lukeman design. The use of thermo-jet torches allowed for rapid, accurate removal of the stone and, in collaboration with Roy

Faulkner, Hancock had the gigantic memorial finished by 1972.

The grandiose neoclassical character and the gigantic size of Mount Rushmore and similar projects call for comment about our seemingly irresistible need to enshrine ideals that are anything but inhuman through overwhelming and inhuman scale. Consider, for example, the 150-foot (45-meter) Statue of Liberty or the Cristo Redentore above Rio de Janeiro. On the other hand, colossi have been built for reasons of vainglory: the Colossus of Rhodes collapsed after one generation; the 120-foot (36-meter) statue of Nero (originally near the Roman Colosseum and providing its name) is long gone. One of the multitude of Egypt's Ramessean statues is described by the poet Percy Shelley as a colossal wreck, "two vast and trunkless legs of stone." Destroyed by nature or by conquerors, such works are at once monuments to our engineering ingenuity and our transience.

Further reading

Shaff, Howard, and Audrey Shaff. 1985. *Six Wars at a Time: The Life and Times of Gutzon Borglum.* Sioux Falls, SD: Permelia.

Smith, Rex Alan. 1985. *The Carving of Mount Rushmore.* New York: Abbeville Press.

Zeitner, June, and Lincoln Borglum. 1976. *Borglum's Unfinished Dream: Mount Rushmore.* Aberdeen, SD: North Plains Press.

Mycenae, Greece

Imposing even as a ruin, Agamemnon's city Mycenae—Homer called it "Mycenae, rich in gold"—stands on a foothill of Mount Euboea between Hagios Elias and Mount Zara near the modern village that still bears its name: Mikínai. Seat of the semilegendary Atreus, it is also rich in tragic myth. Atreus's dynasty was cursed because he fed his brother Thyestes with his own children. His son Agamemnon sacrificed his daughter to gain fair winds to take his war fleet to Troy. When he returned, his wife Clytemnestra killed him; she in turn was killed by her son, Orestes. Except on the southeast, where a steep ravine provided natural fortification, the citadel or acropolis (high city) of Mycenae was surrounded by massive and daunting walls. Parts were of polygonal masonry, with

shaped stones fitted together, and the gates were built of finely dressed ashlar. But most of the defenses were built of "cyclopean" masonry, so named because the later Greeks, unable to accept that humans could have moved such huge blocks, attributed them to the mythical giant Cyclops. The true purpose of such gigantic walls is still debated by scholars: they were certainly defensive, but some suggest they may have been employed more as a *show* of strength. Whatever the case, for engineering audacity and skill they challenge even our modern imaginations.

The generic term "Mycenaean" is used for the Late Bronze Age (Helladic) culture that arose on the Greek mainland around 1650 B.C. and whose powerful, militaristic city-states dominated the region from 1400 until 1100 B.C. Mycenaean navies controlled the Aegean and colonized Crete, Cyprus, the Dodecanese, northern Greece, Macedonia, Asia Minor, Sicily, and parts of Italy. Then they seem to have outgrown their resources, and despite an attempt to secure the Black Sea grain routes by annexing Troy (sometime between 1250 and 1180 B.C.), the Mycenaean culture suffered such attrition that it was easily subsumed by the migrating Dorians a century later.

From its hilltop at an elevation of about 900 feet (270 meters), the citadel of Mycenae commanded a large, fertile hinterland and the Plain of Argos extended before it; the major route between the Bay of Argos and Corinth, Thebes, and Athens to the north passed under its ramparts. There had been neolithic and early Helladic use of the site between 3000 and 2800 B.C., but the earliest significant developments took place in the seventeenth century. Indeed, most of the surviving defenses date from after 1380 B.C., built in three major stages—ca. 1350, 1250, and 1225.

The walls of Mycenae were generally between 15 and 35 feet (4.7 and 10.7 meters) high, rising in places to 56 feet (17 meters); parts of them were as much as 46 feet (14 meters) thick. The earliest circuit (ca. 1350) enclosed the *megaron* (palace) precinct with all its ancillary buildings. About 100 years later the walls were extended to include the main western gate and an older grave pit close to it. Another gate, much smaller but just as cunningly designed

The Lion Gate, Mycenae, Greece (ca. 1250 B.C.), set in the cyclopean masonry walls of the citadel.

for defense, is on the north of the citadel. Around the same time, a tunnel was built in cyclopean masonry, leading to a subterranean spring-fed cistern on the northeast side.

Impressive as they are, the walls of Mycenae pale beside the splendor of the western gate, now known as the Lion Gate (ca. 1250 B.C.), mainly because of the majestic sculpture—the earliest large relief sculpture on the Greek mainland—that crowns it. The gate had a forecourt about 50 feet long by 25 feet wide (15 by 7.5 meters). The 10-foot-square (3.5-meter) opening was formed by four massive stone blocks (a threshold, flanking pilasters, and lintel), averaging about 12 by 7 by 3 feet (3.5 by 2 by 1 meters) in size. The double gates themselves were of bronze-sheathed timber. The remarkable feature was above the lintel. The corbeled triangular opening (known as a "relieving triangle") was invented by the Mycenaeans to divert the huge loads of the upper wall masonry away from the lintel and into the jambs—a major step forward in civil engineering. Here the opening was filled by a relatively thin stone panel bearing a relief carving of two lionesses flanking or adoring a column. The composition evokes many earlier relics found on Crete, and the overt symbolism born of this agrarian culture's emphasis on fertility should not be lost on us. From the Lion Gate, a 12-foot-wide (3.6-meter) road—Homer described the "broad streets of Mycenae"—led via a terraced ramp toward the defensible entrance of the flat-roofed megaron and its associated complex of buildings near the summit. Most of the palace has been lost.

The citadel survived an attack around 1200 B.C., only to be destroyed, possibly by invading Dorians, about a century later. The walls were not pulled down and the buildings outside, found near every Helladic acropolis, were not deserted. It seems that Mycenae was continuously occupied in some form until about 468 B.C., when the small preclassical city built on the ruins of the ancient citadel was destroyed by Argos and its population banished. The city was briefly reoccupied in the third century B.C. A new temple was built at the summit of the acropolis and the city wall repaired. There is some evidence of Roman occupation, but when the Greek traveler Pausanias visited the region around A.D. 160, he found only ruins. Serious archeological investigations began in 1841 and have continued intermittently.

See also Treasury of Atreus

Further reading

Fitton, J. Lesley. 1996. *The Discovery of the Greek Bronze Age.* Cambridge, MA: Harvard University Press.

Mylonas, George E. 1966. *Mycenae and the Mycenaean Age.* Princeton, NJ: Princeton University Press.

Simpson, Richard Hope. 1981. *Mycenaean Greece.* Park Ridge, NJ: Noyes Press.

Mystra, Greece

The ruins of the medieval city of Mystra are 3 miles (5 kilometers) northwest of modern Sparta in the Peloponnese. In 1204 the Fourth Crusade, turned aside from its original purpose by Venetian bribes, sacked Constantinople and established Frankish dominion over Greek territories. Among the most important states they founded was the Principality of the Morea, or the Principality of Achaea, governed from 1210 by Geoffroi I de Villehardouin. In 1249 his second son, Guillaume II de Villehardouin, built a castle atop a steep cone-shaped foothill overlooking the fertile valley of Eurotas and strategically commanding the Taygetos Range to the west and the valley of Laconia to the east.

Over the next few centuries the city of Mystra grew on the slopes below. Its name probably comes from the shape of the hill, which resembled a Myzethra cheese. Mystra, with a population that once exceeded 42,000, has been dubbed the "wonder of the Morea." Like Venice, but for different reasons, it occupies a site that is totally inappropriate for a city, and its construction was a significant architectural achievement.

In 1261 the Byzantine emperor Michael VIII Palaeologus regained Constantinople. The following year, Guillaume II de Villehardouin paid his ransom—he had been captured in 1259—with a number of castles including Mystra, and Michael VIII installed a Byzantine despot. The Villehardouin line survived until 1301, when Philip of Savoy became Prince of Morea. Throughout most of the fourteenth century the principality was in the hands of the Angevin House of Naples, and then controlled by the Venetians. The Byzantines regained it through matrimonial and political alliances and in 1448 Constantine XI Paleologus, the last Byzantine emperor, was crowned at Mystra. For about 350 years after 1460 Turks and Venetians took and retook the city. In 1821 it was among the first places the Greeks liberated from their Turkish oppressors. Ironically, the demise of Mystra was brought about by the foundation of the modern town of Sparta in 1834. The first inhabitants came from the old city; others built the modern village of Mystra.

Mystra has had a tumultuous history, and the different traditions of its occupiers account for its hybridized architecture. In the mid–thirteenth century, the Byzantines' persistent attempts to expel the Franks caused anxiety among the local populace. Many left the Eurotas plain to settle closer to the castle of Mystra. Houses were built on the lower slopes of the hill, and soon churches were constructed, clinging to the mountainside. This precipitous medieval city was surrounded by inner and outer circuit walls, commissioned in 1249 by Guillaume II de Villehardouin, and later repaired and augmented by the Byzantines and the Turks when they occupied the city. The walls were fortified by high rectangular towers, and of course dominated by the castle. They can hardly be described as concentric, because they snaked along contours and plummeted down steep slopes; nevertheless, they contained and defended the city. On its northeast and west sides the craggy hill of Mystra climbs sheer from the narrow valley. The defensive walls divided Mystra into the lower and upper quarters: the urban classes lived in the former, while the aristocracy occupied the latter with its palaces, two- or three-story vaulted mansions, and various administrative buildings. Two heavily fortified gates—the Monembassia and the Nauplia—linked them.

The L-shaped Palace of the Despots, possibly begun by Guillaume II de Villehardouin and built in stages between the thirteenth and fifteenth centuries, occupies an incongruously flat terrace overlooking the Eurotas Valley to the east. The two wings housed many different functions: the private apartments, a palace chapel, an open colonnade, and a large well-lighted hall for assemblies and ceremonies. Just north of the palace stood the mid-fourteenth-century church of Hagia Sofia, a centrally planned funerary chapel for the despot Manouil Katakouzenos. The winding streets of Mystra, as they followed the contours of the hillside, are lined with churches, many built after the metropolitan bishop of Lacedaemonia—the medieval name for Sparta—transferred his cathedra to Mystra. Chief among them is the "mixed architectural type" cathedral: the Metropolis of St. Demetrios (ca. 1309) is a three-aisled basilica at its lower level; the fifteenth-century upper floor, consisting of a women's gallery, is a cross-in-square roofed with five cupolas. Many churches—the thirteenth-century Church of St. Theodore, the Church of the Virgin Evangelistria, and the Peribleptos Monastery (both fourteenth century)—were purely Byzantine in form. Apart from the fifteenth-century Pantanassa Convent, which is still in use, the buildings of Mystra have been reduced to ruins, some by fire, others by being used as quarries when modern Sparta was being built. A few fine frescoes survive; many more have been destroyed.

Extensive restoration work has been undertaken over many years by the Committee for the Restoration of the Mystras Monuments and the Fifth Ephorate of Byzantine Antiquities. Mystra was inscribed on UNESCO's World Heritage List in 1989.

See also Venice, Italy

Further reading

Chatzidakis, Manolis. 1981. *Mystras: The Medieval City and the Castle*. Athens: Ekdotike Athenon.
Runciman, Steve. 1980. *Mistra: Byzantine Capital of the Peloponnese*. London: Thames and Hudson.

N

Nazca Lines
Peru

The Pampa Colorada (Red Plain) is a 37-mile-long (60-kilometer) and 15-mile-wide (24-kilometer) plateau in the coastal desert of southern Peru near the town of Nazca. Across its broad face are carved staggeringly cyclopean patterns, an agglomeration of designs on the earth's surface known as geoglyphs, which portray animals, birds, and other forms, mostly made by removing the dark reddish brown surface to expose a lighter-colored substratum; in some places piled rocks define the enigmatic forms. The challenge presented to the modern imagination by this ancient engineering feat is threefold: its momentous scale and the accuracy of surveying techniques that could project straight lines for miles over irregular terrain are remarkable enough. Beyond them is the uncanny ability of a people whose entire spatial experience was planar, never far above the surface of the earth, to conceive of geometric patterns and representational images whose accuracy and intricacy could be fully appreciated only from high—indeed, very high—above.

The Nazca Lines, as they are called, comprise literally thousands of zigzag, parallel, crossed, or radiating lines: some are 6 feet (1.8 meters) wide, others just a tenth of that. Some stretch for 6 miles (10 kilometers), maintaining their straightness regardless of the uneven topography. There are also simple or complex geometric shapes, including triangles and rectangles, nearly twenty varieties of fantastic birds, a monkey, a spider, a dog, a fish, a tree, and a hummingbird represented. As to their size: the monkey occupies the area of a football stadium; one bird has a 350-foot (100-meter) wingspan; and the spider, among the smallest geoglyphs, has a diameter of 150 feet (45 meters). Together, the lines and figures cover 45 square miles (115 square kilometers). Of course, they are best seen from above and were discovered only when aircraft first crossed the area in the 1930s.

The origin of the lines remains uncertain, although because of their similarity to design motifs on other artifacts, they are attributed to the well-developed Nazca civilization, which flourished between 200 B.C. and A.D. 600. Based on the same evidence, some sources suggest that three successive cultures were responsible for the lines: the Paracas (900–200 B.C.), the Nazcas, and later settlers who migrated from Ayacucho around A.D. 630.

Each culture was agrarian and it is likely that the lines may have been associated with rituals to guarantee a rich crop. On the other hand, the German anthropologist Dr. Maria Reiche, who studied the Nazca Lines for nearly fifty years, believed that they were a vast astronomical calendar, also associated with farming. Studies in the 1980s led others to the conclusion that, while part of elaborate rituals related to fertility, the lines had neither astronomical

Nazca Lines, Pampa Colorada, Peru, date uncertain (perhaps 200 B.C.–A.D. 600). Aerial view of the 300-foot-long (90-meter) monkey carving.

nor calendrical significance. A decade later a new theory emerged: they charted the origins and courses of aquifers—rivers beneath the desert—associated with irrigation farming in the region. In our modern culture of scientism we disengage the rational from the spiritual, and care must be taken to avoid too simple an interpretation of the actions of people whose universe was better integrated. All of the suggestions about the purpose of the Nazca Lines could be accurate.

Even in their own time and place the Nazca Lines were not an isolated phenomenon. Many geoglyphs are to be found throughout South America. Areas with lines and figures very like Nazca's have been studied on the central Peruvian coast between the Fortaleza, Pativilca, and Rimac Valleys. Others have been found in the Viru Valley, on Peru's north coast, and in the Zana Valley, more than 600 miles (1,000 kilometers) north of Nazca. More examples of ground figures and hill figures survive on the other side of

the world. The 370-foot-long (110-meter) White Horse (ca. 500 B.C.) cut into the chalk hills at Uffington is among Britain's most famous, second only to the pre-Christian Cerne Giant in Dorset, a 180-foot-tall (54-meter) human figure, carrying a 120-foot (36-meter) club; there is also the Long Man of Wilmington, Sussex.

As late as 1998 a 2.5-mile-tall (4-kilometer) figure of an aboriginal warrior was discovered carved on the desert floor near Marree in the South Australian outback. It was soon exposed as a hoax, created with the help of satellite tracking equipment and earthmoving machinery. The very fact of the difficulty of making such a figure using modern technology emphasizes more the incredible achievement of the ancient creators of the Nazca Lines.

Further reading

Aveni, Anthony, ed. 1990. *The Lines of Nazca*. Philadelphia: American Philosophical Society.

Reiche, Maria. 1949. *Mystery on the Desert: A Study of the Ancient Figure … near Nazca, Peru.* Lima: n.p.

Reinhard, Johan. 1986. *The Nazca Lines: A New Perspective on Their Origin and Meaning.* Lima: Editorial Los Pinos.

Nemrud Dagi
Turkey

The *hierotheseion* (royal burial precinct) of King Antiochos I of Kommagene (reigned ca. 69–36 B.C.) stands on Nemrud Dagi, the highest point of his domain, near the modern village of Kahta in the southeastern Turkish province of Adiyaman. It has been characterized by UNESCO as "one of the most ambitious constructions of Hellen[ist]ic times." The megalomaniac king reshaped the 7,000-foot-high (2,150-meter) mountain by leveling the rock and filling the artificial platform with huge statues of himself and the gods (whom he claimed as kin); he then ordered a 500-foot-diameter (150-meter), 163-foot-high (50-meter) tumulus (artificial peak) of fist-sized rocks to replace the natural summit. It is believed that his tomb, yet unopened, lies beneath the massive pile of rubble.

Kommagene was a small buffer state between the Roman Empire and the kingdom of Persia. Located between the Amanos Mountains and the upper Euphrates, its capital Samosata commanded a strategic crossing of the great river. Mithradates' father, Ptolemy, used that fact to seize control of the resource-rich area. It became an independent state in 162 B.C. After a brief subjection of the area to the Armenians, in 69 B.C. the Roman general Pompey installed Antiochos I on the throne. About 100 years later King Antiochos IV lost his wars with Rome and Vespasian absorbed Kommagene into the province of Syria.

Antiochos I attempted to establish a new order. His first action was to build a hierotheseion to his father Mithradates Kallinikos I (died 63 B.C.) in the city of Arsameia (now Eski Kale). Its decorations and inscriptions made it clear that Antiochos intended to Hellenize the Kommagenian culture, uniting the Persian Parthian world with the Greco-Roman; in effect, he set out to establish a new religion in which his own assumed divinity loomed large. Nowhere was that more evident than in his own hierotheseion on Nemrud Dagi.

The great tumulus is flanked on the east, west, and north by terraces carved from the mountain; it has been estimated that their creation involved the removal of 7 million cubic feet (200,000 cubic meters) of rock cut away by hand. On the east terrace stood an array of statues of the king and the gods, up to 33 feet (10 meters) high, carved from massive stone blocks mined in a remote quarry. The figures were set in order and identified by inscriptions written in Greek and Persian: Antiochos himself, the mother goddess Kommagene, the father god Zeus-Oromasdes (largest of the statues), Apollo-Mithras, and Herakles-Artagnes. Their faces were finely carved in the late Hellenistic style. At either end, the row of deities was guarded by the royal symbols: an eagle and a lion. At the eastern corner of the terrace stood a pyramidal altar of fire, and various elements around the platform carried carved relief portraits of the illustrious Persian and Macedonian ancestors whom Antiochos claimed as his own. Other relief decoration abounded.

As far as the topography would allow, the west terrace, set some 33 feet (10 meters) lower than the east, was organized in the same way, to much the same purpose: the apotheosis of Antiochos. The syncretized Persian and Greek gods facing east and west on the respective terraces revealed Antiochos's attempted cultural synthesis. One inscription asserted that he had commissioned the site for posterity "as a debt of thanks to the gods and to his deified ancestors for their manifest assistance"; he wanted to set for his people an example of the piety due "towards the gods and towards ancestors."

The north terrace, 269 feet (80 meters) long, was used for assemblies and rituals and also served as a processional way connecting the other terraces. Gigantic stone eagles flanked its entrance. The great tumulus was built on a rocky hill framed by the terraces. According to inscriptions, this was the place where Antiochos ordered that his remains should be buried. He died before his elaborate project was completed, and his son Mithradates neither finished the monumental work nor promoted the religious synthesis begun by his father. The site was abandoned, the last of its priests probably leaving soon after A.D. 72.

Nemrud Dagi was rediscovered in 1881 by one Karl Sester; an 1882–1883 German exploratory expedition followed, as well as a Turkish investigation. The findings of both groups were published, but no more research was carried out until 1938, when Germans F. Karl Dörner and Rudolf Naumann visited the site. Dörner returned after 1951 to work with the American Teresa Goell. In 1984–1985 a Turkish-German restoration team, led by Dörner, reerected the bases of the statues in their places. In 1987 the site was inscribed on UNESCO's World Heritage List, and the following year a 35,000-acre (13,850-hectare) region around Nemrud Dagi was declared a national park. In July 1997 the Turkish government assured the world that the stone heads—all had fallen from their places—would be reset, and measures would be taken to protect the site, not only from natural damage but also from that caused by vandals or just careless tourists. Eighteen months later, the Netherlands-based International Nemrud Foundation received presidential support for a five-year master plan to restore the site, and work commenced at the end of May 2000.

Further reading

Bey, Osman Hamdy, and Osgan Effendi. 1987. *Le tumulus de Nemroud-Dagh*. Istanbul: Archaeology and Art Publications.

Sanders, Donald H., ed. 1996. *Nemrud Dagi: The Hierothesion of Antiochos I of Kommagene*. Winona Lake, IN: Eisenbrauns.

Newgrange
County Meath, Ireland

Newgrange is one of the most notable archeological monuments in Europe. Named in Gaelic Uaimh na Gréine (Cave of the Sun), the great passage tomb stands on a low hillock beside the River Boyne in County Meath, Ireland, about 9 miles (14 kilometers) from the sea. Newgrange was built around 3150 B.C., making it as old as some of the neolithic temples on Malta and much older than the pyramids of Egypt. It is a dramatic testimony to the ancient Celts' scien-

tific and architectural sophistication. Its designers employed great mathematical skills to create such an uncannily accurate astronomical instrument of gargantuan scale. It forms the center of Brú na Bóinne, a region steeped in megalithic culture and ritual. Around it are more than forty prehistoric sites: standing stones, burial mounds, and other passage tombs. Irish mythology identifies Newgrange as the burial place of the high kings of Tara and the home of a preternatural race known as Tuatha de Danainn (people of the goddess Danu); other traditions are attached to the mystical place.

Newgrange is a colossal stone-and-turf tumulus, 1 acre (0.4 hectare) in area and approximately circular in plan, averaging about 280 feet (85 meters) in diameter; the top of its flattish dome is 44 feet (13.5 meters) high. The mound is surrounded by a retaining wall of white quartz and water-washed round granite boulders standing on a foundation of ninety-seven huge curbstones, many of which are decorated with incised patterns of triple and double spirals, concentric semicircles, lozenges, and zigzag lines. It has been estimated that there are some 224,000 tons (203,200 tonnes) of material in the structure. None of the stone is local: the curbstones and those used inside the tumulus were quarried about 20 miles (36 kilometers) from the site; the quartz comes from Wicklow, about 60 miles (100 kilometers) to the south; and the 1,600 granite boulders come from the Mourne Mountains, just as far to the north. All were quarried, transported, dressed, and fitted into place using only stone tools, and without the use of the wheel. The mound was encircled by about 40 widely spaced standing stones, up to 8 feet (2.4 meters) high, in a 340-foot (104-meter) ring. They were probably erected about 1,000 years later. Only twelve survive. The reconstruction as it can now be seen is based on some scholars' interpretation of the position of the quartz layers found during excavations under the direction of Michael J. O'Kelly between 1962 and 1975.

For all its size, the mound encloses very little space. A single low passage, 3 feet (less than a meter) wide, penetrates 62 feet (19 meters) into the interior. The passage is lined with standing stones from 5 to 6.5 feet (1.5 to 2 meters) high and richly decorated with patterns similar to those on the curbstones. There are twenty-two standing stones on one side and twenty-one on the other, supporting a corbeled roof of flat stones. Following the profile of the hill, the floor rises until the passage terminates in a cruciform chamber measuring about 21 by 17 feet (6.4 by 5.2 meters), with a 20-foot-high (6-meter) corbeled roof. Its stones are carved with grooves that prevent rainwater from entering the interior. Three low apses, their walls also carved with intricate geometric designs, open from the central space. Each contains a massive stone basin.

The entrance, crowned with a rectangular opening known as a "roof box," and the passage are exactly designed so that, at dawn on the winter solstice, a shaft of sunlight penetrates to illuminate the central chamber and a triple spiral on one of the great basins. Of course, the tangential sunlight also vitalizes the carvings on the passage walls, bringing life in the depth of winter. The astrophysicist Thomas Ray has calculated that the architecture was not approximately oriented but amazingly accurate. Five millennia ago, as viewed from the inner chamber, the gap in the roof box would have matched almost exactly the sun's apparent width. Ray demonstrates that the first beam would strike the exact center line of the floor in a patch of intense light about 6 feet (2 meters) long and just a few inches wide. Then in the space of twenty minutes it would broaden, narrow again, and withdraw.

Conjecture abounds about the purpose of Newgrange, although the truth is shrouded in mystery. It seems clear that it was more than a tomb. Cremated remains found on the floor originally had been placed in the basins in only two of the recesses; the center one, whose triple spiral is annually illuminated by the rays of the sun, contained no remains. Some sources suggest that it was the focus of religious rites and only occasionally used for burials; others think that its purpose changed over centuries; and still others that it was simply a giant calendar—the least acceptable of all the explanations. Archeological investigation continues. Newgrange is open to the public and the inevitable impact of large numbers of visitors—close to 200,000 a year—is endangering its ancient fabric. Although it remains as weatherproof as ever, the humidity from tourists' breath is a growing threat that the ancient builders could never have foreseen. It is expected that access will be restricted, especially during the summer months.

Further reading

Brennan, Martin. 1980. *The Boyne Valley Vision*. Portlaoise, Ireland: Dolmen Press.

O'Kelly, Michael J., et al. 1982. *Newgrange: Archaeology, Art, and Legend*. London: Thames and Hudson.

Ray, Thomas P. 1989. The Winter Solstice Phenomenon at Newgrange, Ireland: Accident or Design? *Nature* 337: 343–345.

O

Offa's Dike
English-Welsh border

Built by Offa, King of Mercia (A.D. 757–796), the impressive earthwork known as Offa's Dike formed a boundary, albeit discontinuous, between England and Wales. One of the most remarkable structures in Britain, it runs 177 miles (280 kilometers)—7.5 miles (12 kilometers) longer than Hadrian's Wall—from the Dee Estuary in the north to the River Wye in the south. It is now generally agreed that the dike was not so much a fortification as a substantial line of demarcation. No earlier Anglo-Saxon king had unified southern England as Offa did, and with unity came power and wealth. He also formed ties with rulers across the channel and was accepted as an equal by Charlemagne, king of the Franks, with whom he entered a commercial treaty in 796. Even Pope Adrian I treated him with great respect.

Some have suggested that the dike was built to give Mercia command of the approaches to the English lowlands, but parts of it rise as much as 1,300 feet (400 meters) above sea level. The boundary it marked was hardly precise; during Offa's reign there were English communities to the west of it and Welsh communities to the east. Be that as it may, the dike's very presence made a strong statement about separation. There is an apocryphal story that "it was [once] customary for the English to cut off the ears of every Welshman who was found to the east of the dike, and for the Welsh to hang every Englishman whom they found to the west of it."

The dike consists of an earth bank in places 20 feet (6 meters) high with a 12-foot-deep (3.6-meter) ditch on the western side; sometimes their combined width was 60 feet (18 meters). Natural features seemed to be used wherever practicable, but for the most part an earth embankment was built—a total of 80 miles (130 kilometers). Elsewhere it was discontinuous, giving rise to the speculation that, having been initiated late in Offa's reign, it was never finished. On the other hand, perhaps in those locations, the local topography served the same purpose. Welsh historian John Davies writes that Offa's Dike was "perhaps the most striking man-made boundary in the whole of Western Europe." Thousands of workers must have toiled to build it, evidence of Offa's resources and the integrity of his kingdom. In places the straightness of its line for kilometers is evidence of the technical skills of its builders. No written records about the project survive. After 1066 the Norman invaders saw the value of the dike for defense, and many castles and abbeys of early date stand in its shadow on the eastern side.

Offa's Dike is a reminder of persistent Welsh-English antipathy. Although there is everywhere in Britain a challenge to the myth of a "united kingdom," lately evidenced by the Scottish and Welsh elections of 1999, the signs of cultural disintegration are nowhere stronger than in Wales. That is expressed even in language differences. As someone has commented: for the source, "we must look to the mid eighth century, when a long ditch was constructed, flanking a high earthen rampart that divided [the Celts from the Saxons] and which even today marks the boundary between those who consider themselves Welsh [and] those who consider themselves English."

Further reading

Noble, Frank. 1983. *Offa's Dike Reviewed*. Oxford, UK: B.A.R.

Fox, Cyril. 1955. *Offa's Dike: A Field Survey of the Western Frontier-Works of Mercia*. Oxford, UK: Oxford University Press.

Jones, John B. 1976. *Offa's Dike Path*. London: H.M.S.O.

Orders of architecture

To the ancient Greeks, the word "cosmos" conveyed the idea of "the garnished universe," the "world set in order." They believed that creation was the act of a great Demiurge who brought structure and form out of preexistent chaos, an ordering that permeated the physical universe. It was therefore perceptible in nature as a ubiquitous mathematical proportional system, a harmony in everything that could be seen or heard. To be in accord with that harmony, their own creations—music, sculpture, architecture—needed to correspond to cosmic order. Their great architectural achievement was to seek for that truth and express it in the development of three systems of building, each with its distinctive proportions, detail, and form, according to the culture that generated it. Those systems are known as the Doric, Ionic, and—little used by the Greeks—Corinthian orders of architecture. Each comprises a column with a base (in the case of the latter two), shaft, and capital, and a supported entablature, consisting of architrave, frieze, and cornice. Each was governed by an evolving system of proportions, linked to the *module*, the base diameter of the column; each thus imposed architectural order.

The Doric order developed in the regions speaking a Dorian dialect, that is, mainland Greece and the colonies in southern Italy, Sicily, and further west. It is clearly derived from an earlier timber architecture, and when the transition was made from wood to stone in order to produce more appropriate, du-rable buildings for the gods, the form and details, having gained a kind of sanctity, were translated to the new material, right down to the fixing pegs. The sixth-century Temple of Hera (the so-called basilica) at Paestum, Italy, is a well-preserved example of the archaic form, with its squat proportions, coarse moldings, and heavy entablature. The classical quest for cosmic harmony led to refinements of form and detail until the Dorian Greeks achieved what appears to have been a satisfactory conclusion in the proportional balance and visual nuances of the Parthenon, Athens (447–432 B.C.).

The baseless column of the Doric order, rising directly from the temple platform *(stylobate)* made necessary by the uneven terrain had a tapering shaft with twenty shallow flutes separated by sharp arrises. The capital consisted of the convex *echinus* molding crowned with a flat rectangular slab *(abacus)*. The plain architrave of stone blocks, with a molding at the top decorated with raised panels *(regulae)* and round projections *(guttae)*, spanned from column to column. Above it was the frieze, consisting of double-grooved slabs *(triglyphs)* alternating with plain panels *(metopes)*. The metopes and the relief sculptures that usually decorated them were painted in bright colors. The order was crowned by an overhanging molded cornice decorated with flower or figure sculptures.

The Greeks continued to use the Doric order until about the second century B.C. But it presented several difficulties. First, with no base to protect it, the column was subjected to wear and accidental damage. Second, it was extremely difficult to make: not only did the columns taper but they were also carved with a slight swelling *(entasis)* about halfway up to make them *look* straight. Coupled with the need to maintain sharp arrises between the flutes, that demanded very skillful masons' work. Third, the placement of the triglyph was problematic. Because it was impossible to locate one over the center of each column and at the midpoint of the spaces between the columns, the appearance was regarded as unwieldy.

The Ionic order, which was fully developed by the sixth century B.C., was created by Greeks who established colonies along the southwestern coast of Asia Minor (now Turkey). The most remarkable Ionic

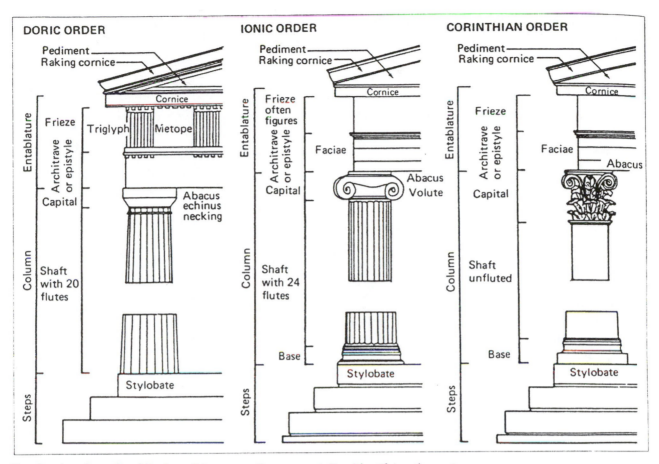

The Greek orders of architecture. Diagrammatic representation identifying the parts.

building in the region was the huge Artemiseion at Ephesus, one of the seven wonders of the ancient world, commenced some time around 400 B.C. By then, the Ionic order had been long established on the Greek mainland, and the Erechtheion (421–406 B.C.) and the Temple of Athena Nike (427–424 B.C.), both on the Athenian Acropolis, are fine examples. It is of more slender proportion than the Doric. The molded base of the column rested on a stylobate, and the shaft, with twenty-four deep flutes (separated by narrow flat surfaces rather than sharp arrises), carried a capital—usually carved from a single block—with symmetrical spiral scrolls (volutes) flanking an echinus molding ornamented with an egg-and-dart pattern and supporting an abacus.

The earliest Ionic capitals had rosettes in place of volutes, and the origins of the spiral pattern are obscure. Suggested sources have been rams' horns, bulls' horns, the foliage pattern of palm trees, and even the helical form of some seashells. Whatever the case,

unlike the Doric, the capital was best viewed from front or back, and that presented a difficulty for designers, especially where the row of columns turned the corner of a building. Then, the outer volute on the corner column was turned outward at 45 degrees to make it "right" from two sides—a compromise, not a best solution.

The Ionic entablature had an architrave comprising three bands, each projecting beyond the one beneath; the highest was narrower than the other two. Above was a continuous frieze, usually encrusted with sculpture, and at the top sat a cornice enriched with dentil moldings, a corona, and a cyma molding.

The Corinthian was the most ornate order of architecture. It was also the latest, reaching its mature form around the middle of the fourth century B.C. Apart from its distinctive and elaborate capital—a sort of inverted bell, covered with carvings of acanthus leaves—it is otherwise very similar to the Ionic order, although the proportions are more slender.

The oldest known example is in the *cella* of the Temple of the Epicurean Apollo at Bassae (ca. 420 B.C.). Among the chief examples are the circular structure known as the Choragic Monument of Lysicrates (334 B.C.), the earliest surviving building with external Corinthian columns, and the octagonal Horologion of Andronicos (also known as the Tower of the Winds), with two Corinthian porches (before 50 B.C.). Both are in Athens. Also in that city was the massive Temple of the Olympian Zeus, started about 530 B.C. and completed by Hadrian in the second century A.D. It was perhaps the most notable of all Corinthian temples; it was certainly the largest. The Corinthian order was seldom used by the Greeks, although it solved the problems that had been presented by the Doric and Ionic orders. However, it was enthusiastically developed in the Roman world.

The Romans copied Greek art and architecture, captivated by the forms rather than the cosmology that generated them. They employed the Corinthian order more than any other, cosmetically modifying it by changing the column base, adding complicated carved embellishments to the cornice, and producing all manner of fanciful variations to the capital, with showy leafage and sometimes grotesque human and animal figures. The so-called Composite order, attributed to the Romans by sixteenth-century writers, was simply a distortion of the Greek precedents, combining Ionic volutes with Corinthian acanthus motifs. The Romans also used the Ionic order but seem to have been too impatient to achieve the refinements of the Doric, inventing their own version. The Roman Doric, used infrequently, was also influenced by a slender column (with a base) developed by the Etruscans. The Tuscan order, known only from the account of the first-century architectural writer Vitruvius, closely resembled the Roman Doric.

The classical orders were eclipsed by the rise of Christianity, although they persisted in vestigial forms. An interest in Vitruvius was awakened in fifteenth-century Italy, and architects, captivated by all things Roman, made archeological studies of ancient ruins and employed the orders, often in an intuitive, uninformed way. Leone Battista Alberti and other more derivative architectural writers, including Serlio, Scamozzi, Vignola, and Palladio, urged the systematic application of the Roman—not the Greek—orders, with pedantic rules of proportion. Later, beyond Italy, the Frenchmen Philibert Delorme and Claude Perrault and the Englishman William Chambers wrote theoretical treatises about the architectural orders. Subsequently, during the artistic period known as the Greek Revival, strict conformity to the proportions of the original Greek models was practiced. Modern architecture had no place for the orders, but with the rise of postmodern architecture in the second half of the twentieth century, they appeared again, often so abstracted and deformed as to be barely recognizable. The American architect Charles Moore even invented one to flank his lively Piazza d'Italia in New Orleans (1977–1978); he named it the Neon order.

See also Erechtheion

Further reading

Mauch, J. M. von, and C. P. J. Normand. 1998. *Parallel of the Classical Orders of Architecture*. New York: Acanthus.

Onians, John. 1988. *Bearers of Meaning: The Classical Orders in Antiquity, the Middle Ages, and the Renaissance*. Princeton, NJ: Princeton University Press.

Summerson, John. 1980. *The Classical Language of Architecture*. London: Thames and Hudson.

Øresund Link
Scandinavia

The ambitious project to construct a fixed rail and road link across the 65-mile (105-kilometer) Øresund Strait was agreed to by the Danish and Swedish governments in March 1991. The resulting 10-mile (16-kilometer) combination of submarine tunnel, artificial island, and bridge, carrying a double-track electrified railroad and four lanes of freeway between Kastrup, Denmark, and Lernacken, Sweden, was officially opened on 1 July 2000. Responding to complex social, cultural, economic, ecological, geological, and technological constraints, the transoceanic international highway is a major achievement of design and logistical skill, demonstrating that high technology and environmental sustainability are compatible.

Each country was responsible for extending its transport system to connect with the link. A/S

Øresund built the Danish land works, while Svensk-Danska Broförbindelsen constructed those in Sweden. The government-backed companies continue to be equal partners in the Øresund Consortium, initially charged with financing and building the link. Since completion, the consortium owns and operates it, funded by tolls charged to cover maintenance costs and to finance loan repayments.

It is reasonable to describe the link in a logical way, crossing from west to east. The Danish land works include about 11 miles (18 kilometers) of railroad from Copenhagen Central Station to the coast and 2.5 miles (4 kilometers) of track exclusively for freight trains. Work was completed in fall 1998, about a year after the Danish freeway system was extended to run beside the railroad. Where it passes through built-up areas, the wide transportation corridor is 23 feet (7 meters) below grade to reduce noise; it is even roofed over in places. Extensive planting—over 350,000 trees and shrubs—makes it less visually intrusive while dampening the traffic noise.

The coast-to-coast link begins with a 470-yard-long (430-meter) artificial peninsula at Kastrup on the Danish shore, built of dredged material from the Øresund seabed. It is the site of the entrance/exit of the 2.5-mile-long (4-kilometer) tunnel that joins Kastrup and an artificial island (also built of dredged material) across the Drogden Channel. The tunnel, the longest immersed tube for both road and rail in the world, is constructed from twenty prefabricated concrete elements, some weighing as much as 61,600 tons (56,000 tonnes). It contains separate well-lit, fire-resistant tubes for each direction of the freeway and the railroad; that is, four in all, as well as a central gallery for maintenance access and emergency evacuation. The components were cast at Copenhagen North Harbor, floated, and towed by four tugs to the tunnel trench about 8 miles (12 kilometers) away. When each element was lowered into place and secured, the trench was refilled around it. The top of the tunnel is about 35 feet (11 meters) below sea level.

Eastbound traffic emerges from the tunnel on a narrow, 2.7-mile-long (4.3-kilometer) artificial island that has been named Peberholm, over which the open highway and the railroad make the transition to the spectacular 4.9-mile-long (7.85-kilome-ter) Øresund Bridge, which gracefully curves across the sea, soaring to a clear height of 186 feet (57 meters) above the Flintrännan navigation channel between the Strait of Kattegatt and the Baltic. The two-level—freeway above, railroad below—cable-stayed High Bridge is supported by four pylon legs. It is 3,570 feet (1,092 meters) long, with a 1,600-foot (490-meter) main span. The High Bridge is connected to Peberholm Island and the Swedish coast by approach bridges, 2.34 and 1.88 miles (3.74 and 3.01 kilometers) long, respectively, carried on a total of 51 reinforced concrete piers. Many of the bridge components—caissons, piers, road decks, and super-structure—were fabricated in Karlskrona and Malmö, Sweden, others in Cadiz, Spain. They were lifted into place by the 9,500-ton capacity (8,600-tonne) floating crane *Svanen* using a satellite navigation system to achieve the necessary precision.

Land works in Sweden include a 4.7-mile (7.5-kilometer) extension of the existing Continental line, connecting the Øresund Link near the coastal town of Lernacken to Malmö Central Station and beyond. That stage of the work was finished in 1999. A new 6-mile (10-kilometer) freeway links Lernacken to the Outer Ring Road east of Malmö. The 2.6-mile (4.2-kilometer) City Tunnel, scheduled for completion by 2007, is planned to connect Malmö Central with the Øresund rail link. New tracks have been built for freight traffic. Freight trains are limited to speeds of 75 mph (120 kph), but the track is designed for high-speed passenger trains traveling at 125 mph (200 kph). The 34-mile (55-kilometer) international commuter journey between Sweden and Denmark takes just 30 minutes.

Environmental awareness characterized the entire project. The link was not allowed to affect water flow in Øresund, so the seabed was partly excavated to offset the impact of the peninsula, the artificial island, and the bridge pylons. Waste from the construction works was strictly monitored, and more than 7.82 million tons (7.11 million tonnes) of spoil from boring, dredging, and digging was used to create the artificial peninsula and the island. Moreover, the work was organized to avoid disruption of natural patterns, such as the seasonal growth of seaweed, the use of wetlands by birds, and the migration of schools of herring.

P

Paddington Station
London, England

Paddington Station, London, terminal of the Great Western Railway linking England's capital with the Atlantic port of Bristol, was designed by engineer Isambard Kingdom Brunel (1806–1859) with the assistance of architect Matthew Digby Wyatt (1820–1877). Built between 1850 and 1854, it was one of the first stations to utilize the iron-arched roof and the ridge-and-furrow roof glazing also employed in Joseph Paxton's Crystal Palace of 1851. It led to further exploitation of the iron arch in stations such as St. Pancras (1863–1865) and to extensive use of the roofing system.

Railroad terminals were a significant nineteenth-century architectural development that added a new building type to the townscape. There were two quite specific types of space required—a head building that housed the pedestrian entrance, ticket sales area, baggage storage, and refreshment and waiting rooms, and an adjacent shed with platforms at which trains and travelers arrived and departed. The building type presented architects with a dilemma since there was no existing morphological or stylistic precedent. Therefore, the designers of the earliest railroad stations merely adopted or adapted conventional building forms, materials, and styles. As the popularity of train travel increased, so did the need for wider station sheds to accommodate more tracks and plat-

forms; because of the limitations of traditional construction technology, structural advances and new materials such as iron and glass offered potential solutions. The relatively new profession of engineers, unrestrained by historicism, took up the challenge and designed sheds that exposed contemporary materials and advances and (most importantly) met their clients' demands. Sometimes they collaborated with an architect, as Brunel did by inviting Wyatt to design Paddington's decorative details. But despite innovation and audacity, the architectural form and esthetic of the railroad sheds was not well received, and they were obscured by a masonry head building that looked back to any of a number of past styles.

Brunel, son of French-born engineer Marc Brunel (1769–1849), served his apprenticeship supervising the construction of his father's Thames Tunnel project (1825–1843). Although work there was in abeyance, he won a commission for the acclaimed Clifton Suspension Bridge near Bristol (1829). In 1833 the promoters of the Great Western Railway appointed him engineer of the bold project to connect London by rail with the west of England. Brunel chose and surveyed the route and prepared plans. Despite outspoken opposition from landowners and rival transport providers, he argued the railroad's merits in lengthy public hearings and before a parliamentary committee. He was highly praised for his persistence and performance when the bill for the Great Western

Paddington Station, London; Isambard Kingdom Brunel, engineer; Matthew Digby Wyatt, architect, 1850–1854. Archival photograph of interior, showing platform sheds.

Railroad was finally approved in August 1835. During the line's construction, Brunel supervised the laying of tracks and the building of bridges, viaducts, and tunnels, including the 2-mile (3.2-kilometer) Box Tunnel, which took about two and a half years to complete. The 151-mile (241.6-kilometer) railroad was finished in mid-1841 with one terminus at Temple Meads in Bristol and the other at Paddington, then a new suburb west of London. Brunel designed the Gothic-inspired Temple Meads Station (1839–1840) with its striking unsupported timber-arch roof spanning 72 feet (21 meters). The first Paddington Station was a temporary structure fitted between the arches of a road bridge. The permanent and grander replacement commenced in 1850 and was finished four years later.

Paddington Station was built in a cutting and without a main facade. The Great Western Hotel (later Great Western Royal Hotel) of 1851–1853 by the architect Philip Hardwick (1820–1890) was next to the shed and served as its head building. Its Italian Renaissance detail, twin Jacobean towers, and mansard roof contrasted sharply with Brunel's airy, cathedral-like structure.

The entire train shed was 700 feet long, 238 feet wide, and 33 feet high (214 by 70 by 10 meters), covered by a triple-arched roof. The central arch, spanning 102 feet (31 meters), was flanked by one of 70 feet (21 meters) and another of 68 feet (20 meters). Two 50-foot-wide (15-meter) transepts integrated the space. In 1916, a fourth shed, spanning 109 feet (30 meters), was added on the northeast side. Originally, in the absence of a southern concourse, a retractable drawbridge provided access to the inner platforms. The roof, ironclad and partly glazed, was carried on slender wrought-iron arches; a cast-iron column bolted to a brick foundation supported every third rib. Wyatt provided the neo-Renaissance-cum-Moorish embellishments to the columns and the sinuous wrought-iron ornamentation in the glazed end-arches. The effect of the spacious train shed interior has been described as "dramatic" and its ambiance as that of a greenhouse.

Paddington Station is now a heritage-listed building under redevelopment by architects Nicholas Grimshaw and Partners as the central London terminal for the express train to Heathrow Airport. A concourse extension, named the "Lawn Area" because it incorporates the site of the station master's former garden, has been covered by a glass roof. Glass and aluminum have been used extensively to enhance the character of Brunel's building. The second stage of the project includes replacement of the 1916 train shed with a new concourse opening to the adjacent Regent's Canal; a prominent transfer structure is also proposed with more buildings, including a 42-story tower block.

See also Crystal Palace; St. Pancras Station; Thames Tunnel

Further reading

Binney, Marcus, and David Pearce, eds. 1979. *Railway Architecture*. London: Oris Publishing.

Bryan, T. 1999. *Railway Heritage*. London: Silver Link Publishing.

Clinker, C. R. 1979. *Paddington, 1854–1979: An Official History of Brunel's Famous London Railway Station in Its 125th Year*. London: Avon-Anglia Publications.

Palace of Minos
Knossos, Crete

Probably the best known of all Cretan architecture, the ruins of the Palace of Minos at Knossos stand near the River Kairatos on the north side of Crete, about 3 miles (5 kilometers) inland, near the modern city of Iráklion. The complexity of its buildings defies verbal description, and its sophisticated planning and lively decoration separate it from its contemporaries anywhere in the world.

Crete is naturally fortified by the sea and was protected by a powerful navy, so the town site was chosen probably for reasons other than defense: an elevated position, a good water supply, access to the coast, and a forest providing building lumber. At the height of its glory, around 1650 B.C., Knossos extended its boundaries. The English archeologist Arthur Evans, perhaps a little generously, claimed that the population reached 80,000.

It is generally thought that Crete was first inhabited in the seventh or sixth millennium B.C. by migrating neolithic farmers from Asia Minor. By around 3000 B.C., they were using bronze tools, and over the next 800 years or so, a sophisticated society emerged that traded with Egypt and invented its own hieroglyphic script. What is known as the Age of the First Palaces has been dated at 2200–1700 B.C. Four large sites have been identified: Phaestos, Mallia, Zakros, and Knossos. Each took the same general form, comprising a number of multistory wings grouped around a central courtyard and built of substantial timber frames, filled with plastered stone blocks. The complex layouts included luxurious living quarters, dining halls, artists' workshops, basement warehouses, and other rooms with a religious purpose. The First Palaces were all destroyed by an earthquake around 1700 B.C., and the so-called Second Palaces rose on their ruins. Over the next two centuries Crete became the preeminent Aegean Bronze Culture site and the first center of European civilization. By about 1580 B.C., through profitable trade, its influence had spread to neighboring islands, the Greek mainland, and beyond. Its dominance was in part due to its location at the crossroads of Africa, Asia, and Europe, and in part to its naval power.

The new palace at Knossos became the center of Minoan government, traditionally regarded as the seat of King Minos. Some ancient writers identify several kings named Minos, and the term may have described the *office* of ruler, like Egypt's pharaoh. The historical basis of this person is impossible to discover. The later Greeks accepted him only in myth, as the son of Zeus and Europa. Myth also has it that he asked the engineer Daedalos to build a labyrinth so cunningly designed that it could be used to imprison the monstrous Minotaur. There is a clear connection with the complicated layout of the second palace. Little is known about their form of government: whether they had a priest-king, priestess-queen, or some other form of ruler gives rise to the question, Was the Palace of Minos *only* a palace?

Archeologists have discovered two main and connected functions of its many parts: economic and religious. We can only speculate. Why the Minoans should have produced an architecture and interior design so

The Palace of Minos, Knossos, Crete, seventeenth century B.C. Detail of the ruined north entrance.

different from their regional contemporaries remains enigmatic. With an area of 211,000 square feet (20,000 square meters) the palace at Knossos was larger and grander than those at Phaestos, Mallia, and Zakros. Perhaps that expressed Minos's domination of other Cretan kings; perhaps the smaller palaces were his alternative residences. Without any obvious fortifications, the palace at Knossos was approached along a ramped road leading to a western outer court, then through a complicated, defensible ceremonial gateway that occupied most of the south wing and gave access to the rectangular central court that formed the nucleus of the plan. The court was surrounded by four multistory wings of varying height. The east wing housed the nobles' quarters, the queen's elegant two-story apartment, workshops, and a shrine. The west wing, also two stories, comprised long, narrow store-

rooms, more shrines, and the throne room; its upper level, reached directly from the gate, housed a number of large halls. The north wing contained a hypostyle hall (now known as the Customs House), a basin for ritual washing, and a square, stepped structure that may have been an open-air theater.

Everywhere, shrines were provided for the worship of a mother goddess, for whom no specific temple was built, and of whom no monumental sculptures were made. Associated with that worship was the double ax *(labrys)*, frescoes of which appear on some of the palace walls. And everywhere connections were made to other parts of the palace and the buildings beyond it, through imposing porticoes, broad stairs, and processional paths. Archeologists have coined evocative (and sometimes speculative) names for the parts of the complex: the Royal Villa, the House of the Frescoes, the Caravanserai, the Unexplored Mansion, the Temple Tomb, the House of the High Priest, and the South Mansion. Knossos was connected to other Cretan centers by paved roads.

Elsewhere in the Mediterranean basin, grand buildings, domestic or otherwise, adhered to a formal axiality of plan that allowed direct and easy movement through them; they seem to have been thought of as maps or patterns on the ground that were simply extruded to form space. By contrast, the Palace of Minos was quite clearly *conceived* in terms of space. There were intriguing changes in floor level. Some rooms were narrow (perhaps as a result of the building materials used); others were wide with central columns. Some were comparatively low; others soared through two stories. Some were windowless and dim; others were lit from above or through light wells. Some were enclosed; others could be opened by means of movable partitions so that fresh air could circulate. A network of conduits, pipes, and drains supplied running water for bathrooms (there was even a flush toilet in the queen's apartment)—a sanitation system that would not be seen again in Europe for 4,000 years. This spatial and technological experimentation was enhanced and enlivened by a celebration of color and decoration. Frescoes adorned the gypsum-faced walls with informal, vivacious scenes in an explosion of deep red, ochre, green, and blue. In the gateway there was a procession of marching youths bearing offerings; in the throne room, a frieze of griffins in a forest; and in the queen's apartment, fish, coral, and frolicking dolphins. Other recurrent themes included the bull and the dangerous Minoan sport of bull leaping. The distinctive downward-tapering Minoan columns, owing their shape to that of the cypress trees of which they were made, were plastered and painted red or black.

The Palace of Minos was partly damaged by earthquakes in about 1600 B.C. and again around 1500. Real disaster struck in the mid–fifteenth century B.C. when Santorini exploded, sending devastating shocks and tidal waves throughout the region. Soon after, Mycenaean raiding parties from the Greek mainland—perhaps Athenians under the legendary Theseus—overthrew Minoan domination of the Aegean. The other palaces were left deserted, but the invaders rebuilt Knossos and occupied it as their seat of government until about 1380

B.C., perhaps even longer. After about 1000 B.C. it was subjugated by the Dorians, and it became a Roman colony in the third century B.C. Surviving as a city well into the Christian era, Knossos later became the see of a bishop. During the Middle Ages it was reduced to a small village named Makrys Toihos.

The Cretan antiquarian Minos Kalokairinos uncovered part of the palace's west wing in 1878. Many tried to extend the excavations, but because the land was then privately owned they were forced to discontinue. When Crete achieved independence in 1898, its antiquities became state property, and two years later Evans initiated systematic excavation. By the end of 1903 almost the entire palace had been exposed. Work continued until 1912, extending the dig to adjacent areas of the ancient city, and was again resumed in 1922–1931. Evans has been criticized by some scholars for the speculative nature of some of his restoration, the accuracy of which has rightly been questioned. After 1941 responsibility for the excavations passed to the British School of Archaeology, and since 1952 conservation work has continued in conjunction with the Greek Archaeological Service.

Further reading

Castleden, Rodney. 1990. *The Knossos Labyrinth: A New View of the "Palace of Minos" at Knossos*. London: Routledge.

Graham, J. Walter. 1987. *The Palaces of Crete*. Princeton: Princeton University Press.

Hood, Sinclair, and William Taylor. 1981. *The Bronze Age Palace at Knossos: Plan and Sections*. London: Thames and Hudson.

Panama Canal
Panama

The Panama Canal, completed in 1914, is a 50.7-mile (81.6-kilometer) passageway through the Isthmus of Panama, connecting Cristobal on the Atlantic coast and Balboa on the Pacific at the narrowest point of the landmass of the Americas. By navigating its three locks, each of which raises or lowers them 85 feet (26 meters), ships can move from ocean to ocean in about twenty-four hours; that saves the several days needed to sail the many thousands of miles around South America. The Panama Canal has been acknowledged as one of the twentieth century's greatest engineering triumphs, which displays the combined skills of an international team of structural, hydraulic, geological, and sanitary engineers.

A canal joining the oceans had been suggested as early as the sixteenth century. The conquistador Hernando Cortés proposed a route across the Isthmus of Tehuantepec, while others favored Nicaragua or Darién. King Charles V of Spain ordered a survey of the Isthmus of Panama in 1523, but although plans were made by 1529, the project lapsed. Several alternative schemes followed, including one of 1534 close to the present canal, but the Spanish then lost interest in the project until the beginning of the nineteenth century. In 1819 the government approved the construction of a canal but the revolt of Spain's American colonies meant that control of potential sites was wrested from her. The new Central American republics took up the idea, but they had to find European or U.S. investment to realize it. The California gold rush in the mid–nineteenth century aroused U.S. interest, and a number of feasibility studies undertaken between 1850 and 1875 suggested two routes: across Panama or across Nicaragua. But the Americans were not the only ones interested.

In 1875 the Frenchman Ferdinand de Lesseps, flushed with the success of his Suez project, first announced his interest in a Central American canal. On 1 January 1880, the Panama Canal was symbolically inaugurated, and a year later French engineers employed by the Compagnie Universelle du Canal Interoceanique arrived at Colon on the Atlantic coast. Construction of a sea-level canal (as opposed to a lock canal) began in 1882 along the route of the 1855 Panama Railroad. But financial mismanagement, the tropical climate, and disease took their toll. The company was liquidated in February 1889 and by the following May all work had ceased. Following a scandal involving charges of bribery, de Lesseps died in France in 1894. In October of that year the Compagnie Nouvelle du Canal de Panamá was established, and work on the canal resumed.

The United States had been involved in canal projects since about 1887 with little success. In 1889 Congress chartered the Maritime Canal Co. to build a canal in either Panama or Nicaragua. Based on cost, the latter location was chosen, but within four years the company failed. At the end of the decade two successive congressional commissions favored the Nicaraguan route, but the French were seeking a buyer for their project. When the asking price dramatically fell from $100 million to $40 million, a canal in Panama became a viable alternative for Congress. The 1902 Spooner Act provided funds to buy the Compagnie Nouvelle and gave the power to negotiate a treaty with Colombia (Panama had been a Colombian province since 1821) over the canal zone. Panama soon declared its independence from Colombia—not without American help—and by early 1904 a treaty was ratified between the new nation and the United States.

The latter's control of the canal project began in May. Work was set to resume under the management of the Isthmian Canal Commission, which decided to build a canal with locks rather than the sea-level waterway favored by de Lesseps. Before work could begin, the disease that had ravaged the French builders had to be overcome. Dr. (later Colonel) William Gorgas identified malaria and yellow fever as the most dangerous, although cholera, diphtheria, smallpox, tuberculosis, and even bubonic plague were not unknown. Gorgas took measures to destroy the mosquito population, while other medical teams treated the additional problems.

John F. Wallace, a civilian, was appointed as chief engineer. Arriving on site in June 1904, he started the excavation at Culebra Cut—a 10-mile (16-kilometer) section through the highest, most difficult terrain on the route—and work continued there for about a year. Eventually, almost 100 million cubic yards (77 million cubic meters) of earth and rock would be removed from the cut, using more than 9,600 tons (8,700 tonnes) of explosives. Dissatisfied with the logistical backup provided, Wallace complained to President Theodore Roosevelt, who replaced the commission with new members. Then, in July 1905, when the engineer again complained about his own working conditions, Roosevelt replaced him, too. He was succeeded by John F. Stevens, another civilian. When Stevens soon stated that he was not "anxious to continue in service," Roosevelt took that as a resignation and U.S. Army Lieutenant (later Colonel) George W. Goethals was appointed in his place. Goethals carried the canal through to completion.

Under his direction, 42,000 men constructed the canal between Colon and Balboa. It comprises the channel (of course), 300 feet (92 meters) wide at the bottom and up to 1,800 feet (550 meters) wide at the top; a massive earth dam at Gatun (to provide water for the locks from what was then the world's largest artificial lake); and a series of three locks at Gatun at the Atlantic end, a set of two more at Pedro Miguel, and another single lock at Miraflores at the Pacific end. Each of the locks is 110 feet (933.6 meters) wide and over 1,000 feet (304 meters) long. The locks and their gates were built in the United States, transported in sections, and concreted together. Water flows in and out through culverts in their walls, and double sets of doors at each end protect the canal. Railroad locomotives move the ships in and out of the locks. A breakwater at the Pacific end, constructed with the spoil from the excavation, prevents silt blockage of the mouth. Defensive works at both entrances maintain security. The widening of the Gaillard Cut was completed in 1970, allowing two-way passage of vessels through the entire waterway.

The Panama Canal was officially opened on 15 August 1914 by the passage of the S.S. *Ancon*, although the first vessel to cross the isthmus was in fact the *Cristobal*. The project was completed on time and under budget by $23 million. American costs over the ten-year construction period totaled $352 million, and it is estimated that the French had earlier spent a further $287 million. More than 80,000 people had been employed to build the canal, and it is surmised that about 30,000 died, mostly through disease. Since 1903 the United States has invested another $3 billion, of which about 65 percent has been recouped through tolls. The Panama Canal is used for almost all interoceanic sea traffic. At first around 2,000 ships a year navigated it; currently, the number is in the order of 15,000 and only the modern supertankers and container vessels are too wide to pass through the locks. Ships pay a toll based on tonnage, and although the cost seems high, it is much less than that of the time and fuel involved in sailing around South America.

The original Hay-Bunau-Varilla Treaty with Panama granted the United States a perpetual lease on a 10-mile-wide (16-kilometer) strip of land—the Canal Zone—flanking both sides of the canal. Probably for military reasons, the United States re-

peatedly interfered in Panama's affairs until 1936, when it finally gave up its right to use troops outside the Canal Zone. Disputes about the canal contract continued until two new treaties were signed in 1977. They provided that the canal would be operated by the Panama Canal Commission, a U.S. government agency appointed to replace the former Panama Canal Company, from October 1979 until the end of 1999. Then control passed to Panama, in the face of vocal opposition from a largely right-wing lobby in the United States that still refers to the waterway as "United States Canal in Panama," believing that the handover presented a serious threat to their country.

See also Suez Canal

Further reading

Jaén Suárez, Omar, et al. 1999. *The Panama Canal*. Paris: UNESCO.

McCullough, David. 1999. *The Path between the Seas: The Creation of the Panama Canal, 1870–1914*. New York: Simon and Schuster.

Wood, Robert E. 1963. *Monument for the World: Building the Panama Canal*. Chicago: Encyclopaedia Britannica.

Pantheon
Rome, Italy

As its name suggests, the Pantheon in Rome was dedicated to many gods. Its seven interior niches housed statues of Apollo, Diana, Mars, Mercury, Jupiter, Venus, and Saturn, and the great dome also had religious significance since it symbolized the heavens. Even now, when stripped of much of its enrichment, the scale and simple geometry of the Pantheon awe the visitor. Moreover, its sophisticated engineering stirs imagination for its ancient engineers. Many of their modern counterparts are at a loss to understand how the structural system worked, much less how it has survived for two millennia. The great Florentine artist Michelangelo Buonarotti concluded that it was the result of "angelic and not human design."

The first Pantheon was built in 27 B.C. for Marcus Vipsanius Agrippa, son-in-law of Augustus Caesar. Except for lower parts of the porch and foundation, it was irretrievably damaged by fire in A.D. 80, and

Pantheon, Rome, architect(s) unknown, ca. 118–128. Gianpaolo Pannini's painting of the interior, ca. 1740.

reconstruction to a slightly different design was commissioned by the Emperor Hadrian. The present building dates from between 118 and 128, although some scholars believe that it was completed under Pius about twelve years later. Lucius Septimius Severus and Caracalla sponsored a restoration in 202. In 609 the building was presented to Pope Boniface IV by the Byzantine emperor Phocas, and it was dedicated as the Church of St. Mary of the Martyrs (now Santa Maria Rotonda) on 13 May.

The great portico was originally approached across a colonnaded rectangular courtyard on the temple's north side. Three rows of 46-foot-high (14-meter) columns support its gable roof, which rises to 80 feet (24 meters). The entablature carries a pediment that once was crowded with bronze relief sculptures of a battle between the Titans and the gods. The front row of eight columns and the second row of four are of Egyptian gray granite; the third row, framing the door and its flanking apses, is of Egyptian red granite. All are crowned with Corinthian capitals carved in white Pentelic marble from Greece. The massive bronze entrance doors are 21 feet (6.3 meters) high. With their fanlights, they were originally gold plated.

The interior of the Pantheon—a single volume—is a 143-foot-diameter (43.3-meter) cylinder upon which rests a hemispherical dome. The total inside height is the same as the diameter. A semicircular apse covered by a hemidome faces the door. Around the wall on each side are three 14-foot-deep (4.2-meter) recesses, alternately rectangular and semicircular in plan and screened from the central space by pairs of 35-foot-high (10.5-meter) marble Corinthian columns supporting entablatures and a deep cornice. The lower part of the wall once was faced with marble and porphyry, the upper with marble pilasters and paneled with antique yellow marble, serpentine, and pavonazetto.

Above it all soars the magnificent coffered dome, the largest in the world until Brunelleschi built his masterpiece on Florence Cathedral thirteen centuries later. Five rows of twenty-eight square diminishing coffers, each recessed in four steps, rise to a central oculus—a circular window open to the sky. Originally the coffers were decorated, perhaps with gold stars on a background of blue. Even when the doors are closed, light enters the vast space through the 19-foot-diameter (8.7-meter) oculus. As the sun moves, a spotlight slowly swings across the interior, illuminating and enriching the colored wall and floor surfaces. Externally the dome was covered with gilded bronze plates, but they were taken to Constantinople in 655 and replaced by lead. Indeed, much of the precious material of the Pantheon has been plundered; the exterior surface of the wall was once veneered with colored marbles. Now the brickwork and the huge relieving arches of the second tier are exposed.

A 14.75-foot-deep (4.5-meter) concrete foundation supports the Pantheon's 20-foot-thick (6-meter) cylindrical brick-faced concrete wall, 104 feet (31.7 meters) high. Roman concrete consisted of lime, pozzolana, and a few pieces of very coarse aggregate. Bricks of various shapes were used as "lost formwork." The wall was made lighter as it rose by using lighter aggregate in the concrete; every 4 feet (1.2 meters) the brick "formwork" was tied together with a through-course of brickwork. It was designed to bear a range of complex stresses and loads, and it seems that instead of working as a solid mass it behaves structurally like a series of massive piers acting as buttresses to resist the thrust imposed by the dome.

Roman concrete domes normally were made of lightweight aggregates such as pumice to reduce their mass and constructed on timber centering supported from the ground. The thickness of the Pantheon dome reduces from nearly 20 feet (6 meters) at the base to about 5 feet (1.5 meters) at the edge of the oculus. The engineers probably knew that such a huge span of unreinforced concrete might not develop enough tensile strength; therefore they used lighter aggregates near the apex and the oculus—really a brick-reinforced compression ring—to minimize lateral thrust. Thus, to halfway up the dome is built as a series of seven stepped rings of concrete with alternating layers of bricks and tufa; probably, each was allowed to develop full strength before the next was placed. The top 30-foot (9.2-meter) section was made in the usual way: alternate layers of 9-inch (7.5-centimeter) pieces of tufa and volcanic slag bonded with mortar. Although the Pantheon did not employ a revolutionary structural system, it represents the high point in Roman concrete technology evolved from well-thought-through construction traditions.

See also Roman concrete construction

Further reading

Ching, Francis D. K. 1996. *Architecture: Form, Space, and Order*. New York: Van Nostrand Reinhold.

Licht, Kjeld De Fine. 1968. *The Rotunda in Rome: A Study of Hadrian's Pantheon*. Copenhagen: Gyldendal.

MacDonald, William L. 1976. *The Pantheon: Design, Meaning, and Progeny*. London: A. Lane.

Parthenon
Athens, Greece

Ralph Waldo Emerson said of the Parthenon, "earth wears no fairer gem upon her zone." Even if that was going a little too far, we certainly may assert that the great temple, built 447–432 B.C., is the high point of Greek Doric architecture. The Greeks' quest for cosmic harmony can be traced in their sculpture and their architecture, especially temples. From archaic shrines—the religious equivalent of the royal *megarons*—the building type underwent a refinement of form and detail until it eventually achieved the proportional balance and visual nuances of the Parthenon. Having achieved perfection in the eyes of its creators, Doric architecture then had nowhere left to go.

Scholars continue to interpret the Athenian Acropolis, differing over the location of buildings long gone. For centuries, successive shrines to the city's patron goddess, Athena Parthenos (the Virgin Athena), were built there, including the archaic Hekatompedon, which may well have been a sacred enclosure open to the sky. Peisistratos (602–527 B.C.) encouraged the Athena cult by commissioning a temple just north of where the Parthenon would later be built. Embellished by his sons after 520, this "old temple" was for a while the only one on the Athenian Acropolis, but it was destroyed when the ragtag Persian armies sacked the abandoned city in 480. Within thirteen years Cimon and Themistokles had cleared away the debris and rebuilt the perimeter wall of the Acropolis.

The Parthenon was commissioned by Perikles, who was the effective ruler of nominally democratic Athens from 461 until 428. The temple was to house a cult statue of Athena made by his friend Pheidias, the most famous artist of the day. Pheidias was appointed general superintendent of Perikles' comprehensive redevelopment of the Acropolis, a fifty-year plan that included the Parthenon, the Propylaea, the Ionic Temple of Athena Nike, and the Erechtheion. All was funded with money collected from Athens' allies—the Delian League—to finance a second war with Persia that never happened. The architects of the Parthenon were Iktinos and Kallikrates, although their exact roles remain uncertain. Probably the latter was responsible for site management and the technical side of construction, executing Iktinos's design.

Overall, the rectangular gable-roofed building was 228 feet (69.5 meters) long, 101 feet (30.9 meters) wide, and 65 feet (20 meters) high. It stood on a three-tiered platform—necessary on the uneven terrain—formed of 20-inch (50-centimeter) steps; the top one formed the floor of the temple. A surrounding colonnade, known as a peristyle, had eight Doric columns at the ends and seventeen along the sides. Each had a base diameter of a little over 6 feet (1.9 meters) and was just over 34 feet (10.4 meters) high. The 22,300 tons (20,300 tonnes) of marble needed for the work was quarried at Mount Pentelicon, about 10 miles (16 kilometers) from Athens. The blocks for the walls and the drums that made up the columns were carefully dressed to form perfect joints, and no mortar was used in the entire building.

The subtlety of refinement that makes the Parthenon a great architectural achievement is, by definition, invisible. There is no truly *straight* line in the entire building, although many may appear to be straight. Minute curves and adjustments were made to create illusions that would refine the gracefulness of the temple. A number of examples will serve to make the point. To the naked eye, a straight-sided column appears narrower halfway up than at the top or bottom; those on the Parthenon had a slight swelling *(entasis)* so they appeared to be straight. Because corner columns were seen against the sky, they were slightly thicker than those seen against a wall; then all would appear to be the same. Even more remarkably, the axes of all the columns were inclined toward the center of the facade—projected, they would meet thousands of feet in the sky—so they would appear to be vertical. And the platform is slightly convex: on the ends it rises 2.375 inches (60 millimeters) toward the center, and about twice that on the sides, because truly horizontal surfaces would have appeared concave to the eye.

The Parthenon (Temple of Athena Parthenos), The Acropolis, Athens, Greece; Iktinos and Kallikrates, architects, 447–432 B.C. Aerial view of ruins from the southeast.

The exterior was painted in bright colors and adorned with sculpture, also painted, made under Pheidias's direction. Ninety-two rectangular panels *(metopes)* above the columns were carved in deep relief with allegorical scenes from various historical and mythological battles: the Trojan War, the Athenians and their enemies, the Lapiths and centaurs, and the gods and giants. Under the peristyle, the walls of the temple were crowned by a 3.25-foot deep (1 meter) continuous frieze, 39 feet (12 meters) above the floor. It portrayed in low relief figures of 350 people and 125 horses participating in the annual Greater Panathenaia, a procession in which the youth of the city accompanied a wheeled ship carrying a new robe *(peplos)* for an ancient wooden statue of Athena. The sculpted procession took two directions, beginning at the southwest corner and meeting above the central eastern door. Contemporary accounts tell us that very high relief sculptures in the east pediment that decorated the gable end narrated the birth of Athena, flanked by other deities, while the

west pediment depicted her battle with Poseidon. Only fragments survive.

The ordinary people were not allowed to enter the Parthenon. All religious ceremonies took place in a courtyard *(temenos)* to the east. Only the priests entered the inner chambers, of which there were two. The smaller *(parthenos* or *opisthodomos)*, reached from the west portico, housed the temple treasury. Its marble-covered timber roof was supported by four slender Ionic columns, perhaps symbolizing the protective role that Athens then enjoyed among the city-states. At the east end was the sanctuary *(naos)*, 98 feet long by 63 feet wide (29.8 by 19.2 meters). Its roof was supported by a two-story, superimposed Doric colonnade, creating aisles on the long sides of the room. Natural light came through the large central door in the east wall.

At the west end of the *naos* was the 40-foot (12-meter) standing figure of Athena Parthenos. She has long since been lost but descriptions survive. Covering a wooden and metal frame, her body-length tunic was of gold plates, and her exposed face, hands, and feet were of ivory; her eyes were made from precious stones. According to the ancient Greek writer Pausanius, her helmet was emblazoned with an image of the Sphinx, and on her breast she wore a head of Medusa carved from ivory. In one hand she held a 6-foot-high (2-meter) statue of Victory and in the other a spear; her shield lay at her feet. Perikles's political opponents spitefully had Pheidias indicted for stealing some of the materials intended for the statue, and the artist was later forced into exile.

In 404 B.C. dominance in the Aegean passed to Sparta after the twenty-eight-year Peloponnesian War, caused in part by Perikles' misappropriation of the Delian League's money. Despite Athens' brief renaissance in the fourth century, Greece came under Macedonian control in 118 B.C. Rome followed Macedon, and by the end of the fourth century Christianity was established as the state religion. Paganism was moribund, and temples, including the Parthenon (which became the Church of St. Mary), were "recycled." Pheidias's wonderful statue was looted and taken to Constantinople. Following the Ottoman invasion of Greece the Parthenon was again converted, this time into use as a mosque. The still intact building was next employed as an ammunition dump during the Turkish-Venetian war. In September 1687 a Venetian cannonball struck the gunpowder, causing an explosion that killed 300 men and reduced the Parthenon to ruins.

The Turks recaptured the Acropolis the following year and began selling antiquities. In 1801 the British ambassador to Turkey, Thomas Bruce, the seventh earl of Elgin, obtained permission to remove "a few blocks of stone with inscriptions and figures," a euphemism that gave him license to pillage the remaining metopes, the frieze, and what remained of the Parthenon's pediment sculptures. Fifteen years later, allegedly at a loss, he sold the "Elgin marbles" to the British Museum. In January 1999 a majority of the European Parliament, as part of a growing international campaign, unsuccessfully petitioned the museum to return the fragments to Greece. The debate continues, not without rancor. The Parthenon's other, more destructive enemy is the atmospheric pollution that plagues Athens. That problem, too, has commanded an urgent international movement to save an outstanding piece of world architecture.

See also The Acropolis; Erechtheion

Further reading
Boardman, John, and David Finn. 1985. *The Parthenon and Its Sculptures*. Austin: University of Texas Press.
Carpenter, Rhys. 1970. *The Architects of the Parthenon*. Harmondsworth, UK: Penguin.
Tournikiotis, Panayotis. 1996. *The Parthenon and Its Impact in Modern Times*. New York: Harry N. Abrams.

Pennsylvania Station
New York City

Pennsylvania Station, between Seventh and Eighth Avenues, New York City, represented the high point of railroad architecture. Built from 1904 to 1910 at a cost of $100 million (about $5.6 billion in today's terms), it was over 30 percent larger than its largest contemporary, Liverpool Street Station in London, England. In its first year of operation 112,000 trains carrying over 10 million passengers passed through Pennsylvania Station. It is not remarkable for its size alone, but also because it epitomized Beaux Arts architecture on the eastern seaboard of the United States just at the time when modernist ideas were challenging it in Europe.

At the end of the nineteenth century, rail transport in the United States was dominated by the rich and powerful Pennsylvania Railroad. It carried more passengers and freight than any other company, servicing about 20,000 stations. It also led in technology, management, and operating practices. But the company had no station in New York; passengers were obliged to reach the metropolis by ferry from the Pennsylvania Railroad terminus in Hoboken, New Jersey. In 1899, the railroad's new president Alexander J. Cassett set about to remedy the situation, and the following year he acquired control of the Long Island Railroad. Direct access to Manhattan was critical, and Cassett planned a terminal there to service both railroads, making use of the tunnel then being built under the East River. Twenty-five acres of real estate, bordered by Seventh and Eighth Avenues and Thirty-first and Thirty-third Streets, was secured at a cost of $10 million. Existing buildings were demolished, and Thirty-second Street from Seventh to Ninth Avenues was closed and incorpo-

Pennsylvania Station, New York City; McKim, Mead and White, architects, 1902–1910. Interior of waiting room, photographed ca. 1911.

rated into the site. The New York architectural firm of McKim, Mead and White was commissioned and the design work for Pennsylvania Station began in 1902. One architectural historian has written that the outcome was "one of McKim's most monumental and moving designs, a giant of a building that still retained a human scale. In catching or meeting a train at Pennsylvania Station one became part of a pageant—actions and movements gained significance while processing through such grand spaces" (Wilson 1983, 211).

The Seventh Avenue entrance was approached through a 780-foot-long (239-meter) Roman Doric colonnade with 35-foot-high (10.6-meter) columns, carrying a low, flat-roofed attic story. Within the colonnade was a row of shops, and at either end

pedimented porticoes led into carriageways for motor vehicles that gave access to the waiting room. The central pavilion, higher than the rest, carried sculptured eagles and figures of women supporting a large clock. The 430-foot (131-meter) facades to Thirty-First and Thirty-Third Streets were relatively plain, but reduced to human scale with attached architectural orders. The main waiting room was probably the most striking part of the building. Based on the ancient Baths of Caracalla in Rome, its 320-by-110-foot (98-by-34-meter) area was roofed with three coffered cross vaults soaring 150 feet (46 meters) above the pink marble floor. The walls, interspersed with giant Corinthian orders at the springing of the vaults, were lined with Italian travertine, and the subtlety of the beautiful stone was brought out by the light that streamed through the huge windows beneath the vaults. The entrance landings at either end of the waiting room were framed with Ionic colonnades. Commuters, dwarfed in the magnificent space, reached the street-level entrances by means of broad stairways. The concourse, about twice the area of the waiting room and down one level from it, was as delicate as the other was massive. It was roofed with barrel vaults of glass framed in a filigree of iron, and therefore flooded with light. The twenty-one railroad tracks were on another lower level, 40 feet (12 meters) below the street.

Excavation work started in summer 1904 and the station was mostly completed by August 1910. At first dubbed the "Manhattan Gateway," Pennsylvania Station soon became the gateway to America. It reached its peak usage toward the end of World War II, with over 109 million passengers in 1944. After that, changes took place in intercity travel. Congress adopted a 40,000-mile (48,000-kilometer) national system of interstate highways in the Federal-Aid Highway Act of 1944; although the roads were not built immediately, that eventually led to the automobile's taking precedence over the train, a situation that was exacerbated by the advent of inexpensive air travel. By about 1955 the railroad was eclipsed as Americans' preferred form of passenger transport.

In 1962, Madison Square Garden purchased the air rights to Pennsylvania Station and in October 1963 began demolition, despite public outcry. All that remains is the underground section; its twenty-one tracks carry 600,000 passengers every day. One positive outcome of the loss of the magnificent building was New York's Landmarks Preservation Law, enacted in 1965. There is also an increased national awareness of the importance of preserving architectural heritage. New York's Grand Central Terminal, a contemporary of Pennsylvania Station, was saved and rehabilitated at a cost of $196 million; the work was finished in 1998. In May 1999 the Metropolitan Art Society announced a $484 million proposal to convert New York's central post office (also designed by McKim, Mead and White), which once faced Pennsylvania Station across Eighth Avenue, into a new Pennsylvania Station. The commission was won by architects Skidmore, Owings and Merrill, and their design was made public early in 2000.

See also Baths of Caracalla

Further reading
Diehl, Lorraine B. 1985. *The Late, Great Pennsylvania Station*. New York: American Heritage.
Middleton, William D. 1996. *Manhattan Gateway: New York's Pennsylvania Station*. Waukesha, WI: Kalmbach Books.
Wilson, Richard Guy. 1983. *McKim, Mead and White Architects*. New York: Rizzoli.

Persepolis
Iran

The ruins of Persepolis (in Persian, Parsa) lie at the foot of Kuh-i-Rahmat (Mountain of Mercy) beside a small river on the Marv Dasht plain of southwestern Iran, about 400 miles (640 kilometers) south of Tehran. Widely held to be one of the greatest architectural complexes of the ancient world, and even claimed to be the most beautiful the world has ever seen, it was probably commissioned by Darius I between 518 and 516 B.C. as the ceremonial center and temporary royal residence of the First Persian (Achaemenian) Empire. Persepolis flourished under later kings. Xerxes I (reigned 486–465 B.C.) built the Throne Hall and the ceremonial gateway. His son Artaxerxes I (464–425) finished the hall, Artaxerxes II (ca. 350) built the so-called Unfinished Palace, and more buildings were added as late as the reign of Artaxerxes III, who died only eight years before the city was looted and burned by Alexander the Great's armies in 330 B.C. Helped by traitors, the Macedonians took Persepolis by surprise, massacred the defenders, and stripped the palaces and the treasury of gold and silver.

The earliest Achaemenian capital was established by Cyrus I at Pasargadae, 48 miles (77 kilometers) to the north of the Persepolis site. Soon the administrative center was moved to Susa, a further 230 miles (370 kilometers) north, which was better placed strategically for dealings with Mesopotamia. Darius I then decided, for his own reasons, to create Persepolis; perhaps he wanted to build a dynastic shrine in the Achaemenian homeland. Or there may have been a political motive. But Persepolis was never a capital, or even a city in any sense of that word. It was established as a venue where the subject nations would pay homage to the Persian kings. There were no temples, and its palaces were for temporary occupation only.

Persepolis stood on a half-constructed, half-natural limestone terrace that measured about 1,475 feet north to south and about 985 feet east to west (450 by 300 meters), rose up to 60 feet (18 meters) above the plain, and was surrounded by a fortified triple wall. Its northern part, with the Gate of Xerxes, the Audience Hall of the Apadana, and the Throne Hall, was the ceremonial precinct, to which access was restricted. The southern part housed the Palaces of Darius (Tachara), Xerxes (Hadish), and Artaxerxes III, the Harem, the Council Hall (Tripylon), treasuries, barracks, and other ancillary buildings such as the royal stables and chariot house.

The main ceremonial approach to the platform was at the northwest corner by a 23-foot-wide (7-meter) monumental stairway of over 100 shallow steps; it was richly carved in low relief with symbols of the god Ahura Mazda and sculptures of people bringing annual tribute to the Achaemenid kings. The stair led to the only entrance to the terrace, the Gate of Xerxes (called the Gate of All Nations), a square hall built of decorated sun-dried brick, its roof supported by four columns. It had three huge doorways, 36 feet high, with double doors of timber sheathed in decorated metal. The southern door opened into a courtyard before the Apadana, the audience hall of Darius and Xerxes. This vast 198-foot-square space—the largest building of the complex—was a forest of thirty-six stone columns, rising more than 60 feet (19 meters) from bell-shaped bases and crowned by capitals decorated with double bulls', lions', human, or mythical horned lions' heads, supporting a roof frame, built of cedar imported from Lebanon.

The processional way through the eastern door of the Gate of Xerxes led visitors to the east before then turning south toward the Throne Hall (also known as the Hall of a Hundred Columns), a 230-foot-square (70-meter) room containing literally 100 columns. Its principal portico, facing north, was flanked by two huge stone bulls, and its eight stone portals were decorated with low reliefs of the spring festival and scenes of the king fighting monsters. On the Persian New Year, Now Ruz (21 March), delegates from the twenty-eight subject nations would pass to the Throne Hall to pay homage and present their gifts and offerings—silver, gold, weapons, textiles, jewelry, and even animals. Later, when the treasury at the southeast corner of the terrace could no longer hold the tributes, the Throne Hall also served to store and display the riches of the Persian Empire.

All these buildings glowed with color: green stucco predominated, and the figures in the relief carvings were brightly painted. In much Achaemenian architecture, mud-brick walls were faced with blue, white, yellow, and green glazed bricks with animal and floral ornaments. The forests of pillars, many of them sheathed in gold and embellished with ivory, were hung with embroidered curtains. Precious stones were used in mosaics.

The long-forgotten site of Persepolis was rediscovered in 1620, and although many subsequent visitors wrote of it, serious investigation did not commence until 1931. James Breasted of the Oriental Institute of the University of Chicago commissioned Professor Ernst Herzfeld of Berlin to excavate and (where possible) restore the remains of the city. Herzfeld (working 1931–1934) and Erich Schmidt (1934–1939) thoroughly documented the extensive ruins. UNESCO declared Persepolis a World Heritage Site in 1979.

Further reading

Schneider, Ursula, comp. 1976. *Persepolis and Ancient Iran*. Chicago: University of Chicago Press.

Wilber, Donald N. 1989. *Persepolis: The Archaeology of Parsa, Seat of the Persian Kings*. Princeton, NJ: Darwin Press.

Pétra
Jordan

Pétra (the name means "rock") in southern Jordan lies about 50 miles (80 kilometers) south of the Dead Sea on the border of the mountainous Wadi Araba Desert. Although there is evidence of earlier occupation of the site, the city was founded around the sixth

El-Deir (aka the Monastery), Pétra, Jordan, third century B.C. This is one of the many rock-hewn buildings in the ancient city; its original purpose is uncertain.

century B.C. as the practically inaccessible capital of the Nabataean Arabs who dominated the region and controlled international trade routes between Asia, southern Arabia, and the markets of the Mediterranean basin. Wealthy and powerful Pétra was partly built, partly carved from the beautiful pink sandstone of its mountain fastness. Its remarkable buildings, representing the hybridization of several cultural sources over almost a millennium, make it one of the great architectural achievements of history. When it was added to UNESCO's World Heritage List in 1985, it was acclaimed as "one of the most famous archaeological sites in the world, where ancient Eastern traditions blend with Hellenistic architecture." Added to that distinction must be the Nabataeans' hydraulic engineering achievements, comprising extensive water-conservation systems and sophisticated measures to avoid flooding of their city.

Following unsuccessful attempts by the Seleucid Antiochus and the Judean Herod the Great to absorb

Pétra into their kingdoms, in 64 and 63 B.C. the Roman general Pompey conquered Nabataea. It remained independent (but taxed), a neutral zone between the desert nomads and Rome's territory. Pétra burgeoned over the next century. The city was wholly Romanized under Trajan in A.D. 106, when Nabataea became Arabia Pétraea. Twenty-five years later, Hadrian renamed the capital Hadriane Pétra and installed Sextius Florentinus as governor. Early in the fourth century, great changes swept the Roman Empire: Christianity was recognized by the state and in 330 Constantine moved his capital to Byzantium, renaming it New Rome (now Istanbul). Pétra, while devastated by earthquake in the mid–fourth century, flourished until late antiquity, after which it began to decline. Its last contact with the Western world until the nineteenth century was in the 1100s, when Crusaders built and briefly occupied a small fortress there.

In 1812 the Swiss orientalist Johann Ludwig Burckhardt learned of Pétra from local Bedouins, and in the years that followed many Europeans visited and recorded it. The romance of the place was irresistible, as the theologian Dean John W. Burgon wrote in 1845: "Match me such marvel, save in Eastern clime, A rose-red city 'half as old as Time.'" Serious archeological investigations began toward the end of the nineteenth century, yielding a number of publications, including Alois Musil's *Arabia Pétraea* (Vienna, 1907–1908) and Brunnow and A. von Domaszewski's *Die Provincia Arabia* (Strasbourg, 1904–1909).

The main access to Pétra is along a trail—originally paved—known as the Alley (Al-Siq); in places it is only about 16 feet (5 meters) wide, but it winds between sheer cliffs hundreds of meters tall. The Alley opens into a large open space, also surrounded by high walls, and directly opposite a building that exemplifies Pétra's unique architecture and is probably the city's most famous building: the Treasury (Al Khazneh). Carved from the pink sandstone, the 100-foot-wide (30-meter) facade soars to 130 feet (40 meters). Its architecture is manifestly Roman influenced; the lower story, its classicism mutated by Nabataean inventiveness, has a pedimented loggia with some freestanding, some attached Corinthian columns. In the attic story, two side pavilions with half-pediments flank a circular central pavilion with a conical roof, crowned with an urn. Decoration and ornament abound. A single door leads to a large square room, precisely excavated from the rock. Typically for Pétra (and just as atypical for Rome), the interior is as frugal as the outside is embellished.

The cliffs near Al Khazneh are honeycombed with cave tombs. The widening Muses Valley (Wadi Musa) leads first to the east, its walls pitted with smaller tombs, past an 8,000-seat Roman theater. Nearby, steps lead westward to the High Place of Sacrifice. At a point where the wadi swings to the north, there rises the King's Wall, a cliff face from which three large structures known as the Royal Tombs have been carved. The first and most impressive is the Urn Tomb, cut deep into the rock and standing upon a two-story arcade. The others are badly weathered, but there are clues to their former glory: the Corinthian Tomb, redolent of Al Khazneh, and the Palace Tomb, modeled on a Roman urban palace.

The elongated "downtown" heart of Roman Pétra, flanking a once paved, colonnaded street along the Wadi Musa, lies about 300 meters north of the King's Wall. The center dates mostly from Nabataea's golden age—say, from about 300 B.C. to A.D. 150—but there are also relics of the Roman and Byzantine Christian cultures, with an underlying Nabataean accent, indicating the ability of the Nabataeans, common to many trading peoples, to assimilate foreign cultural expressions. An open marketplace and public fountain once stood at its southern end. Opposite is the gateway to the sacred precinct *(temenos)* of the freestanding Temple of Dushara, the chief deity of the Nabataeans. The Temple of the Winged Lions, northeast of the gateway, was dedicated to his partner, the fertility goddess Atagartis. A climbing pathway to the northwest leads to the Monastery (El-Deir), a rock-hewn building similar to Al Khazneh, even larger but more crudely detailed. In fact, there are the remains of over 800 structures around Pétra. Most are rock-hewn: besides the temples and tombs, there are other funerary monuments, public baths, houses (of course), public buildings, paved streets, and all the elements of a thriving metropolis.

The Nabataeans were masters of hydraulic engineering. More than sixty water sources served Pétra. A channel was carved into the cliff face of Al-Siq to catch water in the winter and carry it to the many storage cisterns—some as large as 1,500 cubic feet (40 cubic meters)—within the city. A clay pipeline was also built, conveying water about 3 miles (5 kilometers) from the Muses Spring. The cisterns had built-in filtration systems and were linked by a network of channels. It has been estimated that the total supply, used for drinking, cooking, bathing, and even irrigating crops, could support the population of 25,000 for several months.

In 1958 P. J. Parr and C. M. Bennett of the British School of Archaeology began highly objective excavations of the center of Pétra. Subsequently the Pétra/ Jerash Project, a cooperative venture of the Jordanian Department of Antiquities, the Universities of Jordan and Utah, and a team of Swiss archeologists, continued work at the site. In 1998 and 1999 a consolidation and site protection program, part funded by the Samuel H. Kress Foundation and an American Express Award through the World Monuments

Fund, was implemented under the management of Dakhilallah Qublan and Pierre Bikai of the American Center for Oriental Research.

Further reading

Browning, Iain. 1973. *Pétra*. London: Chatto and Windus.

Glueck, Nelson. 1966. *Deities and Dolphins: The Story of the Nabataeans*. London: Cassell.

McKenzie, Judith. 1990. *The Architecture of Pétra*. Oxford, UK: Oxford University Press.

Pharos of Alexandria
Egypt

Taking the name of the long narrow island on which it stood, the Pharos of Alexandria was the most famous lighthouse of antiquity. Situated on a high mound at the end of a long peninsula, 3 miles (5 kilometers) from the city, it was a technological marvel and formed the prototype of the modern lighthouse. Since the sixth century A.D. (when it replaced the walls of Babylon) it has been listed among the seven wonders of the ancient world.

In 323 B.C. Alexander the Great died in Babylon, leaving no heir. Forty years of conflict followed as his generals fought for control of the vast Macedonian Empire. By 280 B.C. three major dynasties emerged: the Seleucids in Asia, Asia Minor, and Palestine; the Antigonids in Macedonia and Greece; and the Ptolemies in Egypt, the wealthiest and most enduring of all, who would reach their peak under Ptolemy II Philadelphos (reigned 285–246 B.C.). His capital was the grand city of Alexandria, designed by Dinocrates, personal architect to Alexander. Around 290 B.C., because of the dangerous sandbanks in the approaches to Alexandria's harbor, Ptolemy I Soter initiated plans for a lighthouse. The work was incomplete when he died five years later. In 281 his son Ptolemy II Philadelphos commissioned the engineer Sostratus of Cnidus to build a great lighthouse on the eastern point of the island of Pharos, reached across a causeway named the Heptastadion. There is a tradition that structural and other calculations were made at the famous Alexandrian library.

Hellenistic accounts like those of Strabo and Pliny the Elder describe a tower faced with glistening white marble, crowned with a bronze mirror whose reflection of the sun could be seen 35 miles (56 kilometers) off shore. Apocryphal accounts claim the mirror was also a secret weapon used to burn enemy ships at sea. Indeed, most descriptions are sketchy, and popular images of the Pharos have been based on an interpretation of coins, terra-cottas, and mosaics published by Herman Thiersch in 1909. The most detailed description of the Pharos was made in 1166 by the Arab traveler Abou-Haggag Al-Andaloussi (to whom Thiersch did not have access), portraying a structure composed of three battered tiers: the lowest was square, 183 feet (56 meters) high, with a cylindrical core; the middle was a 90-foot-high (27.5-meter) octagon with a side length of 60 feet (18.3 meters); and the third was a cylinder 24 feet (7.3 meters) in height. There are few accounts of the interior of the great tower, except to say that it had many rooms and corridors. Fuel for the nightly bonfire was mechanically lifted through an internal shaft. The fire could be seen for about 100 miles (160 kilometers).

Including the foundation pedestal, the lighthouse soared to about 384 feet (117 meters). A wide spiral ramp gave access to the top, where there was a huge statue, possibly representing either Alexander the Great or Ptolemy I Soter in the role of Helios, the sun god. Some later accounts identify the figure as Poseidon, but more recent scholarship suggests it was Zeus Soter (Zeus the Savior). Still others believe there were statues of Castor and Pollux. Whatever its subject matter, the sculpture took the total height above 440 feet (135 meters), about as high as a forty-story office building. The Pharos was the second-tallest building in the world until the Eiffel Tower was constructed 2,000 years later. The monument was dedicated to Ptolemy Soter and his wife Berenice, and an inscription read, "Sostratus, the son of Dexiphanes, the Cnidian, dedicated this to the Savior Gods, on behalf of those who sail the seas."

The Pharos served the mariners of the Mediterranean for about 1,500 years. In A.D. 642, the Arabs conquered Egypt and moved their capital to Cairo. In 796 the upper story of the lighthouse collapsed. Later, Sultan Ibn Touloun (reigned 868–884) built a mosque on the partly ruined tower. In the middle of the tenth century, an earthquake shook Alexandria and caused another 72 feet (22 meters) of masonry to fall. Despite frequent and sometimes extensive repairs being undertaken by the Arabs, earthquakes continued to have a cumulative effect: no fewer than twenty between 1303 and 1323 meant that the Pharos finally toppled some time before 1349. By then Al-Malik-an-Nasir had begun to build a similar lighthouse beside it but the project was halted at his death. Around 1480 the Egyptian Mameluke Sultan Qait Bey built a fortress over its ruins, using its stones for walls.

In the early 1990s the Egyptian government began building a breakwater to protect the fortress from storms. The project was postponed while archeologists from the Egyptian Supreme Council of Antiquities and the French Centre d'Études Alexandrines searched the harbor. Since 1996 they have found over 2,000 objects, columns, capitals, thirty sphinxes, and—most significantly—two colossal statues (one of Ptolemy I and another of a female torso) scattered over more than 5 acres (2 hectares) of the seabed near Alexandria. It is believed that the finds include the ruins of the fabled Pharos.

In September 1998 the U.S.$70 million Alexandria 21st Century Project was announced by the Fondation Internationale Pierre Cardin, claiming the support of UNESCO and proposing to build a 475-foot-high (145-meter) glass-covered concrete lighthouse on the site of the original Pharos. Happily, it came to very little.

Further reading

Clayton, Peter, and Martin Price. 1988. *The Seven Wonders of the Ancient World*. London: Routledge.

Cox, Reg, and Neil Morris. 1996. *The Seven Wonders of the Ancient World*. Parsippany, NJ: Silver Burdett.

Sly, Dorothy I. 1996. *Philo's Alexandria*. London: Routledge.

Pisa Cathedral: The Campanile (Leaning Tower)
Pisa, Italy

The city of Pisa stands on the River Arno in the Tuscan region of northern Italy. Its Piazza dei Miracoli is graced by the most beautiful group of Romanesque buildings in the country: the white marble basilican cathedral

The Campanile (Leaning Tower), Pisa, Italy; Bonanno Pisano and others, architects, 1173–1350. Detail of upper levels; the cathedral is in the background.

(begun 1063); the circular, domed baptistery (begun 1153); and the highly original bell tower (campanile), situated between the apse and the southeastern end of the cathedral's transept and now famous as the Leaning Tower of Pisa. Quite apart from its innovative cylindrical form, the location is remarkable, because bell towers usually stood near the west front of churches. Of course, the tower was designed to be vertical, but it started to lean very early in its construction. Since 1183—only a decade after it was started—repeated and inventive attempts have been made to correct the incline. They continued for 800 years, until the very end of the twentieth century, when modern technology seemed to halt the incremental incline—by then it had reached 10 percent—and save the life of the tower. Taken together, those remedial actions represent a considerable architectural and engineering feat.

The authorship of the tower remains uncertain. Tradition identifies Bonanno Pisano as the architect, but later scholars name Diotisalvi, Biduino, or Guidolotto. Other sources suggest the German Guglielmo of Innsbruck. The Pisa campanile, essentially a hollow cylinder of just over 50 feet (15.5 meters) outside diameter at the base, is 180 feet (55 meters) high. The structure consists of an external wall faced with gray and white San Giuliano limestone ashlar; between it and an inner wall of dressed limestone, 293 steps wind to the top. The continuous facade is divided into six by elegant arcades and crowned (of course) by a cylindrical belfry, a little smaller in diameter than the tower. Throughout, the wall surfaces are inlaid with patterned, colored marble in the Tuscan Romanesque manner, reducing the large building to an appreciable human scale and creating a harmonious unity with the cathedral and the baptistery. There are earlier cylindrical bell towers elsewhere in Tuscany and Umbria, and even in Ravenna, but Pisan historians claim that the tower of Pisa is unique and locally inspired.

Building work commenced on 9 August 1173 under the auspices of the Opera Campanilis Petrarum Sancte Marie (Stonework of St. Mary's Bell Tower). The foundations were set only 10 feet (3.36 meters) deep, on a bed of dry stones. There is no question that poor foundation soil—clay and sand strata interlayered with pockets of underground water—could not support the highly concentrated loads imposed by the tower; this, together with differential sinking of the soils, was the major contributor to the failure of the tower. The problem first appeared within about ten years, probably when the fourth arcade was reached. The building had sunk by more than 1 foot (30 centimeters), causing a lean of about 2 inches (5 centimeters). It was not unique in that: similar problems would be experienced in Holland, for similar reasons. Church towers still standing at Zierikzee and Leeuwarden have substantial leans; both were abandoned, and the architect of the latter committed suicide.

It seems that the Italian builders were more resolute than their Dutch counterparts, and after some delay, work on the tower of Pisa resumed, probably in the first decades of the thirteenth century. Remedial action was taken in two stages as attempts were made to reduce the lean. It is not known how high the building was when in the early 1270s Giovanni di Simone began to correct the inclination by raising one side of the galleries. By 1284, the six gallery levels were completed; then 156 feet (48 meters) high, the tower inclined about 3 feet (90 centimeters) from the vertical. The work was again suspended but the main part of the building was finished by 1319. The belfry, designed by Tommaso Pisano, was in place by 1350.

Major works undertaken in 1838 to save the tower of Pisa changed the proportion of groundwater and resulted in accelerating the inclination; it was only after some time that it settled to become about a millimeter a year. For over a century, eccentric suggestions were made to correct the problem, including a proposal to landscape the surrounding area to slope so that the tower would *appear* to be vertical. Moved to action by the 1989 collapse of the campanile of Pavia Cathedral, the Consorzio Progetto Torre di Pisa (Tower of Pisa Project Consortium) commissioned engineers to stabilize the Leaning Tower, then inclining more than 15 feet (4.6 meters) from true. It was closed to the public in 1990, and rescue work began. After some unsuccessful experiments, the three-part "final" solution was reached in July 1998. The tower was restrained by steel cables while 990 tons (900 tonnes) of lead were stacked against its base away from the direction of lean. Then excess water and mud were pumped from under the tower, allowing it to settle and in effect correct itself. As the tower straightened, the lead counterweights were regularly reduced. The project, with an overall cost of 54 billion lire (U.S.$27 million), was completed by the end of 2000. The result was that the lean was corrected by about 1 foot (30 centimeters), bringing the tower of Pisa to the condition it was in about three centuries earlier.

Further reading
Carli, Enzo, ed. 1989. *Il Duomo di Pisa: Il battistero, il campanile*. Florence: Nardini.

Conant, Kenneth John. 1973. *Carolingian and Romanesque Architecture, 800 to 1200*. Harmondsworth, UK: Penguin.

Pierotti, Piero. 1998. *Una torre da non salvare [How Not to Save the Tower of Pisa]*. Pisa: Pacini Editore.

Pneumatic structures

The most familiar inflated membrane structures are airships, from nonrigid blimps to giant vessels such as the proposed 1,003-foot-long (307-meter) ATG SkyCat cargo lifter with a payload of 2,200 tons (2,000 tonnes). The German firm Zeppelin built several rigid-frame airships between 1900 and 1936, including the famous *Graf Zeppelin*. The new technology had consequences in the building industry. The English aeronautical engineer Frederick W. Lanchester first proposed an air-supported structure in 1917. Immediately after World War II Walter Bird designed and built prototypes of pneumatic domes to house large radar antennae for the U.S. Air Force. Known as radomes, they had many civilian commercial applications and paved the way for a new kind of architecture.

Pneumatic—or air-supported—structures have their form sustained by creating, with the aid of fans, an air pressure differential between the interior of the building and outside atmospheric conditions. The increased air pressure—about the difference between the lobby of a high-rise building and the top floor—is so slight as to be virtually undetectable and causes no discomfort. The structural system enables the achievement of large spans without columns and beams, providing totally flexible interior spaces. Made from laminated membranes such as fiberglass, nylon, or polyester, coated with polyvinyl chloride (PVC) for weather protection, the electronically welded components are tailored to define the building shape. The durability and heat- and light-filtering properties of the membrane are determined by the careful choice of surface finishes and inner lining. Because of its lightness, the air-supported structure is among the most efficient structural forms, combining high-tensile strength materials with the shell form. The capital cost of an air-supported roof is typically up to one-third that of a conventional building; considered on a cost-per-seat basis—they are widely used for sporting venues—the advantage becomes even more obvious.

The United States pavilion at Expo '70 in Osaka, Japan, was one of several air-supported buildings at the fair. At the time, it was the largest structure of its kind ever attempted, a superellipse 265 feet wide and 465 feet long (81 by 142 meters). Its architects were Davis, Brody and Associates, working with designers Chermayeff, Geismar, de Harak and Associates; the engineer was David Geiger. The sloped sides of the pavilion, covered externally with paving tiles, were a 20-foot-high (6-meter) earth berm that supported a concrete ring 1,000 feet long, 4 feet high, and 11.5 feet wide (306 by 1.2 by 3.5 meters). Crisscross steel cables were locked into the ring to retain the roof once it was inflated. The roof of the pavilion was made of a translucent, closely woven fiberglass fabric, coated on both sides with vinyl. The seams were bonded by heat and pressure. Once inflated, the roof behaved almost as predicted.

More ambitious examples were bound to follow. In 1975, the 80,000-seat Silverdome in Pontiac, Michigan, boasted an air-inflated membrane roof measuring 720 by 550 feet (220 by 168 meters). The Hubert H. Humphrey Metrodome in Minneapolis, Minnesota, designed by Ellerbe Becket architects and completed in 1982, provided seating for up to 63,000 spectators under an air-supported roof of Teflon-coated fiberglass more than 10 acres (4 hectares) in area. Its claim to be the biggest air-supported domed stadium in the world was challenged the following year by the B.C. Place Stadium in Vancouver, Canada. In 1988 the Japanese architectural firm Nikken Sekkei and Takenaka, working with engineer David Geiger, designed the Tokyo Dome, with an air-inflated membrane roof of almost 660 feet (201 meters) span. Since about 1990, it seems that there has been a greater demand for sports arenas with openable roofs.

Further reading

Dent, Roger N. 1972. *Principles of Pneumatic Architecture*. New York: Halsted Press.

Vandenberg, Maritz. 1996. *Soft Canopies*. London: Academy Editions.

Pompidou Center (Beaubourg)
Paris, France

The Centre Nationale d'Art et de Culture Georges Pompidou, commonly known as the Pompidou Center, is in the Marais district of Paris. Initially given the working title Beaubourg (after its site), the center was

Pompidou Center (Beaubourg), Paris, France; Renzo Piano and Richard Rogers, architects, 1971–1977. Covered external escalators on the Plateau Beaubourg facade give access to the internal spaces.

formally named for its initiator, French president Georges Pompidou (1911–1974), following his untimely death. In December 1969 he announced an international design competition for a monumental multiuse public library, modern art museum, and contemporary arts center. The winning team, chosen from 681 entries, was directed by two architects—the Italian Renzo Piano (b. 1937) and the Englishman Richard Rogers (b. 1933). They were assisted by Gianfranco Franchini (one of Piano's erstwhile fellow students), and Ted Happold and Peter Rice of the structural engineering firm Ove Arup and Partners. Rice had been site engineer on the Sydney Opera House. Opened in February 1977, the Pompidou Center was a bold and innovative building redolent of the radical but unbuilt schemes of the Archigram group. Renowned for its highly flexible plan and the external exposure of its structure and services, the Pompidou Center was immediately acclaimed as the Parisian symbol of late-twentieth-century high-tech architecture.

In his first year of office, President Pompidou decided that the library proposed by his predecessor Charles de Gaulle should have a broader function. A passionate champion of the arts, Pompidou envisioned a national center that would act as a focus for the extensive array of cultural activities then evident in Paris. He wanted a building complex that would encourage all forms of artistic expression, promote the connection between the arts and social life, and be widely accessible to the person in the street. Flexible and uninterrupted internal spaces were needed to enable exploitation of its multiuse focus and to meet changing needs.

The design team coalesced early in 1971, following Happold's approach to Rogers about the competition; they had collaborated before. Rogers was then partner with Piano, and they saw Beaubourg as an opportunity to correct their firm's bleak work outlook. Piano, based in Genoa, Italy, was an architect and industrial designer. From the mid-1960s he had been experimenting with lightweight shells. One of his projects was the Olivetti plant in Scarmagno, Italy (1968); he also designed the building's components. He met the Italian-born Rogers in London in the late 1960s, and they collaborated on the Italian Industry Pavilion at the 1970 Osaka World's Fair. Rogers had been a member of England's Team 4 with Norman Foster. Over the next few years Piano and Rogers together produced the ARAM Medical Center in Washington, D.C. (1970), the Fitzroy Street Commercial Centre in Cambridge, England (1970), and offices for the B and B Upholstery Co. in Como, Italy (1971–1973). In several aspects—a totally flexible plan, an external steel structure as the basis of the esthetic, and the use of strong color—the latter foreshadowed the Pompidou Center.

The vast Plateau Beaubourg, cleared of dilapidated 1930s housing but surrounded by historic buildings, was selected as the site for Pompidou's new arts complex. It was being used as a car park for the nearby food markets at Les Halles, then in the process of demolition. The architects retained about half the Plateau as a pedestrian open space, intended for meetings and street theater, as well as gardens and sculpture. The new building was pushed to the eastern edge alongside an existing street. Construction began in April 1972, and Beaubourg was inaugurated by President Valéry Giscard d'Estaing on 31 January 1977; it opened to the public two days later. Finished on time and under budget, it cost Fr 993 million (then U.S.$100 million).

All the Pompidou Center's mechanical services are on the street side; on the side facing the open space, external escalators within clear acrylic vaults give access to the building's impressive interiors. To create uninterrupted internal spaces suited to unlimited uses, the steel-framed structure is outside the building, supporting an internal envelope enclosed by a glass skin. Six floors, 532 feet (166 meters) long and 192 feet (60 meters) wide, with movable suspended partitions, house a spectrum of functions, including a library, an art museum, and an industrial design center. There are also spaces for exhibitions, theater, dance, and musical productions, as well as a cinema, lecture and meeting rooms, a restaurant, and (more recently) an Internet café. The Piano and Rogers of-

fice designed the furniture, which was sympathetic to the building's esthetic. Outside, the unavoidably expressed vertical, horizontal, diagonal, and criss-crossed prefabricated tubular steel structural components form grids and lattices to create the architectural composition of the facades. The frame, together with the escalators and circulation walkways, and the service ducts are painted green, blue, red, yellow, gray, or white, according to their function. The building is raised on *pilotis*—doubtless the influence of Le Corbusier—creating a covered undercroft with space for shops. A 700-place car park is provided beneath the Plateau.

The architects' intention was compromised in the realization of Beaubourg for a variety of reasons: the cooler attitude of Pompidou's successor, Giscard d'Estaing; public reaction to the unconventional design; and of course the cost. Fire regulations were also an agent of change because the need to provide fire-isolated sections in such a large building confused the plans for total internal flexibility, achieved (according to the design) by having movable floors and walls. One of the abandoned proposals—it had impressed the competition jurors—was the display of information from the structural frame in the manner of Oscar Nitzchke's proposal for the unbuilt Maison de la Publicité (1932–1935). A lit screen was investigated but costs were prohibitive.

Like the Eiffel Tower (1889), the Pompidou Center, while provoking controversy for its eccentric design, was a highly successful tourist attraction. Daily visitor numbers soon reached 20,000, four times the predicted traffic. Having welcomed some 160 million people in its first twenty years, the Center was closed for renovations from late 1997 until January 2000. Piano and Jean-Francois Bodin, designer of the Musée Matisse in Nice (1987–1993) and the renovated Musée d'Art Moderne de la Ville de Paris (1992–1994), jointly supervised the Fr 576 million (U.S.$92.75 million) project. Essential maintenance including painting was carried out, the building and its infrastructure were modernized, and additional exhibition spaces were created by the removal of offices to an off-site location. A more orderly approach was taken to the use of the internal spaces by rationalizing use patterns. Access to the library from the ground floor was improved and the surroundings enhanced by additional trees.

See also Archigram; Renault Parts Distribution Center

Further reading

Bragstad, Jeremiah O., and Ivan Zaknic. 1983. *Pompidou Center*. Paris: Flammarion.

Silver, Nathan. 1994. *The Making of Beaubourg: A Building Biography of the Centre Pompidou, Paris*. Cambridge, MA: MIT Press.

Pont du Gard
Nîmes, France

The highest aqueduct the Romans ever built, described as "the most daring construction" of its day, supplied the provincial town of Nemausus (modern Nîmes) in Gaul (France). It delivered daily an estimated 44 million gallons (200 million liters) that were distributed through ten mains to the city's baths, fountains, public buildings, and houses. The most spectacular part of it is now known as the Pont du Gard. The description on the UNESCO World Heritage List of this remarkable feat of engineering reads: "The hydraulic engineers and … architects who conceived this bridge created a technical as well as artistic masterpiece."

Aqueducts had been employed more than two millennia earlier in the cities of the Indus valley and Mesopotamia, but the extensive systems that the Romans constructed both at home and abroad were the most sophisticated in the ancient world. The earliest, the Aqua Appia, was a 10-mile (16-kilometer) underground conduit built to serve the city of Rome in about 310 B.C. The Aqua Marcian, built 150 years later, carried water 56 miles (90 kilometers) to the capital; for about a fifth of its length it was supported on arches above the ground. Altogether, ancient Rome was supplied by eleven aqueducts, delivering an estimated 350 million gallons (1.6 billion liters) daily.

By the end of the first century B.C. Nemausus had become a key Roman settlement with a growing population, thought to be close to 50,000. Its water supply was inadequate, and following a visit by the emperor Augustus's son-in-law Agrippa in 19 B.C.,

plans were put in hand to bring water about 30 miles (50 kilometers) from the Fontaine d'Eure springs at Uzès to a stone reservoir near the city. The aqueduct was probably completed during the reign of Trajan (A.D. 98–117); it took over eighty years to build. In all Roman aqueducts, water flowed from source to destination under gravity, so the scale of the project was daunting—mountains had to be tunneled, hollows filled and valleys crossed—and a high degree of precision also was critical: the Uzès spring stood only 57 feet (17 meters) above the reservoir, and the 12-mile (20-kilometer) direct route between them called for 30 miles (50 kilometers) of aqueduct winding through the hilly region. The average gradient was a mere 1 in 3,000; in some places it was as gentle as 1 in 20,000. It was made steeper just before it reached the Gardon Valley, in order to reduce the height of the awesome Pont du Gard across the river about 11 miles (18 kilometers) northeast of Nîmes.

The massive 155-foot-high (47.2-meter) three-tiered bridge spans 920 feet (275 meters) across the valley. It was constructed of locally quarried limestone, finely dressed into ashlar blocks, some weighing up to six tons. The lower two tiers are laid without mortar, the blocks being secured with iron clamps. The 473-foot-long (142-meter) bottom tier has six irregular voussoir arches, 73 feet (22 meters) high, set out to span between stable rock outcrops in the riverbed. The middle tier has eleven 67-foot-high (20-meter) arches; spanning just over 800 feet (242 meters), it carries a 23-foot-high (7-meter), 35-arch arcade that supports the *specus*, a covered rectangular water channel about three feet (1 meter) wide and 6 feet high. That the Romans considered the structure to be utilitarian and "ordinary" is evidenced by the projecting blocks that nobody bothered to dress after the centering that they supported was removed.

The bridge has proven to be extraordinarily strong. Despite the Gardon River being "one of the most treacherous and rapid" in France, the Pont du Gard has resisted its onslaught for two millennia. Sheer mass—the weight of stone has been estimated at 16,000 tons (14,500 tonnes)—is one reason for that, and another is the combined skills of the engineers who designed it and the masons who built it. More-

Pont du Gard, Nîmes, France; engineer(s) unknown, completed ca. 100. The projecting stones were used to support the temporary wooden centering.

over, its slightly convex line and shaped piers combine to resist the current and whatever debris the river sweeps down, even when the entire bottom tier is entirely immersed, as sometimes happens; the foundations stand on bedrock; and its arches are much thicker than was then customary.

The simplicity of construction of the Pont du Gard provides an excellent demonstration of the Roman technological development par excellence: the voussoir arch. They learned much from their Etruscan forebears, including the exploitation of the arch, that enabled them to develop a new kind of architecture. When a flat stone beam (lintel) spans between upright supports (posts), the construction method is called trabeated. If the span is too great or the lintel too heavy, tensile or stretching forces will cause it to crack on the bottom. Posts needed to be closely

placed to avoid such failure. The arch and its extensions—the vault (arches placed side by side) and the dome (intersecting arches)—allowed the use of compressive materials (the only durable kind available) to enclose space, spanning huge distances without the need for intermediate supports. That technique, known to the Greeks, Egyptians, and Mesopotamians, had not been exploited by them but the Romans applied it with genius to build such bridges as the Pont du Gard and to create vast interior spaces.

The wedge-shaped stones used to construct arches are called voussoirs. Normally, they were cut roughly to shape in the quarry and accurately dressed on the site. Centering (a temporary timber framework), itself strutted from the ground or from projections in the lower levels of the structure, carried the arch until it could support itself. The first voussoirs were placed at the springing (the part of the pier carrying the arch) and others added until a keystone locked the arch at the center.

Some scattered remains of the aqueduct survive: a 55-foot-long (17-meter) three-arched bridge near Bornègre; a few exposed sections of the subterranean channel; another raised section near Vers; sections of tunnel; and the Pont de Sartanette, spanning about 100 feet (32 meters) over a small valley. The Pont du Gard itself continued in use until the ninth century, after which it was abandoned. Some of its stones were subsequently plundered by builders with less noble projects in mind, but generally it has survived human intervention. The bottom tier has long been used as a pedestrian thoroughfare, and in the Middle Ages stones were removed from one side of the piers so that laden mules could pass. In 1702 local authorities began renovations, attempting to close cracks, filling in ruts and building corbels to help support the road. The removed stones were replaced, and in 1743 a new bridge was built beside the lowest tier, widening the "roadway" for vehicular traffic. Overawed by the majesty of the structure, Napoléon III ordered its restoration (1855–1858), and subsequent projects have consolidated the mighty piers and arches.

As noted, the Pont du Gard was given World Patrimony status by UNESCO in 1986, and two years later the Regional Council of Gard began a historical and ecological protection program. In 1991–1992, fretted stones in the bottom level were replaced and waterproofing was improved; similar work continues on the second tier. Currently, the monument attracts more than 1 million visitors annually. There is growing concern that the Pont du Gard will be further threatened as those numbers increase—an example of the tension that everywhere exists between conservation and tourism.

Further reading

Aicher, Peter J. 1995. *Guide to the Aqueducts of Ancient Rome.* Wauconda, IL: Bolchazy-Carducci.

Fabre, Guilhem, et al. 1992. *The Pont du Gard: Water and the Roman Town.* Paris: Presses du CNRS.

Hauck, George. 1988. *The Aqueduct of Nemausus.* Jefferson, NC: McFarland.

Pontcysyllte Aqueduct
Denbighshire, Wales

At Pontcysyllte (Welsh for "connecting bridge") the Llangollen Canal crosses the River Dee at a height of 120 feet (36 meters) by means of a breathtaking aqueduct that marches more than 1,000 feet (306 meters) over the valley on nineteen slender, tapering (and partly hollow) masonry piers. Built between 1795 and 1805, the graceful structure remains the highest navigable canal aqueduct ever built. Besides its own weight it supports 1,680 tons (1,524 tonnes) of water. Pontcysyllte Aqueduct has been correctly identified as "one of the heroic monuments of the Industrial Revolution," and the novelist Walter Scott asserted that it was the finest work of art he had ever seen.

Part of Britain's 2,000-mile (3,200-kilometer) canal network, the spectacular Shropshire Union Canal—really a collection of canals built by various companies at different times—stretches 67 miles (107 kilometers) from the English Midlands town of Wolverhampton to the River Mersey. The original ambitious plan to link the Mersey and Severn Rivers was never achieved. The first stage, the Chester Canal from the River Dee in Chester to the town of Nantwich, was completed in 1779. Between 1796 and 1806 the Ellesmere Canal (later renamed the Llangollen Canal), fed largely by the Dee at the Horseshoe Falls, was built to connect Wrexham's ironworks and collieries with Chester and Shrewsbury. Joining the Chester Canal to Ellesmere Port in the Mersey Estuary, it reached as far as Llantisilio near Llangollen in North Wales. It is probably the most beautiful canal in Britain.

At Pontcysyllte the Llangollen Canal is channeled over the Dee Valley in a 12-foot-wide (3.6-meter) trough made of cast-iron plate and supported by four arched iron ribs spanning 44 feet (13.6 meters) between the tall piers. The water level comes to within a few inches of the top of the trough, and because a towpath is cantilevered over the surface, the navigable width for long boats and barges is reduced to about 7 feet (2 meters). There is a safety railing on the towpath side, but on the opposite side there is a sheer drop to the valley floor, 120 feet (36 meters) below.

Pontcysyllte Aqueduct, Denbighshire, Wales; Thomas Telford and William Jessup, engineers, designed and constructed 1795–1805. A canal boat can be seen crossing high above the Dee Valley.

The courageous decision to use a relatively unknown material followed hard upon Thomas Farnolls Pritchard's graceful Ironbridge at Coalbrookdale, Shropshire. Tradition continues to attribute the design of the Pontcysyllte Aqueduct to the Scots engineer Thomas Telford, who indeed claimed most of the credit, but more recent evidence suggests that, since he was working under the direction of William Jessup, the latter played a major part in the project. The financial backers of the Llangollen Canal approached Jessup in 1791, seeking an engineer "of approved character and experience." Jessup prepared the earliest working drawings, he gave expert evidence before the parliamentary committee, and he first proposed an aqueduct with an iron trough. Many consider Jessup to be the greatest canal and river navigation expert of his day.

The 70-foot-high (21-meter), 400-foot-long (120-meter) aqueduct at Chirk (also with a cast-iron trough but supported with conventional masonry); the 1,200-foot (360-meter) Darkie Tunnel; and the Whitehurst Tunnel, all built by Jessup and Telford between 1796 and 1801, also form part of the Llangollen Canal. The spoil from their excavations was used to construct the huge earth embankments of the Pontcysyllte Aqueduct. The structure has been listed for many years as a historic monument and it was among thirty-five sites nominated by the British government in April 1999 for inclusion on the UNESCO World Heritage List.

See also Ironbridge, Coalbrookdale

Further reading

Caswell, Lionel T. 1958. *Thomas Telford.* London: Longmans Green.

Hadfield, Charles, and A. Skempton. 1979. *William Jessup, Engineer.* Newton Abbot, UK: David and Charles.

Wilson, E. A. 1975. *The Ellesmere and Llangollen Canal: An Historical Background.* London: Phillimore.

Postmodern architecture

Simplistically, postmodern architecture emerged in the 1960s as a reaction to the Modern Movement that had commanded world architecture since the mid-1920s. Its theories were first expounded by the American architect Robert Venturi and realized in his Chestnut Hill Villa of 1962. Within less than a decade, designers were willfully denying the pervasive geometrical glass boxes that Henry-Russell Hitchcock and Philip Johnson had dubbed the International Style. Ornament (which the modernists had once equated with crime), color, and texture were again accepted, rather embraced, by architects. Historical precedents were revisited and often transposed into the language of twentieth-century technology to become a new visual language, an architectural patois. Eclecticism, for years a pejorative term, became a basis for design. And at first it was *just* design, because architects made more drawings and models than buildings. Although it is difficult to choose from a plethora of examples, among the icons of this new way of making architecture were Philip Johnson and John Burgee's AT&T Building in New York (1978–1984) and Michael Graves's "flamboyantly decorative" Portland Public Service Building in Oregon (1980–1983). As one commentator has observed, postmodernism became "the style of choice for developers of commercial buildings" everywhere. It has the same kind of stylistic anonymity of "globalness" as the Modernism it replaced.

Johnson (b. 1906) received a degree in architectural history from Harvard in 1930 and immediately became the first director of the Department of Design at the New York Museum of Modern Art. In 1940, inspired by the work of the Dutch modernist J. J. P. Oud, he returned to Harvard and emerged with an architectural qualification four years later. He worked alone and with others and became widely known from the early 1950s for his puritanically modernist buildings—some consider him a clone of Ludwig Mies van der Rohe—such as the Seagram Building in New York (with Mies, 1958) and the Glass House (1962) in New Canaan, Connecticut, until he formed a partnership with Burgee in 1967. Johnson then renounced Modernism (he had castigated Oud for doing that in 1946) and converted to postmodernism. His final artistic

Sony Building (formerly AT&T Building), Madison Avenue, New York City; Philip Johnson and John Burgee, architects, 1978–1984.

position was as an anti-postmodernist, leading the English architectural historian Dennis Sharp to opine that Johnson was philosophically fickle, with "more interest in [architectural] style than in substance."

Be that as it may, the AT&T Corporate Headquarters at 550 Madison Avenue, New York City, is a milestone in the development of twentieth-century architecture, the first postmodern skyscraper and a key building in the popularization of postmodernism. The 600-foot-high (184-meter), bland rectangular prism covers its site. Perhaps in reference to nineteenth-century skyscrapers, perhaps to a classical column, the main facade is divided into three parts: an entrance at the base, a tall shaft of identical floors, and a wide band of windows near the building's crown. The base, which originally enclosed a public open space, includes portals of epic proportion. A central 110-foot-high (33-meter) arch, surmounted by oculi, is flanked by three 60-foot-high (18-meter) rectangular doorways. Some critics suggest it borrows from Alberti's Sant' Andrea in Mantua, of 1472–1494. Unlike the featureless window-walls of modernist office towers, the shaft is sheathed in pink granite, and the fenestration is designed (like the early skyscrapers) to express the steel structural frame beneath.

The most controversial feature of the AT&T Building was the 30-foot pediment, ostensibly to mask mechanical equipment on the roof. Many regarded it as kitsch, and critics immediately dubbed it "Chippendale" because it evoked the work of the eighteenth-century English cabinetmaker. Indeed, the epithet was applied to the entire building, and Johnson interpreted the bestowal of a nickname as complimentary; otherwise he described his building as a "neo-Renaissance essay on the use of stone." In January 1992 the building was leased to the Sony Corporation.

Graves (b. 1934), one of the most honored twentieth-century architects, trained at the University of Cincinnati and Harvard. His early practice was limited to mostly domestic buildings. Among the notable examples are the Hanselmann house at Fort Wayne, Indiana (1967); additions to the Alexander house at Princeton, New Jersey (1971–1973); and the Crooks house, also at Fort Wayne (1976).

The Portland Public Service Building was the first of his large-scale projects to be realized. With subsequent commissions including the Humana Building at North Carolina State University (1982–1985), the San Juan Capistrano Public Library (1981–1983), and extensions to the Newark Museum (completed 1989), it placed him beside Venturi and Denise Scott-Brown, Frank Gehry, and Charles Gwathmey in the hall of champions of American postmodern architecture and design.

The freestanding fifteen-story municipal office building on Southwest Fifth Avenue, Portland, Oregon, houses the municipal Building, Planning and Design Review departments. It was the winning entry in a design-and-build competition sponsored by the city fathers. Johnson, as adviser to the jury and the client, was influential in securing the commission for Graves over the other shortlisted designs by Arthur Erickson and the Mitchell-Giurgola partnership.

Built on an entire 200-foot-square (61-meter-square) city block in the urban precinct, it is flanked by the city hall and county courthouse buildings on two sides, and a transit mall and a park on the others. To emphasize the association with other local government functions, Graves deliberately organized the facades in what he described as a "classical three-part division of base, middle or body, and attic or head," an approach that Johnson adopted for the AT&T Building. Described as a "wildly innovative and controversial postmodern landmark," the hefty building, rising from a heavy four-story base, has facades of diverse designs, clad with strongly colored tiles—brown, blues, and terra-cotta—against an ivory background. The square windows are relatively small, and they puncture the walls at regular intervals, another denial of the glass curtains of a decade or so before. The symmetrical park front has two huge seven-story pseudocolumns with boxy, floor-height capitals and flutes evoked by vertical bands of windows. Above the central main entrance there is a 40-foot (12-meter) hammered-copper sculpture of "Portlandia" (the female figure on the city seal) by sculptor Raymond Kaskey; it was added in 1985 at Graves's initiative. Around the corners, the facade is adorned at the tenth-floor level with a stylized swag of blue ribbons, made of concrete: on one, they hang sedately in place; on the other they appear to be blowing in the wind. Inside the building, Graves used the same colors as the exterior (a decision that provoked some criticism); he also designed the furnishing textiles and other details for the offices. Since 1995, the building's structural problems have become evident and are worsening. Despite costly repairs, the building may soon become unsafe to use.

Further reading

Graves, Michael. 1999. *Michael Graves: Selected and Current Works*. Mulgrave, Victoria: Images Publishing Group.

Jencks, Charles. 1991. *The Language of Post-Modern Architecture*. New York: Rizzoli.

Marder, Tod A., ed. 1985. *The Critical Edge: Controversy in Recent American Architecture*. Cambridge, MA: MIT Press.

Schulze, Franz. 1996. *Philip Johnson: Life and Work*. Chicago: University of Chicago Press.

Potala Palace
Lhasa, Tibet

The thirteen-story, 380-foot-high (117-meter) Potala Palace rises from sheer walls on a cliff named Marpo Ri (Red Hill), 130 meters above Lhasa, the capital city of what is now the Tibet Autonomous Region of the People's Republic of China. The 1,200-foot-wide (360-meter) complex of stone and timber buildings contains literally thousands of rooms with a total floor area of 154,000 square yards (130,000 square meters). Its shrines and the tombs of eight Dalai Lamas make it a focus of pilgrimage for Tibetan Buddhists. The walls, some up to 16 feet (5 meters) thick, were reinforced against earthquakes by backfilling with molten copper; their building stones were carried to the site by pack animals and slaves to construct a skyscraper on the roof of the world, nearly 12,000 feet (3,600 meters) above sea level. Wide stone stairways climb steeply from the city below. As one writer has commented, the Potala Palace was an achievement comparable to building the pyramids.

Legend has it that the craggy Marpo Ri was the site of a cave used as a religious retreat by the first emperor of a unified Tibet, Songtsan Gampo (A.D. 617–665), who also introduced Buddhism to the country. He ascended the throne when he was just thirteen, and in a twenty-year reign he established a powerful empire, with his armies ranging from northern India eastward to China and westward to Turkey. He moved his capital from Yarlung to Lhasa, and in 637 he built a palace, Kukhar Potrang, on Marpo Ri for his Chinese bride, Wen-Ch'eng. During his successor's reign, much of that building was destroyed by Chinese invaders, and its size and character are unknown. However, what little remained was subsumed into a new structure when the existing Potala Palace was initiated. There are two chapels in the palace, the Chogyal Drubphuk and the Phakpa Lhakhang, said to date from Songtsan Gampo's time.

By the end of the seventeenth century the ridge of Marpo Ri was crowned by a line of towering buildings that seem to be at one with the outcrop. The Potrang Karpo (White Palace) was the first phase of the new palace, commissioned in 1645 by the "Great Fifth" Dalai Lama, Ngawang Lobsang Gyatso (1617–1682). He moved his official residence from the Ganden Palace at Drepung (then the largest monastery in the world) to the Potala. The White Palace was completed by 1653. The central pavilion, known as the Potrang Marpo (Red Palace) and flanked on both sides by the White Palace, had not been started when he died, so the monks kept his death a secret for fourteen years while the work was finished! The Red Palace, and it is indeed dark red, was added between 1690 and 1694 by a workforce reputed to have consisted of 7,000 workers and 1,500 skilled artisans. The massive building in the so-called City of the Sun thus became the official winter seat of the Dalai Lamas and their extensive entourage. It was also the seat of government of the theocracy of Tibet. The name Potala, probably after the South Indian mountain sacred to Siva, dates from the eleventh century.

The Red Palace housed four meditation halls, thirty-five chapels, shrines, assembly halls, and the gem-encrusted golden stupas marking the tombs of the fifth to the thirteenth Dalai Lamas (except the sixth). In the western wing of the White Palace was the private cloister of the Dalai Lamas, the Namgyal Monastery, home to more than 150 monks, while its eastern wing contained government offices, a school for officials, and the National Assembly's meeting halls. There were also repositories for ancient religious books and manuscripts, arms and armor, and the treasures accumulated by the monastery over centuries. The lower levels were a maze of storerooms. In 1922 many of the major rooms in the White Palace were renovated, and two whole stories were added to the Red Palace. The agglomeration of buildings at the foot of the Red Hill, once a village named Sho, was the location of government offices and the headquarters of the Tibetan army.

A Communist Chinese army of 84,000 invaded Tibet in October 1950, unchallenged by the rest of the world. The troubled decade ended with a March 1959 uprising, the consequent toppling of the Tibetan government, and the voluntary exile of the

Potala Palace, Lhasa, Tibet; architect(s) unknown, 1645–1694.

Dalai Lama and 100,000 of his compatriots. In the suppression that has followed, it is claimed that more than 1 million people have been killed. The conqueror's policy of resettling Chinese in Tibet means that the Tibetans have become a minority in their own land, where Chinese has been made the official language. The Potala Palace, although its south and north sides were targeted by Liberation Army artillery in 1959, suffered minimal damage. Unlike many monasteries and shrines—about 6,000 have been destroyed—it was left untouched by the Red Guards during Mao Tse-tung's infamous cultural revolution. Nevertheless, many of its treasures were removed to China, and some later appeared on foreign markets.

Paradoxically, in 1990 the Chinese government sanctioned a 35 million yuan (then U.S.$4.2 million) restoration and renovation of the Potala Palace as a five-year plan. The budget subsequently blew out to 53 million yuan (U.S.$6.4 million), but it seems that Beijing considered it money well spent if it could mollify the resentment of the Tibetan religious hierarchy and attract tourists to Lhasa. The whole project, worthy in itself and praised by international conservation bodies, was swathed in propaganda. The Chinese government has also trebled the annual maintenance budget since the restoration was completed in August 1994. The following December the Potala Palace was inscribed on UNESCO's World Heritage List.

Further reading

Guise, Anthony, ed. 1988. *The Potala of Tibet*. London: Atlantic Highlands.

Nan, Hui. 1995. *The Potala Palace of Lhasa*. Beijing: Chung-kuo shih chieh yü ch'u pan she.

Q

Qosqo, Peru

Qosqo ("navel" or "center") in southern central Peru was once the ancient capital of the Inkan Empire. Continuously occupied for three millennia, the oldest living city in the Americas perches 11,150 feet (3,400 meters) above sea level in the Andes Mountains. Strategically located, Qosqo reached out to the entire Tahuantinsuyu (Land of the Four Quarters) by means of an extensive road network. In the days of its glory, the city boasted about 100,000 houses and somewhere between 225,000 and 300,000 citizens, many of whom lived in the neighboring farmland. The population compares with modern Rochester, Jersey City, or Anaheim. It was remarkable for its physical planning, its social organization, and the gold-festooned buildings of massive masonry that adorned it.

Farmers and herdsmen of the Marcavalle culture established permanent settlements in the Qosqo Valley around 1000 B.C. The Chanapata followed 200 years later, and successive groups—Qotakallis, Sawasiras, Antasayas, and Wallas—also occupied the site for about six centuries from A.D. 600. There is a tradition that Inkan Qosqo was founded some time in the eleventh or twelfth century by the legendary king Manco Cápac. What is clear is that under the ninth ruler, Pachacútec (reigned 1438–1463), it became a thriving urban center and the hub of the far-flung empire's religious and administrative life.

Its ascendancy lasted until 1533, when Pizarro's conquistadors entered the city. The invaders corrupted its name to Cuzco—meaning "hypocrite" or "humpback." To further diminish its power, within two years the Spanish established Lima as the new capital of Peru. In 1536 Manko Inka led his armies against them, and a protracted bloody war followed. But within forty years, the last emperor, Tupaq Amaru I, was defeated, captured, and beheaded in Qosqo.

The plan of Qosqo, like that of all Inkan cities, had several determinants. First was the cosmology of the builders, who framed it within imaginary lines governed by Pachamama (Mother Earth) and the Apus and Aukis (spirits of the mountains and valleys). Second, creating a balance, was the pragmatism of an agrarian people who had a habit of optimum land use (so that even city streets were narrow). Third, formal rules of "symmetry, opposition, repetition and subordination" constrained relationships between elements of the urban design. Basically, Qosqo comprised two parts: the *hawan* (upper sector) to the north and the less important *uran* (lower sector) to the south. The second division, into four, reflected the Tahuantinsuyu, and twelve neighborhoods were created by dividing each of the four into thirds. Each neighborhood was again divided into three. Tradition attributes the city plan to Pachacútec, and there is some evidence to support a second tradition that the central part of his capital was based upon the shape of a puma (considered sacred by the Inka) crouching over the Saphi River. That stream was diverted through a paved canal crossing Qosqo's central plaza—in every way the city's heart.

Urban life was focused on the plaza. The great open space, paved with flagstones, was divided into two by the Saphi Canal, one part providing the setting for the Inkas' principal rituals and ceremonies. That was surrounded by the most important buildings, including the sprawling, low palaces of the rulers and their extended families. Built of dressed stone, or at least stone faced, many were brightly painted; some had marble gates. The other part of the square was for public gatherings and celebrations. Near its center stood a platform *(Usnu)* from which the emperor and other dignitaries could speak to the people. Among Pachacútec's improvements to city center were the Coricancha (a courtyard once covered in gold) and the Temple of the Sun, also encrusted with gold plates and flanked by the trappings of the priesthood: cloisters, dormitories, gardens, and terraces, all "sparkling with gold." What was beauty to the Inkas was merely wealth to the rapacious Spaniards and therefore quickly plundered after 1533.

Within the context of public buildings, something must be said of the Inkas' superlative stonemasonry skills. Like the Spaniards, we marvel at the enormous granite or andesite boulders—some were almost 30 feet (9 meters) high—weighing hundreds of tons, that were transported great distances from quarries and without the benefit of the wheel. They were carved and dressed, for the most part, with stone chisels, although there is evidence of some bronze tools being used. Whatever the case, the bond known as Imperial Inkan masonry—medium-size stones laid in regular horizontal rows on very thin clay beds—was dressed with such accuracy that it was difficult to see the joints or slip even a knife blade between them.

From Qosqo's central plaza, four main streets led to the high roads to the Four Quarters and formed the base of the divisional structure already described. Long, narrow, straight streets, all paved with cobbles of Rumiqolqa basalt, followed a regular, right-angled grid. Along the streets, covered channels carried a clean water supply. In contrast to the carefully dressed stone of the palaces, houses on the perimeter of Qosqo were built of random rubble or mud brick (adobe) and lined with painted clay stucco. Their steeply pitched, timber-framed roofs were skillfully thatched with *ichu*, the local wild grass.

Following a series of abortive uprisings between 1780 and 1815, Peru was finally emancipated from Spanish colonial rule in 1821. In the meantime, because of the local dialect, the Spanish name for the city had been changed to Cusco. The Inkan metropolis had been overlaid by three centuries of colonial architecture. Nevertheless, in 1933 Cusco was recognized as the "Archaeological Capital of South America," and fifty years later it was inscribed on UNESCO's World Heritage List. By the late twentieth century, a strong nationalistic movement pressed for reversion to the original name, and in 1990 the municipality officially adopted "Qosqo" and the 1993 Peruvian Constitution declared it to be the Historic Capital of the country.

See also Inka road system

Further reading

Burland, Cottie Arthur. 1967. *Peru under the Incas*. New York: Putnam.

Gasparini, Graziano, and Luise Margolies. 1980. *Inca Architecture*. Bloomington: Indiana University Press.

Hemming, John, and Edward Ranney. 1990. *Monuments of the Incas*. Albuquerque: University of New Mexico Press.

Queen's House
Greenwich, England

The Queen's House on the edge of the Royal Park at Greenwich near London was designed by Inigo Jones—probably the greatest of all English architects—early in the seventeenth century. It was a major architectural feat because it represented, all at once and in a single building, the introduction of a new kind of architecture in the face of a well-established and reactionary building industry.

Before Jones (1573–1652) stepped on her architectural stage, England had been trying for almost a century to come to terms with the new forms of the Italian Renaissance. Henry VIII's attempts to bring Italian craftsmen to England had been resisted by

The Queen's House, Greenwich, England; Inigo Jones, architect, 1617–1635. The park front. The colonnade marks the former London-Dover road; the building between the house and the Thames is the Royal Naval Hospital, designed by Christopher Wren.

his subjects, and his later breach with most of Catholic Europe had stemmed the inflow of artistic ideas. The cultural standoff was maintained through Elizabeth I's long reign and well into the seventeenth century. Anything of the Renaissance that *did* reach England came, often in clumsy caricature, through northern European pattern books, and attempts to use supposedly Italian details in English architecture generated the epigram, "The Englishman Italianate is the devil Inkarnate." Single-handedly, Jones changed that.

His early life is obscure, but in 1603 he was working for the Earl of Rutland. Two years later Anne of Denmark, James I's queen, asked him to design scenery and costumes for a royal masque at the Palace of Whitehall. In 1611–1612 he briefly held the office of Surveyor to the Crown Prince, Henry, and shortly after his master's death, he was promised the position of Surveyor of the King's Works. The following

year Jones traveled in Italy with the Earl of Arundel and visited Venice, Vicenza, Bologna, Florence, Siena, Rome, and Genoa. He was impressed with modern Italian architecture and especially the country houses designed by Andrea Palladio (died 1580). He bought a copy of Palladio's *The Four Books of Architecture*, published in Italian in 1570. Soon after returning to England Jones succeeded Simon Basil as Surveyor.

His first royal architectural commission was for the Queen's House for Anne of Denmark. James I often went down from London to Greenwich (perhaps for fear of the plague) where Pleasaunce Palace stood on the site of the present Royal Naval College. Anne wanted a villa linking the palace garden and the Royal Park, which were divided by the main road between Deptford and Woolwich. Jones built the house with a two-story wing on each side of the road, joined at the upper level by a bridge, making it possible to pass from the palace gardens into the park without crossing the thoroughfare. When Anne died in 1619 work was halted. The basement and unfinished ground floor walls were covered with straw to protect them from frost, and a decade passed before work resumed.

In 1629 James's son Charles I gave the house to his queen, Henrietta Maria, and Jones completed it for her. By 1635 the outside was almost finished. Apart from its ingenious siting, the house was un-English in a number of ways, most notably for its carefully proportioned H-shaped plan, that contrasted with the rambling layout of contemporary English houses. Spatial organization within the Queen's House was symmetrical, geometrically laid out in keeping with the principles of visual harmony set down by Palladio. The Great Hall at the building's core was a 40-foot (12-meter) cube, the pattern of its marble floor matching the geometrical composition of the ceiling panels. From one corner of the hall, the so-called Tulip Stair—the first cantilevered staircase in England and of the kind recommended by Palladio—led to the king's and queen's separate apartments on the upper floor. Each suite comprised rooms planned to fit the court routine: a presence chamber, anteroom, privy chamber, antechamber, bedchamber, inner closet, and outer closet. A loggia on the south side of the house looked out across the Royal Park.

Another major departure from convention was the outside appearance of the house. The park front had a loggia in the center of the second story, and the proportions of solids and voids can be related to Palladio's Palazzo Chiericati at Vicenza (1550–1580). The riverfront had a central full-height projection to relieve its flatness; a horseshoe stair led from the palace garden to the podium on which the Queen's House stood. The building was crowned with a balustrade. The plain upper walls were set above a ground floor with regular, deep recessed joints. The stories also had windows of different heights, but in the eighteenth century the ground floor windows were lengthened. All is not what it appears, because the house is built of brick covered with white stucco in imitation of stone and prompting the alternative name "the White House." Although Jones believed that the outside of buildings should be "solid, proportionable to the rules, masculine and unaffected," the interiors were a different matter, and the Queen's House was lavishly decorated and fitted out.

The ceiling panels of the Great Hall, showing Peace surrounded by the Muses and Liberal Arts, were painted in 1635 by the Italian father-daughter team Orazio and Artemisia Gentileschi. Henrietta Maria furnished the rest of her house so opulently that an impressed visitor exclaimed that it "far surpasseth any other of that kind in England." But the fact was there *was* no other of that kind in England.

The interior was possibly incomplete when civil war erupted in 1642. When the king's houses were seized by Parliament in the following year, Jones's surveyorship was terminated. In 1645 he was arrested and his property confiscated; that was put right a year later. The king was executed in 1649, and Jones died (some say of grief) in 1652. Anti-Catholic feelings compelled the queen to flee the country. Following the restoration of the monarchy in 1660, Charles II intended to live in the Queen's House while building a new palace, but Henrietta Maria (now Queen Mother) moved in and remained until her death in 1689.

Jones's student and nephew John Webb undertook the restoration of the house in 1662, following his uncle's meticulous documentation and adding two bridges to make the plan of the upper floor into a perfect square. In 1690 the Queen's House became the residence of the Ranger of Greenwich Park, and in 1708 the ground floor windows and original casements were altered, spoiling Jones's careful design. The house was painstakingly restored in the 1980s.

It is difficult for us to grasp how innovative, even alien, the white, classical Queen's House would have appeared in Stuart England. Inigo Jones had categorically departed from every English precedent, and his design was regarded by one critic as "some curious device," because no one understood the theory upon which his architecture was based. His lead would not be followed for a hundred years. His architectural feat was achieved for a number of reasons: first, he was a new kind of architect, with royal patronage; second, he was no slave to fashion but had a thorough commitment to the principles that underlay Italian Renaissance architecture; and third, he was a practical man with consummate drafting skills that allowed him to communicate exactly what he required of the craftsmen, although they were unfamiliar with his kind of architecture.

Further reading

Charlton, John. 1976. *The Queen's House, Greenwich*. London: H.M.S.O.

Summerson, John. 1966. *Inigo Jones*. Harmondsworth, UK: Penguin.

Tavernor, Robert. 1991. *Palladio and Palladianism*. New York: Thames and Hudson.

R

The Red House
Bexley Heath, England

Designed for William Morris in 1859 by his friend and coworker Philip Webb, the Red House in the London suburb of Bexley Heath has been called "a cornerstone in the history of English domestic architecture." Much more than that, although in one sense a piece of eclectic architecture, it was a milestone in the way that architects designed houses, making the house to fit the occupant, rather than (as had been the case) forcing the occupants to fit the house: the earliest glimpse of functionally constrained design. Early in the twentieth century the German critic Hermann Muthesius recognized it as "the first house to be conceived as a whole inside and out, the very first example in the history of the modern house." At that moment, the ideas behind it were taken up and developed by the American architect Frank Lloyd Wright and fed back into the European Modern Movement.

The now-famous English social reformer, designer, novelist, and poet William Morris (1834–1896) originally intended to become a Church of England priest. While at university he decided to devote himself to art. He then worked briefly for the Gothic Revival architect G. E. Street, but influenced by the Pre-Raphaelite painters Edward Burne-Jones and Dante Gabriel Rossetti, soon turned, also briefly, to painting. In 1857 he met Jane Burden, one of Rossetti's

models, and two years later they were married in Oxford. Morris was financially independent—his annual income of £900 was substantial—and in summer 1858, while on a rowing holiday in France, he decided to build a house at Upton in Kent, southeast of London. He commissioned the architect Philip Webb (1831–1915), with whom he had worked in Street's office, to design a house "very medieval in spirit" and "in the style of the thirteenth century." Webb resigned his position and commenced work on his first building as an independent architect.

Named for its brick walls and clay tile roofs, the Red House was indeed medieval in spirit. It rejected the formal aspects of the fashionable Gothic and the exotic Italianate for the familiar vernacular—"home-grown" English domestic architecture, emphasizing the spirit, not the letter, of a medieval past that Morris and his friends viewed with wistful longing. The house was simple and (remarkably for its day) free from architectural ornament. One writer has commented, "Form was more important than decoration. Outside it [had] steeply tiled roofs, long ridge-lines, tall chimney-stacks and steeply recessed porches. Inside it had plain tiled floors, a simple open staircase and large wooden dressers" (Bradley 1978, 26). The house was begun in 1859 and completed within a year.

Rossetti thought it "more of a poem than a house … but an admirable place to live in too." That livability

The Red House, Bexley Heath, England; Philip Webb, architect, 1859–1860. Detail of exterior.

was because Webb designed the Red House to suit the Morrises' needs. At that time, houses were planned according to usually symmetrical, formal geometries, and their occupants were obliged to tailor their daily lifestyle to the limits imposed by the architecture. The Red House's L-shaped plan, partly enclosing a courtyard, was unconventional and informal, and rooms were disposed according to the way in which the Morrises intended to use them. The east wing of the first floor contained the kitchen and service rooms; its north wing had a waiting room and a bedroom. The dining room was at the northeast corner, off the large central hall. The main entrance was at the north end of the hall; at the other, an open oak staircase led to the second floor, which housed bedrooms, servants' quarters, and a drawing room at the northeast corner, over the dining room. There was a large L-shaped study at the western end. Morris's biographer Fiona MacCarthy asserts that the house was the symbolic point of departure for his crusade against the Industrial Age.

Much later in life Morris reflected that, because he could not find appropriate furniture and furnishings in English shops, he decided "with all the conceit of youth" (he was twenty-five) that he would design and make them for himself. With Webb, Jane Morris, and his Pre-Raphaelite friends he decorated the house; they painted medieval-style murals, constructed furniture, sewed embroideries, wove tapestries, and designed wallpapers and stained-glass windows. In fact, the Red House and its contents—"a temple to art and craft"—were at the very foundation of the English Arts and Crafts Movement, both literally and philosophically.

Morris then realized that his interest lay in the decorative arts, and the success of the Red House collaboration provoked the formation of Morris, Marshall, Faulkner and Co. in 1861; the firm's brochure described it as "Fine Art Workmen in Painting, Carving, Furniture and the Metals." The partners were Morris, Burne-Jones, Rossetti, the painter Ford

Madox Brown, Philip Webb, Charles James Faulkner, and Peter Paul Marshall, a surveyor. The Red House housed the firm's workshops, and they continued to design and manufacture all the types of artifacts that had been produced for it. In November 1865 Morris, because of lack of money (the firm was mismanaged), moved his family to new Queen Square, Bloomsbury, London, to share premises with the firm. The Red House was sold.

It was owned as of 2001 by Mrs. Doris Hollamby. She and her architect husband Edward (died 1999) bought it in 1952, after it had been used for offices of the National Assistance Board during World War II. The Hollambys undertook to accurately restore the house, which is now open to the public. An association known as the Friends of Red House has been formed to support the Red House Trust, in order to physically maintain the house, its garden, and orchard, and secure its "long-term future."

Further reading

Bradley, Ian. 1978. *William Morris and His World*. London: Thames and Hudson.

Hollamby, Edward, and Charlotte Wood. 1991. *Red House: Bexley Heath, 1859*. New York: Van Nostrand Reinhold.

MacCarthy, Fiona. 1995. *William Morris: A Life for Our Time*. New York: Knopf.

Muthesius, Hermann. 1979. *The English House*. London: Crosby Lockwood Staples.

Reichstag
Berlin, Germany

The restored Reichstag in Berlin, designed by the London architectural firm of Foster and Partners, epitomizes a new kind of architecture—one that respects the physical and cultural environment and takes account of the past while assuming responsibility for the future.

The institution known as the Reichstag was set up in 1867 by the German Chancellor Otto von Bismarck to allow the bourgeoisie to have a role in the politics of the new empire, a confederation of princely states under the King of Prussia. From 1871 the Reichstag met in a disused factory until a neo-Renaissance building (1882–1894) was created for it by the Frankfurt architect Paul Wallot. After the reunifica-tion in 1990, the new Germany's Parliament, comprising the two houses known as the Bundestag and Bundestat, made Berlin the capital of the Federal Republic of Germany in June 1991. It also voted, by a small majority, to move its own seat from Bonn to Berlin, locating it in the historic building.

The monument was in a sorry state and held memories of the failure of the Weimar Republic and the disastrous Third Reich. Before the notorious Berlin Wall came down, it was cut off from the old center, just outside the boundary; now it is in the middle of the city. The Reichstag building had been patched up in the cold war years, and the facades and the interior underwent desultory restoration in the 1960s. It was used as a historical museum between 1958 and 1972, and spasmodically for meetings of the West German Parliament. In June 1992 an international architectural competition was held to restore the Reichstag, and eighty architects submitted proposals.

Following some debate and a second stage of the competition among the three shortlisted entries, Foster and Partners were awarded the commission in July 1993. The consulting engineers were Leonhardt Andra and Partner, the Ove Arup Partnership, and Schlaich Bergermann and Partner. The Foster partnership originally proposed a huge mesh canopy supported on columns to enclose Wallot's building and extend it into the Platz der Republik. Axel Schultes and Charlotte Frank's urban plan for the Spreebogen district of Berlin, the result of a contemporary competition, set the framework for new buildings and called for a rebriefing and consequent changes to the design. Building work began in July 1995 and the new Reichstag was opened in April 1999; it cost DM 600 million (approximately U.S.$330 million).

According to the architects, their final design was constrained by four factors: the history of the Reichstag, which in its earliest days had symbolized liberty; the day-to-day processes of the Parliament; questions of ecology and energy; and (naturally) the economics of the project. Because Wallot's building was to be preserved as far as possible, the Reichstag is a living historical museum that frankly shows the scars of its past—pockmarks caused by shells, charred timber, and Russian graffiti from the post–World War

Reichstag, Berlin, Germany; Foster and Partners, architects, 1993–1999. Interior of glass dome, showing the inverted cone used for climate and lighting control.

II occupation are all left visible. Because it was believed that the processes of democracy should be transparent, Wallot's formal west entrance was reopened to serve for all users of the building, politicians and public alike. The great steps lead to a tall, top-lit narthex; on entering, the visitor is confronted by a glass wall that defines the lobby; beyond that, another transparent partition gives a view into the parliamentary chamber. Members of the public may occupy public balconies or follow interlocking spiral ramps to a viewing deck that looks down into the chamber from within the cupola. The functional needs of the Parliament required the demolition of many of the accretions of the earlier refurbishment.

Visually and structurally, the design is dominated by a new glass-and-steel hemispherical cupola at the center of the restored building, which replaces and evokes the war-damaged original dome, removed in 1954. But the cupola is more than an esthetic or symbolic choice. At its center a curving, inverted cone of mirrors reflects daylight into the plenary chamber. The cupola is fitted with a movable sunscreen: in summer it tracks and blocks the sun to prevent overheating of the interior; in winter it is set aside to allow warming sunshine to penetrate into the building. The cone also acts as a convection chimney; fresh air enters the building through air shafts and rises through the floor of the chamber. As it heats up it is drawn into the cone, and an extractor expels it from the building. An aquifer at a depth of 100 feet (30 meters) stores cold water that is circulated through pipes in the Reichstag's floors and ceilings in the summer. Warmed in the process, the water is then pumped into another subterranean lake, 1,000 feet (300 meters) beneath Berlin. At that depth it retains its heat, and in winter the process is reversed to heat the building. The Reichstag power plant that drives the pumps is fueled by renewable grape seed oil. In the 1960s the restored Reichstag emitted 7,700 tons (7,000 tonnes) of carbon dioxide a year; the new building emits 440 tons. Germany has been a world leader in energy conservation, and the building that now symbolizes national unity fittingly exemplifies that mind-set.

Further reading

Foster, Norman, et al. 1999. *Rebuilding the Reichstag*. Woodstock, NY: Overlook Press.
Pawley, Martin. 1999. *Norman Foster: A Global Architecture*. New York: Universe.

Reinforced concrete

Concrete is a combination of small aggregate (sand), large aggregate (gravel), a binding agent or matrix, and water. Historically, lime was used as a matrix, mostly for mortars that had no large aggregate. In 1774 the British engineer John Smeaton added crushed iron-slag to the usual quicklime-sand-water mix, making the first modern concrete for the foundations of the Eddystone Lighthouse off the English coast. Fifty years later, a new matrix was discovered. Portland cement, a calcium silicate cement made with a combination of calcium, silicon, aluminum, and iron, is the basis of modern concrete. In 1824, the English stonemason Joseph Aspdin made it by burning (on his kitchen stove) finely ground limestone and clay, then grinding the combined material to a fine powder. It was named for its original use in a stucco that imitated Portland stone. However, the burnt clay yielded silicon compounds that combined with water to form a much stronger bond than lime. It was to revolutionize the architectural and engineering world.

For the next thirty years or so, plain concrete, because of its tremendous compressive strength (resistance to crushing), was used for walls. Sometimes it replaced brick as fire-resistant covering for iron-framed structures. Reinforced concrete, developed first by the French, combines concrete's compressive strength with the tensile strength (resistance to stretching) of metal—at first iron and later steel—reinforcing bars or wire. The first person to employ such construction was the Parisian builder François Coignet, for drainage and building works throughout the 1850s. His own all-concrete house in Paris (1862) survives.

The new material was also investigated by the Parisian market gardener Joseph Monier, who was granted a patent in 1867 for garden pots made of cement mortar reinforced with a cage of iron wire. Over the next decade, he built concrete water-storage tanks, patented several ideas for bridges, and promoted reinforced concrete for floors, arches, railroad ties, and bridges. Not being an engineer, he was not permitted to build bridges in France, so he sold his patents to German and Austrian contractors Wayss, Freitag and Schuster, who built the first reinforced concrete bridges in Europe.

Monier exhibited his plant pots at the 1867 Paris Exposition, where they caught the attention of the builder François Hennebique (1842–1921), who then looked for applications in the building industry. Beginning with reinforced concrete floor slabs in 1879, by 1892 he had developed a complete building system of columns, beams, and floors, which he applied to an apartment building in Paris. Having patented the system, he wound up his contracting firm to become a consulting engineer. The structural and esthetic implications of monolithic reinforced structures were staggering. Because concrete is in effect a liquid, the only limitations placed upon the plasticity of architectural shapes lay in the buildability of the formwork (known as shuttering) that held it in place while it set. The remaining major technological step was taken by yet another Frenchman, Marie Eugène Léon Freyssinet (1879–1962), who perfected prestressed concrete, allowing the construction of even greater spans.

It was left to Auguste Perret (1874–1954) to address the problem of a suitable esthetic for the new material, because (of course) there was no historical precedent. Although he never became an architect in the strictest academic sense, Perret was one of the pioneers of French modern architecture, in whose work we see the first rational expression of reinforced concrete. Perret developed Hennebique's structural system, but while the latter had disguised his buildings' reinforced concrete skeletons with masonry, from as early as 1903 Perret—he called himself a "builder in reinforced concrete"—began frankly expressing them as part of the architecture. That approach reached its pinnacle in his war memorial church of Notre Dame du Raincy, Paris, of 1922–1923, claimed by architectural historian Peter Collins to be the most revolutionary building of the first quarter of the twentieth century. Raincy was a model for other churches by Perret, including St. Therese, Montmagny (1925); a chapel at Arcueil (1925); and St. Joseph, Le Havre (begun in 1950). Perret's carefully designed shuttering, producing "off-form" surfaces that needed no further finishing work, inspired the so-called Brutalist architecture movement, mostly British, of the late 1950s, as well as Japanese architecture right to the end of the twentieth century.

Beyond what may now be described as orthodox reinforced concrete construction, a number of engineers—the German Ulrich Finsterwalder, the Italian Pier Luigi Nervi, and the Spaniards Eduardo Torroja y Miret and Felix Candela—pushed the versatile material to its technological limits, developing the cantilever and the thin shells that may be regarded as the ultimate concrete form.

See also Airplane hangars; Maillart's bridges; Roman concrete construction; Shell concrete

Further reading

Collins, Peter. 1959. *Concrete: The Vision of a New Architecture.* London: Faber and Faber.
Idorn, Gunnar M. 1997. *Concrete Progress: From Antiquity to Third Millennium.* London: Thomas Telford.
Straub, Hans. 1952. *A History of Civil Engineering.* London: L. Hill.

Renault Distribution Center
Swindon, England

High-technology (usually contracted to "high-tech") architecture was a movement born in the 1960s and sustained through the 1980s. It sought to express zeitgeist—the spirit of the age—defined by its followers as resting in the technological advances of industry, communications, and travel, including aerospace developments. These advances offered an alternative building approach. "High-tech" architects produced machinelike structures of flexible plan, applying lightweight materials such as sheet metal, glass, and plastic to innovative structural techniques; they employed easily assembled, sometimes mass-produced, building components. Usu-

ally, the structure was made explicit (often reinforced by colorful paintwork). Sometimes the services were exposed. Constructed between 1980 and 1983, the Renault Distribution Center at Swindon, about 100 miles (160 kilometers) southwest of London, was and continues to be regarded as the archetypal high-tech building.

The brief called for a building that established a progressive corporate identity and stood out from the featureless industrial shed typical of the area. It was to suit multiple functions, be quickly constructed, and capable of later extension. Architect Norman Foster of Foster Associates, together with the engineering firm of Ove Arup and Partners, responded with a design for a visually arresting, structurally self-explicit building that dominated what has been described as an otherwise bleak landscape.

Foster was born in 1935 in Manchester, England. He trained as an architect and town planner at the University of Manchester (1956–1961) before undertaking a master's degree in architecture at Yale University (1961–1962). Soon after, following a brief period with Richard Buckminster Fuller, he returned to England to set up practice with his wife Wendy and Richard and Su Rogers. They worked as Team 4 between 1964 and 1967 until the partnership was superseded by Foster Associates, now Foster and Partners. In the late 1960s Foster and Richard Rogers made a significant architectural statement at Swindon with the Reliance Controls Factory (1967), one of the last Team 4 projects. Its elegance, use of off-the-shelf components, exposed steel structural bracing, metal cladding, wall-high glazing, flexible plan, and focus on improved employee working conditions all challenged the conventional wisdom about industrial buildings. Reliance Controls was an early example of high tech.

Almost two decades later the Swindon landscape was again confronted—this time by the unconventional form of the Renault Center. Brilliant yellow (Renault's corporate color) cable-stayed tubular steel masts supported a reinforced polyvinyl chloride (PVC) membrane roof that covered spaces for spare parts warehousing, visitor reception, distribution and regional offices, vehicle showroom, after-sales maintenance training, and staff dining. Its expressive, detailed outline and functional, worker-friendly spatial arrangements were characteristically Fosteresque, balancing the high-tech approach with client and social needs. In its marketing literature, Renault enthusiastically reproduced images of its marqueelike center, regarding it as the quintessence of its corporate image.

When presented with plans for the sloping 16-acre (6.5-hectare) site, the local authority consented enthusiastically to the unexpected design and to the proposed 67 percent land coverage (the usual limit was 50 percent). The prefabricated rectangular building was formed as a series of suspended modules—forty-two in total—comprising 52-foot-high (16-meter) masts, connected to pin-jointed portal frames. Each module measured 91 feet (24 meters) square and was 25 feet (7.5 meters) high at the edge and 31 feet (9.5 meters) in the center. As extensions were required, modules could be unbolted and new ones added. Initially, thirty-six modules were devoted to warehousing, the rest located at the narrower end of the site where the building tapered to a generous entry and porte cochere.

The fully exposed, repetitive mast arrangement flowed graciously beyond the external walls, glazed for showroom and dining but sealed elsewhere with steel skins. Ample natural lighting was achieved by clear glass panels inserted where the mast pierced the roof membrane and by a louvered roof light at the apex of each module; the louvers could be opened for ventilation. The building was centrally heated and lit according to the function of the space. Foster Associates designed the furniture.

The Renault Distribution Center has been described as ushering in the firm's Hong Kong and Shanghai Bank headquarters (1979–1986) in Hong Kong, also noted for its extrinsic structural expression. However, unlike the sprawling Renault building, the bank headquarters is a soaring triple-layered tower (the tallest forty-one stories) with immense tubular steel trusses from which the floors are suspended. Many consider this Foster's magnum opus.

Further reading

Abel, Chris. 1991. *Renault Centre: Swindon, 1982*. New York: Van Nostrand Reinhold.

Davies, Colin. 1988. *High Tech Architecture*. New York: Rizzoli.

Sudjic, Deyan. 1986. *Norman Foster, Richard Rogers, James Stirling: New Directions in British Architecture*. London: Thames and Hudson.

Retractable roofs

The Houston Astrodome in Texas, opened in 1966, was the first stadium with a roof over the playing area. It set a trend for sports fields for the next twenty years. Its roof, designed to resist 135-mph (216-kph) winds, has a clear span of 642 feet (196 meters); it is 208 feet (64 meters) high at the apex. It was not, however, the first arena to have a roof. It was predated by almost 2,000 years by the Flavian Amphitheater in Rome, better known as the Colosseum. The Colosseum measured 620 by 510 feet (189 by 156 meters), and the perimeter of the fourth story had stone brackets supporting wooden masts from which an awning (velarium) was suspended across the interior to shield spectators from the sun. The velarium was not fixed; teams of sailors handled the rope-and-pulley system that allowed it to be opened and closed depending on the weather.

The Toronto SkyDome, designed by architects Rod Robbie and Michael Allen and inaugurated in June 1989, was the first modern stadium with a fully retractable roof. SkyDome provides 2 million square feet (186,000 square meters) of usable floor space for up to 30,000 spectators. The 8-acre (3.24-hectare), 11,000-ton (10,000-tonne) roof rises 282 feet (86 meters) above the field level. It consists of a fixed panel and three movable panels, framed with steel trusses and covered with a polyvinyl chloride (PVC) membrane laminated to an insulated steel sheet, moving on a system of tracks and bogies. The roof can open in twenty minutes to uncover the entire field area and over 90 percent of the seating. Since SkyDome, many similar structures have developed the new technology that enables very large buildings, once considered static, to become (at least in part) flexible—an architectural feat.

Amsterdam Arena, the Netherlands, was opened in September 1996, the first retractable roof stadium in Europe. The stadium is 540 feet (165 meters) wide and 770 feet (235 meters) long; the roof, soaring 255

Amsterdam Arena, the Netherlands; Rob Schuurman, architect, completed 1996. Aerial view, with the roof closed.

feet (78 meters) above the playing field, consists of two movable panels that retract across the short span. The designer was Rob Schuurman. Bank One Ballpark in Phoenix, Arizona, designed by Ellerbe Becket, was completed in early 1998. Two 200-horsepower motors open or close the retractable roof over the 48,000-seat stadium in under five minutes. Each half of the roof consists of three movable trusses that telescope over a fixed end truss. Either side can be opened to any position, independently of the other. The 52,000-seat Colonial Stadium in Melbourne, Australia, was opened in 2000. Its 540-foot-span (165-meter) retractable roof, employing a lightweight space-truss structure, opens or closes in less than eight minutes. Other arenas, such as the Sports Park Main Stadium of the Oita Prefecture, Japan, and Miller Park, Milwaukee, Wisconsin, were completed in 2001. The former, designed by Kurokawa Kisho Architectural Urban Design and the Takenaka Corporation, has a retractable 895-foot-diameter (274-meter) hemi-spherical steel-framed shell roof; Miller Park has a seven-panel roof.

See also Colosseum (Flavian Amphitheater)

Further reading

Crawford, Bob. 1998. *Diamond in the Desert: A Pictorial Visit to Bank One Ballpark*. San Francisco: Woodford.

O'Malley, Martin, and Sean O'Malley. 1994. *Game Day: The Blue Jays at SkyDome*. Toronto and New York: Viking.

Petersen, David C. 1996. *Sports, Convention, and Entertainment Facilities*. Washington, DC: Urban Land Institute.

Roman concrete construction

Concrete is made by mixing broken stone or gravel and sand (aggregate), a bonding agent, and water, and allowing the mixture to harden through chemical process into a solid mass. So-called cementitious materials had been used in ancient Egypt about 3,000 years earlier and later by the Chinese, Minoans, and Mycenaeans, but this synthetic stone—a new building material—was developed and exploited by the Romans from about the third century B.C.

Ambrose advised his protégé Augustine: "When in Rome, live as the Romans do; when elsewhere, live as they do elsewhere." Throughout the Roman Empire, the architecture they built was a weighty presence imposed upon the subject peoples—a consistency probably more to serve the colonizers, isolated from the familiar things of home, rather than for the colonized. Throughout history architecture has provided a social anchor for migrant peoples. The Roman way was to come, to see, to build, and there was, especially in the days of imperial expansion, a need to build quickly and in a familiar way. That was made possible by the use of concrete.

The Romans used concrete *(opus caementicium)* for all parts of their structures: foundations, walls, and roofs. It was made by combining pozzolana (a volcanic earth found in many places in southern Europe) with lime, broken stones, bricks, tufa, and sometimes pumice. Such a mixture could set even underwater. Lime was obtained by crushing limestone or seashells, or sometimes replaced by gypsum as a binding agent. The Romans placed a very dry mix of pozzolana and wet lime, layer for layer, over rock fragments, and carefully tamped it into place. Its structural strength depended upon what is now called the water-cement ratio: the higher the proportion of bonding agent to water, the stronger the concrete. The combination of a dry mix and thorough consolidation made the material extremely durable.

At first, concrete was limited to places where it would not be seen. For foundations, it was placed between wooden form boards that were stripped once the mixture had hardened. For building above the ground, its brutal appearance, once the formwork was removed, presented an esthetic problem. Because the many advantages—strength, versatility, economy, availability, and speed of erection—more than offset that single disadvantage, the Romans simply used more presentable materials to face the concrete, usually as a "lost" formwork. For example, as late as 20 B.C. the architectural theorist Vitruvius recommended building two face walls of squared stone *(opus quadratum)*, 2 feet (0.6 meter) thick, tying them together with iron cramps, and filling the cavity with tamped concrete. But that was for prestigious buildings, and a number of alternative wall constructions had already been developed.

From around 200 B.C., slabs of volcanic tufa were used as permanent facings; far more common was the technique known as *opus incertum*, which employed small, random pieces of tufa, carefully packed together. Over time the shapes were made increasingly regular, and by about 50 B.C. 4-inch-square (10-centimeter) pyramidal tufa blocks were employed. Set diagonally, their sharp apexes penetrated about 10 inches (25 centimeters) into the concrete infill, providing an excellent bond. Because of its netlike appearance, the method was called *opus reticulatum*. By then, fired clay bricks were also being used for facing. Over the next two centuries the predominant technique was *opus testaceum*, flat slice-of-pie-shaped bricks, tied at intervals with bonding courses through the wall. The late empire saw a further variation, called *opus mixtum*, consisting of alternate courses of brickwork and small squared stones. It is stressed that these systems provided only a presentable *surface*: the real work of the wall was done by the immensely strong concrete mass, which normally supported innovative superstructures, also made of concrete.

Unfinished concrete was not only unattractive to the eye but it also presented an architecture that was, to the Roman mind, inappropriate in appearance. They therefore covered it, whether brick faced or not, with a variety of decorative surfaces: stucco (a mixture of marble dust and lime) perhaps 3 inches (7.5 centimeters) thick in up to five layers, and molded, patterned, painted, and sometimes veneered with mosaics of marble and even glass tesserae. The most important buildings had marble veneers, held in place by bronze pins and nonstructural architectural orders applied as pilasters or half-columns that masked the concrete structure and reduced the visual scale.

Roman public architecture existed to move *in* and *through*, rather than *around*. Such urban buildings as the *thermae* (bathhouses) and *basilicas* (law courts) demanded interior spaces uncluttered by columns that could accommodate huge gatherings of people. In order to achieve vast interior spaces, the Romans exploited the semicircular arch, a technology inherited from their Etruscan forebears. The arch and its three-dimensional extensions, the vault (a prismatically extended arch) and the dome (a rotated arch), could span large distances without intermediate supports. With characteristic directness, Roman engineers found expedient solutions. A small rectangular room could be covered by a semicircular barrel vault carried on continuous parallel side walls. A square room could be roofed by a cross vault (two barrel vaults placed at right angles), supported by piers at the corners and allowing the space to be lit from all sides. Larger rectangular spaces could be enclosed by a procession of such vaults built side by side. A polygonal space received a hemispherical dome and an apse a half-dome, carried on drums above the base walls. By using concrete for these roof structures, the Romans enclosed volumes that would not be equaled for over 1,000 years.

The simplest barrel vaults consisted of a series of parallel brick arches cross-tied as in opus testaceum and filled between with concrete; that is, the concrete was packed into brick compartments. The whole structure was supported by wooden centering until the mortar had set. Other vaults and domes were directly formed in mass concrete. The technique had two main advantages: once the centering was designed and placed, it employed unskilled labor, and it enabled complex plan forms to be roofed without the cost of dressed stone construction. Often, the weight was reduced by using hollow clay boxes or even wine jars, especially in the groins of cross vaults; alternatively, vaults were lightened by forming recesses or coffers in their undersides. Domes were generally much thicker toward their base and therefore appeared externally as inverted saucers, while inside they were hemispherical. As they rose, lighter materials, such as pumice—a stone that floats—were used for aggregate.

In order to underline the achievement of the Roman engineers and architects, it is helpful to consider the size of some concrete structures; a couple of examples may suffice. The concrete barrel vaults that spanned the 76-foot-wide (23-meter) side aisles of the fourth-century-A.D. Basilica of Maxentius in Rome were 8 feet (2.45 meters) thick. The 142-foot (43-meter) concrete dome of the Pantheon, also in Rome, is 4 feet thick (1.3 meters) at its apex and 20 feet at its base. The massive loads of these roofs were carried to the ground through huge piers or thick walls, and their horizontal thrusts resisted by buttressing elements integrated with the architectural design.

See also Baths of Caracalla; Pantheon; Reinforced concrete

Further reading
Adam, Jean-Pierre. 1994. *Roman Building: Materials and Techniques*. Bloomington: Indiana University Press.
Hamey, L. A., and J. A. Hamey. 1982. *The Roman Engineers*. Minneapolis: Lerner.
Sear, Frank. 1989. *Roman Architecture*. London: Batsford.

Royal Albert Bridge
Saltash, England

The Royal Albert Bridge at Saltash, completed in 1859, was Isambard Kingdom Brunel's last bridge and probably his finest work. Certainly, it was one of the great engineering feats of the nineteenth century, because (it is widely agreed) of its size, its economy of design, its revolutionary superstructure, and not least because of the way in which Brunel solved difficult logistical problems. It was one of the first bridge projects on which compressed air was used to allow underwater foundation work to proceed.

Dividing Cornwall from the rest of England, the tidal reaches of the River Tamar were once a major maritime thoroughfare. The twelfth-century port of Saltash lies on the west shore of the Tamar Estuary near the English Channel coast, nearly facing Plymouth on the opposite side. A railroad into Cornwall, the county in the extreme southwest of England, was first proposed in 1844. The Cornwall Railway Company was formed in 1845, and it successfully applied for the necessary act of Parliament to provide either a steam ferry to transport trains across the 1,100-foot-wide (336-meter), 85-foot-deep (26-meter) river or to build a bridge. The project was delayed because the Admiralty was concerned about restricted access to the Devonport naval base, close to Saltash. Finally, in 1852 Brunel's proposal for a bridge with two main

Royal Albert Bridge, Saltash, England; Isambard Kingdom Brunel, engineer, 1852–1859. View of the bridge under construction, showing the prefabicated tubular trusses that formed the main spans (1859 photograph).

spans was adopted because a single pier in the river would offer least hindrance to water traffic. During the construction the plans were changed for financial reasons; Brunel designed the bridge for a single-track railroad. The authorities demanded a clearance of 100 feet (31 meters) under the bridge at high tide.

The Royal Albert Bridge is 2,240 feet (683 meters) long. Each of the two main spans is 455 feet (140 meters), and the 17 side spans of the long, curving approach viaducts vary between 70 and 90 feet (21 and 28 meters). Brunel first proposed a single-span bridge but because of difficult ground conditions changed the design.

Brunel found a firm base on rock in the middle of the river for the center pier, at a depth of more than 87 feet (27 meters) below high-water mark. Debris had to be cleared to expose a good foundation. To that end, a 95-foot-tall (29-meter) iron cylinder, 35 feet (11 meters) in diameter, was fabricated onshore. A dome was constructed about 20 feet (6 meters) above its lower end, and a 4-foot-wide (1.2-meter) gallery, divided into 11 compartments, was built around the cylinder below the dome. The cylinder was floated into position and sunk to the riverbed in June 1854. Compressed air was fed into only those compartments where men were working, obviating the need to supply it to the whole space under the dome all the time. The foundation was cleared, and the rock was leveled with a 16-foot-thick (5-meter) base layer. By the end of 1856 the circular granite center pier was completed to a height of 12 feet (3.7 meters) above river level.

Four hollow octagonal cast-iron columns, 10 feet (3 meters) in diameter and stiffened by cross-bracing, rise from the center pier to the same height as the tapering masonry piers at the ends of the approach viaducts. Two columns support each of the huge main trusses. Those trusses were fabricated on the riverbank. Each comprises a curved, wrought-iron elliptical tube 16.75 feet (5.1 meters) wide—constrained by the single-track railroad—and 12.25 feet (3.7 meters) high, forming a flat arch that carries the weight of the superstructure. The arch is connected to massive catenary iron chains at eleven equidistant points by pairs of vertical standards, braced by diagonal bars; the chains support the girders under the railroad deck, 110 feet (34 meters) above high-water mark.

Beginning on 1 September 1857, the first 1,200-ton (1,016-tonne) truss was floated into position on four pontoons. Through the combined efforts of 500 men on shore and on five vessels at strategic points in the river, it was put into place with great accuracy. As the masonry pier progressed, the truss was raised a little at a time by hydraulic jacks. By July 1858 it had reached its full height, and the second was ready for floating into position. The process was repeated for the second truss. At their landward ends, the trusses are carried by piers, with arched openings through which the railroad passes.

The bridge was opened by Queen Victoria's consort, Prince Albert—hence the name—on 3 May 1859, just a few months before its creator, Brunel, died. Its construction made possible a continuous rail journey between London and Truro. A branch line to Falmouth opened in 1863 and was later extended to the new docks then being built. A neighboring suspension bridge carrying the A38 road over the Tamar was completed in 1961. Early in 1998 the Royal Albert Bridge was refurbished. The £1.2 million (U.S.$1.75 million) project involved cleaning back the paintwork to bare metal and repainting and replacing much of the timber deck, all without unduly disrupting the rail services.

See also Clifton Suspension Bridge

Further reading

Binding, John. 1997. *Brunel's Royal Albert Bridge: A Study of the Design and Construction.* Truro, UK: Twelveheads Press.

Hay, Peter. 1973. *Brunel: His Achievements in the Transport Revolution.* Reading, UK: Osprey Publishing.

Vaughan, Adrian. 1993. *Isambard Kingdom Brunel, Engineering Knight-Errant.* London: J. Murray.

The Royal Pavilion
Brighton, England

The Royal Pavilion, Brighton (1817–1822), "a grand oriental fantasy" with Indian domes and minarets and Chinese interiors, is a fascinating example of the diverse architectural styles allowed in the Regency period, which was otherwise dominated by refined neoclassical architecture. Two elements were necessary for its realization: an esthetically adaptable architect—in this case, John Nash (1752–1835)—and a client powerful enough to get what he wanted—the Prince Regent (later King George IV, 1762–1830). More importantly, it is probably the first attempt by any architect, freed from classical and Gothic precedents, to use cast iron to make legitimate architecture.

George, Prince of Wales, first visited the coastal resort of Brighton (then Brighthelmstone) in 1783. He was already deep in gambling debts, a heavy drinker, and a notorious womanizer. In 1784 he again visited Brighton and in the same year fell in love with the twice-widowed Maria Fitzherbert. When she refused to become his mistress he agreed to marry her, but secretly, because English law prohibited royalty from marrying Catholics. Two years later his comptroller, Louis Weltje, obtained from Thomas Kemp, Member of Parliament for Lewes, a three-year lease with an option to purchase on a timber house facing the sea at Brighton. He relet it to the prince, undertaking to rebuild it. Between May and July 1787 the architect Henry Holland enlarged and converted the modest but "respectable" farmhouse to the Marine Pavilion, a double-fronted Palladian affair with a domed Ionic rotunda. Maria was provided with a nearby villa.

In 1795, attempting to persuade Parliament to pay his accrued debts of £650,000, the prince entered a

The Royal Pavilion, Brighton, England; John Nash, architect, 1817–1822. General view, photographed ca. 1875.

political marriage to his cousin, Caroline of Brunswick. After a daughter was born in January 1796 the royal couple lived apart. Caroline returned to Germany and George to the arms of Mrs. Fitzherbert. He moved his court to Brighton, where he planned the next stage of his evolving seaside house. Henry Holland was engaged on the other side of the country, so his assistant P. F. Robinson supervised the mutation of the Marine Pavilion's interiors into a Chinese palace. The prince bought imported Chinese furniture and porcelain and even costumes, and from about 1802 chinoiserie interiors were executed by the firm of John Crace and Sons.

A year earlier George had directed Holland to design Chinese exteriors for the house, and in 1805 Holland's successor William Porden made similar plans. The prince's imagination was then seized by a new house, Sezincote in Gloucestershire (architect Samuel Cockerell), in the "Moghul" style. In 1807 Humphrey Repton, the landscape architect for Sezincote, was commissioned to design an Indian exterior for the Marine Pavilion. Although George

liked the result, he could not afford to build it. But he had Porden construct a huge circular stables and riding house west of the house; crowned with a central cupola, it was in the Indian style.

In February 1811 King George III again retreated into madness, and the prince was appointed Regent. Breaking his promises to the Whigs, he supported the incumbent Tory government. He also disappointed Repton, who expected to complete the pavilion when his client had sufficient funds. Instead, in 1812 the commission was passed to James Wyatt. But he died in September 1813, and there was another hiatus until January 1815, when George capriciously engaged Repton's sometime partner John Nash, who had designed London's Regent Street and Carlton House Terrace (both 1813) for him.

The eclectic Nash swathed the exterior of Holland's building in a mixture of pseudo-Moghul and neo-Gothic detail; internally, he changed it beyond recognition. The house is symmetrical about a long north-south axis. A central porte cochere on the west front enters through a vestibule, across a long gallery into a domed salon, flanked with drawing rooms at the "back" of the house. The prince's apartments are in the northwestern wing of the first floor; visitors' apartments occupy the southwestern wing. Above are "small but elegant" bedrooms. The banqueting hall, its opulent decor designed by Robert Jones and Frederick Crace, is in the southeastern corner. From a cluster of plantain leaves at the center of its 45-foot-high (14-meter) blue saucer dome hangs a gigantic gilded dragon gasolier. The walls are decorated with Chinese motifs in brilliant colors. The music room, also by Jones and Crace, is at the other end of the first floor. Its domed ceiling is formed by gilded scallop shells above an octagonal cornice, supported by dragon brackets; the walls and drapes are gold and crimson. It is possible to describe the disposition of spaces, but their scale and extravagant splendor must be experienced.

The roof is a fantastic agglomeration of domes, tentlike roofs (over the banqueting hall and music room), chimneys disguised as minarets, and "oriental" finials. Nash adventurously used cast iron in the structure and the interior and exterior details. The huge central ogee dome over the salon, and the other roofs, as well as the bases of the chimneys and pinnacles, are framed in the new material. The extremely ornate palm-tree columns supporting the roofs of the drawing rooms are also cast iron, as are the four simple, slender columns with copper palm-leaf capitals that carry the central roof lantern in the kitchen. The double staircases at either end of the long gallery are cast iron, too, but their balustrades and other details (like the wall mirrors throughout the house) are disguised with paint as bamboo, befitting the Chinese mood. Outside, iron was used for the elegant lattice tracery on the east front of the music room.

The diarist John Wilson Croker visited the Royal Pavilion before it was finished and declared it an "absurd waste of money," accurately prophesying, "[it] will be a ruin in half a century or more." The affairs of state limited George IV's visits to his extravagant "pleasure dome." His dissolute lifestyle eventually overtook him; addicted to alcohol and laudanum, he began showing signs of insanity, and shortly before his death he became increasingly reclusive. His successor, William IV (reigned 1830–1837), used the pavilion, but Queen Victoria acquired a summer home on the Isle of Wight. She sold the pavilion to the people of Brighton in 1850 after moving its furnishings to Buckingham Palace. It was put to various uses until, following World War II, interest in it was revived. Beginning in 1980 the storm- and fire-ravaged building was refurbished by the Brighton Borough Council, and Queen Elizabeth II returned most of the original furnishings. The Royal Pavilion, its splendor restored, was reopened to the public in 1990.

Further reading

Dinkel, John. 1983. *The Royal Pavilion, Brighton*. New York: Vendome Press.

Goff, Martyn. 1976. *The Royal Pavilion, Brighton*. London: Michael Joseph.

Summerson, John. 1980. *The Life and Work of John Nash, Architect*. Cambridge, MA: MIT Press.

S

Sagrada Familia (Church of the Holy Family)

Barcelona, Spain

The 328-foot-tall (100-meter) spires of the Church of the Sagrada Familia dominate the skyline of Barcelona, the chief city of Catalonia, in northeastern Spain. This unique church, which, in the tradition of the medieval cathedrals of Europe, remains unfinished more than a century after it was started, is one of the great pieces of world architecture. Its fantastic forms defy our vocabulary and confound any attempt at stylistic classification. It marks the fin de siècle rejection of historical revivalism—perhaps it is the last *true* Gothic church—but unlike the willful forms of the contemporary Art Nouveau (a category to which some historians have consigned it), it is respectful of the past in its local context and the broader sphere. To repeat, it is unique.

Around 1874, José María Bocabella y Verdaguer (1815–1892), the proprietor of a religious bookshop and cofounder of the reformist Society of Devotees of Saint Joseph, initiated a proposal to build a votive church, the replica of the basilica at Loreto, Italy. Members of the society were solicited for funds, but the money raised was not even enough to buy land in Barcelona. At the end of 1881, 5 acres (2 hectares) of land were bought in the city's outlying "new town" on the Muntanya Pelada, near the Gran Via Diagonal. Changing his mind, Bocabella commissioned the diocesan architect Francisco del Villar y Lozano, who produced a church of neo-Gothic design. The foundation stone was laid about a year later, but the building had not progressed far when del Villar fell out with the administration and resigned. Bocabella's son-in-law offered the lapsed commission to Juan Martorell, technical supervisor of the project. He declined, recommending his 31-year-old erstwhile assistant, Antoni Gaudí y Cornet (1852–1926), a fiercely nationalistic Catalan, who took over in November 1883.

The son of a coppersmith, Gaudí studied at Barcelona's Escola Superior d'Arquitectura, graduating in 1878. Not a particularly good student, he nevertheless established a busy practice. His early work, particularly the Casa Vicens (1878–1880) in Barcelona, attracted the attention of the wealthy industrialist Count Eusebio Güell, who became his patron. For Güell he designed, amongst other works, the Palacio Güell (1885–1889) and Park Güell (1900–1914). Both are fine examples of the sensuous, free-curving, and richly decorated architecture for which Gaudí became admired by his European contemporaries. Of course, he brought the same celebration of form to the Sagrada Familia.

Francisco del Villar had quit the project when the walls of the crypt, the chapels, and part of the pillars of his prosaic church were built. The crypt is therefore neo-Gothic, structurally and esthetically, but

Sagrada Familia (Church of the Holy Family), Barcelona, Spain; Antoni Gaudí y Cornet, architect, from 1883 (unfinished). East front, with the Portal of the Nativity.

Gaudí modified it as much as he could and surrounded it with a moat. Completing its vaulting in 1887, he turned to the apse (he preferred to finish one section of the church before addressing the next). Although he was constrained by the completed foundations, Gaudí otherwise abandoned del Villar's design and replaced the neo-Gothic buttresses with sloping columns. In 1893, with the apse still incomplete, and in the face of criticism (because others believed that the west front, facing the city, was more important), he turned his attention to the east front, where he wanted to celebrate Christ's nativity. Gaudí proposed four 330-foot-high (100-meter) towers for each front of the Sagrada Familia. The towers of the evangelists on the east facade are the last built under his supervision; their tips, glorious with glazed color ceramic, were finished after his death. The huge Portal of the Nativity, enriched by sculpture, is the only one of the four planned portals completed in his lifetime. All these elements eloquently express the spirit of Gothic architecture while celebrating their own ebullient originality. In 1917 Gaudí completed designs for the west front with its Portal of the Passion of Christ. The south front was to have the Portal of the Last Judgment. Space does not permit a description of the church; in any case, words would fail to create an image. The building that one of Gaudí's friends called "a marvelous, budding flower" must be personally experienced.

Because the project was privately financed, building work was intermittent. After 1908, Gaudí committed himself exclusively to the church as designer, construction supervisor, and fund-raiser. Becoming increasingly reclusive, he eventually moved to accommodations on the building site. He became obsessed with the idea that his church could redeem Barcelona from what he saw as the evils of secularism. In 1922, the architect Teodoro de Anasagasti Asensio proposed that the church become a public work, to be financed by the state. On the afternoon of 7 June 1926, Gaudí was struck by a trolley car while crossing a street near his beloved Sagrada Familia. He died in the hospital three days later without having regained consciousness. He left only sketches of the overall project, as well as drawings and scale models of various parts. Supervision was taken over by his associate Sugañes, who completed the east portals in 1935; he was working on the vestries when the Spanish civil war broke out, and he died in the conflict. In 1936 an anticlerical mob overran the Sagrada Familia, burning the plans and destroying the models. Further building activity was halted until 1952, when architect Lluis Bonet rebuilt the models.

In the face of debate over whether the church should remain uncompleted as a monument to Gaudí, construction began again in 1979, closely following his plan. Funded by private donations and the sale of tickets to increasing numbers of visitors—there were 1.2 million in 1999—the work proceeds. The main nave has been under construction since 1986 under the supervision of architect Jordi Bonet. It is expected that the church will be covered with its very complicated irregular vaults by 2010, although the exterior of the roof will still be unfinished. The construction council is optimistic that the Sagrada Familia will be completed in only fifty years.

Since 1992 there has been a movement among Barcelona's Catholic hierarchy to effect Gaudí's eventual canonization; it was given fresh impetus in 1998, and in April 2000 the Diocesan Beatification Process was officially opened. The city of Barcelona has declared 2002 the International Year of Gaudí to commemorate the 150th anniversary of his birth.

Further reading

Bergós Massó, Juan. 1999. *Gaudí, the Man and His Work*. Boston: Little, Brown.

Burry, Mark. 1993. *Expiatory Church of the Sagrada Família: Antoni Gaudí*. London: Phaidon Press.

Zerbst, Rainer. 1988. *Antoni Gaudí i Cornet: A Life Devoted to Architecture*. Cologne: Taschen.

St. Chapelle
Paris, France

St. Chapelle, at 6 boulevard du Palais, is now surrounded by the Palace of Justice on the Ile de la Cité, Paris, near Notre Dame. It was built as a palatine chapel for King Louis IX of France (known as St. Louis, reigned 1226–1270) between 1242 and 1247, and consecrated on 26 April 1248. During Louis IX's reign, Gothic architecture in France entered the *rayonnant* phase, its name derived from the radiating spokes of the large rose windows that characterized the style. Refining the stone-framed architecture of the age, architects further reduced the amount of solid wall in favor of expansive traceried stained-glass windows. The masonry that remained was in the form of narrow but very thick buttresses that dealt with the thrusts imposed by vaulted stone ceilings. St. Chapelle, with its luminous glass curtains, represents the highest degree of this structural refinement and is probably the most beautiful surviving example of the French Gothic of any phase.

In 1239 Louis IX purchased (at extravagant cost) a number of relics of the crucifixion of Christ from his bankrupt cousin, Jean de Brienne, the Emperor of Constantinople. The most important of them was the crown of thorns; there was also a piece of iron from the lance used by the soldiers and the sponge

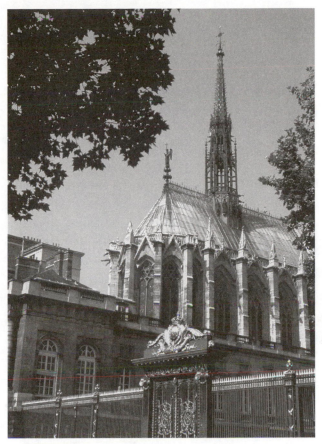

St. Chapelle, Paris, France; architect(s) unknown, 1242–1247. East end and spire, beyond the Palace of Justice.

on which Jesus was offered sour wine. From de Brienne's successor, Baudouin II, Louis bought a piece of the true cross. To purchase them and fashion a reliquary—a bejeweled chest that was destroyed during the French Revolution—it is said that Louis spent two and a half times what it cost to build St. Chapelle. Soon after acquiring the relics, he commissioned a private chapel within the royal palace on the Ile de la Cité to hold them. There is some debate about the identity of the architect; many sources identify Pierre de Montreuil, who had worked on Notre Dame, Paris, and St. Denis, but St. Chapelle may have been the work of Robert de Luzarches or Thomas de Cormont.

The building in fact houses two chapels. The lower, entered from the courtyard, was dedicated to the Virgin Mary and was for the use of servants of the royal household. It is relatively low—its vaults are 22 feet (6.6 meters) high—and rather dimly lit. Two small spiral staircases within the walls connect it to the upper chapel, for which it may have been designed as a foil; certainly, there is a breathtaking contrast in the quality of the respective spaces. The official access to the upper chapel, which was dedicated to the Holy Crown and the Holy Cross and reserved for the use of the sovereign, was by a gallery directly linking it with the royal apartments. Entering through a sculpture-enriched double portal, the visitor is greeted by an explosion of color and light. Fifteen lofty stained-glass windows, rising 65 feet (20 meters) from just above floor level to the gilded arches of the vaults, fill the entire area between the buttresses—in total, 6,600 square feet (620 square meters)—to create a space that has been described as "Gothic architecture at its most daring and successful" and "a cage of Light." The windows were restored in the nineteenth century after the depredations of the French Revolution. About 65 percent of them date from the thirteenth century; together, they depict more than 1,130 scenes from the Old Testament and the life of Jesus.

St. Chapelle was burned in 1630 and was rebuilt. During the Revolution it stood in danger of demolition but was saved, though damaged. It was then used as an archives store until 1837, but in 1846 a twenty-year restoration program, "almost amounting to renewal," was initiated. The architects Félix Duban, Jean-Baptiste Lassus, Émile Boeswillwald, and E. E. Viollet-le-Duc replaced the roof and the stair and redecorated the interiors. The building is now a museum.

Further reading

Bony, Jean. 1983. *French Gothic Architecture of the Twelfth and Thirteenth Centuries.* Berkeley, CA: University of California Press.

Bottineau, Yves. 1967. *Notre-Dame de Paris and the Sainte-Chapelle.* Chicago: Rand McNally.

Grodecki, Louis. 1975. *Sainte-Chapelle.* Paris: Caisse nationale des monuments historiques et des sites.

St. Denis Abbey Church
St. Denis, France

The Abbey of St. Denis is situated in a small municipality (now a suburb) of the same name, about 4 miles (6.4 kilometers) north of Paris. Its thirty-sixth abbot, Suger (1081–1151), commissioned the present church from about 1140. It is a milestone in the history of architecture because, like Durham Cathedral in England, it has in it the seeds of a new way of building for Europe: the highly inventive structural system that we know as the Gothic. In particular, Suger's choir at St. Denis, the first application of pointed arches in a major building, marks one aspect of the transition from the Romanesque style, which was quite hobbled by the use of round-headed arches; that is, the transition from wall architecture to framed architecture.

Denis, first bishop of Lutetia, and his missionary companions were martyred in 258, and buried at St. Denis. When the persecutions ended in the fourth century, a small chapel was built that became a popular shrine for pilgrims by the end of the sixth century. The Merovingian king Dagobert founded a Benedictine monastery there in 630, replacing the chapel with a large basilica and enriching the new royal abbey. He also bestowed many rights and privileges on the little town, not least the honor of building his tomb. Eventually, the abbey was to house seventy royal sepulchers. Charlemagne, king of the Franks, commissioned a new church in 750 and much of the earlier building was subsumed. Systemic reforms were introduced by Abbot Hilduin (815–830; ca. 831–840) during his second term of office, and the Abbey of St. Denis, because of the relics it held, grew in significance and prosperity. In about 1127 Suger assumed the position of abbot, to which he had been elected in Rome five years earlier.

Between 1123 and 1127, as adviser to Louis VI (reigned 1108–1137), he was engrossed in affairs of state but soon after he set out to thoroughly reform his monastery, first of all establishing a more rigorous discipline for the monks and dealing with its financial problems. Then he turned to the building. The old abbey church had been completed in 775, and by the middle of the twelfth century it had become dilapidated; from 1135 Abbot Suger initiated an extensive renovation program. His motives have been widely discussed by historians; it is clear that he was moved by religious and esthetic sensibilities, but because St. Denis was the royal abbey (and thus a symbol of royal power), its renovation was also a political statement at a time of unrest in France. In fact, the only loyal region to Louis VI was the Ile-de-France, and it was in the king's interest to patronize the rebuilding of the church.

Suger wrote an account of his renovation program titled *A Little Book on the Consecration of the Church of Saint Denis.* The first major phase was the reconstruction of the west facade and the narthex: "dismantling a certain addition said to have been built by Charlemagne we ... vehemently [enlarged] the body of the church, tripling the entrance and doors, and erecting tall, worthy towers." Only the northeastern tower survives. The new monumental west front, built in the dour Romanesque style, had a high rose window above the central portal, admitting more light into the church. But the critical part of Suger's work—his architectural feat—is in the choir at the east end of the church.

The new choir was built over the ninth-century crypt. Its seven chapels, radiating from a semicircular ambulatory and integrated with it, formed the first example of the distinguishing element of French Gothic architecture, the *chevet*. Supported on slender cylindrical columns standing on square bases, it comprises ribbed stone vaults in which a regular network of pointed—not semicircular—arches carries thin panels. It is likely that the idea of the pointed arch (an Arab invention) was brought back to France by masons who had accompanied the First Crusade (1095). The device made the height of the arch independent of its span and allowed a much more accurate structural frame to be developed, in which loads could be gathered at columns and sideways thrusts resisted by buttresses. The walls, freed from their load-bearing function, could then be thinner and penetrated by more and larger windows. What was first done at St. Denis was developed and refined to produce the luminous interiors of the thirteenth century, like St. Chapelle in Paris (1248).

Suger, impressed with fifth-century ideas, was fascinated by the role of light in churches. God is light, and his creation, ordered by his light, praises him by reflecting light back to him. That applied to inanimate things, such as precious stones and stained glass, as much as to people. Therefore his additions to St. Denis were lit by glorious stained-glass windows and "wor-

thily painted with gold and costly colors." The choir was a repository for the relics and remains of St. Denis and was afforded the most extravagant treatment, provided with the help of Abbot Suger's close friend Louis VII, who came to the French throne in 1137. The pulpit was made of sculptured ivory tablets decorated with figures wrought from copper. The morning altar was of black marble, with sculptures of the martyrdom of St. Denis in white marble, and the high altar was surrounded by gold sides enriched with precious stones, with figures in relief. Suger declared himself so overcome by the sight of it that he thought he was no longer on Earth, but near Paradise. He insisted that "we must do homage also through outward ornaments ... with all inner purity and with all outward splendor." Not all his contemporaries agreed, and there was a protracted debate between Suger and Bernard of Clairvaux, Abbot of Cluny, about the propriety of opulence in the house of God. The choir was completed in 1144, having taken three years and three months. To achieve it, masons and other craftsmen had been gathered from all over France. But none was acknowledged in Suger's report of the project; he claimed all the credit for himself, having a memorial stone inscribed "Bright is the noble work enlarged in our time, I, who was Suger, having been leader while it was accomplished."

Although he was "eager ... to follow up on [his] successes," by the time the choir was built the abbey's funds had been depleted. The nave and transepts were not rebuilt until the fourteenth century, under the supervision of the mason-architect Pierre of Montreuil. After the Council of Trent (1545–1563), the abbey was placed at the head of a body of ten other monasteries; these were joined in 1633 to the Benedictine Congregation of Saint-Marui. Although its monastic buildings were reconstructed, the great church remained intact. About sixty years later, Louis XVI suppressed the abbot's office. The abbey was eventually dissolved during the French Revolution, when the church was vandalized. In the nineteenth century it was restored by the architect E. E. Viollet-le-Duc and is now a national monument. Today its greatest threat comes from air pollution.

See also Chartres Cathedral (Cathedral of the Assumption of Our Lady); Durham Cathedral; St. Chapelle

Further reading

Crosby, Sumner McKnight, and Pamela Z. Blum. 1987. *The Royal Abbey of Saint-Denis: From Its Beginnings to the Death of Suger, 475–1151.* New Haven, CT: Yale University Press.

Gerson, Paula Lieber, ed. 1986. *Abbot Suger and Saint-Denis: A Symposium.* New York: Metropolitan Museum of Art.

Rudolph, Conrad. 1990. *Artistic Change at St.-Denis: Abbot Suger's Program and the Early-Twelfth-Century Controversy over Art.* Princeton, NJ: Princeton University Press.

St. Geneviève Library
Paris, France

The St. Geneviève Library in the place du Panthéon, Paris, was designed in 1843 by Henri Labrouste (1801–1875) and built between 1844 and 1851. It is the first public building to have a frankly exposed structural iron frame. Wrought iron and cast iron, used to great structural and esthetic effect in engineering works since the late eighteenth century, were still widely regarded as unsuitable for legitimate architecture (except for decorative details like balustrades or ornamental hardware), simply because the classical and medieval styles that informed contemporary design provided no precedent for the manner of their use. That was despite their many advantages: they were incredibly strong in compression, noncombustible, and inexpensive; moreover, they could be prefabricated and mechanically fixed, thus avoiding "wet work" and minimizing the labor involved in making good, which represented—and continues to represent—a major cost in building. Besides all this, their properties allowed the combination of visual lightness and natural illumination.

The library held the collection of books and documents, covering all aspects of theology, the arts, and science, from the former Abbeye de Sainte-Geneviève, founded in the sixth century. The Augustine monastery had been reformed by Abbot Suger in 1148, after which it had maintained a library and a school for copyists. The library was reestablished in 1624 and by the mid–nineteenth century it held manuscripts and about 120,000 books. In February 1842, it was decided to relocate the collection, which since before the French Revolution had been housed on the top floor of former abbey buildings, to the former Collège de Montaigu. It was moved in October when appropriate alterations had been made to the college and would remain there for eight years while a new building, approved by the Chamber of Deputies in 1843, was completed.

The new library was to be part of an urban renewal scheme around the Panthéon, which before the Revolution had been the Church of Sainte-Geneviève (1758–1789), designed by Jacques Germain Soufflot and completed by J. P. Rondelet. In 1844, planning regulations provided for the creation of the broad rue Soufflot and the erection of a town hall that would face the library across the place du Panthéon and echo Soufflot's Law Building of the University of Paris. Labrouste was immediately given the commission, and his designs, also sympathetic with the proportions of Soufflot's building, were accepted late in 1843. A Parisian, Labrouste had studied architecture at L'École des Beaux-Arts from 1819 to 1824. After winning the Grand Prix, he continued at the French Academy in Rome until 1830, developing notions of "romantic rationalism" before returning to Paris to set up his own atelier. The foundation stone of the Library of St. Geneviève was laid in August 1844, and the main construction work was completed in 1847. The building was officially opened on 4 February 1851.

The St. Geneviève Library hybridizes neo-Renaissance and rationalist architecture, the first through its composition and decorative details, the second by the refreshing expression of internal organization and function in the design of the facades. The two-story rectangular building, measuring 278 by 69 feet (85 by 21 meters), follows a straightforward architectural program: a comparted ground floor houses library stacks and rare book storage areas that flank a central foyer. A stair leads to a vast reading room that occupies the entire upper story. That airy space is lit by forty-six huge high-level, arched windows, nineteen on each long side and four at each end. Across its width, it is divided by a central spine of elegantly slender cast-iron Ionic columns standing on short piers. The columns, braced longitudinally with filigree arches, support decorative openwork cast-iron round arches, care-

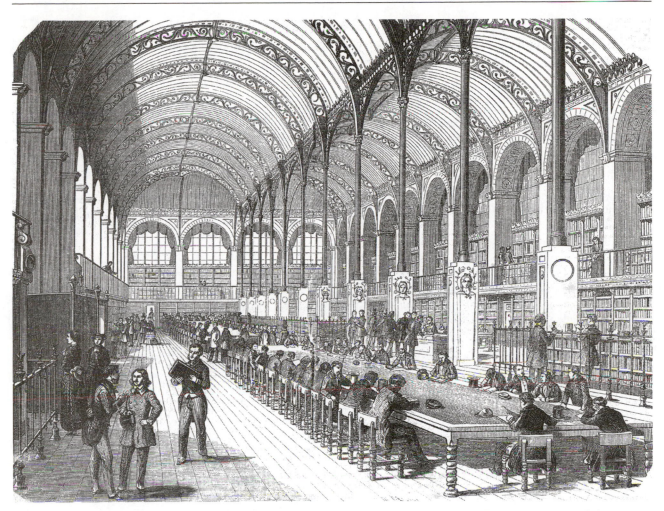

St. Geneviève Library, place du Panthéon, Paris, France; Henri Labrouste, architect, 1843–1851. Interior of the reading room on the upper floor, from an 1852 engraving.

fully designed by Labrouste in consultation with one Calla, of the Val d'Osne foundry in Paris. The transverse arches in turn directly carry barrel vaults formed by plaster reinforced by iron mesh on a network of joists. Originally a flat ceiling had been intended, but Labrouste's inventiveness and the flexibility of the General Council of Civil Buildings made possible, after prototypes were built and tested, this first and beautiful expression of an iron structure in "serious" architecture. However, it would be a long time before the architectural profession came to terms with the new material.

Rather a disappointment in the light of the reading room, the library's relatively modest facades incorporate Renaissance elements, such as the virtual arcade of almost continuous windows on the upper level. The narrowness of the pilasters between them indicates the iron columns within and contrasts with the massive masonry piers that define the corners of the structure. A frieze of swags below the prominent string course that marks the upper floor level is almost the only applied decoration. Initially, Labrouste proposed a garden along the place du Panthéon facade "to distance the noise of the public thoroughfare" and prepare visitors to his library "for due reverence." Lack of space prevented its realization.

The architect's estimation of future space needs, questioned by the client during construction, quickly proved insufficient. Much-delayed major renovations were made in the early and mid–twentieth century. In 1930, the reading room was redisposed to increase the number of places from 400 to over 700; a small extension was built in 1954, followed by a larger one in 1961. Although Labrouste's library did not sit well with other buildings on the site, its planning made it possible to add urgently needed public spaces, offices, and stores. The Library of St. Geneviève again demands general reorganization, as technology and usage patterns change.

Further reading

Gargiani, Roberto. 1997. "Ornament and Construction in the Library of Ste-Geneviève, Paris, 1839–1850." *Casabella* 61 (May): 60–73.

Middleton, Robin. 1999. "The Iron Structure of the Bibliothèque Sainte-Geneviève as the Basis of a Civic Decor." *AA Files* (Winter): 33–52.

Zanten, David van. 1987. *Designing Paris: The Architecture of Duban, Labrouste, Duc, and Vaudoyer.* London and Cambridge, MA: MIT Press.

St. Katharine Dock
London, England

Toward the end of the twentieth century, because of technological changes in world shipping, the St. Katharine Dock area near London's Tower Bridge was forced to alter its function after more than a thousand years as a trade center. That adaptation of building use foreshadowed a universal trend in which former warehouses became (usually luxury) apartments. For that reason, and because of the model cargo handling and storage design that it represented in the nineteenth century, St. Katharine Dock is worthy of a place in any list of architectural achievements.

The Saxon King Edgar (reigned 959–975) granted 13 acres (about 5 hectares) on the site to several knights. Because they were permitted to use the land for profit, that gift laid the foundation for foreign trade. In 1125 the property and its small dock passed to a convent and a hospital was established; twenty-five years later Queen Matilda endorsed the Royal Foundation of St. Katharine. Wharves were later built along the tidal inlet, and the area became known as St. Katharine Dock late in the sixteenth century. It was in the eighteenth century that the Thames was changed from a relatively quiet river into the major commercial thoroughfare into the heart of London, center of

world trade. Some sources claim that up to 800 vessels at a time were moored in the Pool of London, and market forces generated the 8.5 square miles (22 square kilometers) of the city's Docklands. Made at great social cost, the transformation was empowered by legislation: the West India Dock Act of 1799 authorized the first enclosed docks; the London Dock Act followed, authorizing a dock at Wapping. Between 1802 and 1806, the West India, London, and East India Docks became operational. The St. Katharine Dock Act was passed in 1825, and the following year the St. Katharine Dock was opened between the London Dock and the Tower of London.

The engineer Thomas Telford was commissioned to build the new dock and faced the constraint of a very small site. Assisted by Thomas Rhodes, he designed a unique system of two connected basins—the East Dock and the West Dock—that provided maximum wharf frontage. They were linked to the Thames through a 180-foot-long, 45-foot-wide (45-by-14-meter) lock with three gates, and steam engines maintained the dock water level above that of the river. One of the largest engineering projects ever seen in London, at times employing 2,500 men, St. Katharine Dock took two years to construct. The neoclassical dock offices and especially the six-story warehouses, designed by the architect Philip Hardwick, were also revolutionary. Constructed virtually at the wharfside, they enabled cargo to be unloaded directly from ship to storage, saving time and reducing pilfering. Because they backed on the access roads, the warehouses also did away with the need for a boundary wall around much of the dock. The West Dock warehouses were completed by 1828 and those around the East Dock a year later. The largest were 470 feet long and 140 feet deep (144 by 43 meters). In all of them, squat, cast-iron Doric ground-floor columns supported vaulted brick and iron floors and the superstructure; the walls were of gray London bricks. The window frames were also of cast iron. St. Katharine Dock was officially opened on 25 October 1828.

Although it was celebrated as an engineering, architectural, and commercial triumph, not everyone would have agreed. The London *Times* reported that the acquisition and clearing of the land for this "mag-

Warehouses at St. Katharine Dock, London, England; Philip Hardwick, architect; Thomas Telford, engineer, 1828–1829.

nificent speculation" led to the demolition of 1,250 houses and tenements and the displacement of 11,300 residents. Freeholders and leaseholders were compensated, but not the dispossessed. And when philanthropic housing schemes were finally introduced in the 1840s, the benefits were only for people with regular incomes.

The dock originally handled exotic, valuable international cargoes. The company provided incentives for fast turnaround of vessels, and with competitive charges and low land-transportation costs, it enjoyed early prosperity. In 1852, Hardwick's associate George Aitchison built two more warehouses. At their peak they were capable of "accommodating

120 ships besides barges and other craft." But London had overdeveloped new docks, and St. Katharine's share of the traffic soon declined until it was left with only coastal and cross-channel traffic. Financial difficulties followed, and in 1864 the St. Katharine and London Dock Companies merged to take over Victoria Dock. Between the two World Wars, because ocean vessels were too large to berth there, St. Katharine Dock became a convenient storage facility close to the city of London for goods that were carried by lighters from vessels downstream. The dock and warehouses were badly damaged by incendiary bombs in the Blitz of 1940, and were never fully repaired. Upriver docks were far too small to deal with container vessels, and in the 1960s the Port of London Authority began to locate its functions at Tilbury.

In 1968 the Greater London Council (GLC) closed St. Katharine and London Docks and announced its plans to replace them with houses. The warehouses were in use until 1969, when the GLC acquired the site and invited redevelopment proposals. Seven developers presented plans for a trading and leisure facility, and the project was awarded to the Taylor Woodrow Property Company. They proposed (and built) a mix of privately and publicly owned apartments, a five-star hotel, a trade center, office buildings, restaurants, shops, and a marina, with a total of about 5,000 tenants and residents.

Further reading

Beckett, Derrick. 1987. *Telford's Britain*. Newton Abbot and North Pomfret, UK: David and Charles.

Foster, Janet. 1999. *Docklands: Cultures in Conflict, Worlds in Collision*. London: University College London Press.

Ledgerwood, Grant. 1985. *Urban Innovation: The Transformation of London's Docklands, 1968–1984*. Aldershot and Brookfield, UK: Gower.

St. Pancras Station
London, England

Built between 1863 and 1865 for the Midland Railway, St. Pancras Station has been described as the epitome of the railroad buildings that evolved following advances in iron technology in the second half of the nineteenth century. It was one of a number of London stations, including Victoria and Charing Cross, erected during the 1860s railroad boom, when national and international travel was becoming more popular. St. Pancras established Midland's footing in the capital; coming as it did after other companies had erected their London terminals, it was deliberately intended to impress by its scale and architectural style. Its substantial train shed, designed by company engineer William Henry Barlow (1812–1902) with R. M. Ordish, achieved the widest single-arch span then built. This daring engineering accomplishment was unrivaled. Several years later, a grand Victorian Gothic Revival hotel and terminus building was added to the front of the shed. Named the Midland Grand, it was designed by the eminent Gothic Revival architect George Gilbert Scott (1811–1878) and constructed between 1868 and 1876.

St. Pancras was built next to King's Cross Station (1851–1852), the Great Northern Railway's terminus designed by architect Lewis Cubitt. The dissimilar approach to the design of each station reveals a dilemma of the age—the functional station building was celebrated as an engineering triumph and a demonstration of technological and structural progress but was not popularly, or professionally, accepted as "real" architecture. King's Cross exposed its function—a yellow brick facade with two arched windows flanking a central clock tower was effectively integrated with the sheds behind. The elaborate Midland Grand Hotel hid St. Pancras Station as if to suggest that only the former was acceptable for public display. The station proper was a mere shed, an industrial structure symbolic of a building type that did not fit neatly into the accepted definition of architecture. That divide between architecture and engineering would persist for several decades.

The St. Pancras train shed was an immense 700 feet (213 meters) in length. The roof, framed of wrought-iron trussed-lattice arches at 30-foot (9-meter) intervals, spanned 243 feet (74 meters), rising to a height of 100 feet (30 meters). The side walls were masonry, supported on masonry piers. Three-inch-diameter (76-millimeter) tension rods tied the feet of the arches below platform level. Iron and glass clad the roof frame. The train platform was at second-floor level, and the floor below was designed to

The train shed, St. Pancras Station, London, England; William Henry Barlow, engineer, 1863–1865. Photographed ca. 1868.

store beer barrels, the length of the barrels determining the spacing of the columns that supported the platform.

The uninterrupted roof arch created a spacious and dramatic interior, but being tied at its feet, it did not have freedom of rotation at that point, making the structure statically indeterminate. This dilemma was overcome after the development of the hinged arch, employed with spectacular effect in the Galerie des Machines at the Paris Exposition of 1889.

In the heyday of railroad travel, the Midland Grand Hotel was amongst the most luxurious in London. Scott designed it at the peak of an illustrious career; two of his other works were then in progress—the Albert Memorial (1863–1875), Kensington Gardens, and the Foreign Office (1868–1873), Whitehall. There were over 300 lavishly decorated and amply furnished rooms opening off broad corridors. Hydraulic lifts were provided. One of the hotel's features was a spiral staircase supported on exposed iron beams. The hotel at St. Pancras was a landmark for the picturesque composition of its various Gothic elements—tracery windows, turrets, gables, pinnacles—and for the clock tower in the style of Big Ben. Its curved facade suited the site and the materials selected by Scott—red Nottingham bricks, red and gray granite, and beige stone—added to its distinctiveness. This colorful building stood out against the city's smoky pall and was a refreshing addition to the rather dull surroundings. The total cost of the shed and hotel was about £1 million.

In 1935, when the cost of refurbishing the hotel proved too high, it was closed and soon, renamed St. Pancras Chambers, converted to the Midland

Railway Company's offices. Many of its original features were removed or concealed. Threatened with demolition in the 1960s, it was saved as a significant example of Victorian Gothic architecture and assigned Grade 1 listed status by English Heritage. However, it was vacated in the 1980s due to inadequate fire standards. London and Continental Stations and Property, the current owner of the building (2001), has undertaken extensive structural and external and internal restoration works. These have revealed original decoration such as mosaics, ornamental ceiling panels, and stenciling. The railroad shed survives intact and remains a bustling London terminus.

See also Galerie des Machines (Gallery of Machines); Paddington Station

Further reading

Meeks, C. L. V. 1956. *The Railroad Station*. New Haven, CT: Yale University Press.

Radford, John B. 1983. *Midland Line Memories: A Pictorial History of the Midland Railway Main Line*. Kent, UK: Midas Books.

Simmons, Jack. 1968. *St. Pancras Station*. London: Allen and Unwin.

St. Paul's Cathedral
London, England

St. Paul's Cathedral in the city of London, created by the astronomer, mathematician, and designer Sir Christopher Wren (1632–1723), is the crowning work in the large oeuvre of one of the greatest English architects of his time, perhaps of all time. With it, English architecture regained the tradition of construction that it had developed for 400 years, and that had been displaced temporarily by Italian theories of proportion and emphasis upon appearance. Although it clearly drew upon classical and Italian models, Wren's great church was primarily concerned with space and the structural systems that achieved it.

The earliest church on the site was a wooden structure built in A.D. 604 by King Ethelbert of Kent for Mellitus, first bishop of the East Saxons. It burned down in 675 and was replaced by Bishop Erkenwald in 685, only to be destroyed by Viking raiders seven years later. Again rebuilt, it was again destroyed by fire in 1087. A new Norman church, now known as Old St. Paul's, was completed in 1240 after 150 years in the building. It was consecrated in 1300. A Gothic choir was added by 1313, and the following year a 489-foot (150-meter) spire was completed.

By the beginning of the seventeenth century, the cathedral had fallen into disrepair and disuse. In 1633 Inigo Jones, Surveyor of the Royal Works, was instructed to restore it. He had renovated the transepts and nave in the "modern" classical style and added a classical portico to the west front, when in 1642 England was divided by the civil war. Cromwell's Roundhead troops commandeered St. Paul's and deployed the nave as cavalry barracks, stabling their horses in the chancel. They dismantled Jones's scaffolding and sold the timber. During the Commonwealth (1649–1660) little changed and the building was being used as a public market. But in 1662, after the monarchy had been restored, King Charles II undertook its reinstatement as a cathedral. Temporary repairs were made to the choir so that services could recommence, and a Royal Commission was established to determine the structural condition of the building. In 1663 Wren was asked to make proposals for the restoration. His plan, which recommended the continuation of Jones's program and included his first design for a dome, was accepted on 27 August 1666. A week later the Great Fire swept through two-thirds of London, destroying over 13,000 houses and nearly ninety parish churches. Old St. Paul's was affected, and at first Wren thought that a new church could incorporate the existing nave walls. But in 1668 some of the masonry collapsed, and it was decided to start again. The Norman church was demolished.

In 1668 Wren was commissioned to design the new St. Paul's, and in 1669 he was installed as Surveyor to the King's Works. His original scheme, which can still be seen in the "First Model" of 1670, was approved by the king. By 1673, however, the design was rejected by the church because it "was not splendid enough." Wren responded with a spectacular alternative presented to the conservative dean and chapter of St. Paul's. It showed a domed church in the shape of a Greek cross (in which all arms are of equal length). It, too, was rejected by the clergymen, who thought it too modern and too Roman. Wren

St. Paul's Cathedral, London, England; Sir Christopher Wren, architect, 1675–1710. View from the northeast.

therefore compromised his ideal design to develop an elongated version—that is, a Latin cross—demonstrated in the so-called Great Model, which gave the churchmen what they wanted: a traditional English church with a spire. Completed in 1675, this Royal Warrant design, endorsed by the king, bore the interesting qualification that gave Wren the liberty to make "variations, rather ornamental than essential." But Wren knew, as did Charles II, that he in fact had complete license, and he changed the proposed church so much that it bore little resemblance to the Warrant design. He omitted the spire, reduced the length of the nave by three bays, increased the size of the dome, and raised the aisle walls. Construction work began as soon as the first contracts were signed in July 1675.

As it was built, St. Paul's Cathedral combined a floor plan based on traditional English churches with architectural elements and decorations drawn from Wren's seven months' experience in Europe when he fled the plague in 1661. Some writers identify specific Italian Renaissance and Baroque buildings, all in Rome, as sources of inspiration: Borromini's Church of St. Agnese, Bramante's Tempietto in the cloister of S. Pietro in Montorio, and Pietro da Cortona's Santa Maria della Pace. The dome is modeled on the ancient Pantheon in Rome. By 1694 the masonry of the choir was finished, and in 1697 the first service was held in the cathedral. Wren sought the best artists and craftsmen to work on details of the church: the master wood-carver Grinling Gibbons carved the choir stalls, and the wrought-iron sanctuary gates were the work of Jean Tijou. But after twenty-two years of building, there was no dome. Wren's salary was halved until the work was completed. St. Paul's was finished in 1710, making it the first English cathedral to be completed while the original architect was still alive.

Christopher Wren represents a new kind of architect, or more correctly, the reappearance of the old kind of architect, who was interested in architectural theory but also in the practical issues of design, process, and structure. He was, in short, a modern architect. Although a few of his contemporaries probably had some grasp of theory, none had the training or background that allowed them to develop it in any scientific way. Wren took an active role in the construction of his great cathedral, personally hiring and supervising the workmen, auditing and approving the accounts for materials and labor, and visiting the site each Saturday. Nowhere is his commitment to making architecture better seen than in the brilliant construction of the great dome, then the second largest in the world. It is in fact a triple dome. The outer hemispherical shell is timber framed and sheathed with copper. The inner saucer dome, 111 feet (34 meters) in diameter, begins at 173 feet (53.4 meters) above the floor at the crossing and is decorated with scenes from the life of St. Paul by Sir James Thornhill. At its center an oculus admits light to the interior from the lantern above. Therein lies the genius of Wren: in order to transmit the tremendous load of the stone lantern, reaching to 355 feet (109 meters) above the ground, he constructed a cone of brickwork hidden between the inner and outer domes. Wren is buried in St. Paul's Cathedral, and his epitaph stands upon his tomb; it reads in Latin *Si monumentum requiris, circumspice*—If you seek his monument, look around.

See also Pantheon

Further reading

Downes, Kerry. 1988. *Sir Christopher Wren: The Design of St. Paul's Cathedral*. Washington, DC: American Institute of Architects Press.

Gray, Ronald. 1982. *Christopher Wren and St. Paul's Cathedral*. Minneapolis: Lerner Publications.

Hart, Vaughan. 1995. *St. Paul's Cathedral: Sir Christopher Wren*. London: Phaidon Press.

St. Peter's Basilica
Vatican City, Italy

St. Peter's Basilica is the central place of the Roman Catholic Church. From its inception, it took 225 years to complete. No fewer than sixteen architects were responsible for it, under the patronage of twenty-two popes. Nevertheless, the great building presents a degree of integrity, of harmony (perhaps helped by the mellowing passage of the centuries) that might seem improbable given its heterogeneous and sometimes philosophically conflicting sources; that ultimate unity of form and detail is in itself no small architectural feat.

In A.D. 323, the first Christian Roman emperor, Constantine the Great (died 337), commissioned a magnificent basilica on the Vatican Hill, south of the River Tiber. It was built with difficulty on the sloping site, its altar supposedly above the spot where St. Peter was believed to have been buried around A.D. 64, and dedicated to him. Twelve centuries passed from the building of Constantine's basilica to the first phase of its demolition.

Between 1309 and 1377, for political reasons, the papal residence was at Avignon, France. Rome became derelict; according to some sources, packs of wolves roamed the streets. Its churches were neglected, and the old St. Peter's descended into decay, its walls leaning and its frescoes encrusted with dust and grime. With the popes again in residence, around the middle of the fifteenth century Rome succeeded Florence as the center of the Italian Renaissance, and in 1452 Pope Nicholas V (reigned 1447–1455) commissioned the architect Bernardo Rossellino (1409–1464) to build a new apse for St. Peter's west of the old one. Rossellino, who had already restored the church of San Francesco, Assisi, and many other buildings in Italy, proposed to surround the choir and transept, continuing the elongated Latin-cross plan. But only the tribune and foundations had been built when Nicholas V died and work stopped. Pope Paul II (1464–1471) passed the project to Giuliano da Sangallo in 1470, but none of the subsequent three popes pursued it.

Early in 1505 the warrior-pope Julius II (1505–1513) was considering what form his own tomb might take. The sculptor Michelangelo Buonarotti designed an imposing monument, but it called for an appropriate setting. Julius decided to rebuild St. Peter's, and late in 1505, a competition was held for the design. The winner was Donato Bramante, who, inspired by the ancient Pantheon in Rome, proposed a Greek cross (all of the arms of which are equal), with towers at the corners and a central dome raised on a drum. Julius laid the foundation stone on 18 April 1506. Despite the theological and esthetic arguments for a centrally planned church, the Greek cross was impractical for the Roman liturgy and thus unacceptable to the clergy. Bramante lengthened one arm to form the traditional Latin cross. Much of Julius II's money was diverted to wars with the French, and when his architect died in 1514, only the four main piers of St. Peter's were completed. They determined all further developments.

Julius's successor, the Borgia Leo X (1513–1521), commissioned Rafaello da Urbino, assisted by Giuliano da Sangallo and Fra Giocondo da Verona. The latter two modified Bramante's plan to a slightly elongated central nave with three aisles on either side. They died in 1516 and 1515, respectively, and Rafaello simplified their plan, seemingly to little effect. After his death in 1520 the new architects, Antonio da Sangallo the Elder, Baldassarre Peruzzi, and Andrea Sansovino, "without fixed plans and [attempting] all manner of experiments," revived arguments about a Greek cross versus a Latin cross. Seething conflicts in southern Europe impeded the work until Pope Paul III (1534–1549) engaged the architect Antonio da Sangallo the Younger, who revived Bramante's plan. Da Sangallo died in 1546. The following year Paul III offered the post of Prefect of Works to the 71-year-old Michelangelo, who accepted but refused payment for his work. He adapted Bramante's original plan, reducing the floor area by 40 percent, and started designing the three apses and the dome, which he made possible by strengthening the central piers. By 1564, when the great artist died, one apse and the drum of the dome as high as the entablature were completed. He was succeeded by Pirro Ligorio (better remembered as a coin counterfeiter than as an architect) and Giacomo da Vignola.

Pope Gregory XIII (reigned 1572–1585) next placed Giacomo della Porta in charge of St. Peter's. Assisted by Domenico Fontana, he subtly altered Michelangelo's design for a hemispherical dome, building a taller, more pointed structure based on Brunelleschi's Florence duomo. Work began in 1587, and the dome was completed under the patronage of Sixtus V (1585–1590). His successor, Gregory XIV (1590–1591), commissioned the lantern. Working for Paul V (1605–1621), Carlo Maderna gave the final building the form of a Latin cross by extending the nave to the east by building three more bays. He also won a 1607 competition for the design of the main facade, completed in 1626, and added an extra bay at each end to support bell towers. Pope Urban VIII solemnly consecrated the church on 18 November 1626.

But the project was not yet finished. Only one bell tower was built (and later demolished), to a 1637 design of Gian Lorenzo Bernini, the last architect to contribute to St. Peter's. For Pope Alexander VII (1655–1667), Bernini put final touches to the sumptuous basilica (the baldachin beneath the dome and St. Peter's Chair) and between 1656 and 1660 laid out the vast elliptical piazza, whose colonnaded arms open to welcome all to the bosom of Mother Church.

In 1984, the U.S. Catholic organization the Knights of Columbus collaborated with Reverenda Fabrica di San Pietro to restore the main facade. St. Peter's Basilica, with a floor area of 3.7 acres (1.5 hectares), was the largest church in Christendom until 1989, when its size was exceeded by the mimetic and extravagant Church of Our Lady of Peace (commenced 1986) in Yamoussoukro, Côte d'Ivoire, Africa.

Further reading

Bruschi, Arnaldo. 1977. *Bramante*. London: Thames and Hudson.

Hibbard, Howard. 1971. *Carlo Maderna and Roman Architecture, 1580–1630*. London: Zwemmer.

Stinger, Charles L. 1998. *The Renaissance in Rome*. Bloomington: Indiana University Press.

St. Pierre Cathedral
Beauvais, France

Beauvais is capital of the French *departement* of Oise, north of Paris. It already was an important center in pre-Roman Gaul. The Romans called it Bellovacum, and tradition has it that Lucianus, Maxianus, and Julianus founded Christianity there at the cost of their lives in about A.D. 275. Beauvais became a countship in the ninth century. Power passed to the bishops in 1013, although the date of the foundation of the diocese is unknown. The first cathedral was built in the tenth century alongside the Romanesque church still known as the Basse Oeuvre, and dedicated to Peter, the patron saint of Beauvais. That cathedral was damaged by fire in 1180 and again in 1225, and reconstruction was undertaken. The second cathedral, never finished, is regarded by many as the most ambitious structure in Gothic architecture, one of the wonders of medieval France.

Because it is among the last Gothic churches, the architects of Beauvais Cathedral were able to draw widely on the experience of other builders. The High Gothic phase (1225–1232)—a five-aisle church with wide transepts and towers—was commissioned by Bishop Miles de Nanteuil. But little was built before his funds were exhausted. Changes were made when work resumed in 1238 for Bishop Robert de Cressonsac; although more modest than de Nanteuil's project, his church (had it been completed) would have been much grander than most of its contemporaries. But more significant changes were yet to be made. The final phase was built for Bishop William de Grez from about 1247.

In order to allow more light to enter the churches, the High Gothic mason-architects pushed the structural boundaries to the limit by increasing the height of the vaults. Their architecture is distinguished by its emphasis on verticality and the apparent slenderness of the structural elements. In the *rayonnant* style—the name comes from the spokes of its characteristic rose windows—which became popular during the reign of Louis IX (1226–1270), the emphasis on height was displaced by the refinement of the masonry frame; the consequently larger window area produced the same net effect as greater height: more light penetrated into the church. Beauvais Cathedral synthesizes the High Gothic and the *rayonnant*. The piers were more widely spaced, thus reducing masonry and gaining more stained glass. Moreover, by superimposing a tall clerestory on the already lofty arcade of the choir, William's architects added more than 16 feet (5 meters) to the height of the building, whose soaring vaults then reached 157 feet (48 meters)—about three and a half times their span.

The choir was completed in 1272. Its vaulting collapsed only twelve years later. When it was rebuilt (to the same height) between 1337 and 1347, additional piers placed between the existing ones strengthened the structure and allowed the builders to replace the quadripartite vaults with a more conservative sexpartite system. The flying buttresses were reinforced at the same time, and iron tie rods were introduced for extra security. Transepts were added in 1500–1548, and in 1558–1568 a 495-foot-tall (151-meter) tower was built over the crossing. It collapsed in 1773 and was never rebuilt. Even today, Beauvais Cathedral is without a nave.

Debate continues about the reasons for the collapse of the choir. One theory suggests that the vaulting system was underdesigned, with too widely spaced piers, while another blames uneven and too rapid settlement of the foundation soil beneath the highly concentrated loads that weakened the buttresses and superstructure. Yet another attributes the failure to interruptions to the work and changes to the plans when the building was partly completed. Current experience presents a plausible alternative: the buttresses swayed (as they still do) in the gale-force winds that come off the English Channel, affecting the overall stability of the church. It seems likely that the choir ceiling fell through the combined effect of all these factors.

The town of Beauvais was heavily bombed in 1940, and after World War II it was reconstructed to the original plan. Although the cathedral escaped the bombing, its structural stability is still compromised, for all of the reasons stated. Some well-meaning efforts at preservation have only exacerbated the problem. Informed and urgent action is necessary to save the highest medieval church in Europe. Beauvais Cathedral was included on the World Monuments Watch 2000 List of 100 Most Endangered Sites; in June of that year a grant to meet the engineering costs related to a structural-modeling project was received from the Samuel H. Kress Foundation.

See also Chartres Cathedral (Cathedral of the Assumption of Our Lady)

Further reading

Bony, Jean. 1983. *French Gothic Architecture of the Twelfth and Thirteenth Centuries*. Berkeley and Los Angeles: University of California Press.

Murray, Stephen. 1989. *Beauvais Cathedral: Architecture of Transcendence*. Princeton, NJ: Princeton University Press.

San Paolo fuori la Mura (St. Paul's outside the Walls)
Rome, Italy

Perhaps the most demanding question that can be asked of any architect is to invent a building to suit a new purpose, and the provision of an adequate, even seemly, answer is indeed an architectural feat. From the beginning of the fourth century A.D., congregational worship by large numbers of people needed a hall, and the Roman basilica—a civil law court—became one model for churches in western Europe. The early Christian architects ingeniously combined the vast, articulated open spaces of the basilica with the familiar layout of the Roman *domus* to produce a new architectural type: the basilican church. San Paolo fuori la Mura, begun soon after 314 and completed in the thirteenth century, although completely rebuilt after 1823, is the clearest example, simply because the others have been "modernized" over the years.

The first 250 years of Christianity were marked by intermittent brutal persecution, so the places where the faithful gathered were necessarily unobtrusive. At first the religion generated no specific architectural forms, its underground congregations meeting in larger houses *(domus)* or even literally underground in the catacombs near Rome. In A.D. 313 the Edict of Toleration granted legal recognition to Christianity and the emperor Constantine initiated a program to build meeting places for the emancipated believers. The church leaders and their architects had a problem: was there an appropriate form for a building whose God did not "dwell in temples made with hands"? For associative reasons, pagan temples provided an inappropriate precedent; neither did their spatial organization suit the church's developing liturgy.

The plan of the first basilican churches reflected the social organization of the domus, creating a house for the family of God. And in terms of a gathering place to worship the King of Kings, what could surpass the basilica—from the Greek word for royalty? The standard plan, linear in every respect, included an atrium, or forecourt; a narthex, or porch; and a long nave flanked by single or double side aisles. Every feature of construction and decoration drew the worshiper's eye immediately to the altar in a polygonal or semicircular apse at the eastern end. The space was achieved with the simplest and lightest of structural systems: slender parallel walls, either solid from the ground or carried above lofty arcades, supported trussed timber roofs (usually obscured by a coffered ceiling). Avoiding the need to cope with lateral thrusts, the architects created vast ethereal spaces that still provoke admiration, even awe.

In Rome the established Christian foci—often the sites of martyrdoms—became the locations of these huge basilicas: San Giovanni in Laterano (311–314), San Pietro (begun 324), Santa Maria Maggiore (432–440), and San Paolo fuori la Mura. Between the nave and the apse each had a short transept, an oblong hall crossing the nave to form a T-shaped cross. Typically, columns, marble panels, dressed masonry, and even roof tiles from pagan temples were used as instant building materials. Many smaller basilicas were built, such as Santa Sabina, Rome, and Sant' Apollinare in Classe, Ravenna, both of the fifth century.

Tradition has it that the apostle Paul was executed in Rome between A.D. 64 and 68 and that his tomb in the Via Ostia was conserved by a woman named Lucina who owned the surrounding land. There may have been an oratory on the site as early as 103, and the place was certainly marked by a *cella memoriæ* at the beginning of the third century. Around 314 Constantine had transformed it into the basilica that was rebuilt and enlarged in 386 by the Consul Sallustrius on the orders of the emperors Valentinian, Theodosius, and Arcadius. The mosaic decoration was completed under Honorius and Theodosius's daughter, Galla Placidia. The church of San Paolo fuori la Mura reached its height from the thirteenth century onward.

The medieval worshiper would have approached across the atrium, seeing the magnificent color of Pietro

San Paolo fuori la Mura (St. Paul's outside the Walls), Rome; architect(s) unknown, ca. 314–386. Interior, looking east.

Cavallini's mosaics on the west façade. The eleventh-century bronze doors were made in Constantinople by Staurakios of Chios for Hildebrand; inlaid with niello work, they portrayed saints and martyrs. Once inside, the vast space—nearly 440 feet long by 217 feet wide (132 by 65 meters), with a coffered ceiling 100 feet (30 meters) above the marble floor—would have been seen, but hardly taken in, all at once. Eighty fluted columns of pavonazzetto and Parian marble divided the 82-foot-wide (24.6-meter) nave from the flanking double aisles. Twenty-four of them were perfectly matched; the remainder came from various pagan temples. High on the nave walls panels were decorated with frescoes of the lives of St. Peter and St. Paul and other biblical subjects; below them and above the arcades, mosaic portraits of the popes led the eye

inexorably to the east end. There, a huge triumphal arch, with scintillating gold and colored mosaics of biblical subjects, was carried on two 50-foot (15-meter) marble Ionic columns. Framed by the arch, the tomb of St. Paul (1285) stands under a Gothic baldachin by Arnolfo di Cambio; behind it, the apse was "all aglow with mosaic." San Paolo fuori la Mura was a glorious, wondrous space.

The building has had a checkered history. There have been damaging earthquakes and simple neglect when the area near the Tiber became malarial. Because it was outside the walls it was exposed to invaders: the Langobards plundered it in 739 and Saracens in 847, leading Pope John VIII belatedly to build a citadel in 872. Through all this, it preserved its original character and form more than any church in Rome for nearly 1,500 years. It was almost totally destroyed by fire in July 1823 through the carelessness of a workman. The whole world, Catholic and non-Catholic alike, contributed to its restoration by the architect Luigi Poletti; for example, the Khedive of Egypt sent pillars of alabaster, and the Tsar of Russia gave malachite and lapis lazuli for the tabernacle. San Paolo fuori la Mura was reconsecrated in December 1854. As it stands today, the church articulates by its structure the brilliance of the early Christian architects.

Further reading

Sanderson, Warren. 1993. *Early Christian Buildings: 300–600*. Champlain, NY: Astrion.

Ward-Perkins, J. B. 1994. *Studies in Roman and Early Christian Architecture*. London: Pindar Press.

White, L. Michael. 1997. *The Social Origins of Christian Architecture*. Valley Forge, PA: Trinity Press.

Seikan Tunnel (Seikan Tonneru)
Japan

After two decades of planning and construction, the 33.5-mile-long (53.85-kilometer) Seikan submarine tunnel was opened to traffic on 13 March 1988. Part of a railroad between Aomori City and Hakodate City, it links Honshu, the main Japanese island, with Hokkaido to the north, passing under the 459-foot-deep (140-meter) Tsugaru Strait. The tunnel runs 328 feet (100 meters) beneath the ocean bed for 14.5 miles (23.35 kilometers); thus, at 787 feet (240 meters) below sea level, it is the deepest railroad line in the world. The journey between the terminals takes two and a half hours. It has been called "one of the most formidable engineering feats of the twentieth century."

Before the tunnel was built, ferries provided transport between Honshu and Hokkaido. After a typhoon sank a ferry boat in the Tsugaru Strait in September 1954, killing 1,430 people, public pressure for the construction of a safer crossing increased. Plans were already under way: geological studies of the area had been started as early as 1946, and the Tsugaru line among those planned in Japan's Railway Construction Act of August 1953. In April 1964, almost immediately after the newly established Japan Railway Construction Corporation took over research operations from Japan National Railways, a basic plan was called for, and exploratory inclined shafts were sunk on either side of the strait by early 1966. Five and a half years passed before the minister of transport approved the final plan, allowing construction to commence.

Excavation began on the underwater section of the tunnel first, with a 16.35-foot-diameter (5-meter) pilot tunnel, which was started from both ends, meeting in the middle in January 1983. Next followed a service tunnel of similar size. The latter, about 100 feet (30 meters) from the rail tunnel and connected by regularly spaced junctions, now serves as a maintenance and escape route. The main tunnel's entrance section was commenced in August 1982.

The main tunnel, with an internal width of 32 feet (9.7 meters) and a clear height in the center of almost 26 feet (7.85 meters), was slowly constructed through the seismically unpredictable seabed by drilling and blasting (nearly 3,300 tons, or 3,000 tonnes, of explosives were used); the nature of the geology made the use of a tunnel-boring machine impracticable. The submarine section of the tunnel was completed in March 1985, and eighteen months later the two-track electrified rail system was in place. Commercial rail services for passengers and freight began in March 1988. There are two stations in the tunnel: Tappi Kaitei on Honshu and Yoshioka Kaitei on Hokkaido, just on the respective coasts.

Because the Seikan Tunnel's signal system is based on an uninsulated rail circuit, rails were welded together to form a single rail—the longest in the world—running nearly the length of the tunnel. This provides a much smoother ride for commuters. The tunnel was designed with high-strength rails, large-radius curves, and minimal gradients to enable possible future use by Japan's famous Shinkansen (bullet trains). For the time being, the cost of extending the Shinkansen has proved prohibitive. Moreover, the Seikan Tunnel had cost 1.1 trillion yen (U.S.$7 billion) to build—almost twelve times the original estimate. Much of the budget blowout was attributable to inflation, because the ten-year estimated construction time had, through accidents and other circumstances, extended to twenty-five. The delays had another effect: ironically, the convenience offered to travelers by the Seikan Tunnel has now been rivaled by quicker and cheaper air travel between Honshu and Hokkaido.

Further reading

Nihon Tetsudo Kensetsu Kodan Sapporo Koji Jimusho. 1990. *Tsugaru kaikyosen kojishi. Seikan Tonneru.* 2 vols. Hakodate-shi, Japan: Nihon Tetsudo Kensetsu Kodan.

Yutaka, Mochida. 1994. *Seikan Tonneru kara ei-futsu kaikyo tonneru e: chishitsu, kishitsu, bunka no kabe o koete.* Tokyo: Chuo Koronsha.

Semmering Railway
Austria

The 26-mile-long (41.8-kilometer) Semmering Railway climbs through an altitude of 1,400 feet (439 meters) over the Semmering alpine pass, at an elevation of 2,930 feet (898 meters), between Gloggnitz and Mürzzuschlag in southeastern Austria. Designed in 1843 and built between July 1848 and July 1854, it is still in use. The railway has been described in UNESCO World Heritage List documents as "one of the greatest feats of civil engineering of [the] pioneering phase of railway building." Moreover, it "represents an outstanding technological solution to a major physical problem."

The railway was designed by the Austrian engineer Karl von Ghega (1802–1860) as part of a double-track connection between Vienna and Trieste, Italy. In May 1842 the section of line between Vienna and Gloggnitz was opened. Just then, von Ghega, who had been appointed inspector of the Southern State Railway, was undertaking a study tour in England and North America. In August an imperial edict called for the extension of the line over the high Semmering alpine pass, and by the end of the following January von Ghega had prepared three proposals, offering the alternatives of steam locomotive (eventually chosen), cable inclines, and atmospheric railway for the consideration of Ermengildo von Francesconi, Director General of State Railways. Little action was taken on the Semmering Railway, and in October 1844 a section of track between Graz and Mürzzuschlag was completed, still leaving a gap in the line over the

mountains. Political vagaries and reservations about a project that had no precedents—no one anywhere had built a railway over mountains—meant that von Ghega's plans were shelved for four years. Then, prompted or panicked by a revolution in Vienna in spring 1848, the government was moved to urgent action.

Construction of the Gloggnitz-Payerbach and Mürzzuschlag-Spital sections of the line started in summer 1848. Each day, up to 5,000 laborers commuted from Vienna. At the peak of the project about 20,000 men and women were employed. They built sixteen arched masonry viaducts (some of them two stories high) with a total length of a little under 1 mile (1.5 kilometers) and excavated 15 tunnels with a total length of 3 miles (4.8 kilometers). By means of these structures and over 129 bridges, the track climbed along its winding, precipitous path, carved from the faces of the mountains entirely by hand and without the use of explosives. The track was lined with stone station buildings, fifty-five two-story houses for linesmen, and thirty-two timber-framed signal boxes, as well as workshops and engine halls.

Von Ghega had not convinced everyone that a steam locomotive railway would work, and for three years the project proceeded in the face of criticism from his professional peers, who preferred the use of fixed steam engines and cable inclines. They were silenced in September 1851 when a competition was held on the line. Twenty-two years earlier in England, a similar competition had established the suitability of steam locomotive power across flat country. Von Ghega's conviction that normal vehicles could be used on the Semmering was vindicated when the four participating locomotives easily passed the test of pulling 154 tons (140 tonnes) at 7 mph (11.4 kph) on the steepest gradient. The Austrian engineers Wilhelm Freiherr von Engerth and Fischer von Röslerstamm were charged with developing the first triple-coupled locomotives, and they were built in Germany and Belgium in 1852.

The Semmering Railway was inaugurated in October 1853. The budget had been exceeded by a factor of four. Goods traffic started in May 1854, followed in July by a scheduled passenger service. Of course, a rail trip through mountain country was a totally new experience and many made it simply for the pleasure of watching the grandeur of the landscape. The journey, pulled by any of the twenty-six mountain locomotives in service, took a few minutes over two hours. The entire Vienna-Trieste line was opened at the end of July 1857.

The high elevation of the Semmering main tunnel caused problems because of freezing. Around the turn of the century, attempts were made to heat it with gas burners, and it was actually closed off with doors between trains. Flooding was also a difficulty, and after the particularly bad 1946–1947 winter, several hundred wagon loads of ice had to be excavated from the tunnel. A second tunnel was completed in March 1952 to overcome the difficulty. The Vienna-Gloggnitz line was electrified seven years later, and a four-year program was launched to upgrade tracks and buildings and electrify the Semmering section. The modifications were finished by May 1959. Coupled electric locomotives now transport up to 1,100 tons (1,000 tonnes) over the pass. Journey time, for either freight or passenger trains, is just forty-two minutes. Maximum speeds are limited to 37.5 mph (60 kph) on the north ramp and 44 mph (70 kph) on the south. The Semmering Railway continues to function well in the face of major technological change since it was first built, a testimony to the "detailed and well-founded planning" of von Ghega. However, the increased traffic has called for constant maintenance. The Semmering Railway, preserved in its nearly original form, has come under the Austrian Law of Protection of Monuments since 1923, a status confirmed by the Federal Office of Monuments as recently as March 1997. It was inscribed on UNESCO's World Heritage List a year later.

Further reading

Niel, Alfred. 1960. *Der Semmering und seine Bahn*. Vienna: Ployer.

Winn, Bernard C. 1987. *Railways Revisited: A Guide to Little-Known Railways in Austria and Germany*. Merced: Incline Press.

Shell concrete

In 1919 Dr. Walter Bauersfeld of the Carl Zeiss optical works in Jena, Germany, proposed a planetarium.

Following his 1922 success with a 52-foot-diameter (16-meter) iron-rod dome built on the roof of the company's building—the first lightweight steel structural framework in the world—Bauersfeld consulted the structural engineers Dyckerhoff and Widmann about a larger version. Then, together with their designers Franz Dischinger and Ulrich Finsterwalder, he built the world's first lightweight thin-shell concrete dome for Zeiss's sister company, Schott and Partners. It was 131 feet (40 meters) in diameter and only 2.4 inches (6 centimeters) thick. The new structural technology, honed in later structures, made possible clear spans of lighter weight than had ever been imagined. Because concrete shells depend on configuration rather than mass for their strength, and because they exploit the fact that concrete is essentially a fluid, they have been characterized as the ultimate concrete form.

Some of the most exciting examples have come from Spanish engineer-architects. Eduardo Torroja y Miret (1899–1961) was perhaps the most innovative engineer of the early twentieth century, notable for shell concrete roof designs that employed continuous surfaces and eliminated the need for ribs. Three examples should suffice. Torroja's first thin-shelled concrete roof was for the Market Hall in Algeciras, Spain (1933–1934), designed in conjunction with the architect Manuel Sanchez. The low-rise dome, supported at six points on its perimeter, spans 156 feet (48 meters); it is only 3.5 inches (9 centimeters) thick. In 1935, working with the architects Carlos Arniches Moltó and Martín Domínguez Esteban, Torroja produced the folded plate roof for the grandstand at the Hipódromo de la Zarzuela, Madrid. The cantilever on the graceful structure is 73 feet (22 meters). In the same year, with architect Secundino Zuazo, he designed a 180-foot-long (55-meter) vaulted roof over an indoor pelota court in Madrid. Its two intersecting parallel vaults, of 40 and 21 feet (12.2 and 6.4 meters) radius, respectively, span a total of 98 feet (30 meters); the general thickness of the vast shell is a mere 3.14 inches (8 centimeters). With José Maria Aguirre, in 1934 Torroja founded the Instituto Técnico de la Construcción y Edificación in Madrid to "develop new uses and theories for reinforced concrete"; after his death in 1961, the institute was renamed the Instituto de Ciencias de la Construcción Eduardo Torroja. In 1952, the editors of *Concrete Quarterly* had claimed with some accuracy, "Torroja has extracted the utmost from his chosen material, concrete; no other material could have given such structures, no other designer has."

Another Spanish-born architect, a generation after Torroja, earned the nickname "the Shell Builder." Felix Candela Outeriño (b. 1910), known best simply as Felix Candela, graduated from Madrid's Escuela Superior de Arquitectura in 1935 but did not practice until he emigrated to Mexico in 1939. First working for others in Acapulco, he moved to Mexico City and set up his own practice, specializing in the design and construction of thin tensile concrete shells. Because this was a new way of building, Candela acted as architect, structural engineer, and contractor, even training the construction workers himself. He began building beautiful churches such as Medalla de la Virgen Milagrosa (1953–1955), Nuestra Señora de la Soledad Chapel (1955), and San José Obrero (1959), all in Mexico City, and the Open Chapel in Lomas de Cuernavaca (1958). The roofs and sometimes the walls are noteworthy for their seamless concrete construction, often only 1.07 inches (4 centimeters) thick. Candela stated, "It is the shape that matters." He insisted that "the shell must be stable and of a shape which permits an easy way to work. It should be as symmetrical as possible because this simplifies its behavior" (Faber 1963, 199). To this end, he frequently made use of the hyperbolic paraboloid, a form that made the construction of timber formwork easy because it is generated only from straight lines. The best example can be found in Los Manantiales Restaurant of 1958 in Xochimilco near Mexico City; the thin concrete shell structure that encloses its radial plan is based on eight hyperbolic paraboloid segments. Critics have remarked that in Candela's work both design and structure have been sharpened to the finest edge, imparting "a new dynamism" to architecture.

Shell concrete was promoted in North America by a single expatriate European engineer, the Viennese Anton Tedesko (1902–1994). From 1930 to 1932 he collaborated with the shell concrete pioneers Dischinger and Finsterwalder, and two years later he moved to the Chicago offices of Roberts and Schaefer, for whom he worked until 1967, designing what have been described as "watershed buildings," including the Sports Palace (1936) in Hershey, Pennsylvania. Its ribbed-barrel, shell concrete roof spanned 255 feet (78 meters) and covered an area measuring 230 by 340 feet (70 by 104 meters). During World War II he designed seaplane hangars in San Diego for the U.S. Navy and later produced the main terminal building at Lambert International Airport, St. Louis, roofed with four bays of huge groin vaults. It was mainly through Tedesko's evangelism that shells became respected in the United States.

According to engineering historian David Billington, the other U.S. "thin-shell" visionary engineer was Jack Christiansen of Seattle, who was chief designer of the city's Kingdome (1976, demolished 2000). The building was not technically a dome; its 9-acre (3.63-hectare) roof was formed of hyperbolic paraboloids—22 rib arches with a 5-inch-thick (13-centimeter) concrete shell spanning between them. Kingdome was the last (and biggest) major thin shell built in the United States; it was preceded by the New Orleans Exhibition Hall of 1968 (on which Christiansen worked) and Denver's Paraboloid (1969), created by I. M. Pei and Tedesko. After experimentation in the sixties and seventies, such large shells were eclipsed by the newer technologies of air-supported structures and steel-framed retractable roofs.

That is not to say that the concrete shell is passé. Since the early 1960s the structural engineer Heinz Isler (b. 1926) has designed more than 1,000 shells, mostly in his native Switzerland. His elegant structures, inspired by natural forms, challenge those of Candela for elegance. Instead of mathematically calculating the forms, Isler designs by experiment, using catenary models in much the way that Antoni Gaudí did almost a century before. Notable among his earlier works are the Wyss Garden Center at Solothurn (1961), a thin, double-curved geometric shell; the Sicli Company Office building (1969–1970) in Geneva; and the Bürgi Garden Center at Camorino (1973).

See also Airplane hangars; Airship hangars; Jahrhunderthalle

Further reading

Chilton, John. 2000. *Heinz Isler*. London: Thomas Telford.
Faber, Colin. 1963. *Candela, the Shell Builder*. New York: Reinhold.
Fernández Ordóñez, José A. 1999. *Eduardo Torroja: Engineer*. Madrid: Ediciones Pronaos.

Shibam
Yemen

Surrounded by a 23-foot-high (7-meter) mud-brick wall, the Yemeni city of Shibam lies at the southern edge of the Rub'al-Khali Desert at the junction of several wadis and the Hadramawt Valley. Popularly known as the "Manhattan of the Desert," this city of about 7,000 inhabitants has more than 500 earthen high-rise houses, up to twelve stories high—the world's oldest skyscrapers—neatly contained in a quarter-square-mile (half a square kilometer) rectangle. The city has been there for at least 1,800 years, but most of these remarkable dwellings date from the sixteenth century. Inscribed on UNESCO's World Heritage List in 1982, Shibam was described as "one of the oldest and best examples of urban planning based on the principle of vertical construction."

Shibam was on the caravan route of the incense trade and replaced Shabwa as the capital of the Hadramawt in the third century A.D. Since then, it has enjoyed several periods of religious, political, and especially commercial power. Because it stands on a hillock barely 100 feet (30 meters) above the floor of the deep Hadramawt wadi, Shibam has been victim to floods, and it was partly destroyed by water in 1532. Thus flood protection is among the reasons given for the traditional form of its unique high-rise houses; others include the need to conserve agricultural land (the city is surrounded by groves of date palms), the desire to gather patriarchal families under one roof, and, more pragmatically, at least in earlier times, to accept the protection afforded by the perimeter wall.

Many of the towers of sun-dried, straw-reinforced mud brick taper upward, perhaps to ensure greater

The city of Shibam, Hadramawt Valley, Yemen. Aerial view; some parts of the "skyscrapers" date from ca. 1200.

stability. The external walls, up to 130 feet (40 meters) high, are partly load bearing. They are set on stone foundations, supported by a framework of timber posts and beams and a large internal stairway built of stone. At ground-floor level the walls are from 4 to 6 feet (1.3 to 2 meters) thick. Interior surfaces are finished with a skim coat of lime plaster to which egg whites are added, producing a highly polished surface. The exteriors are stuccoed with a mixture of clay and chopped straw. High above the narrow lanes that separate the buildings, the living quarters on the upper floors are joined by networks of covered flyovers, so that neighbors can meet without needing to climb up and down stairs. The uppermost floors are usually covered with a thick layer of impermeable white stucco that not only protects from the rain but also reflects much of the solar radiation.

Ventilation is an important constraint upon the design in the relentless desert heat. The crowding of the tall buildings within such a small urban area produces some degree of self-shading. The walls are pierced by regular rows of narrow windows, rooms often having upper and lower sets that from outside create an illusion of more stories than there really are. Openings are closed with wooden *mashrabiyas* and doors, some as much as 800 years old, elaborately carved with open geometric patterns to encourage natural ventilation. The highest row of openings allows hot air to flow from the interior of the building in the summer.

Some inhabitants of Shibam have been forsaking these traditional houses for modern dwellings that line the highway to Suyan, about 30 miles (19 kilometers) away, producing a version of urban sprawl. Without maintenance, the skyscrapers in this "Manhattan of the Desert" are deteriorating; many are cracking and showing symptoms of collapse. Indeed, over the last decade, more than thirty have succumbed to the elements. Despite UNESCO's listing and the self-conscious effort of the local people—rigorously applied building codes insist upon traditional designs, materials, and techniques—to conserve its character, the unique architectural landscape of Shibam is in imminent danger of disappearing.

Further reading

Lewcock, Ronald. 1986. *Wadi Hadramawt and the Walled City of Shibam*. Paris: UNESCO.

Stark, Freya. 1983. *The Southern Gates of Arabia: A Journey in the Hadhramaut*. Boston: Houghton Mifflin.

Shwedagon Pagoda
Yangon, Myanmar

The most spectacular building in Yangon (formerly known as Rangoon) is the Shwedagon Pagoda, a great bell-shaped, solid brick *stupa* covered with an estimated 55 tons (50 tonnes) of gold. It rises 368 feet (112 meters) on Theinguttara Hill, above the city. The sixteenth-century English adventurer Ralph Fitch wrote that "it is of a wonderful bigness, and all gilded from the foot to the top.... It is the fairest place, as I suppose, that is in all the world; it stands very high...." The base of the pagoda, nearly 1,500 feet (460 meters) in perimeter, is surrounded by over seventy sculpture-enriched smaller shrines. It may be approached from four directions, and in the sixteenth century the gates to its three tiered terraces opened from long avenues lined with fruit trees. Although the tourist literature justifiably claims it to be the highest pagoda and the largest golden monument in the world, it is an architectural feat if for no reason other than its size and economic value. It is a thing of great glowing beauty, and a high point in the development of Buddhist architecture in Southeast Asia.

The Shwedagon Pagoda's origins are immersed in myth. Tradition asserts that it has stood on its hill for 2,500 years, although archeologists believe it to be about 1,000 years younger. But Theinguttara Hill had long been sacred because of the relics of three earlier Buddhas buried there: the staff of Kakuthan, the filter of Gawnagon, and the waistcloth of Kassapa. The legend describes how two brothers met Guatama Buddha, who entrusted them with strands of his hair to be enshrined in Shwedagon. With divine help, they and King Okkalapa discovered the holy hill. To guard all four relics, consecutive pagodas of silver, tin, copper, lead, marble, iron, and gold were built one upon the other. The pagoda was damaged by earthquakes on at least eight occasions between 1564 and 1919, but rebuilding and enhancement by successive kings

Shwedagon Pagoda, Yangon, Myanmar; architect(s) unknown, ca. 1455–1462. Detail of main gilded shrine.

caused it to grow from the original 66 feet (20 meters) to its present height.

Although it was in the insignificant Lower Myanmar town of Dagon, it seems that Shwedagon emerged as a major shrine during the Mon Kingdom of Hantharwaddy toward the end of the fourteenth century A.D., when King Binnyau undertook repairs to its structure. Between 1455 and 1462, Queen Shinsawpu built the terrace, balustrade, and encircling walls and gilded the pagoda, by then raised to a height of 302 feet (92.3 meters), with her body weight in gold leaf. Further changes were made in the eighteenth century.

In 1755 King Alaungpaya, patriarch of the Kon-Baung dynasty, reunited the whole of Myanmar. He recognized the strategic importance of Dagon and renamed it Yangon ("the end of strife"), which it retained until the British occupation of Burma in 1851. Shinbyushin, king of Ava, extended the Shwedagon Pagoda to its present height in 1774 and crowned it

with a new golden *htidaw*—a seven-tiered umbrella. His son Singu regilded it four years later. In 1786 the top half of the building was brought down by a violent earthquake. The pagoda's present form dates from its rebuilding at that time. A major renovation was ordered by King Mindon in 1871, when another new htidaw of layered gold, copper, steel, and zinc was erected. No major maintenance was then undertaken until 1999.

When the periodical gilding was being applied at the end of 1998, workers discovered the serious deterioration of King Mindon's htidaw. Under the direction of the Committee for All-Round Perpetual Renovation of Shwedagon Pagoda, a 700-strong team of laborers and artisans set about to preserve the failing structure. Ngi Hla Nge of Yangon Technological University supervised the engineering work. At the end of April 1999 a new htidaw, consisting of a steel frame covered with gold and alloys and studded with nearly 8,000 precious stones, was hoisted into place; at its pinnacle there is a 76-carat diamond. Most of the cost was met by donations of cash, gold, and jewelry. The restoration was completed by the end of 1999.

Further reading

Aung Than, U. 1957. *Shwedagon*. Rangoon: Government Printer.

Fitch, Ralph. 1625. *The Voyage of Master Ralph Fitch, Merchant of London*. London: n.p.

Moore, Elizabeth, Hansjorg Mayer, and U. Win Pe. 1999. *Shwedagon: Golden Pagoda of Myanmar*. Bangkok: River Books.

Taik, Aung Aung. 1989. *Visions of Shwedagon*. Bangkok: White Lotus; Berkeley, CA: Independent Scholars of Asia.

Sigiriya (Lion Mountain)
Sri Lanka

Sigiriya (Lion Mountain), about 130 miles (210 kilometers) from Colombo in central Sri Lanka, is a ruined ancient stronghold built on a sheer-sided rock pillar. It rises 1,144 feet (349 meters) above sea level and 600 feet (180 meters) above the surrounding plain. On the summit King Kasyapa I (reigned A.D. 477–495) built a palace. Together with the surrounding gardens, it is the best-preserved first-millennium city in Asia, combining symmetrical and asymmetrical elements, changes of level, and axial and radial planning. The central rock is flanked by rectangular precincts on the east 234 acres (90 hectares) and the west 104 acres (40 hectares), all surrounded by a double moat and three ramparts. The city plan, based on a square module, extends 2 miles (about 3 kilometers) from east to west and over 1,000 yards (1 kilometer) from north to south, with precincts set aside for hunters, scavengers, foreigners, and even heretics. There were separate cemeteries for high and low castes, hostels, and hospitals. As well as the city within the inner and outer ramparts, suburban houses spread beyond the walls. Sigiriya demonstrated a sophisticated level of urban design at a time when Europe was in its Dark Ages.

The origins of this remarkable architectural achievement are obscured by legend. A romantic if grisly tradition has it that Kasyapa murdered his father Dhatusena and usurped the throne. Seven years later, full of "paranoia, arrogance and delusions of divinity," he forsook the Sinhalese capital Anuradhapura and built his palace on Sigiriya. Of course it was defensible, although in the event, that was of little consequence. In A.D. 495 his half brother Moggallana, the rightful heir, led an army against him and Kasyapa came down from his fastness to fight. When his forces were routed, Kasyapa cut his own throat. An alternative version has more historical support. Dhatusena was frustrated in his quest for the imperial title that Sri Lanka's rulers traditionally held as protectors of Buddhism, because the king of Java would not relinquish it. A priest advised Dhatusena that if he reigned from the summit of a rock, a palace in the sky, he could win that higher status. When Dhatusena named Moggallana as his successor Kasyapa fled to India, returning to invade Sri Lanka seven months later. Anticipating losing the battle, his father killed himself, and Kasyapa entered Anuradhapura to seize the throne. When Kasyapa discovered Sigiriya, there was a monastery in the lower levels of the rock. He built a new place for the monks before he started work on the fortress, commissioning a Sinhalese architect, Sena Lal, to complete the work.

The entire summit of Sigiriya, nearly 3 acres (1.2 hectares) in extent, was once surrounded by an outer wall on the very rim of the cliff. An ancient guidebook, the *Sihigiri Vihara Suvarnapura*, describes a colonnaded mansion set in landscaped gardens made only for the use of King Kasyapa and his queen. The palace garden had terraces and rock-cut pools replete with aquatic flowers. They were fed by a mechanical pumping system from a lake at the base of the rock that kept them full of water even in the dry season. The west and south slopes of Sigiriya were terraced with houses for members of the royal household; on the west there were also two flights of stairs to the summit and a theater with seats hewn from the rock.

The spectacular feature that gave the Lion Rock its name was the ceremonial entrance, approached from the west through an elaborate water garden surrounded by a moat and across a flat piece of terrain known as the Plateau of Red Arsenic. The covered brick-and-timber stair leading to the summit was reached through the mouth of a brilliantly colored lion, built with brick and limestone and towering 45 feet (14 meters) against the granite cliff. All that remains of the lion are two gigantic paws with talons unsheathed, and a mass of broken brickwork. Images of Dhatusena and Kasyapa are painted on the rock above the lion's head. The stairway rises past hollows to a gallery about halfway up, enclosed by a 10-foot-high (3-meter) polished plaster wall—the Mirror Wall—and then to a small cave about 45 feet (14 meters) higher, in which there are paintings of half-naked women. There is some debate about their identity: they could be *apsaras* (heavenly maidens), courtiers, or even Kasyapa's concubines. Of the original 500 portraits, only 19 survive.

Below Lion Mountain, a western precinct was entered across the inner moat, through a timber-and-brick gatehouse. Its water gardens were symmetrically planned, with cleverly designed hydraulic systems for horticulture, agriculture, surface drainage, and erosion control. The pools and cisterns (some 400 of them) were fed from the Sigiri Maha Wewa, an artificial lake stretching for 8 miles (13 kilometers) from the foot of the rock. There were ornamental and recreational water courses and even cooling systems using underground terra-cotta conduits. A miniature water garden inside the western precinct's inner wall consisted of water pavilions, pools, cisterns, court-yards, conduits, and water courses, many designed as cooling devices with carefully considered visual and aural effects. The largest water garden had a central island linked to the "mainland" by cause-ways that formed four L-shaped pools with different water levels, designed for different functions. A narrow fountain garden was flanked by four moated islands, oriented perpendicular to the central axis of the water garden; their summer palaces were reached by bridges cut into the rock. An octagonal pond marked the point where the water garden met the asymmetrical boulder garden, set at a higher level, with its sinuous paths and natural boulders. A massive masonry wall ran from the octagonal pond to the bastion on the southeast, where wide brick walls connecting a series of boulders surrounded a rock-cut throne. Not all the gardens have been excavated and described.

When Kasyapa died, Moggallana seized Lion Rock and promptly deserted it as the capital, fixing the seat of government at Anuradhapura. Rediscovered during the British occupation of Sri Lanka, the archeological reserve and historic site of Sigiriya was declared a World Heritage Site in 1982. Conservation work continues. In 1990 an area of 12,600 acres (about 5,000 hectares) around Sigiriya was declared a wildlife sanctuary. The UNESCO-sponsored Central Cultural Fund has begun restoring Sigiriya's water gardens, and the Sigiriya Conservation Policy means that the gardens will be stripped of all introduced plant species, leaving only the ancient flora.

Further reading

Bandaranayake, Senake. 1999. *Sigiriya: City, Palace, and Royal Gardens*. Colombo: Ministry of Cultural Affairs.
Wijesinghe, Piyadasa. 1997. *Sigiriya and the God-King of Sri Lanka*. Colombo: ANCL Book Publishing Project.

Skellig Michael
Ireland

Skellig Michael (Sceilig Mhichil), or Great Skellig, is the larger of a pair of forbidding limestone pinnacles—the other is Small Skellig—jutting from the Atlantic Ocean about 7 miles (12 kilometers) off the Valentia peninsula at the southwest tip of Ireland.

The ruins of Christian monastic buildings, some dating from the seventh century, on the island of Skellig Michael, Ireland.

Skellig Michael, only 44 acres (17 hectares) in area, is dominated by two crags, one of 712 feet (218 meters) and another of 597 feet (183 meters). On top of the latter, reached via steep, winding stairways cut from the rock, there is an artificial platform with a cluster of six circular drystone huts *(clochans)*, two boat-shaped oratories, some stone crosses, and a cemetery—all that remains of a monastery established, possibly by St. Fionán, sometime in the sixth century A.D. and called "the most westerly of Christ's fortresses in the Western world." This monastic foundation on Skellig Michael was one of many in which the Christian tradition was preserved as the rest of western Europe was plunged into a dark age. As the art historian Kenneth Clark once put it, European civilization escaped by the skin of its teeth.

Moreover, the island's remarkable architecture, although small, was an achievement of inventiveness and effort seldom seen in history.

In the sixth century, just as the monastery was being established, St. Columbanus described the Irish as people "living on the edge of the world." Hermits who chose to settle on such islands as Skellig Michael subjected themselves to long periods of total isolation, while through lives of prayer and contemplation they searched in silence and solitude for God. In winter Skellig Michael was, and still is, completely inaccessible. Paradoxically, monasticism was the vehicle for the eastward spread of Celtic Christianity. It was established in Scotland and the north of England by Columba, Ninian, Wilfrid, and Aidan.

The platform was reached by any of three zigzagging stairways—one with 670 steps—from different points at the base of the island. The monks built them by carving the rock and by carrying and placing thousands of flat stones. The terracing at the top was achieved, probably over decades, by constructing massive drystone retaining walls and filling behind them. On these level places the reclusive churchmen built their huts, using a flat-stone, corbeled technique already thousands of years old. The successive courses of the circular buildings, laid without mortar and with outward-sloping joints to drain the rainwater, gradually diminished in diameter, closing the building to form a pointed dome—a "beehive" dome. The 6-foot-thick (almost 2-meter) walls and roof were thus integrated into a single entity, providing living quarters and storage. The monks grew vegetables, probably in soil imported from the mainland, supplemented by a diet of seabirds, gulls' eggs, and fish. They traded eggs, feathers, and seal meat with passing boats for grain, tools, and animal skins, from which they made vellum.

Vikings plundered the Skellig Michael settlement four times between 812 and 839 but the little community survived. Some rebuilding took place in 1860, and it is likely that many of the surviving buildings on the island are from that period. Around 1000 the chapel was added to the monastery, using stone from nearby Valentia Island. Sometime in the twelfth century, the monks withdrew to the Augustinian priory at Ballinskelligs on the mainland.

Skellig Michael became and remains a pilgrimage site. In the nineteenth century, two lighthouses were built on it, one of which remains functional. From 1986 the Irish Office of Public Works carried out restoration of the Skellig monastic site, and the buildings were added to UNESCO's World Heritage List in 1996.

Further reading

Horn, Walter, et al. 1990. *The Forgotten Hermitage of Skellig Michael*. Berkeley: University of California Press.

Lavelle, Des. 1993. *The Skellig Story: Ancient Monastic Outpost*. Dublin: O'Brien Press.

Waal, Esther de. 1999. *Every Earthly Blessing: Rediscovering the Celtic Tradition*. Harrisburg, PA: Morehouse.

Skyscrapers
Chicago

Only seldom for ideological, political, or pragmatic reasons has a society called for a new building type. Ecclesiastes asserts "There is nothing new under the sun," and most human endeavor is characterized by building upon what we already have, going "back a bit to make [ourselves] fit for going further along the way [we] seek to explore." Even the first Christian basilicas of the fourth century A.D., despite a desperate search for a new form, drew upon precedents. A complex network of constraints lay beneath the invention of the tall commercial building, the modern skyscraper that originated in Chicago. The conception of this building type was an architectural feat.

By about 1870 the United States was becoming an urban industrial nation and Chicago, more than any other city, was the focus of that change. What had a few years earlier been a frontier town was transformed into an internationally significant industrial metropolis as the cattle, grain, and lumber trades flourished and manufacturing activity grew. Between 1850 and 1870 the population increased tenfold to 330,000 and the city covered 18 square miles (46 square kilometers) on the Lake Michigan shore. Redevelopment and continual rebuilding provoked Mark Twain to comment that the city was "never the Chicago you saw when you passed through the last time."

The Reliance Building, Chicago, Illinois; Daniel Burnham and John Wellborn Root, architects; Edward C. Shankland, engineer, 1890–1894.

Most of its buildings were of timber construction. Fire was a perennial problem and there were over 600 in 1870. The unprecedented conflagration that started on 8 October 1871 within 30 hours swept through 73 miles (117 kilometers) of streets, killing about 300 people and leaving 100,000 homeless. It caused more than $192 million damage—a third of Chicago's total property value—and many fortunes were lost. The theoretical and structural innovations applied to rebuilding the city were in the vanguard of an emerging industrial architecture, but the Great Fire was only a catalyst that enabled the compounding of elements already present.

Capitalism sired the skyscraper. The architecture that we call Chicago School sprang from the frugal pragmatism of speculative clients with an eye on the "bottom line." Compelled by commercial interests, not least soaring land prices (a 600 percent increase from 1880 to 1890), architects were constrained to create a new building type, and within a couple of decades Chicago became the urban wunderkind of the Western world, where, as someone observed, "Old World assumptions were overthrown by New World realities as the past was discounted, the present glorified, and the future eagerly anticipated." European visitors recognized the downtown Loop as a nucleus built around the quest for profit and a functional esthetic, and carried their discoveries to audiences at home.

The ten-story Montauk Block (1882), designed by Daniel Burnham and John Wellborn Root for Brooks Brothers of Boston—who insisted that it must be "for use and not for ornament"—was the first building to be dubbed "skyscraper." Like the Rookery (1885–1886) and Monadnock Building (1889–1891) by the same architects, it was of conventional load-bearing construction. The unsuitability of the technique is underlined by the fact that the sixteen-story Monadnock's external brick walls were 6 feet (1.8 meters) thick at ground level, wasting valuable floor space.

The economic imperative to find a better way was matched by technological potential. What soon developed was (typically) a tall office building with a metal structural skeleton, entirely covering a relatively enclosed site and whose large windows provided adequate daylight and ventilation. Access to the upper floors was by means of an electric elevator, a wonder of the age.

William Le Baron Jenney pioneered the metal structural frame essential to the development of the skyscraper. The evolution of his ideas may be seen in the "simple, glass-enclosed cage" of the first Leiter Building (1879), followed by his nine-story Home Insurance Building (1884–1885), the first multistory iron structural frame ever erected. Earlier buildings had used frame construction internally but their external walls had been self-supporting; the Home Insurance Building carried the envelope on the frame, heralding the potential of true skeleton construction.

Iron had long been used for architectural ornament and such incidentals as hinges and door handles, but not as structure. In the Industrial Revolution, engineers, unhindered by the esthetic formalism embraced by architects, were first to turn cast and wrought iron to such applications as bridges, aqueducts, railway sheds, and conservatories. When architects *did* use iron, it was out of sight or in such frivolous buildings as John Nash's Royal Pavilion at Brighton (1818–1821). Joseph Paxton's iron-and-glass Crystal Palace (1851), although a temporary structure not regarded as "real" architecture, demonstrated the potential of prefabrication, and Henri Labrouste's National Library in Paris (1862–1868) showed how the metal column offered both structural flexibility and a new esthetic of proportion. Eiffel and Boileau's Bon Marché department store in Paris (1867) further underlined how iron and glass fit the articulation of commercial spaces. The lessons were taken up in the United States. Since 1854 James Bogardus had been experimenting with a metallurgical architecture; taking his lead, commercial buildings with cast-iron fronts and even cast-iron frames proliferated in American cities between 1850 and 1880.

Iron had disadvantages: in fire, it failed at relatively low temperatures, and it had very low tensile strength. The former was easily addressed, and well-established techniques existed: columns and beams were simply encased in a fire-resistant material. The use of steel, readily available in large quantities of predictable strength since the Siemens-Martin open-hearth process had been developed in 1875, overcame the second problem.

Burnham and Root's Rand McNally Building (1889–1890), a complete steel-framed commercial building, freed the skyscraper from its dependence on masonry walls and created "the plan and structure of the [modern] urban office block." They followed it closely with the fourteen-story Reliance Building (1890–1894). Its initial four-story stage, designed by Charles Atwood and the structural engineer E. C. Shankland, is claimed to be the first example of the comprehensive system known as Chicago construction. It consisted of a riveted steel frame with plaster fireproofing, carrying hollow-tile flooring on steel joists. The bay windows (dubbed "Chicago windows") had a central pane of fixed plate glass with opening lights at the sides and spandrel panels of terra-cotta. The concrete foundations extended 125 feet (37.5 meters) beneath the ground. That fact points to another important technological necessity that literally underpinned the skyscraper: the development of an appropriate foundation design method that dealt with the concentrated loads imposed by tall, framed buildings.

Efficient and safe systems of vertical transportation were also necessary. Steam-powered traction elevators had been used in Britain since 1835, and the German Werner von Siemens first applied an electric motor to a rack-and-pinion elevator in 1880. Motor technology and safe control methods evolved rapidly, and seven years later an elevator was built in Baltimore that moved the cage by means of a cable wound on a drum. The Otis direct-connected geared electric elevator was first used commercially in New York City in 1889. The technology of the skyscraper had been established. But what of an esthetic appropriate to the new architecture?

Although Jenney had pioneered structural innovation, he had less success with the architectural expression of the framed building. He began to address the esthetic questions in his second Leiter Building (1889–1891), Fair Store Building (1890–1891), and Manhattan Building (1889–1891) but better answers were provided by others. Some sources cite the Marshall Field Wholesale Store (1885–1887) by Henry Hobson Richardson as the stylistic model for the Chicago School. The skyscraper was a distinctly American building. The Industrial Revolution was a source of significant change in Western social structure but many nineteenth-century architects were slow to accommodate that change, wallowing in a morass of historical revival styles. Not so the French theorist E. E. Viollet-le-Duc (1814–1879), who recognized that new materials were essential as a means to a modern architectural end, and he insisted that they be used in accordance with their properties and honestly expressed in the form of the building. His widely published theories had a marked influence on the Chicago School architects just when the United States was beginning to realize that it was different from the Old World. That revelation was expressed by Walt Whitman, Ralph Waldo Emerson, and the sculptor Horatio Greenough, who in 1852 had called for a home-grown American architecture in terms that would be echoed (albeit in a French accent) in the theories of Viollet-le-Duc.

The Borden Block (1879–1880) by Dankmar Adler and Louis Sullivan was one of the first buildings to repudiate solid wall or heavy pier construction. Its narrow vertical piers allowed the maximum penetration of natural light to the interior. Ornament, while never completely rejected, gradually freed itself from historical precedent and became integrated with the making of the architecture, subordinated to frank structural expression, the needs of the building's users, and the nature of materials. The economy of form of the mature skyscraper can be seen in the Marquette Building (1894–1895) by Holabird and Roche, in which narrow piers and recessed spandrels frame large rectangular windows. It is a little ironic in an essay about Chicago architecture that Adler and Sullivan's Wainwright Building (1890–1891) in St. Louis, Missouri—their first work that exclusively used metal framing—is probably the best example of the skyscraper esthetic. Sullivan clearly expressed the external elements according to the idea set out in his tract *The Tall Building Artistically Considered*. He insisted that there should be a base (the public floors), a shaft (any number of identical upper floors), and a capital (the pronounced cornice crowning the composition). He denied that this articulation of the tall building form reflected the classical column, but the connection is inescapable. As noted, all human endeavor is characterized by building upon what we already have.

See also Curtain walls

Further reading

Condit, Carl W. 1964. *The Chicago School of Architecture*. Chicago: University of Chicago Press.
Huxtable, Ada Louise. 1984. *The Tall Building Artistically Reconsidered: The Search for a Skyscraper Style*. Berkeley: University of California Press.
Morrison, Hugh. 1998. *Louis Sullivan, Prophet of Modern Architecture*. New York: W. W. Norton.
Turak, Theodore. 1986. *William Le Baron Jenney: A Pioneer of Modern Architecture*. Ann Arbor, MI: UMI Research Press.

Snowy Mountains Scheme
Australia

The Snowy Mountains Scheme, one of the largest engineering and construction projects in the world, extends over 2,700 square miles (7,000 square kilometers) in Australia's Snowy Mountain Range. The "Snowies," as they are known, form part of the Australian Alps, a southern extension of the Great Dividing Range that stretches parallel to the east coast from northeastern Queensland to Victoria. The highest peaks reach about 7,250 feet (2,300 meters). The government-financed scheme is complex but conceptually straightforward. Aqueducts and dams collect melted snow and rainwater from the upper reaches of the Snowy River and its tributary the Eucumbene, store it in reservoirs, and then divert it westward via underground tunnels. On the way, it falls 2,625 feet (800 meters) and passes through a series of power stations that generate 3,740 megawatts of electricity—approximately 16 percent of the generating capacity of southeast Australia—for the Australian Capital Territory, New South Wales, and Victoria. The water is finally released to augment irrigation along the vast inland Murrumbidgee-Murray River system for use by New South Wales, Victoria, and South Australia. The Snowy Mountains Scheme, begun in October 1949 and completed on time and under budget in 1972, involved the construction of 16 dams, 7 hydroelectric power stations, about 90 miles (145 kilometers) of tunnels, and 50 miles (80 kilometers) of aqueducts. The capital cost was $A820 million. In 1967 the American Society of Engineers listed it among the seven engineering wonders of the modern world. Thirty years later, the American Society of Civil Engineers recognized the scheme as an International Historical Civil Engineering Landmark, ranked with the Panama Canal and the Eiffel Tower.

As early as 1884 there were proposals for using the Snowy's waters to supplement the inland rivers and relieve frequent drought conditions. In the early twentieth century, when Canberra was chosen for the site of the national capital, the river's potential for hydroelectric power production was also considered. After protracted negotiations between the federal government and New South Wales and Victoria, in 1947 a joint federal-states technical committee was established to investigate the latent economic value of the Snowy River. The Snowy Mountains Hydroelectric Power Act was passed two years later, and an agency was created to investigate, design, and construct the Snowy Mountains Scheme.

There were not enough workers in Australia to achieve the task, so the Snowy Mountains Authority recruited migrant workers from over thirty countries. They made up nearly 70 percent of the 100,000 people engaged on the scheme during the entire construction period; the size of the workforce peaked at 7,500 in 1969. Belgian, British, French, Italian, Japanese, Norwegian, Swiss, and U.S. companies were contracted for parts of the work, as engineering technology was imported from international centers of specialized engineering excellence. In all, about forty major contracting firms and many smaller companies built the Snowy Mountains Scheme.

The Guthega Dam and Power Station was the first stage of the project. Built by Norwegian contractors between 1951 and 1954, Guthega Power Station became operational in April 1955. The scheme's biggest reservoir, Lake Eucumbene (with nine times the volume of Sydney Harbour), feeds the two main sections of the scheme, the Snowy-Tumut Development and the Snowy-Murray Development, through underground tunnels. Construction began on the Tumut project in 1954 and lasted until 1973. The water it diverts to the Murrumbidgee River flows via tunnels to Tumut 1 and Tumut 2 underground power stations and to Tumut 3, the largest station in the system. The Murray project, which sends water to the Murray River, was commenced in 1961. Dams and tunnels were built at Jindabyne and Island Bend. The Murray project comprises Geehi Dam, Murray 1 and Murray 2 Dams and Power Stations, Khancoban Dam, and a pumping station at Jindabyne. To construct and operate the scheme, about 500 miles (800 kilometers) of new roads were built, and nearly 190 miles (300 kilometers) of existing public roads were widened and improved.

New engineering practices were developed during construction. Rock bolting was used to strengthen the walls of the subterranean tunnels and power stations in the Tumut project. The Snowy Scheme was among the first enterprises to use computers for engineering design. Other gains included an increased knowledge of rock mechanics, advances in concrete and cement technology, improved quality of steel manufacture and welding techniques, improved diamond-drilling techniques, and advanced corrosion-protection paint systems. Workers set and reset records for hard-rock tunneling.

The Snowy Mountains Scheme also saw sweeping changes that improved the performance of the Australian construction industry: an internal supervisor training program, recognition of the workers' role, and improvements to working conditions and industrial relations. But there were social and personal costs as well: some farms and grazing lands were acquired and flooded by new lakes and reservoirs, and the mountain townships of Adaminaby and Jindabyne were moved to new sites. Although the scheme had a relatively excellent safety record and despite major efforts to provide safe working conditions, 121 men died building it. In 1981 a memorial was dedicated to these men in the town of Cooma.

Since 1991 a systematic multimillion-dollar upgrade of the turbines, generators, and pumps has been in progress, and a new integrated control system has been installed. The Snowy Mountains Hydro Electric Authority, based at Cooma, is presently (2001) responsible for maintenance of the huge network on behalf of the Snowy Mountains Council, formed in 1958 to direct the scheme. Plans are in hand to corporatize the authority, after which the council and the scheme will be managed by a new corporation, Snowy Hydro Limited.

Further reading

Collis, Brad. 1990. *Snowy, the Making of Modern Australia*. Sydney: Hodder and Stoughton.

Unger, Margaret. 1989. *Voices from the Snowy*. Sydney: NSW University Press.

Wigmore, Lionel. 1968. *Struggle for the Snowy*. Melbourne: Oxford University Press.

Solomon's Temple
Jerusalem, Israel

No archeological remnant of Solomon's Temple survives. The Bible provides descriptions, and since it is generally believed that the architectural style was constrained by regional influences, the biblical account is augmented by knowledge of contemporary buildings in the region. It is very possible that it was the most expensive structure ever built, because the gold alone, valued at 2001 prices, was worth something in the order of U.S.$62 billion. Cost aside, the temple is an architectural achievement because during the seven-year course of its construction no "sound of hammer, axe, or any other tool [was heard] at the building site." The level of organization required to prefabricate every component of such a large and complicated stone building was a remarkable achievement.

King David united the Israelite tribes and extended the national boundaries. He ruled for seven years from Hebron, then moved the seat of government to Jerusalem—he had captured the former Jebusite town around 1000 B.C.—and reigned for another thirty-two years. The Ark of the Covenant, the nation's most sacred object, was moved there around 955. Jerusalem stood on the south side of Mount Moriah, where Abraham had been prepared to sacrifice his son Isaac, and the mountain's summit was the site David chose for a future temple to replace the portable Mishkan. Because David was a warlike man, the God of Israel forbade him to build the temple; the task was reserved for his son Solomon. Nevertheless, David made the plans, provided many of the resources, and enlisted foreign stonecutters, masons, carpenters, and all kinds of metalsmiths for the work.

After his father's death, Solomon contracted with Hiram, king of Tyre, to provide cedar and cypress timber in return for grain and oil. He also raised a labor force of 30,000 Israelites and rostered them to go to Lebanon to fell trees, directed by the Sidonians. Rafts of the timber were floated down the coast, and the logs were dressed into boards and delivered to the temple site. Four years later the building work began. Solomon also employed 80,000 men to shape huge blocks of stone in the mountain quarries and another 70,000 to haul materials to the site and perform the general laboring work. There were 3,300 foremen. None of this takes account of the number of people needed to feed such a workforce, nor the resources to produce the food, estimated by some sources to be 4,950 tons (4,500 tonnes) every month.

Solomon extended Jerusalem northward of the original city to include the summit of Mount Moriah, where he built his palace complex. The temple was probably part of that precinct. It stood in its own courtyards, to which the public was admitted only according to status, immediately north of the palace. Although the designers would have avoided the religious icons of their pagan neighbors, architecturally the building was probably a typical Phoenician temple, even in aspects of its decoration. Its furnishings were similar to those in the Mishkan, although more sumptuous, and in principle it had the same spatial organization. It stood on a raised platform, and its porch was approached by a flight of ten steps.

The temple itself was 90 feet long, 33 feet wide, and 45 feet high (about 27 by 9 by 13.5 meters); as one commentator remarks, in size more like a church than a cathedral. A 15-foot-deep (4.5-meter) porch (ulam) stretched across the eastern end. On either side of the only entrance stood two freestanding hollow cast bronze pillars. Each was 27 feet high and about 6 feet in diameter (8.25 by 1.8 meters), with walls 3 inches (7.5 centimeters) thick. Their capitals, in the shape of lilies, were decorated with chains of pomegranates. Solomon named the south pillar Jachin ("Yahweh will establish your throne forever") and the north Boaz ("In Yahweh is strength").

The body of the temple was enveloped on three sides with a three-story annex, which had a separate entrance and contained storerooms and others for the use of the many priests and attendants who served on a rostered basis. The side rooms were connected to the walls of the temple by beams resting on corbeled blocks, and the levels, each 7.5 feet (2.3 meters) high, were connected by spiral stairways. The two major rooms of the temple were paneled from floor to ceiling with carved cedar boards; the floors were boarded with cypress. Every interior surface—walls, ceilings, floors, and roofs—was sheathed in gold, embellished with figures of angels, palm trees, and open flowers. The 60-foot-long (18-meter) outer room, known as the Holy Place (hekhal), was entered directly from the porch; it was accessible only to the priests. The inner room was a 30-foot (9-meter) cube at the west end of the temple, separated from the Holy Place by an embroidered curtain and carved doors of olive wood, overlaid with gold. This Most Holy Place *(debir)* could be entered only by the high priest once a year on the Day of Atonement (Yom Kippur). Within, the Ark of the Covenant was placed and overshadowed by two 15-foot-high (4.5-meter) olive-wood statues of guardian angels, also covered with gold, with outspread wings meeting above the ark.

The Temple of Solomon was dedicated in 953 B.C., accompanied by the offering of 22,000 oxen and 120,000 sheep; needless to say, a great public feast followed. Between 604 and 597 B.C. the Babylonian Nebuchadnezzar II took the Jews into exile and stripped the great building of its fabled treasures. He had it destroyed in 586 B.C. About fifty years later Cyrus of Persia, who had conquered Babylon, allowed the Jews to return to their homeland, and the Second Temple was completed by 515 B.C. For five centuries, although enjoying occasional and short-lived freedom, Jerusalem was successively occupied by the Macedonians, Egyptians, and Seleucids. Roman rule began in 64 B.C., and early in that period the Jewish puppet king Herod the Great (reigned 37–34 B.C.) rebuilt and enlarged the Second Temple. It was totally destroyed by the armies of Titus in A.D. 70, never to rise again. Today its great rectangular platform is occupied by the Islamic Dome of the Rock.

See also Mishkan Ohel Haeduth (the Tent of the Witness)

Further reading

Backhouse, Robert. 1996. *The Kregel Pictorial Guide to the Temple.* Grand Rapids, MI: Kregel Publications.

Bible, The. See especially 1 Kings., chapters 1–7; 1 Chron., chapter 22.

Statue of Liberty
New York City

Originally titled Liberty Enlightening the World, the colossal statue on Liberty Island in New York Harbor stands nearly 307 feet (93.5 meters) high. It represents a woman of pre-Raphaelite appearance, draped in voluminous robes and crowned with a spiked diadem. Her right hand raises a flaming torch at arm's length; her left carries a book emblazoned with "July 4, 1776"; broken chains are strewn around her feet. The towering figure alone is over 152 feet (46 meters) high, her right arm is 42 feet (12.8 meters) long, and her head measures 28 feet (8.5 meters) from neck to diadem. She weighs about 250 tons (227 tonnes). Conceived for other reasons, the Statue of Liberty has become since 1903 first a symbol of U.S. immigration and then a universal icon of emancipation, as well as the most recognizable emblem of New York City, perhaps even of the United States.

Primarily, the monument may have been intended to express *French* republican ideals when France was enduring Napoléon III's repressive rule. The idea was first discussed in 1865 in the Paris home of the scholar Edouard de Laboulaye, who admired the wealthy industrializing nation that had just emerged from the crucible of a civil war. De Laboulaye believed that a monument expressing the idea of liberty would sustain the republican principle in France while strengthening ties with the United States, and he shared his thoughts with a ready listener, the sculptor Frédéric Auguste Bartholdi. But it was not until 1871—in which Napoléon III was deposed—that Bartholdi crossed the Atlantic to offer the statue as a gift from the French people to the American people. Although it would celebrate, on the centenary of the conflict, friendship established during the War of Independence, the gesture coincided with the perceived need to reinforce republicanism in France, where monarchist sentiment survived.

Bartholdi chose New York Harbor, a major gateway to the United States, as a site with the necessary symbolic value. He is said to have been an "academic sculptor driven by two obsessions: liberty and immensity." Impressed with the Ramessean colossi of Egypt and accounts of the Colossus of Rhodes, he set out to design a work of overpowering scale to stand at the harbor entrance. It was agreed that the American people would finance construction of a pedestal and France would pay for the statue and its construction in the United States. The plan was announced late in 1874.

Bartholdi produced a design by 1875. His 4-foot (1.25-meter) clay model was scaled up to produce

300 full-sized plaster sections, from which wooden forms were crafted. The 452 pieces of the statue's outer skin were made by hammer dressing 0.1-inch-thick (2.5-millimeter) sheets of Norwegian copper against the forms. The structural support for the flimsy envelope was designed by the engineer Gustave Eiffel, already reputed for his metal structures, in consultation with the architect E. E. Viollet-le-Duc. A central wrought-iron pylon carried a secondary framework of flexible iron bars to which the copper skin was riveted. This dual construction—a light-weight skin on a substantial skeleton—would not only withstand high wind loads but also safely respond to temperature changes. Lack of money slowed progress, but various fund-raising programs meant that Liberty Enlightening the World was completed and assembled in Paris in June 1884. She stood on public display for six months.

The granite pedestal, designed in 1877 by the French-trained American architect Richard Morris Hunt, was constructed in the courtyard of Fort Wood on Bedloe's Island under the direction of the engineer Charles P. Stone. It stood just under 150 feet (45 meters) high on a mass concrete foundation, 90 feet (27 meters) square and 53 feet (16 meters) deep. In the United States, despite art shows, theatrical galas, auctions, and even prizefights, finance was not forthcoming. The statue waited in Paris while the American Committee looked for $100,000 to complete the pedestal. In January 1885 the statue was dismantled into over 300 pieces, packed in 214 crates, and shipped on the frigate *Isere*, arriving in New York Harbor in June. The funds for the pedestal were finally in hand by August, and the work was finished eight months later. It then took four months to reassemble Liberty Enlightening the World, and the dedication took place on 28 October 1886.

Because her torch was a navigational aid, the statue was first managed by the Lighthouse Board. In 1901 administration was transferred to the War Department, and in 1924 the great figure was declared a national monument. In 1956 Bedloe's Island was renamed Liberty Island, and in 1965 neighboring Ellis Island, site of a large immigration station, became part of the Statue of Liberty National Monument. Between 1983 and 1986 a $140 million rehabilita-

tion project saw French and American craftsmen repairing failed rivets and replacing the rusted iron core with stainless steel. They strengthened the right arm and replaced the old flame, which had been lit from inside, with a gold-plated copper flame lit by reflection, as Bartholdi had originally intended.

As part of the 1880s fund-raising effort the poet Emma Lazarus wrote a sonnet titled "The New Colossus." In 1903 it was inscribed at the main entrance to the pedestal, and its words embody the meaning that the Statue of Liberty has held for immigrants for over a century:

> Here at our sea-washed, sunset gates shall stand
> A mighty woman with a torch, whose flame
> Is the imprisoned lightning, and her name
> Mother of Exiles. From her beacon-hand
> Glows world-wide welcome; her mild eyes
>   command
> The air-bridged harbor that twin cities frame.
> "Keep ancient lands, your storied pomp!" cries
>   she
> With silent lips. Give me your tired, your poor,
> Your huddled masses yearning to breathe free,
> The wretched refuse to your teeming shore.
> Send these, the homeless, tempest-tost to me,
> I lift my lamp beside the golden door.

Further reading

Dillon, Wilton S., and Neil G. Kotler, eds. 1994. *The Statue of Liberty Revisited: Making a Universal Symbol*. Washington, DC: Smithsonian Institution Press.

Price, Willadene. 1959. *Bartholdi and the Statue of Liberty*. Chicago: Rand McNally.

Trachtenberg, Marvin. 1986. *The Statue of Liberty*. New York: Penguin.

Stockton and Darlington Railway
England

The Stockton and Darlington line, the world's first public railroad, was opened on 27 September 1825. As well as carrying coal, the train drawn by "Locomotion No. 1" had about 550 passengers, most of them in coal wagons but some in a carriage named *Experiment*. The steam railroad was to change the course of history. There is little in the modern world that it did not affect, including our perception of time and distance, the pattern of settlement, the relationship between labor and industry, and not least the urban and rural landscape.

George Stephenson (1781–1848), born in Wylam in northeast England, invented the steam locomotive. At the age of fourteen he went to work at Dewley Colliery, becoming an engine man in 1802. Within a decade, his thorough knowledge of engines won him the post of engine wright at Killingworth Colliery, whose manager allowed him to experiment with steam-powered machines. In 1814 Stephenson built his first locomotive: *The Blutcher*, which could haul 33 tons (30 tonnes) uphill at 4 mph (6.4 kph), differed from contemporary engines in that the gears drove the flanged wheels; he later modified it so that the connecting rods directly drove the wheels. Over the next five years Stephenson built sixteen engines at Killingworth, mostly for local use. In 1819 the colliery owners asked him to construct an 8-mile (13-kilometer) railroad between Hetton and the coastal town of Sunderland. Employing locomotives, fixed engines, and cables and inclined planes, it was the first railroad that did not use animal power.

The Stockton and Darlington Railway dates from October 1767, when a few entrepreneurs sought to improve links between the West Durham coal mines and the seaport of Stockton-on-Tees. First (and fashionable) thoughts were for a canal system, and the navigator James Brindley was asked to suggest possible routes. The estimated cost was prohibitive and the scheme lapsed for forty years. It was revived in 1810 by the Tees Navigation Company, whose "New Cut" of the River Tees greatly reduced the length of waterways between coal mines and coast. By then, horse-drawn railroads were becoming popular, and with interest shown by Darlington businessmen, both alternatives were investigated. Again, a consultant recommended a canal; again, the project was abandoned because of the cost. Besides, parochial disputes over the route could not be resolved.

In November 1818 a committee at Darlington, convened by Edward Pease, a retired wool merchant, resolved to apply for an act of Parliament to construct a rail- or tramway to join Stockton and the collieries in West Durham. Although the estimated cost exceeded

that of the canal system, when the committee issued its prospectus for building a horse-drawn railroad, the public response was overwhelming. On 19 April 1821 Parliament authorized the project, and the Stockton and Darlington Railway Company was formed. On the same day, Pease met with Stephenson, who had already walked along the proposed route and who suggested that the company should consider steam locomotion. He invited Pease to visit Killingworth Colliery, where the businessman saw *The Blutcher* in action. In January 1822 Stephenson was appointed chief engineer of the company. Another act of Parliament authorized it "to make and erect locomotive or moveable engines to facilitate the carriage of goods, merchandise and passengers."

Work began on the track in May 1822. Stephenson used 15-foot (4.57-meter) lengths of malleable wrought-iron rails, developed in the previous year by the Bedlington engineer John Birkinshaw. They were carried on cast-iron chairs and laid on wooden blocks for part of the railroad and stone blocks for the remainder. The gauge was set at 4 feet 8 inches, and later eased 0.5 inch to allow freer running of the rolling stock. That became the standard gauge of British and other railroads. On 27 September 1825, the 26.75-mile (43-kilometer) Stockton and Darlington line was ready; most of the railroad had easy gradients, but there were a few very steep places, with inclines of up to one in thirty, which would present severe challenges to the locomotive.

In 1823 Pease, Thomas Richardson (another committee member), Stephenson, and his son Robert had formed Robert Stephenson and Company, the world's first locomotive building firm. In September the Stockton and Darlington Railway Company placed an order for two high-precision steam engines at £500 each. They recruited Timothy Hackworth, foreman smith of the Wylam and Walbottle Collieries, to supervise the Newcastle-upon-Tyne workshop. *Locomotion No. 1*, similar to those Stephenson had built at the collieries, was finished just two years later. She had two vertical 9.5-inch-diameter (240-millimeter) cylinders with a 24-inch (610-millimeter) stroke; four wheels coupled by two side rods had a wheelbase of a little over 5 feet (about 1.5 meters). She weighed 11 tons (10 tonnes).

On 26 September 1825 *Locomotion No. 1* stood at Shildon Lane End, coupled to twelve wagons of coal, another filled with sacks of flour, and the single eighteen-seat passenger carriage, *Experiment*. Tickets had been issued to about 300 people, but somehow 450—according to one account, 553—found room, many by sitting atop the coal. The long train, driven by George Stephenson, preceded by a horseman carrying a flag and with a cortege of twenty-four horse-drawn wagon loads of workmen and others in its wake, moved off to the applause of a great crowd of onlookers. Over the low gradients, for the first few miles the engine pulled its 110-ton (100-tonne) load at speeds reaching 12 mph (19 kph). After a couple of stops for mechanical problems, the train reached Darlington; the first 8.5 miles (13.6 kilometers) were covered in 65 minutes; at one point it achieved a speed of 15 mph (24 kph) "with perfect safety." At Darlington, six coal wagons were uncoupled and their contents distributed to the poor. With two more wagons carrying a brass band in tow, *Locomotion* resumed her journey. Three hours, 7 minutes later she reached the Stockton terminus. A salute was fired from cannon on the company's wharf, and the band rendered "God Save the King." That evening, a celebration dinner was held at the Stockton Town Hall. The feasibility of a steam-locomotive railroad had been made evident to almost 50,000 people, and future success was guaranteed.

Further reading

Emett, Charlie. 2000. *The Stockton and Darlington Railway: 175 Years*. Stroud, UK: Sutton.

Hoole, Ken, et al. 1975. *Rail 150: The Stockton and Darlington Railway and What Followed*. London: Eyre Methuen.

Kirby, M. W. 1993. *The Origins of Railway Enterprise: The Stockton and Darlington Railway, 1821–1863*. New York: Cambridge University Press.

Storm Surge Barrier
Rotterdam, the Netherlands

More than half the Netherlands lies below sea level, and the little country is protected from flooding by about 750 miles (1,200 kilometers) of dikes. The process of global warming and the consequent rise in sea levels will challenge their adequacy, and many of them will need to be raised and reinforced. The extensive Deltaworks project, completed in 1986, secured the province of Zeeland by sealing off its sea inlets. Its northern neighbor, South Holland, remained under threat. Responses to disastrous floods in 1953 had included plans to raise the dikes in the region, but by the 1970s there was public resistance to a scheme that entailed demolishing many historic precincts. The alternative was the construction of a movable storm surge barrier in the man-made approach to Rotterdam Europoort. It is the busiest harbor in the world, and an average of ten ships pass through the New Waterway every hour. Technological and economic feasibility studies led to the construction of the Storm Surge Barrier, one of the engineering marvels of the late twentieth century. Otherwise known as the Maeslant Kering, it is located between the Hook of Holland and the town of Maassluis, a little under 4 miles (6 kilometers) from the North Sea. Built at a cost of 1 billion guilders (U.S.$500 million), it was opened on 10 May 1997.

In response to the Dutch government's call for submissions, the Bouwkombinatie Maeslant Kering consortium's tender was accepted from among six competitors. Contracts were signed in October 1989, and the first pile for the hinge foundation was driven in November 1991. The barrier has a guaranteed life of 100 years. It consists of a pair of 50-foot-thick (15-meter) hollow, arc-shaped steel gates, each 73 feet (22 meters) high and 700 feet (210 meters) long and weighing 16,500 tons (15,000 tonnes). Each is attached by means of 795-foot-long (238-meter) latticed steel arms to a steel ball joint seated in a massive concrete socket on the riverbank. The 33-foot-diameter (10-meter) ball joints each weigh 760 tons (690 tonnes) and work with a tolerance of 0.04 inch (1 millimeter). The figures are almost meaningless, but in terms of comparative size, each half of the barrier—the gate, the two three-dimensional trusses, and one ball joint—weighs as much as two Eiffel Towers.

Normally, the gates are "parked" in docks in the banks. If a water surge of 10 feet (3.2 meters) above a set acceptable maximum is anticipated, a central computer instructs the automated control system to activate the barrier, and water is pumped into the parking docks. When the hollow gates start to float and the water level in the dock reaches that in the New Waterway, the dock gates are opened. The "locomobiles"—their name explains the function—on top of the gates push them horizontally out of the dock into the Waterway. The gates meet in the middle, not quite touching. They are then flooded and slowly sink to the concrete sill on the bottom of the Waterway, 56 feet (17 meters) down. The ball joints move in different directions following the gates' movements: horizontally (when the gates are floated out) and vertically (upon submersion). The gates must be able to ride with the waves when being closed and opened. In a fierce storm, the water could hit the barrier with up to 33,000 tons (30,000 tonnes) force. The loads on the structure are transmitted to the 57,000-ton (52,000-tonne) triangular concrete foundations of the ball joints. After the storm the gates are floated again and driven back into the dock by the locomobiles. The dock gates are closed, and the dock is pumped dry. From the time the computer registers the need to close the barrier until the gates are in place, the operation takes nine and a half hours. After the storm, it takes two and a half hours to return the gates to their docks.

It is expected that the barrier will have to be closed (on average) once every ten years, but changes in sea levels over the next half-century may double that. Seawater can enter the Europoort area freely through other waterways, and a supplementary dike-reinforcement program is being implemented, with a further defense known as the Europoort Barrier acting to support that in the New Waterway.

See also Afsluitdijk; Deltaworks

Further reading

BMK (Bouwkombinatie Maeslant Kering). 1997. *Sluitstuk van de Deltawerken; Stormvloedkering Nieuwe Waterweg*. The Hague: Ministerie van Verkeer en Waterstaat.

Kerssens, P. J. M., et al. 1989. "Storm Surge Barrier in the Rotterdam Waterway: A New Approach." *Delft Hydraulics*: 335–342.

Suez Canal
Egypt

The Suez Canal, an artificial waterway across the Isthmus of Suez in northeastern Egypt, connects Port Said on the Mediterranean coast with Port Tawfiq on the Gulf of Suez, an inlet of the Red Sea. The 101-mile (163-kilometer) canal has no locks, making it the longest of its kind, sea level being the same at both ends. Because it exploits three natural bodies of water—Lake Manzala in the north; Lake Timsah, almost exactly at the midpoint; and a chain known as the Great Bitter Lake in the south, accounting for about 18 percent of its length—it does not follow the shortest possible route. For most of the canal, traffic is limited to a single lane, but there are passing bays, as well as two-lane bypasses in the Great Bitter Lake. A railway on the west bank runs parallel to the canal from end to end. It took a force of an estimated 1.5 million Egyptian laborers, often working under appalling conditions, eight years to dig the Suez Canal; more than 125,000 lost their lives. In every way, the project is comparable with the architectural and engineering feats of pharaonic Egypt.

In fact, the idea of a navigable link between the Mediterranean and Red Seas dates from dynastic Egypt. Earlier canals joined the Red Sea to the Nile, with obvious economic advantages for the land of the Nile. The first, said to have been commissioned by Ramses I around 2000 B.C., linked the Red Sea and the Nile, and a second component was formed by a branch of the Pelusian River that extended to the Mediterranean. Other sources claim that the first canal was constructed in the reign of Tuthmosis III (1512–1448 B.C.), and still others that Necho II (reigned 610–595 B.C.) initiated it, but lack of maintenance meant that it later became unnavigable. Whatever the case, the Persian king Darius I (558–486 B.C.) ordered the work to be completed. His canal, linking the Gulf of Suez to the Great Bitter Lake and the lake to the Nile Delta, remained in good repair through the Macedonian era. It was redug in the time of the Roman emperor Trajan (A.D. 53–117) and again by the Arab ruler Amr Ibn-Al-Aas. When a trade route around Africa was discovered, it again fell into disuse until about 1800. About then, Napoléon Bonaparte's engineers proposed a shorter route to India by digging a north-south canal through the Isthmus of Suez. But they wrongly believed that there was a difference in sea level of about 32 feet (10 meters), an error that undermined the feasibility of the project. The Egyptian khedive Mohammed-Ali (reigned 1811–1848) showed little interest in the scheme, and it lapsed for almost half a century.

On 15 November 1854 the French diplomat and engineer Vicomte Ferdinand Marie de Lesseps, who had long championed a canal across the isthmus, approached Egypt's new ruler, his old friend khedive Said, with a plan privately devised by a French engineer in Mohammed-Ali's service. A. Linant de Bellefonds proposed a canal between Suez and Peluse, crossing the Great Bitter Lake and Lake Timsah. By the end of the month de Lesseps was granted a decree allowing him to dig the canal and manage it for ninety-nine years. A second agreement, signed in January 1856, ensured that the Suez Canal would be open to shipping of all nationalities and accessible for a transit fee. By December 1858 the Frenchman established the Compagnie Universelle du Canal Maritime de Suez, and shares were quickly bought by investors from all over Europe and the Ottoman Empire. Construction work started on the Suez Canal in April 1859.

Even during its construction, the canal was at the center of a political storm because of its critical military and economic importance. Before work started, the British—and particularly the prime minister, Lord Palmerston—were afraid that the French project would threaten their interests in India, and they tried to have the khedive's decree set aside. When that failed, they used political pressure in an attempt to have the digging stopped, only six months after it had started. Such interference continued into the 1860s, after Said was succeeded by the khedive Ismael, who was persuaded to sell his shares in the Compagnie Universelle to Britain, making it the largest single shareholder.

The Suez Canal was completed in August 1869. In the course of its construction, three new towns had been built—Suez, Ismailia, and Port Said—and millions of hectares of farmland had been created. Built at immense human and economic cost (about $330 million in modern values), it was officially opened at Port Said by the French empress Eugénie on November 1869. The great waterway originally had a width of 72 feet (22 meters) at the bottom and 190 feet (58 meters) on the surface. The channel was 26 feet (8 meters) deep. It has been enlarged and deepened many times.

A couple of incidents have highlighted, at great economic cost, the Suez Canal's critical strategic importance. In July 1956 Egypt's President Nasser, in response to the British, French, and U.S. refusal of loans for the Aswan High Dam, nationalized the canal. That provoked the so-called Suez Crisis, beginning with the British and French invasion of Egypt in October. The Egyptians scuttled forty ships that were then in the canal. By the following March the United Nations had prevailed upon Egypt to clear and reopen the waterway. Ten years later, following the Six-Day War, when Israel occupied the Sinai Peninsula, the canal was again closed to shipping. The Egyptians reclaimed it after the 1973 Arab-Israeli War, and, cleared of mines and obstructions, it was reopened in 1975.

In 2001, about 6 percent of the world's seaborne trade passes through it. Of course, it dramatically reduces the east-west voyage distance for vessels; for example, the route between Tokyo and Europoort in the Netherlands is only three-quarters of the distance of that around the Cape of Good Hope. The canal is a major source of income for Egypt, and the Suez Canal Authority continually makes improvements to it. The world's largest bulk carriers—vessels that are 1,600 feet long and 230 feet wide (500 by 70 meters), with drafts up to 70 feet (21.4 meters)—can now navigate the Suez Canal. The duration of the passage is normally about twelve hours. Its present annual traffic capacity is over 25,000 vessels. Tanker traffic has declined, mostly because of competition from the 200-mile (320-kilometer) Sumed oil pipeline between the Gulf and the Mediterranean.

See also Panama Canal

Further reading

Farnie, D. A. 1969. *East and West of Suez: The Suez Canal in History, 1854–1956.* Oxford, UK: Clarendon Press.

Kinross, Patrick Balfour. 1969. *Between Two Seas: The Creation of the Suez Canal.* New York: Morrow.

Rogers, W. G. 1978. *The Real Suez Crisis: The End of a Great Nineteenth Century Work.* New York: Harcourt Brace Jovanovich.

Sultan Ahmet Mosque
Istanbul, Turkey

The deeply religious Ottoman sultan of Istanbul Ahmet I (reigned A.D. 1603–1617) was enthroned at the age of fourteen. Six years later he commissioned his architect Sedefkar Mehmet Agha to build a mosque that would compete for size and splendor with the sixth-century Byzantine church of Hagia Sofia. A site was chosen facing the church across what is now Sultanahmet Square, and Ayse Sultan, whose palace stood on it, was duly compensated for its demolition. Construction started in 1609 on the Sultan Ahmet Mosque, also known as the Blue Mosque, probably the greatest achievement of Ottoman architecture.

Its architect had been a pupil of Sinan, considered by many to be the best architect of the early Ottoman Empire. Mehmet Agha worked in the tradition of his former master, and one of the precedents for his design was Sinan's Suleymaniye Mosque (1550–1557) on the west bank of the Golden Horn. The other was Hagia Sofia itself, on which the Suleymaniye Mosque was based anyway. All, Islamic or Christian, grew around the same major element: an almost square, vast central space crowned with a dome. The Sultan Ahmet Mosque occupies an area of 209 by 235 feet (64 by 72 meters). Its central dome, 77 feet (23.5 meters) in diameter and reaching a height of 140 feet (43 meters), is carried on pendentives above four pointed arches, themselves supported on round, fluted piers. The central structure is stiffened by a hemidome on each of its four sides and by cupola-covered piers at the corners; then, in the manner of much Byzantine and Ottoman architecture, the loads and thrusts are transmitted to the ground by a cascade of flanking ancillary structures.

Sultan Ahmet (Blue) Mosque, Istanbul, Turkey; Sedefkar Mehmet Agha, architect, 1609–1616. Aerial view.

In front of the mosque stands a wide courtyard, enclosed by an openwork wall and entered on three sides through any of eight monumental gateways with bronze doors. The marble-paved inner court, with a central domed fountain for ritual ablutions, is surrounded by an arcade of slender columns of pink granite, marble, and porphyry, each bay-roofed with a cupola. Four marble minarets with pointed spires rise from the corners of the mosque; and two others, not as tall, at the outer corners of the court make the building, with six minarets, unique in Istanbul. They have a total of sixteen balconies, from which the muezzin calls the faithful to prayer, honoring the sixteen Ottoman sultans.

Around the mosque was the extensive *kulliye*, a collection of buildings and functions including the Imperial Lodge *(hunkar)* on its north side, a hospital, a caravansary, a primary school, public kitchen and service kiosk, a bazaar for the trades guilds, two-storied shops, and a college *(medrese)*.

The architectural excellence of the Blue Mosque lies not in its structural ingenuity, because it was in fact highly derivative, nor in its challenge to the grandeur of Hagia Sofia, because it was much smaller than the ancient church. Rather, Sultan Ahmet's building is remarkable for the splendor of its extraordinary decoration, especially the beautiful blue tiles that give the mosque its alternative name. Daylight is admitted through no fewer than 260 carefully placed windows, once glazed with stained glass, and when conditions are right the interior of the mosque is endowed with an ethereal blue haze. These tiles—there are more than 21,000 of them—were produced in nearby Iznik just when the industry was enjoying its highest level of achievement. There is an unsubstantiated tradition that the production of so many hand-decorated tiles completely exhausted the ceramicists, and the Iznik workshops began to decline. The tiles are painted with traditional floral and plant motifs, including roses, carnations, tulips, lilies, and cypresses, all in soft shades of green and blue on a white ground, and they cover the interior walls and piers to about a third of their height. The stunning effect of tiles and light is enhanced by other decorative details, including painted floral and geometrical arabesques on the domes and upper parts of the walls, although these are now for the most part modern replicas of traditional seventeenth-century designs. The graceful calligraphy everywhere is the work of Ameti Kasim Gubari. The wooden doors and window shutters, designed by Mehmet Agha, are inlaid with shell, mother-of-pearl, and ivory, and the pulpit *(mimbar)* as well as the niche indicating the direction of Mecca (mihrab) are both made of white Proconnesian marble, fine examples of Ottoman stone carving.

Sultan Ahmet I died of typhus only a year after his mosque was finished, and his nearby tomb and that of his wife Kosem Sultan was completed by his son Osman II.

See also Dome of the Rock (Qubbat As-Sakhrah); Masjed-e-Shah (Royal Mosque)

Further reading

Goodwin, Godfrey. 1971. *A History of Ottoman Architecture*. Baltimore: Johns Hopkins University Press.

Stierlin, Henri, and Anne Stierlin. 1998. *Turkey: From the Selçuks to the Ottoman*. New York: Taschen.

Vogt-Goknil, Ulya, and Eduard Widmer. 1966. *Living Architecture: Ottoman*. London: Oldbourne.

Sydney Harbour Bridge
Australia

The Sydney Harbour Bridge, irreverently known as "the coat hanger" to Sydneysiders, is the largest, although not the longest, one-bow bridge in the world. It crosses from Dawe's Point on the downtown side to Milson's Point on the North Shore, and its realization was a remarkable economic accomplishment in the years of the Great Depression—using labor-intensive technology, the project employed 1,400 men—as well as being one of the twentieth century's major engineering feats.

Before 1932, the only connections between the city center and the residential suburbs on the North Shore were ferries or a circuitous 12-mile (20-kilometer) road route that crossed five bridges over narrow inlets of the extensive harbor. The notion of a single bridge surfaced from time to time during the nineteenth century, but serious thought was not given to the project until the 1890s. In January 1900, on the eve of the Australian Federation, the New South Wales government invited tenders for the design of a North Shore bridge across the harbor. Twenty-four proposals were received but all were rejected; only one had involved a single-arch bridge.

In 1912 the civil engineer Dr. John Job Carew Bradfield was appointed chief engineer of Sydney Harbour Bridge and Metropolitan Railway Construction, and he developed the basic design with Department of Public Works engineers. Bradfield recommended an arch bridge with granite-faced pylons. In 1922 the government held an international design competition, for which six companies entered twenty schemes. The winning tender was one of six submitted by Dorman Long and Co. of Middlesborough, England. It proposed a single arch that would be built from both shores and supported by cables until it joined at midspan. The complex structural calculations for fabrication and erection were made by consulting engineers Ralph Freeman of the London firm Sir Douglas Fox and Partners and G. C. Imbault. The British firm of Sir John Burnet and Partners was named consulting architect.

The acquisition of land for the approaches and the construction yards called for the demolition of 800 houses, and their occupants were displaced without compensation. Construction began in 1923. The bridge (including the railroad line) took eight years to build. The contractors, with Lawrence Ennis as site manager, established workshops at Milson's Point where the steel components were fabricated; almost 80 percent of the steel was imported from England. The 39-foot (12-meter) reinforced high-grade concrete foundations were set in the rock upon which the whole of Sydney stands, and 118-foot-long (36-meter), U-shaped tunnels were excavated to anchor the 128 steel cable restraints that would temporarily support each side of the bridge. Once the approach spans were built, in November 1929 work started on the 1,650-foot (503-meter) single-span hinged arch. It supports the bridge deck, the hinges at either end transferring the massive load to the foundations at either end. The two halves of the arch were erected to meet in the middle. The components were transported from the workshops on barges, positioned, hoisted by two 640-ton (580-tonne) electric creeper cranes, and assembled as the cranes traveled toward the center of the bridge. For about ten months the halves gradually reached out for each other across the harbor, finally meeting in August 1930. The 160-foot-wide (49-meter) steel deck, carried 194 feet (59 meters) above sea level, was then commenced, working outward in both directions from the center of the bridge to save time moving the cranes. The decking was fixed within nine months. The last of the 6 million rivets was in place late in January 1932. In February ninety-six railroad locomotives were used to test the bridge. The four 278-foot-high (85-meter), Moruya granite–faced concrete pylons, each standing on the gigantic pads that support the pins of the arch, have no structural role; they are there for esthetic reasons, to define the ends of the main span.

The official opening on 19 March 1932 attracted large crowds. The state premier John T. Lang was about to declare the bridge open when ex-Army Captain Francis De Groot spurred his horse forward and slashed the ribbon with his saber. The total cost of the bridge, in today's terms, was about $A12 billion—some sources put it considerably higher. Recovered by tolls on automobiles, it was amortized by 1988. The Sydney Harbour Bridge is an essential rail and road artery in and out of the downtown area. When it was first opened, average daily traffic (in both directions) was about 10,900 vehicles. Toward the end of the century, that figure had approached 152,000. The bridge carries eight vehicle lanes, two train lines, a footway, and a cycleway. The growing pressure of traffic volumes was somewhat relieved after August 1992, with the opening of the four-lane, 1.44-mile (2.3-kilometer) Sydney Harbour Tunnel, close to the bridge. Built at a cost of $A738 million, it offered a ten-minute reduction of crossing time at peak hour. Providing an eastern bypass of the city, it is reputed to save 3 million gallons (13 million liters) of fuel a year.

Further reading

Mallard, Henri. 1976. *Building the Sydney Harbour Bridge*. Melbourne: Sun Books.

Nicholson, John. 2000. *Building the Sydney Harbour Bridge*. St. Leonards: Allen and Unwin.

The Story of the Sydney Harbour Bridge. 1967. Sydney: N.S.W. Department of Main Roads.

Sydney Opera House
Sydney, Australia

The Opera House stands on Bennelong Point, which reaches out into Sydney Harbour, close to the famous bridge. It was designed by the Danish architect Jørn Utzon and engineered by the English firm of Ove Arup and Partners. Fourteen years in the building (1959–1973), the Opera House is one of the internationally recognized icons of Australia; it is also internationally acknowledged by architects, critics, and the wider public as one of the greatest pieces of architecture of the twentieth century, perhaps of any century.

At the end of World War II Sydney, Australia's largest city, had no satisfactory venue for musical performances apart from its city hall. Orchestral concerts were held there, but opera was out of the question; neither were other theaters suitable. From about 1950 a number of influential people, led by Eugene Goosens, chief conductor of the Sydney Symphony Orchestra and director of the New South Wales (NSW) Conservatorium of Music, lobbied the state's Labor government to build a performing-arts center. Their efforts were rewarded late in 1954 when semiderelict industrial land at Bennelong Point was selected as the site for the project.

The following year an international architectural design competition for a performing-arts center was announced—the name "Opera House" was later popularly and inaccurately attached to the building. The competition called for a complex with a 3,500-seat auditorium for opera, ballet, and orchestral concerts and a 1,200-seat theater for drama and smaller musical recitals. There was no budgetary constraint in the design brief. More than 700 architects applied and over 230 entries were received from thirty-two countries. In January 1957 Utzon's success was announced in controversial circumstances. The judging panel comprised two Australians: Professor H. Ingham Ashworth of the University of Sydney and the NSW government architect Cobden Parkes; and from overseas, Professor Leslie Martin of Cambridge University, England, and the famous American architect Eero Saarinen. There is a persistent legend that, because it ignored competition guidelines, Utzon's entry was discarded by the others before Saarinen's late arrival in Sydney. He then reviewed the discarded designs and persuaded his fellow judges that Utzon must win. Other sources claim that Ashworth and Martin had already warmed to the Dane's proposal. Whatever the case, the judges' report observed with prophetic accuracy that Utzon's drawings conveyed "a concept of an Opera House which is capable of becoming one of the great buildings of the world." He was awarded first prize of $A10,000 and commissioned as architect.

Soon the NSW government established a public appeal for the building. When it became clear that the subscription fund would not even approach the estimated cost of $A7 million, the government introduced the Opera House Lotteries. By 1975 they would raise 99 percent of the final cost of $A102 million.

The Opera House stands on a concrete podium projecting into Sydney Harbour, supported on 580 concrete piers founded 82 feet (25 meters) below water level. The podium houses the service areas, dressing and rehearsal rooms, minor theaters, and the box office. The harborside walkways, the raised platform around the building, as well as exterior and interior walls, stairs, and floors are faced with pink-brown reconstituted granite from Tarana, NSW. Rising from the platform are the three sets of roof shells, faced with custom-made white ceramic tiles from Sweden. The smallest shell covers a restaurant; the other two, set slightly off parallel, contain the two major performance spaces—the opera theater and the concert hall—and their foyers. The roof shells are assembled from nearly 2,200 precast concrete elements, each weighing up to 16.5 tons (15 tonnes), held together by steel cable. The two main auditoria, which are actually separate structures under their roofs, are framed with steel and faced inside and outside with brush box and white birch plywood. The open ends of the shells are glazed with laminated, tinted plate glass in bronze frames. The five theaters—there is also a drama theater, playhouse, and "The Studio"—and almost 1,000 rooms cater to the performing arts, cinema, exhibitions, and conventions. There are two other large auditoria, a reception hall, four restaurants and six bars, as well as a library, dressing rooms, rehearsal studios, and offices. About 3,000 events are presented each year, attracting audiences of 2 million.

Sydney Opera House, Sydney, Australia; Jørn Utzon, architect; Ove Arup and Partners, engineers, 1959–1973. Night view from the harbor; the Sydney Harbour Bridge is silhouetted in the background.

Work started in March 1959 with a workforce from Australia, Italy, Spain, Germany, Holland, and Chile. By 1963 the podium was completed and the roof vaults had been started. The steep sail shapes originally proposed by Utzon were, according to the engineers, unbuildable. So by October 1961 he revised them as parts of a sphere, enabling time and cost savings. Cost blowouts began to cause the client concern, and in 1966 a newly elected conservative government in NSW made Utzon the undeserving scapegoat in what was really a political intrigue. It withheld part of his fees, forcing his unwilling resignation. That issue divided the Australian architectural profession. Some architects held protest marches in Sydney; others abandoned their calling, and the Victorian Chapter of the Royal Australian Institute of Architects blackballed Utzon's replacement. Not so the NSW chapter, and in April 1966 a team of local architects, Peter Hall, Lionel Todd, and David Littlemore, was appointed to complete the Opera House under the direction of the government architect. Inside the building, they obscured Utzon's monumental vision by their thoroughly pedestrian design. The thousands of drawings and many models that showed his ideas for the interiors were ignored; almost all the models somehow disappeared after he left Australia.

In the hands of government architects, the completion of the Opera House took longer and cost more than might have been expected under the Dane's supervision. By September 1973 the roof vaults were completed and work commenced on the glass walls and the interiors, and outside, the promenade and approaches. The building was officially opened by Queen Elizabeth II in October 1973.

The Opera House remains a politicians' plaything. In 1998, the NSW Labor government agreed with Utzon, then about 80 years old, that he would have control of all design decisions in a proposed ten-year, $66 million renovation of the building. About the same time, a proposed submission was put forward for formal recognition by UNESCO as a World Heritage Site. It was titled "Sydney Opera House in its Harbour Setting" because of the imminent changes to the interiors. Nevertheless, Australia's ultraconservative prime minister John Howard refused to pass on the document to UNESCO because it might "complicate" those alterations. Despite its rough political passage, as the editors of *Architecture Australia* commented in December 1999, "The original inspiration and genius [seen in the Sydney Opera House] has transcended the ensuing conflicts and controversy to produce an astonishing manifestation of triumphant imagination and spirit."

Further reading

Baume, Michael. 1967. *The Sydney Opera House Affair*. Camden, NJ: Thomas Nelson and Sons.

Drew, Philip. 1995. *Sydney Opera House: Jørn Utzon*. London: Phaidon.

Futagawa, Yukio, and Christian Norberg-Schulz. 1980. *Sydney Opera House: Sydney, Australia, 1957–1973 by Jørn Utzon*. Tokyo: A.D.A. Edita.

T

Taj Mahal
Agra, India

The Taj Mahal, India's most recognizable icon, was built on the banks of the River Jamuna at Agra by the Mughal emperor Shah Jahan (reigned A.D. 1628–1666), in memory of his beloved wife Arjumand Banu Begam, known as Mumtaz Mahal ("Elect of the Palace"), who died in childbirth in 1631. There is a tradition that, on her deathbed, she entreated her husband to build a tomb that would preserve her name forever. The funerary mosque, faced with white marble, was completed in 1653 after twenty-two years in the building. When it was inscribed on UNESCO's World Heritage List in 1983, the Taj Mahal was acclaimed as "the most perfect jewel of Moslem art in India and … one of the universally admired masterpieces" in the world.

The symmetrical square mausoleum stands on a marble plinth above a 186-foot-square (59-meter) red sandstone platform. A 58-foot-diameter (18-meter) pear-shaped dome soars 213 feet (65 meters) above the octagonal central space—a two-story memorial chamber housing the bejeweled cenotaphs of Mumtaz Mahal and her husband. In fact, the real coffins lie in an unpretentious crypt below. Originally, the interior was opulently furnished with Persian carpets and articles of gold, all dimly lit by dappled sunlight that shone through intricately carved marble lattices in the drum of the dome. Two massive silver doors closed the entrance on the south side. At both levels there are eight interconnected anterooms: the four on the sides are rectangular, and the four corner spaces, also crowned with domes, are octagonal. A 163-foot (50-meter) tapering marble-faced minaret stands at each corner of the plinth. The building is praiseworthy for its composition, balance, and massing.

Closer scrutiny reveals that its surfaces, inside and out, are enriched with flower patterns of inlaid semiprecious stones using a technique known as *pietra dura*, as well as panels and bands of calligraphic inscriptions from the Koran. Each piece of decorative work is in itself a jewel, but all are ingeniously integrated in a complex but harmonious whole. Throughout, the builders made exacting adjustments of line and surface to ensure that the Taj Mahal *looked* right. For example, the plinth is slightly convex, so that it appears to be horizontal. The walls are slightly inclined for the same reason. Such refinement extended to the detail of the decoration: the calligraphic inscriptions are more widely spaced as they rise, to appear uniform when viewed from below. It has been said that the tomb was "built by giants and finished by jewelers."

The Taj Mahal was not the creation of an individual. Its overseeing architect was probably the Persian engineer-astrologer Ustad Ahmad, but there were about forty other specialists whose skills combined to

Taj Mahal, Agra, India; Ustad Ahmod (?), architect, completed 1653. Garden front.

fully realize his designs. So, while Ismail Khan from Turkey *built* the dome, Qazim Khan of Lahore cast its 30-foot (8.2-meter) gold finial. The master sculptor and mosaicist was a Delhi lapidary named Chiranji Lal, and the master calligrapher was Amanat Khan from Shiraz. Muhammad Hanif directed the masons, and Mir Abdul Karim and Mukkarimat Khan of Shiraz were what today would be called project managers. The construction employed 20,000 North Indian forced workers, as well as craftsmen from South India, Baghdad, Baluchistan, Bukhara, Shiraz, Syria, and Persia. There is a tentative view—most likely a myth—held by some Italian scholars that the Taj Mahal was designed by the Venetian Geronimo Veroneo. But Veroneo, a goldsmith, did not arrive in Agra until 1640, when the work was already half finished. It is far more probable that he was employed on part of the decoration, among the international band.

Many of the materials were as exotic as the men who worked them. Although the red sandstone was quarried locally, marble was imported from distant Makrana in Rajasthan. A 10-mile-long (16-kilometer) earthen ramp was constructed through Agra, and it is said that the marble blocks were hauled along it to the building site by a team of 1,000 elephants. Bullock carts were also used. More than forty varieties of gemstones, corals, and rare seashells were gathered from all over the region—Afghanistan, Badakhshan, Burma (now Myanmar), Egypt, Tibet, Turkestan, and even the depths of the Indian Ocean.

The great mausoleum stands at the north end of a walled enclosure measuring 1,902 by 1,002 feet (580 by 305 meters), designed in the Charbagh style by Ali Mardan Khan, one of Shah Jahan's courtiers. It faces the domed three-story main gateway of red sandstone at the south end of the long axis, the vista emphasized by a reflecting channel flanked by two avenues of cypresses. The entrance symbolizes the gate to Paradise, and indeed the square garden, divided into four quarters by the waterways, conjures the Garden of Paradise with its rivers of water, honey, milk, and wine. At the central meeting place of the channels is a marble tank representing the celestial Pool of Abundance, exactly placed to reflect the Taj Mahal in its still waters. Each of the four sections created by the waterways is subdivided into four smaller squares and then into four again—sixteen flower beds defined by raised stone paths. An ingenious system of reservoirs and underground earthenware and copper pipes carries water from the river to be fed to the pools and fountains and to irrigate the garden.

Two other buildings at the ends of the garden's transverse axis complete the Taj Mahal ensemble: a red sandstone mosque to the west is reflected by an identical "rest house" to the east. The latter is known as the *jawab* (answer), indicating that it probably was included simply to provide symmetry in the architectural composition, rather than for any practical function. There is no place within the walls from which the serene mausoleum cannot be seen.

The white marble of the Taj Mahal has been yellowed by automobile emissions, acid rain, and industrial pollution. In April 1997 India's Supreme Court enforced earlier orders for almost 300 nearby metal foundries to stop using coal for fuel or risk being closed down. There is now a 62-mile (100-kilometer) "safety zone" around the monument. The other threat to the fabric is the breath of up to 3 million visitors each year, which raises humidity and causes rusting of the iron cramps that hold the marble facing in place. In 1998 the French Rhône-Poulenc Foundation, UNESCO, and the Archaeological Survey of India began a three-year joint program to improve the water tightness of the Taj Mahal, undertake antifungal treatment, and extend research into stone-preservation technology.

The Taj Mahal and its romantic and poignant story have inspired poets everywhere. The Indian Rabindranath Tagore called it "a teardrop on the cheek of time," and the Englishman Edwin Arnold saw it as "not a piece of architecture, as other buildings are, but the proud passions of an emperor's love wrought in living stones."

See also Jantar Mantar

Further reading

Lall, John S. 1994. *Taj Mahal and the Saga of the Great Mughals*. Delhi: Lustre Press.

Mabbett, Hugh, and Fiona Nichols. 1989. *In Praise of the Taj Mahal*. Wellington, NZ: January Books.

Nath, R. 1985. *The Taj Mahal and Its Incarnation*. Jaipur, India: Historical Research Documentation Programme.

Temple of Amun: The Hypostyle Hall
Thebes, Egypt

On the east bank of the Nile at Thebes, 440 miles (700 kilometers) south of the site of modern Cairo, stood the most extensive temple complex in ancient Egypt. From the time of the New Kingdom (1550–1069 B.C.), the northern end of this religious compound (near the modern village of Karnak) was dominated by the great temple devoted exclusively to the worship of Amun-Ra, "King of the Gods" and one of the most important Egyptian cults. Like all dynastic temples, it provided a platform upon which the pharaoh enacted rituals to ensure the annual flooding of the river and maintain life on earth. The idea of a virtual universe raised Egyptian architecture from the function of shelter to a metaphysical plane. It was, in that sense, an architecture of altered states. Other major temples in the complex were consecrated to Amun's wife, Mut, and Monthu; there were smaller shrines for Khons, Oper, and Ptah. The hypostyle—the word is derived from the Greek, "resting on columns"—of the Temple of Amun is the building's most remarkable feature, ranked by many among the world's architectural masterpieces.

The temple was built over twelve centuries under the patronage of many pharaohs; the last additions were made in the Ptolemaic period (ca. 332–30 B.C.). The walled precinct was approached from the west along an avenue of ram-headed sphinxes. A gateway between two massive, battered pylons (never finished) gave access to an open forecourt, measuring about 230 by 260 feet (70 by 80 meters). In its northwest corner stood the Temple of Seti II, and close to its southeast corner the Temple of Ramses III. The courtyard provided the setting for public ceremonies and festivals. The processional way continued along the main axis of the temple complex, and beyond the eastern gateway, through a second pair of pylons, was the hypostyle hall. It was in effect a forest of 134 columns, covered in painted bas-reliefs, filling a room 338 feet wide by 170 feet deep (102 by 53 meters).

Temple of Amun, Thebes, Egypt; architect(s) unknown, 1408–1225 B.C. Ruins of the hypostyle hall, looking west along the central axis.

Its central corridor, defined by twelve gigantic sandstone columns, formed a processional way into the inner parts and sanctuary of the temple, where only the kings and priests could go—"a preparatory passage from this world to the next" (Smith 1990, 14).

Amun-Ofis III commissioned the central columns on the hall's east-west axis in about 1408 B.C. They were 33 feet (10 meters) in circumference, and their papyrus-flower capitals towered 69 feet (21 meters) above the floor, supporting the stone roof slabs of the central nave. Seven aisles on either side of the hall were defined by a total of 122 stone columns with papyrus-bud capitals; these columns were 27.5 feet (8.4 meters) in circumference and 43 feet (13 meters) high. The difference in the general ceiling height and that of the central nave allowed for the provision of clerestory windows formed with vertical stone slats that allowed air and some light to enter the hypostyle hall; most of the space was kept in mysterious shadow. The brilliantly colored, painted decoration of the interior walls and columns was initiated by Ramses I (reigned 1292–1290 B.C.), continued by Seti I (1290–1279), and completed by Ramses II (1279–1225). The external walls are decorated with battle scenes.

To the east of the hypostyle hall, another pair of pylons led to a narrow central court, and yet another pair (although somewhat smaller) to the transverse hall that subsumed the earliest sanctuary. A new sanctuary was built by Philip Arrhidaeus (323–316 B.C.), brother of Alexander the Great. Southward from the transverse hall, a walled enclosure at right angles to the main axis led through a succession of four pairs of pylons to the Temple of Mut and from there to the temples at Luxor 2 miles (3 kilometers) further south.

The Temple of Amun is presently endangered because of foundation failure and erosion of the sandstone at the base of walls through the Nile's annual flooding. Funded in part by the U.S. National Endowment for the Humanities, the Institute of Egyptian Art and Archaeology from the University of Memphis, Tennessee, is recording the inscriptions in the great hypostyle hall.

Further reading

Cenival, Jean-Louis de. 1964. *Living Architecture: Egyptian*. New York: Grosset and Dunlap.

Schwaller de Lubicz, R. A. 1999. *The Temples of Karnak: A Contribution to the Study of Pharaonic Thought*. Rochester, VT: Inner Traditions.

Smith, G. E. Kidder. 1990. *Looking at Architecture*. New York: Harry N. Abrams.

Temple of the Inscriptions
Palenque, Mexico

The modern town of Palenque is about 100 miles (160 kilometers) east of Villahermosa and 30 miles (48 kilometers) south of the Gulf of Mexico in the southern Mexican state of Tabasco. The ruins of the ancient Maya city are a little to the south, perched in dense forests on a shelf carved from the Sierra Oriental de Chiapas and overlooking the basin of the Usamacinta River. The Spanish conquistadors named it Palenque because of the surrounding wooden palisade, but its Mayan appellation remains unknown. The city reached its greatest glory between A.D. 600 and 800, and Mayan stone architecture found its highest expression there in such imposing buildings as the several pyramidal temples, the palace, and of course the towering Temple of the Inscriptions.

In December 1784 the mayor of Palenque, Jose Antonio Calderon, accompanied by an Italian architect named Antonio Bernaconi, found the ruins of the rumored lost city. The first serious investigation of the ancient site was made eighteen months later by Don Antonio del Rio, whose Mayan laborers cleared the jungle around the palace. Having demolished its interiors in search of gold, he found none. But his accurate report, published in 1822, motivated further study of Mayan civilization. Meanwhile, in 1807 the Spanish commissioned one Guillermo Dupaix to investigate ancient Mexican sites, and his draftsman, José Luciano Castaneda, made drawings of Palenque that remained unpublished until 1834.

The Mayan civilization dates from 700 B.C.; within a millennium it extended from Honduras, Guatemala, and El Salvador through the lowlands south of Oaxaca into the Yucatan Peninsula. Architecture and all the arts flourished between A.D. 300 and 800. In this Classic Period their great cities were built upon the remains of earlier settlements. Their civilization began to collapse around 750, and after about 830 construction and development had come to a halt. In another 120 years the cities south of Oaxaca were suddenly abandoned. Many reasons have been given: climatic change, food shortages caused by overpopulation, epidemic disease, invasion, or a peasant revolution. In 1542 the Spanish conquest completed its demise.

Because of the Maya's highly developed agricultural technology and the fertility of the region, an estimated five months a year were free from farmwork. The ruling classes shrewdly used this social surplus to build cities, pyramids, and temples, without resort to draft animals or the wheel. They could also support a separate class of specialist artists and craftsmen. The architecture of Palenque epitomizes the Classic Period's Western regional variant. The city

Temple of the Inscriptions, Palenque, Mexico; architect(s) unknown, ca. 650–700.

became an important population center under Pacal (reigned 615–683). In 647 he built the so-called "Forgotten Temple," upon which the Temple of the Inscriptions is modeled. And he commissioned other major buildings, including the Temple of the Inscriptions and the extensive palace, with its four-story square tower. His sons, Chan-Balum (reigned 684–702) and Kan Xul, completed the temple and their father's tomb. The Temple of the Inscriptions, the highest in Palenque, stands on a 75-foot (23-meter) limestone-faced pyramid in front of a steep hill on the southern edge of the central plaza. Like most of Palenque's ceremonial buildings, the pyramid was painted deep red. Mayan architecture was distinctive for its vivid colors; interiors and exteriors were both painted red, yellow, green, and white. Nine stepped terraces, each with a molded cornice, ascend to the base of the temple itself, which is reached by a steep stairway at the center of the northern side. Covered by a traditional mansard roof and crowned

with a roof comb, the temple building was about 40 feet (12 meters) high. Its front had five portals separated by piers, decorated with more inscriptions and images of stately figures in feathered headdresses in stucco reliefs. Inside, two corbel-vaulted chambers contained (among other decoration) the three panels of 620 glyphs—they set out Pacal's dynastic history—for which the temple is named.

Beginning in 1948 the Mexican archeologist Alberto Ruz Lhuillier excavated a concealed stairway leading 80 feet (24 meters) from the floor of the temple to a 30-by-13-foot (9.2-by-4-meter) vaulted burial chamber, 5 feet (1.5 meters) under the pyramid. It contained Pacal's magnificently decorated sarcophagus. Because the coffin is too large to have been carried down the narrow stair, it seems likely that the crypt was built before the pyramid. A long stone tube joined it to the temple floor, providing a way for the king's soul to reach the living world. The similarities between this tomb and Egypt's pyramids, even in such detail, deserve remark. Despite much popular speculation, serious scholars have dismissed suggestions of a connection.

Adding the Palenque precinct to its World Heritage List in 1987, UNESCO noted that this "prime example of a Mayan sanctuary of the classical period" through the "elegance and craftsmanship of the construction, as well as the lightness of the sculpted reliefs … attest[s] to the creative genius of [Mayan] civilization." Fewer than 10 percent of Palenque's almost 500 buildings have been excavated. Work continues at other parts of the site under the aegis of the multidisciplinary Palenque Project, jointly undertaken by an international group of archeologists and other specialists from the Pre-Columbian Art Research Institute and Mexico's Instituto Nacional de Antropologia y Historia.

Further reading

Abrams, Elliot M. 1994. *How the Maya Built Their World*. Austin: University of Texas Press.

Schele, Linda, and Peter Mathews. 1998. *The Code of Kings: The Language of Seven Sacred Maya Temples and Tombs*. New York: Scribner.

Stierlin, Henri, and Anne Stierlin. 1997. *The Maya: Palaces and Pyramids of the Rainforest*. Cologne and New York: Taschen.

Tension and suspension buildings

Historically, post-and-beam construction and the arch (with its three-dimensional extensions) were regarded as the only ways to build. Both were constrained by a belief in the necessary permanence of architecture. Because the only available durable materials—masonry of various kinds—were strong in compression, structural systems exploited that property. A third way of building, the structure that used stretched filaments and membranes, was limited to short-life buildings like Arab tents because it was made from nondurable materials—wood and animal or vegetable fibers. Two events would change that: the advent of structural steel and reinforced concrete after 1865, and synthetic membranes, after about 1950, combined to create architectural opportunities. Together, these new means to support and enclose provided architects with extensive esthetic possibilities; at last they were able to use the long-span systems that had long been available to bridge builders.

The German architect-engineer Frei Otto advocated tensile-stress construction in a 1954 book *Das hangende Dach (The Hung Roof)*. The following year the Australian architects Barry Patten and Angel Dimitroff of the firm Yuncken Freeman began design work for the Sidney Myer Music Bowl (completed 1959), an open-air theater in the city of Melbourne. Its structure comprises a network of steel cables supporting a roof of aluminum-plywood sandwich panels. The main cable is anchored on each side of the open end to large reinforced concrete blocks and draped between two 75-foot-high (22.8-meter) steel masts, 110 feet (33.5 meters) apart. Secondary cables span about 195 feet (60 meters) from the primary cable to converge on a reinforced concrete ground anchor, and the transverse tertiary cables that carry the roof cladding are draped over them and fixed to individual ground anchors. This early experiment in suspended roof construction is presently (2001) undergoing extensive restoration.

Another early and important innovator was the Finnish-American architect Eero Saarinen (1910–1961). In 1958 he designed, in conjunction with structural engineer Fred Severud, the 3,300-seat David S. Ingalls Ice Hockey Rink at Yale University. Saarinen used a 300-foot (92-meter) concrete arch that undulates (supposedly with the grace of a skater) on the long axis, crossing the arena and the entrance. From this structural backbone, which reaches 76 feet (23 meters) at its highest point, steel cables stretch to the building perimeter and carry a wood slat and an aluminum roof. Saarinen's second essay in tensile construction, hailed as the third most significant building in U.S. history, was the dramatic main terminal at Dulles International Airport, Chantilly, Virginia, built in 1962. Assisted by the structural engineer Joseph Vellozzi, Saarinen covered the 50-foot-high (15-meter) glass-walled space with a curved roof of insulated precast concrete panels covered with a ply membrane, "high at the front, lower in the middle, slightly higher at the back." It is carried on catenary steel cables stretched 141 feet (43 meters) between massive, upward-tapering concrete piers that slant outward as

Main terminal building, Dulles International Airport, Virginia; Eero Saarinen, architect, 1962.

they rise. Recent additions completed in 1991 by Skidmore, Owings and Merrill followed Saarinen's monumental original design.

At the same time as Saarinen, architects in Asia and Europe were extending the new technology. In 1960 the Japanese Kenzo Tange designed stadiums in Tokyo for the 1964 Olympic Games. The larger has a plan comprising two offset semicircles with their open ends elongated to points. Two reinforced concrete columns support a gracefully curving roof made up of a system of steel cables that carry a membrane of enameled steel plates. Tensile-stress construction was put to a different use by the Italian engineer Pier Luigi Nervi, who is better known for his reinforced concrete structures. His paper mill (1961–1962) in Mantua, Italy, has a floor area of 86,000 square feet (about 8,000 square meters) and houses the large machinery needed for paper manufacture. In anticipation of future expansion, the client needed a clear area of 525 feet (160 meters). Nervi responded by designing a suspended roof structure consisting of a steel deck suspended from 164-foot-high (50-meter)

reinforced concrete towers by steel cables; the central span is 535 feet (164 meters) and the two side cantilevers are 140 feet (43 meters).

Tensile-stress construction was perhaps epitomized in the Millennium Dome, built at Greenwich near London to house part of England's Year 2000 celebrations. Despite the fact that the exhibition it housed was mediocre (one critic called the show "vain, vapid, fatuous, inane and … patronizing") and the project a financial disaster, the building itself was spectacular indeed. The dome, with 20 acres (8 hectares) of ground floor area—about twice the size of the Georgia Dome in Atlantic City—was designed by the Richard Rogers Partnership of architects and consulting engineers Buro Happold. The joint main contractors were Britain's largest building firms, Laing and Robert McAlpine. The primary structure of the 1,050-foot-diameter (320-meter) dome consists of 72 radial cables of high-strength steel suspended from twelve 325-foot-tall (100-meter) structural masts. The clear span between the masts is 650 feet (200 meters), and the cables carry a steel net that in turn supports a translucent membrane of Teflon-coated fiberglass at a height of 165 feet (50 meters) at the center.

See also Bedouin tents

Further reading

Nervi, Pier Luigi. 1957. *The Works of Pier Luigi Nervi.* New York: Praeger.

Power, Mark. 2000. *Superstructure.* London: Harper-Collins Illustrated.

Temko, Allan. 1962. *Eero Saarinen.* New York: Braziller.

Thames Tunnel
London, England

The Thames Tunnel was designed by the French-born engineer Marc Isambard Brunel and supervised by his son Isambard Kingdom Brunel between 1825 and 1843. The approximately 1,200-foot-long (365-meter) structure runs under the River Thames between Wapping on the north bank and Rotherhithe on the south in east London. Originally used for foot traffic, it now forms part of the London Underground system. The first tunnel ever built through the soft ground under a river, and the forerunner of modern tunneling techniques (including that used for the Channel Tunnel), it is widely recognized as a landmark feat of civil engineering.

In 1798 one Ralph Todd had tried to build a tunnel under the Thames, further downstream between Gravesend and Tilbury, but the venture failed for financial and other reasons. In 1807, Robert Vazie and Richard Trevithick also attempted to construct a timbered tunnel. In January 1808 the river broke through. The hole was plugged with clay, the tunnel cleared, and work restarted, but a similar accident occurred a month later, when excavation was almost complete. Fearing that more clay dumping would endanger shipping, the authorities called a halt to the project. By about 1815 most of London's docks were near Rotherhithe, while industry was springing up around Wapping. The nearest fixed crossing was London Bridge, about 2 miles (3.2 kilometers) upstream, and by about 1815 nearly 4,000 people a day were ferried in small boats; goods had to be taken on a costly, time-wasting detour over the bridge.

In 1818 Marc Brunel patented his tunneling shield, enabling tunnels to be excavated by a technique commonly employed in coal mines, that is, by sinking vertical shafts and digging from within the shield. The 88-ton (80-tonne) cast-iron structure, built by Henry Maudley, consisted of twelve 3-foot-wide (0.9-meter) sections, each with three compartments in which a man could work. It had a closed face, and at the front, angled jacks held horizontal timber boards in place. The tunneler removed the boards one by one and dug out 4.5 inches (11 centimeters) of soil, replacing the jacks against the board in a forward position. When the entire face had been excavated, the shield assembly was edged forward by the use of screw jacks, and brickwork lining was built behind it. Most contemporary tunnels were built using the "cut and cover" method, that is, by digging a deep trench and roofing it; clearly such a technique was impossible underwater.

Marc Brunel secured financial backing to form the Thames Tunnel Company, and work started in March 1825. Until 1828 his son Isambard Kingdom Brunel, was engineer in charge of construction. The project, undertaken in perilous and testing conditions, did not proceed without problems: there were several roof falls and at least five floodings (in the second, the younger Brunel almost drowned), ten workers lost their lives, and others suffered injury and disease. The work was suspended between 1828 and 1835 because of failing finances, and Marc Brunel was almost bankrupted. The Thames Tunnel was completed in April 1840, but it was not until March 1843 that Queen Victoria officially opened it to the public. Estimated to take three years to build, it had taken eighteen.

The brick structure had a rectangular outer profile measuring 22.25 by 37.5 feet (6.8 by 11.5 meters), pierced by a pair of side-by-side horseshoe inner tunnels lined with terra-cotta tiles. The tubes were horizontally linked by 64 brick-lined passages with classical arches at about 18-foot (5.5-meter) centers that were cut through the partition wall between them. It was little more than 20 feet (6 meters) below the riverbed in places, passing through mostly alluvial deposits. Seepage water ran to sumps in the tunnel floor and was pumped to the surface. The red brick Rotherhithe Engine House, now listed as an ancient monument, was designed by Marc Brunel for the steam-driven pumps. Because there was no money left for road ramps, the Thames Tunnel, while designed for horse-drawn traffic, was originally used as a pedestrian thoroughfare. The only means of access were stairways down the 80-foot-diameter (25-meter) shafts that had been dug on each side of the river to lower the tunneling shield. A newspaper of 1830 offered visitors an adventure: "The Public may view the tunnel every day, from seven in the morning until eight in the evening, upon payment of one shilling for each person," adding that the tunnel was "lighted with gas, [was] dry and warm, and the descent [was] by a safe and easy staircase." Other sources tell a different story, asserting that the 70-foot-deep (21-meter) access stairs and the "dark and dank conditions" soon made it unpopular.

In 1865 it was bought by the East London Railway, and by December 1869 rail tracks were laid to link Wapping to the London, Brighton, and South Coast Railway at New Cross Gate. The oldest section of the London Underground system, it was first used by the East London Line in 1884. In May 1996 Taylor Woodrow Construction commenced refurbishment, building new fiber-reinforced, sprayed concrete linings in its twin bores. The work was completed in spring 1997. Heritage conservation laws required four of the cross passages and part of the tiled finish of the main bores to be preserved.

See also Channel Tunnel

Further reading

Falk, Nicholas. 1975. *Brunel's Tunnel and Where It Led.* London: Brunel Exhibition Project.
Lampe, David. 1963. *The Tunnel: The Story of the World's First Tunnel under a Navigable River Dug beneath the Thames, 1824–42.* London: Harrap.

Theater of the Asklepieion
Epidaurus, Greece

Every modern visitor to the fourth-century-B.C. Theater of the Asklepieion at Epidaurus marvels at its remarkable acoustics. The tearing of a slip of paper, a whisper, or the sound made by a struck match in the orchestra can be heard with perfect clarity everywhere in the theater, even at the very top, 250 feet (80 meters) distant. The theater epitomizes the skill of the ancient Greeks in the creation of a building type. That fact was already recognized in antiquity, when the traveler Pausanias praised its symmetry and beauty. The building is generally attributed to Polykleitos the Younger, and features of the design suggest an original date of around 300 B.C.

For about eight centuries the Asklepieion was the preeminent healing center of the classical world. The cult of the god Asklepieos was active in the region as early as the sixth century B.C., and such was its success that the original sanctuary of Apollo Maleatas became too small for public worship. The fame of the place led to financial prosperity. In the fourth and third centuries B.C. an ambitious program to create monumental religious buildings was implemented: first the temple and altar of Asklepieos, the *tholos*, and the *abaton*, and a little later the Hestiatoreion, the baths, the palaestra, and the theater.

The theater, whose overall diameter was 387 feet (118 meters), was built in two stages: the orchestra, the lower section of seating, and the stage building (*skene*, from which our word "scenery" comes) were constructed first. Originally, the actors worked from the orchestra, with a *thymeli* at its center where the leader of the chorus stood. It was a full circle of 67 feet (20 meters) diameter that was not changed in the extension nor during later restorations. Acoustical problems in earlier Greek theaters were reduced when the actors moved from the orchestra to a raised stage, placing them in direct line of sight and sound with the audience. By the time the Theater of the Asklepieion was completed, the actors' masks, said to have been introduced in the mid–sixth century B.C., had become quite large; evidence suggests that the trumpet design of mouth openings helped to better direct their voices. However, the excellent audibility to an audience seated around three-fifths of a circular orchestra, in the open air, was reached by the ancient builders largely by means of the architecture alone.

The seating area *(theatron* or *koilon)* for 6,200 spectators comprised twelve tiers of thirty-four rows of broad limestone benches divided by stairways. The

Theater of the Asklepieion, Epidaurus, Greece; Polykleitos the Younger, architect, ca. 300 B.C. A permanent architectural *skene* once stood beyond the circular orchestra.

longest distance from the orchestra was 193 feet (58 meters), and to achieve better acoustics, only eight rows of benches of the central eight tiers were set out so that their radii extended to the center. The radii of the tiers on either side met closer to the skene, and their curves were larger. Nevertheless, all the rows look like segments of concentric circles. The lowest row of seats *(proedria)*—ringside, as it were—with back supports, was reserved for honored guests such as officials and priests.

The second phase of building took place in the middle of the second century B.C. The theatron was extended upward with twenty-three stairways giving access to twenty-two tiers, each with twenty-one rows of benches, divided from the original seating by a passage reached by rear ramps at either end of the theater. That brought the number of seats to 12,300. The skene was also extended, probably to two stories. In its completed form the lower level in-

cluded a main room with four columns on its long axis and two small square rooms at each end. The facade of the *proskenium* had a series of fourteen Ionic engaged half-columns framing painted wooden panels. Ramps on either side led to a single-level, 10-foot-wide (3-meter) stage through one door of the monumental gates at the two entrances. A second door led through the *parodoi*, from which the chorus entered the orchestra and the audience reached the lower rows of seats.

Enjoying a renaissance in the second century A.D., the Sanctuary of Asklepieos survived until the sunset of antiquity. But the spread of Christianity in Greece eventually led to its abandonment and consequent deterioration. The theater later collapsed during an earthquake. Systematic excavations (1879–1926) by the Greek Archaeological Society under P. Kavvadias led to the first restoration in 1907, involving the western gate of the theater and the adjacent retaining wall.

Work resumed intermittently between 1942 and 1963, and digs have been continuous since 1974. Since 1985 the Committee for the Preservation of the Epidaurus Monuments has been conducting research, and the restoration of several buildings including the gate of the theater's west parodos is at various stages of completion. In 1988, the Asklepieion was added to UNESCO's World Heritage List.

Further reading

Lawrence, A. W. 1996. *Greek Architecture*. New Haven, CT: Yale University Press.

Martin, Roland, and Henri Stierlin. 1967. *Living Architecture: Greek*. New York: Grosset and Dunlap.

Tomlinson, R. A. 1983. *Epidauros*. London: Granada.

Three Gorges Dam, Yangtze River
People's Republic of China

The Three Gorges Dam on the Yangtze River near Chongqing in China's central Hubei Province is the largest hydroelectric project in history, with twenty-six generators designed to deliver over 18,000 megawatts, 11 percent of the nation's needs. Started in 1994 and scheduled for completion by 2014, it will provide electricity to rural provinces and facilitate flood management and improved navigation for the upper Yangtze. The controversial dam has been widely criticized within and outside China as a socially and environmentally harmful project.

The 3,450-mile (5,500-kilometer) Yangtze is the world's third-longest river. Midway on its journey from the Tibetan Plateau to the East China Sea, it passes through a 120-mile-long (193-kilometer), exquisitely beautiful stretch known as the Three Gorges—the precipitous Qutang, Wuxia, and Xiling— one of China's most scenic regions. It will be submerged through the building of the dam.

The huge dam—five times wider than the U.S. Hoover Dam—will create a 575-foot-deep (176-meter) reservoir nearly 400 miles (640 kilometers) long and an average of 3,600 feet (1,200 meters) wide. According to official Chinese sources, the lake will completely inundate 2 cities, 11 counties, 140 towns, 326 townships, and 1,351 villages; other figures are consistently and considerably higher. About 59,000 acres (23,800 hectares) of rich agricultural land and numerous—some experts say nearly 1,300—important archeological sites will be lost, and an estimated 1.98 million people will be displaced and relocated. Other critics claim the project will increase the risk of earthquakes and landslides. It will also threaten fish stocks and such endangered species as the Yangtze dolphin, the giant panda, and others. As of 2001, the published estimated cost was U.S.$27 billion; the budget has soared from U.S.$10.57 billion in 1992 to a figure that unofficial sources place around U.S.$76 billion.

The dam was proposed as early as 1919 by Dr. Sun Yat Sen (Sun Yi Xian), and it was revived when the People's Republic of China exploded into being in 1949. Chinese and international engineers and scientists were involved in planning and design. Despite public opposition, *The Three Gorges Project Feasibility Study Report* emerged in May 1989 and became a major issue in the Tiananmen Square incident in June, after which Premier Li Peng, mostly for political reasons, became the scheme's principal sponsor. Following more feasibility studies, in April 1992 the National People's Congress approved construction, but about a third of the delegates either abstained or voted against it. The project remains the focus of a political tussle, the outcome of which will shift the balance of power in China's Communist Party.

The main parts of the Three Gorges Project are the dam, the powerhouses, and the navigation facilities. The 7,550-foot-wide (2,310-meter) concrete gravity-type dam is over 600 feet (175 meters) high; its 1,580-foot-long (483-meter) spillway is located in the middle of the original river channel, flanked by intake dam and nonoverflow dam sections. If the dam is finished, two powerhouses will be built at the toe of the intake dams, one on each side; there will be fourteen generator units in the left powerhouse and twelve in the right, connected to fifteen transmission lines to Central China, East Sichuan, and East China. The completed ship lock on the left bank will consist of two-way, five-step flight locks, through which 10,000-ton (9,100-tonne) barges will be able to pass. The one-step vertical ship lift will be able to raise a 3,000-ton (2,700-tonne) vessel.

Responsibility for all aspects of the construction and the eventual management of the project is vested

Three Gorges Dam, Yangtze River, People's Republic of China, commenced 1994. Photograph taken during construction.

in the state-owned China Yangtze Three Gorges Project Development Corporation, established in September 1993. Most of the cost is being met from within China, mainly through the Three Gorges Construction Fund, loans from the State Development Bank, and power revenues from the Gezhouba Hydropower Plant and (when Phase 1 is completed in 2003) from the Three Gorges project itself. Foreign financial institutions are conservative because of the dam's ecological implications; for example, the U.S. Export-Import Bank opposed the dam in May 1996, responding to recommendations of the National Security Council. They refused to guarantee loans to U.S. companies tendering for work on the dam. Some finance comes from Canada and Germany.

Excavation for the dam's foundations were in progress by mid-1993, and the project was formally opened in December. The Yangtze was dammed in November 1997 and diverted through a channel to drain the building site on the riverbed. A new high-way and airport were built, as well as apartment buildings for the 18,000 workers employed on the project. In January 2000 official Chinese sources claimed that the Three Gorges Dam project was on schedule, and work was accelerating. As a last word, it must be added that some engineering experts have warned about the eventual success of the project.

The promised power generation, flood control, and improved navigation all depend on the Three Gorges Project Development Corporation solving the potential problem of sedimentation in the reservoir. Because of the hugeness of the dam, there is no experience on which to draw to predict the rate of sedimentation, and it may seriously reduce the project's life and effectiveness.

Further reading

Qing, Dai. 1998. *The River Dragon Has Come!* New York: M. E. Sharpe.

———. 1994. *Yangtze! Yangtze!* Edited by Patricia Adams and John Thibodeau. London: EarthScan.

Slyke, Lyman P. van. 1988. *Yangtze: Nature, History, and the River*. Reading, MA: Addison-Wesley.

Timgad, Algeria

The Roman town now known as Timgad was founded in A.D. 100 on command of the emperor Trajan (reigned 98–117) and named Colonia Marciana Trajana Thamugas for his sister. It was built on a high plateau north of the Aurès Mountains in Algeria (then Numidia), 94 miles (150 kilometers) south of the modern town of Constantine. The Third Augusta Legion, effectively the Roman police force in North Africa, was garrisoned nearby, and Timgad, designed for veterans, was the archetypal Roman colony. The regular well-ordered layout became one of the principal sources of city plans in Europe and the New World from the fifteenth century until the beginning of the twentieth. It is therefore significant in the development of Western architecture and urban design. In turn, the inspiration for Timgad comes from the Roman army encampment, the *castrum*.

Perhaps because it was easy to set out, or perhaps because it suited military purposes, the right-angled grid formed the structure of the castrum, which might have served as a garrison for months or even years during a campaign. Two main streets, the Via Principia and the Via Praetoria, intersected at the legion's command post. Both extended through fortified gates beyond the enclosing ditch and palisade. Many permanent towns later grew from a castrum; for example, most English cities with "chester"—Winchester, Silchester, Dorchester—in their name have such origins, and in places like Chester the Roman plan is easily discernible. Even when they were building on a completely new site, Roman civil engineers used a similar grid (orthogonal) plan.

The hills around Timgad were once savanna grassland, and the mountains bore conifer and evergreen forests. Numidia was fertile and productive, and the prosperous province of Africa, of which it formed part, became known as the granary of Rome, furnishing a large proportion of the capital's barley, figs, and olives. Large olive presses have been found in and near Timgad. The town was essentially a 1,000-foot-square (300-meter) walled enclosure with gates on three sides. Its two main streets, a north-south *cardo* and a colonnaded east-west *decumanus maximus*, formed a T-junction almost at the center of the town, at the paved, colonnaded forum. At the western approach to the decumanus, just outside the wall, stood the tripartite Arch of Trajan, dating from the third century A.D. The plan was divided into 132 square blocks *(insulae)*, separated by 16-foot-wide (5-meter) paved streets, underneath which sewers ran. The insulae contained 400 houses, some of which were large enough to occupy a whole block, as well as shops and taverns. Timgad boasted twenty public buildings. In the forum there was the *curia* (council chamber), the *basilica* (town hall), a temple, shops, and a public lavatory. Just south of the forum, hollowed out of a low hill, stood a theater designed to hold up to 4,000 people. Other buildings, such as a library and public baths—there were seventeen of them, some quite extensive—were dotted around the town, both within and outside the walls. An aqueduct supplied water to Timgad from a spring about 3 miles (5 kilometers) away. There were also two large markets, one at either end of the decumanus. The large Temple to Jupiter Capitolinus stood outside the walls at the southwest corner of the town.

Almost inevitably, Timgad outgrew its walls. Some estimates put the eventual population at 15,000. The first extensions seem to have followed the orthogonal pattern of the original plan, but soon rather untidy development of "extensive but casually laid-out 'suburbs'" occurred along the arterial roads to the neighboring colonies. Timgad was a center of Christianity in the fourth century and became a stronghold of the Donatist heresy. With the decline of the Roman Empire it lost much of its significance. It was successively despoiled by the invading Vandals in the fifth century, the Berbers in the early sixth, and the Arabs in the eighth. Timgad remained in Arab hands until the French annexation of Algeria in 1830. It was forgotten until the end of the nineteenth century, and protected by isolation and the encroaching sands of the Sahara; then investigations were begun by the French architect Albert Ballu for the Service des Monuments Historiques de l'Algérie. The ancient city was inscribed on UNESCO's World Heritage List in 1982.

Timgad, Algeria; architect(s) unknown, founded 100. Ruins of the colonnaded *decumanus,* leading to the third-century Arch of Trajan.

See also Hippodamus of Miletus

Further reading

Manton, E. Lennox. 1988. *Roman North Africa*. London: Seaby.

Picard, Gilbert, and Yvan Butler. 1965. *Living Architecture: Roman*. New York: Grosset and Dunlap.

Ward-Perkins, John B. 1988. *Roman Architecture*. New York: Electa/Rizzoli.

Tower Bridge
London, England

Tower Bridge (1886–1894) is immediately and universally recognizable as an icon of London. Even during its construction, it was nicknamed "Wonder Bridge"—the largest, most advanced bascule bridge ever seen, employing hydraulic power on a scale never before attempted. Spanning the Thames between the Tower of London and Bermondsey, it was built to ease the traffic congestion at London Bridge, which was then (apart from ferries) the only means of crossing at the east end of the growing city. By the early 1870s the problem had become untenable, and travelers faced delays of several hours. Numerous petitions urged either the widening of London Bridge or construction of another crossing, and in 1876 the Corporation of the City of London formed a "Special Bridge or Subway Committee." It mounted a public design competition to find a solution.

The brief specified that river traffic must not be disrupted during or after construction, that there must be 140 feet (43 meters) clearance for ships, and (perhaps prompted by the problems of Brunel's Thames Tunnel of 1843), the road approaches must be negotiable for vehicles. There were over fifty entries, presenting a spectrum of ideas: steam vehicle-ferries, tunnels, low-level swing bridges, and high-level bridges with long road approaches or elevators for the street traffic. One entry suggested a railroad on the river bottom with very tall cars that could cross above the high-tide mark. In 1878 the city architect Sir Horace Jones (1819–1887) conceived the idea of a low-level bascule bridge—a double drawbridge with a road deck made up of two hinged arms that could be raised and lowered to accommodate both vehicular and river traffic. The bascule, or draw-span, bridge, although on a much smaller scale, had been common in Europe since the Middle Ages. It was not until October 1884 that Jones offered a developed design for a hydraulically operated openable bascule bridge, spanning between masonry-faced towers and connected to the respective banks by single-span suspension bridges. The following year an act of Parliament authorized the Corporation of the City of London to build the bridge. It took eight years to complete and involved five major contractors, employing nearly 450 construction workers.

Work started in April 1886. Horace Jones was appointed architect but died later in the year, and in 1887 his assistant George Daniel Stevenson took over the design. Sir John Wolfe Barry (1836–1918), son of the famous architect Sir Charles Barry, was appointed engineer. Because of provisions in the Tower Bridge Act about the size of the piers and the width of clearway in the river, the foundation builders faced a difficult and protracted task. Under both massive piers they sunk twelve permanent wrought-iron caissons, combined to follow the profile of the pier, into the watertight London clay 20 feet (6 meters) beneath the riverbed. The caissons and the spaces between were filled with concrete; working in temporary caissons, the masons built the brickwork and stone piers up to a level 4 feet (1.2 meters) above high water. Despite its Gothic Revival appearance, Tower Bridge is a steel structure. Behind the Cornish granite and Portland stone masonry of each of its 293-foot (90-meter) towers, the real work is done by four octagonal 5.5-foot-diameter (1.7-meter) steel columns, resting on granite slabs and secured to the piers with huge holding-down bolts. The towers are connected by two high-level cantilever steel footways, each with a suspended span, 143 feet (44 meters) above high water.

In terms of its operation, the bridge combines the suspension and bascule types. Including the approaches it is 0.5 mile (0.8 kilometer) long; it supports a 35-foot-wide (11-meter) carriageway, flanked by two 12.5-foot-wide (4-meter) sidewalks. Between the abutments the Thames is 880 feet (270 meters) wide, and the central opening span is reached from either direction by a 270-foot (82.5-meter) suspension span, carried on parallel latticed steel trusses

Tower Bridge, London, England; Horace Jones and George Daniel Stevenson, architects; John Wolfe Barry, engineer, 1886–1894, from a 1900 photograph.

(chains) slung between the main towers and landward towers, and thence to shore anchorages. Tie bars at the same level as the walkway join the pairs of chains to complete the structure. The opening span of the bridge was its most revolutionary feature. It consists of two counterweighted hydraulic bascules, each constructed of four parallel 160-foot-long (49-meter), evenly spaced girders. When lowered, each bascule cantilevers to span exactly half of the 200-foot (61-meter) opening; they can be raised to an angle of 86 degrees.

The client wanted the bridge to be in context with its surroundings, including the eleventh-century Tower of London. Jones had originally proposed brick facings, but Stevenson later changed the specification to stone backed with brickwork. This "facadism" drew fire from *The Builder* magazine in 1894, which called the bridge "the most monstrous and prepos-

terous architectural sham that we have ever known." The exposed steelwork was originally painted chocolate brown.

Tower Bridge remains the only movable bridge of the twenty-nine that now span the Thames. Its steam-powered hydraulic system has been superseded by an electric-powered mechanism. In 1982, while still functioning as a bridge (some 40,000 vehicles cross it every day), it was made into a museum that welcomes about half a million visitors annually. When it gave access to the Pool of London, Tower Bridge opened more than 500 times every month; that is presently the annual number of openings.

Further reading

Godfrey, Honor. 1988. *Tower Bridge*. London: J. Murray.
Welch, Charles. 1894. *A Short Account of the Tower Bridge ... with a Description of Its Construction by J. Wolfe Barry*. London: n.p.

Treasury of Atreus
Mycenae, Greece

Between 1400 and 1200 B.C. Mycenae was the most powerful ancient Greek city-state. Its ruins now stand above the Plain of Argolis in the Peloponnese, near the modern village of Mikínai. The fortified city was the seat of the tragic, semilegendary Atreus, whose dynasty was cursed because he fed his brother Thyestes with his own children. To gain fair winds to take his warships to Troy, Atreus's son Agamemnon sacrificed his daughter to the gods. After the Trojan War, Agamemnon's wife Clytemnestra killed him before she herself was murdered by her son Orestes. Mycenae was defended by walls so massive that the later Greeks attributed their construction to the mythical giant Cyclops. Outside the walls to the south and flanking the main approach to the citadel lies the mid-thirteenth-century-B.C. tomb—one of nine similar structures—that the second-century-A.D. Greek trav-

eler Pausanias identified as the Treasury of Atreus. Although it is neither a treasury nor directly linked with Atreus, the building is an architectural feat, the epitome of the *tholos* (beehive) tomb. One historian observes that 1,000 years passed before the Greeks put "such technical perfection … at the service of such a grandiose architectural design" (Norwich 1984, 135). The Treasury of Atreus demonstrates first and more than any other building the effectiveness of the corbeled dome, a structural technique that recurs in European architecture, for example, in the sixth-century Hagia Sofia in Constantinople and the fifteenth-century Florence Cathedral dome.

Mycenaean tholos tombs were first used early in the fifteenth century B.C. They were probably reserved for only the highest echelons of society—kings and their immediate families. Others, even if aristocratic, were usually buried in chamber tombs, such as those found within the Mycenaean citadel. The tholos com-

The Treasury of Atreus, Mycenae, Greece; architect(s) unknown, mid–thirteenth century B.C. The dromos, looking west to the door.

prised a domical underground burial chamber *(thalamos)*. The dome was formed by corbeling, that is, by laying each successive masonry course to a slightly smaller diameter than the one below, tapering to a capstone until the structure was closed. The remains of the dead were either laid on the thalamos floor or interred in pits or shafts beneath it. The thalamos was approached through a relatively long and narrow passageway *(dromos)* open to the sky and entered through a doorway *(stomion)*. Normally, but not always, the tholos was set into an excavated hillside that was backfilled when the building was finished. As opposed to the post-and-beam constructions then widely used to house the functions of the living, the corbeled dome was an essentially stable structure and an excellent way to roof large areas without intermediate supports. Since most of the forces in the combined wall-roof were vertical, the sideways thrust inevitably imposed by true dome construction was minimized, reducing the need for lateral restraint.

The dromos of the Treasury of Atreus, opening to the east, is over 120 feet long and 20 feet wide (37 by 6.1 meters). Its sides are lined with huge blocks—each course is almost 3 feet (900 millimeters) high—of hammer-dressed ashlar masonry. The 34-foot-high (10.5-meter) stomion is built of sawn ashlar. Its exterior was once enriched with relief sculpture, long since plundered, in red and green-gray marbles from quarries near Kyprianon in Laconia. The great upward-tapering doorway was flanked by two slender half-columns with cushion capitals, much like those seen in the Palace of Minos at Knossos, supporting a frieze. Over the door are two lintel blocks; the weight of the inside block has been estimated at 132 tons (120 tonnes). Above it, the opening known as the relieving triangle—a structural device that deflected the loads from the lintel—was masked by a thin panel, carved in relief. The door opens directly into the thalamos, 48 feet (14.6 meters) in diameter and 44 feet (13.5 meters) high, its vault formed of 33 regular courses of masonry. Some blocks bore a metal decoration, probably of hammer-dressed ashlar, originally decorated with gilded bronze rosettes evenly spaced in parallel horizontal rows. The Treasury of Atreus also has a rectangular side chamber about 20 feet square (6 meters) and 16 feet (5 meters) high, opening off the main chamber and cut into the natural rock of the site.

See also Florence Cathedral dome; Hagia Sofia; Mycenae, Greece

Further reading

Burg, Katerina von. 1987. *Heinrich Schliemann: For Gold or Glory?* Windsor, UK: Windsor Publications.

Mylonas, George E. 1966. *Mycenae and the Mycenaean Age*. Princeton, NJ: Princeton University Press.

Norwich, John Julius ed. 1984. *The World Atlas of Architecture*. Boston: G. K. Hall.

U

Unité d'Habitation
Marseilles, France

It has been accurately claimed that Le Corbusier's most influential late work was his Unité d'Habitation in Marseilles. The eighteen-story apartment building, universally admired by architects but unloved by the people who live in it, is the first realization—it was followed by three others elsewhere in France and one in Berlin, Germany—of the famous Swiss architect's theories of urban design formulated twenty years earlier. It also expressed the socialist housing ideals of CIAM (in English, International Congresses of Modern Architecture), developed after 1928, and spawned imitations throughout the world. It was, although perhaps for the wrong reasons, an important architectural event.

Between the early 1920s and the end of World War II, Le Corbusier's most significant work—albeit theoretical—was in urban planning. In published plans like *La Ville Contemporaine* for a population of 3 million (1922), the *Plan Voisin de Paris* (1925) that proposed replacing the historic city with eighteen superskyscrapers, and a spate of "classless" *Villes Radieuses* of 1930–1936, he expounded ideas diametrically opposed to the low-rise cities projected by the international Garden City movement. Le Corbusier was not the first with such notions. For example, Hendrik Wijdeveld's projected Amsterdam 2000 of 1919–1920—a radial city in a greenbelt around old Amsterdam, with towering hexagonal apartment blocks, each for 2,000 inhabitants—anticipated the Swiss by at least a couple of years. Le Corbusier, building on Tony Garnier's concept for the Cité Industrielle and the forms of Antonio Sant' Elia's Cittá Nuova, proposed a complex vertical city with separated functions (living, working, recreation) and a carefully designed transport infrastructure. That he physically resolved into what was almost a diagrammatic plan with an orthogonal grid with wide green open spaces between high-rise buildings that had separate access for vehicular and pedestrian traffic. It was simple but soulless and—luckily—never built.

After World War II, the government of France's Fourth Republic established a program to house the thousands of people displaced by the conflict. The program was managed by Raoul Dautry, an admirer of Le Corbusier. In 1946 he commissioned the architect to build a prototype of his Vertical City on the edge of Marseilles. Le Corbusier seized the chance to give substance to his beliefs about social integration and the relationship between a high-density housing scheme and the wider community. L'Unité d'Habitation (literally, Housing Unit), characterized by one writer as a utopian "tower in the park," was the result. The houses are stacked in a block, leaving the remainder of the site for open space to serve the whole community.

Completed in 1952, the huge concrete building—it is 440 feet long, 79 feet wide, and 170 feet high (136 by 24 by 52 meters)—stands on massive 23-foot-high (7-meter) *pilotis*. It houses 1,600 people in 337 apartments; Le Corbusier called them "superimposed villas." There are twenty-three apartment types, varying from single-person studios to those for families of ten people; almost all have two-story living rooms. The Unité was designed as a concrete-and-steel frame into which the precast concrete apartment units were inserted; as the architect put it, it was "a wine rack into which the living spaces [were] inserted as bottles." The units have a mezzanine level and double-height living spaces that open through glazed walls into deep balconies that create the building's interesting facades by forming a sunscreen, the *brise-soleil* that Le Corbusier and others would employ elsewhere. The sides of the balconies are painted in bright primary colors. The offset units form a series of interlocking L shapes, with a corridor at every third floor. Because there are windows at both ends of each unit, there is good natural light and ventilation. Lead sheets in the party walls provide sound insulation.

Le Corbusier believed that collective organization, while protecting the individuality of the family unit, would also integrate that unit into the Unité's society. Therefore, on the seventh and eighth floors he created a shopping and service mall, with food stores, a bookshop, a laundry, a café, a post office, and even hotel rooms, expecting that the services provided would identify his building as a neighborhood. He also included a "roof garden" to serve as a community center. The large, barren space, admittedly with spectacular views but anything but a garden, is dominated by two colossal concrete ventilation shafts. There is a children's playground full of rather forbidding concrete forms and space for a swimming pool, a nursery and a gymnasium.

Le Corbusier built several versions of the Unité d'Habitation; all are isolated buildings and not part of a group. The same building was reproduced at Nantes-Reze (1952–1957) and Briey-en-Foret in the *département* of Meurthe and Moselle (1963). An unrealized development plan for Meaux (1957–1959) in the *département* of Seine and Marne proposed five Unités. The Berlin Unité (1956–1959), built on a hilltop at Reichssportfeldstrasse, has 530 larger apartments on 17 floors, but no communal amenities and oppressive dark corridors. The last Unité was built with André Wogenscky at Firminy-Vert in the *département* of Loire in 1964–1967. It has 414 apartments but was never fully occupied. Renovation work was undertaken by the architect Henri Ciriani in 1995, and the roof was restored by Jean-François Grange-Chavanis in 1995–1996.

The Unité d'Habitation at Marseilles was seized upon as a prototype for collective housing and was frequently copied (largely without much attention to its underlying sociopolitical reasons for being). Le Corbusier's sublime idea, his ideal, whether right or wrong, was reduced to Unité clones, inexpensive high-density housing like London County Council's Alton West estate at Roehampton (1952–1959) and the Bijlmermeer (1966–1969), a complex of high-rise apartments southeast of Amsterdam. Others appeared further afield, including in the new city of Brasília. Almost everywhere, they are associated with social problems, and many became multistory versions of the slums they were built to replace.

See also Brasília; CIAM (International Congresses of Modern Architecture); Garden city idea

Further reading

Futagawa, Yukio. 1972. *Le Corbusier, L'Unité d'Habitation, Marseilles, France, and L'Unité d'Habitation, Berlin, West Germany*. Tokyo: A.D.A. Edita.

Jenkins, David, and Peter Cook. 1993. *Unité d'Habitation Marseilles—Le Corbusier*. London: Phaidon.

U.S. interstate highway system

The Dwight D. Eisenhower System of Interstate and Defense Highways, inaugurated in June 1956 by the Federal-Aid Highway Act, is a 41,000-mile (66,000-kilometer) network linking 90 percent of the major cities whose population exceeds 50,000 and many other urban centers in the mainland United States. The bill earmarked $25 billion to be spent between 1957 and 1969, and the system was to be completed by 1972. Sinclair Weeks, then secretary of commerce, somewhat extravagantly claimed it to be "the greatest public works project in history." In 1994 the

American Society of Civil Engineers placed it among the nation's "seven wonders." Five years later a construction executives conference included it with the top ten construction feats of the twentieth century. It noted that it is "the world's most comprehensive system of roadways," adding, "Its vast scale, rapid development and subsequent advancements have changed the lives of everyone who travels its byways [or] receives goods shipped over its … expanses."

The concept of superhighways was adopted from Europe. The world's first freeway—the 11.8-mile-long (19-kilometer) Avus experimental two-lane highway was built in Berlin, Germany, and completed in 1921. Italy constructed its first autostrada in the early 1920s, including the 80-mile (130-kilometer) road from Milan to the north Italian lakes. Germany's first four-lane autobahn was opened between Düsseldorf and Opladen in 1929. From 1933 Hitler's Nazi government extended the system by creating work to restore the deeply depressed economy. As the fatherland prepared for war, linked north-south and east-west highway networks were obviously strategic, and other *Reichsautobahnen*—between Cologne and Bonn, and Frankfurt and Darmstadt—followed. These early freeways had no shoulders, narrow medians, and even cobblestone exit ramps. By the end of World War II the German network covered over 1,300 miles (2,100 kilometers); by 2000 it comprised about 7,250 miles (11,600 kilometers), reaching most of the reunited nation.

In the depressed 1930s, President Franklin D. Roosevelt recognized that the construction of a national road system would create jobs. But action was first resisted by Congress and then postponed because of the imminent entry of the United States into the European war. Roosevelt foresaw that the postwar repatriation of U.S. servicemen could increase unemployment and perhaps renew economic depression. A highway program would be a hedge against that. Several pieces of legislation littered the years between 1938 and 1956. The Federal-Aid Highway Act of 1938 recommended a 26,000-mile (41,600-kilometer) "inter-regional" network, and a 1944 act authorized the creation of a 40,000-mile (64,000-kilometer) "National System of Interstate Highways," but no extra funds were authorized to build it. Construction started in August 1947, but in the absence of federal money, many states rejected the scheme. In the early 1950s, in the face of cold war, the need for an interstate highway system was perceived to be very urgent. A further Federal-Aid Highway Act of 1952 provided a pittance—only $25 million a year for 1952 and 1953; the federal government would halve construction costs with the states. When Dwight D. Eisenhower became president in 1953, only $955 million had been spent on 6,000 miles (9,600 kilometers). A 1954 act went a little further, offering $175 million annually, with the federal government prepared to meet 60 percent of costs.

Eisenhower the soldier was aware of the strategic importance of highways, and he had experienced the speed and mobility afforded by Germany's four-lane Autobahn. In January 1954 he announced that he would "protect the vital interest of every citizen in a safe and adequate highway system." Without it, the nation was faced with not only inadequate means to respond to catastrophe or attack but also a greater number of road injuries and deaths and the immense cost of traffic delays, highway-related litigation, and the inefficient movement of goods. In June 1956 Eisenhower signed the act that launched "the biggest peacetime construction project of any description ever undertaken anywhere."

The act ensured that the federal government would pay 90 percent of costs for the national project. In order to recoup that expenditure, the related Highway Revenue Act of 1956 increased taxes on gasoline and motor fuels and the excise on new tires, besides imposing new taxes on retreaded tires and a weight tax on heavy vehicles. The states would meet the remainder of the cost: they would build the interstate highways, and would own and operate them. The new law also called for uniform minimum design standards to accommodate projected 1975 traffic volumes; that was later modified to twenty years after construction. The standards included such factors as uniformity of design, access limited to interchange ramps, and no at-grade highway and railroad crossings, a constraint that entailed the construction of more than 55,000 grade-separation bridges. Initially, two-lane stretches and at-grade intersections were allowed on low-usage sections, but revised 1966 legislation required all interstate highways to have at least four lanes (two in each direction). Rest areas would line the highways, but all commercial establishments, including gas stations, were prohibited. The 1966 revisions also allowed existing bridges, tunnels, and toll roads to become part of the system provided they met its strict standards and enhanced transport integration.

There is unresolved rivalry over which segment was first built: the states of Pennsylvania, Missouri, and Kansas all claim the first segment of the interstate system, but their entries were either started before the 1956 act or were upgrades of existing roads. Originally expected to take thirteen years, construction took nearly forty, some of the delay being due to changes in plan, as well as environmental and social issues. In 1990 the project, finally extended to just under 42,800 miles (68,500 kilometers), was renamed the Dwight D. Eisenhower System of Interstate and Defense Highways. The network, completed in 1995, represents under 2 percent of total road mileage in the United States, but it accounts for almost 25 percent of all national travel miles. A 1958 estimate put the total cost of construction at $37.6 billion; that rose to $130 billion, mostly because of inflation.

The Dwight D. Eisenhower System of Interstate and Defense Highways has met at least its economic goals, a success that may be measured by its projected usage being reached ten years earlier than predicted. During construction it provided job opportunities throughout the country and enhanced the national economy. It allows the decentralization of commerce and industry, makes intercity travel much easier, and increases the mobility of the workforce. Indeed, some estimates claim it has already returned, in terms of economic productivity, six times its construction cost. Other statistics suggest another major socioeconomic benefit: dramatic reductions in the number of road fatalities and injuries, just as Eisenhower predicted.

Further reading

Cox, Wendell, and Jean Love. 1996. *The Best Investment a Nation Ever Made: A Tribute to the Dwight D. Eisenhower System of Interstate and Defense Highways*. Washington, DC: American Highway Users Alliance.

Rose, Mark H. 1990. *Interstate Express Highway Politics, 1941–1989*. Knoxville: University of Tennessee Press.

Seely, Bruce E. 1987. *Building the American Highway System: Engineers as Policy Makers*. Philadelphia: Temple University Press.

Weiner, Edward. 1992. *Urban Transportation Planning in the United States: An Historical Overview*. Washington, DC: Office of the Secretary of Transportation.

V

Vehicle Assembly Building, John F. Kennedy Space Center
Merritt Island, Florida

The U.S. National Aeronautics and Space Administration (NASA) was founded in 1958 with a brief to plan and conduct nonmilitary aeronautical and space activities and to develop international space programs. The 140,000-acre (56,658-hectare) John F. Kennedy Space Center on Merritt Island near Cape Canaveral, Florida, was originally established to support the Apollo lunar landing project. It is now operated by NASA as the main U.S. launching site for satellites and spaceflights. In terms of volume, the Vehicle Assembly Building (VAB) at the Space Center is the second largest building in the world, exceeded only by Boeing's 747 aircraft factory in Seattle, Washington. Originally used to assemble Apollo and Saturn space vehicles, it was later modified to serve space-shuttle operations. It is an architectural feat because of its overwhelming size, but more because it was a building type without historic precedent. The new and difficult architectural design problems it presented (and addressed) have been clearly stated:

> The design of the assembly building had to allow for stacking the (110-meter) Apollo-Saturn space vehicle on top of its 14-meter-high movable launch platform.... To handle the stages of the vehicle, bridge cranes had to span 45 meters and lift 121 metric tons to a height of 60 meters.

The architect-engineers faced complex problems, particularly since the structure had to be capable of withstanding hurricane winds. To make room for...three or four vehicles of this size simultaneously required an enormous building [with] four high bays or checkout areas, each big enough to handle all stages of the Saturn V and the spacecraft...assembled in an upright position ready for launch.... Additional low bays would accommodate preliminary work on single stages. (Benson and Faherty 1978, chapter 11).

In August 1962 URSAM, a New York consortium (architect Max Urbahn; structural engineers Roberts and Schaefer; mechanical and electrical engineers Seelye, Stevenson, Value and Knecht; and foundation specialists Moran, Proctor, Mueser and Rutledge) was commissioned to undertake design studies for what was then called the Vertical Assembly Building (VAB). They employed retired Col. William Alexander as project manager, and by the end of October preliminary designs were submitted.

The VAB—the acronym now indicates *Vehicle* Assembly Building—covers 8 acres (3.25 hectares). It is 716 feet (218 meters) long and 518 feet (158 meters) wide and consists of a 525-foot-tall (160-meter) "high-bay" section with four vehicle assembly and checkout bays; a 210-foot-tall (64-meter) "low-bay" section with eight stage-preparation and checkout cells; and a four-story Launch Control Center that houses

Vehicle Assembly Building, John F. Kennedy Space Center, Florida; URSAM consortium, architects and engineers, 1962–1965. Aerial photograph.

display, monitoring, and control equipment for check-out and launch operations. It connects to the high-bay area through an enclosed bridge. The simple box shape chosen by URSAM yielded a strong, stable building at minimum cost. Design was dictated by resistance to wind loads, adaptability to changes in space technology, ease of extension, and efficient work areas. The layout of the building allows access from each high bay to a "transfer aisle," along which the completed and checked shuttle components are moved for assembly, and finally the 3,000-ton (2,700-tonne) crawler transporter—larger than a baseball infield—carries the launch platform and 185-foot (57-meter), 6,000-ton (5,400-tonne) shuttle vehicle on the five-hour journey to the launch pad, 2 miles (3.2 kilometers) away.

URSAM designed the steel-framed VAB for winds of 125 mph (200 kph) and limited side sway, a critical factor, to less than 6 inches (15 centimeters). Space-vehicle launches, although remote from the building, create massive shock waves and noise, so the designers faced it with insulated aluminum panels; natural light was provided by impact-resistant, translucent plastic panels. There were other unique problems created by the size of the building: the air-conditioning system (that needed for the office, laboratory, and low-bay area is equivalent to 3,000 houses), the high-bay doors, and the lifting mechanisms that had to cope with huge loads. The mobile launcher for the Apollo vehicles needed a 450-foot-high (139-meter) opening, 147 feet (45 meters) wide at the base and 72 feet (22 meters) at the top. The doors have seven leaves at the top of the opening and four motor-driven horizontal-sliding leaves at the bottom. Among the largest doors ever made, they weigh between 33 and 72 tons (30 and 66 tonnes). The vehicle assembly processes necessitated two 270-ton (250-tonne) bridge cranes that could span almost 150 feet (45 meters), with a hook height of 460 feet (141 meters); the loads they impose upon the foundations are enormous. The structural engineers designed a bed of 4,225, 16-inch-diameter (40-centimeter) open-end steel pipe piles, driven 160 feet (49 meters) to bedrock.

In May 1963, the American Bridge Division of U.S. Steel signed a $23.5 million contract for supplying and erecting more than 49,500 tons (45,000 tonnes) of structural steel for the VAB frame completion by 1 December 1964. Six weeks later Blount Brothers Corporation of Montgomery, Alabama, contracted to construct the foundations and floors of the VAB for $8 million. Early in October, bids were invited for general construction: completion of the VAB, site works, roads and utility installations, the VAB utility annex, and the Launch Control Center. The Corps of Engineers, which oversaw all the contracts, set the completion date at 1 January 1966. The bidding was won by a consortium of three California companies—Morrison-Knudsen Company, Perini Corporation, and Paul Hardeman—for almost $63.4 million. The South Gate combine soon took over the VAB structural steel contract and Colby Cranes Manufacturing Company's contract for the bridge cranes. The consolidated Morrison-Knudsen, Perini, and Hardeman contract was then $88.7 million. The "topping-out" ceremony for the VAB took place on 14 April 1964, although "bricks and mortar" work continued well into 1965.

Further reading

Benson, Charles D., and William Barnaby Faherty. 1978. *Moonport: A History of Apollo Launch Facilities and Operations*. N.p.: NASA.

Harland, David M. 1998. *The Space Shuttle: Roles, Missions, and Accomplishments*. Chichester, UK and New York: Wiley.

Jenkins, Dennis R. 1996. *The History of Developing the National Space Transportation System: The Beginning through STS-75*. Indian Harbor Beach, FL: Dennis R. Jenkins.

Venice, Italy

Venice is one of the world's densest urban places—a compression of churches, great and small houses, and other buildings crowded around hundreds of *piazzi* and *campi*, little relieved with planting and having only two public gardens. Floating on a cluster of more than 100 low islands about 2.5 miles (4 kilometers) off the Veneto region of the Italian mainland, the historical center of this remarkable city is surrounded by the shallow, crescent-shaped *Laguna Véneta* (Venetian Lagoon) and permeated by a network of over 150 canals, 400 bridges, and countless narrow streets known as *calli*. It is protected from the Adriatic Sea by the Pallestrina, Lido, and Cavallino littorals, a total of 30 miles (48 kilometers) of narrow strips of sand with seaward entrances to the lagoon. In fact, Venice is built in the sea, hardly a suitable place for a city, and it therefore provides a remarkable example of how humanity rises to meet a challenge.

Why did the city's founders choose such a location? When he became sole ruler of the Huns in A.D. 446, Attila set out to extend his domain from the River Rhine across the north of the Black Sea to the Caspian Sea. With the Franks and Vandals, five years later he attacked western Europe, only to be driven back by Roman and Visigoth armies. In 452 he invaded Italy, displacing entire communities, many of which fled to islands along the Adriatic coast, then inhabited only by hunters and fishermen. When Attila withdrew a year later the refugees returned to the mainland—but not all. Some historians identify this relocation as the key to the eventual foundation of Venice. After Attila's death in 453, the Lombards rose to dominate what is now Hungary. Around 568 their king Alboin led an army of Lombards, Gepids, Sarmatians, and others into Italy, overrunning much of the Veneto. He would soon conquer Milan and the Po Valley; Tuscany would follow and, by 575, Rome. The people of the Veneto had again retreated to the lagoons. Because the Lombards remained in Italy, the refugees no longer had homes to which they could return and they remained on the islands. Late in the seventh century their numbers were augmented by more exiles from the harsh Lombard rule.

In the lagoon, a loose confederation of communities emerged, owing allegiance to Byzantium. Each had its economic, religious, and organizational distinctives because it governed islands whose population originated in a specific part of the Veneto. By 726 the Iconoclastic movement—a religious phenomenon demanding the destruction of holy images—reached the Byzantine outposts in Italy. Although the rest of the Eastern Empire was loyal to the Orthodox Church, these Italian communities were bound to Rome. Prompted by the pope, they briefly asserted independence from Byzantium, only to think better of it later—except Venice. The Venetians elected Orso Ipato as doge (leader) in 727, the first head of a polity that would last almost 1,100 years, the most enduring republic in history. When Orso's son Teodato succeeded him in 742, the seat of government was moved to Malamocco on the Lido, and Venice was recognized as an independent city within the Byzantine Empire.

In 755 the pope urged the Frankish king Pepin the Short to invade Italy, ending Lombard rule; they were finally defeated in 773 by his successor Charlemagne. Charlemagne's son Pepin II sent a force against the islands of the lagoon in 810. It overran Chioggia and Pallestrina, the southernmost littoral island, before turning on Malamocco. Although the Franks were repelled with heavy losses, the confederation moved its capital to islands near the center of the lagoon that were protected from naval attack by sandbars. Formed by sediment from the Brenta River (the Grand Canal marks its former course), those islands were known as Rivo Alto or *Ri'Alto* (high bank). After the Franks withdrew, the capital remained there, and 828 saw the establishment of the city that has been known for eight centuries as Venice, with its famous Rialto bridge.

Venice was built on its unlikely clutch of islands by gradually reclaiming land from the lagoon or by forming new land behind seawalls and dikes, backfilled with soil brought by boat from the mainland. Timber—oak and pine for piles and larch for the boards—was cut in the northern Veneto forests and floated across the lagoon. Multiple rows of piles were driven into the hard clay substrata under the muddy islands. In this way the natural waterways between them were turned into defined canals, and new ones were formed by blocking the ends, excavating the waterway, forming a bed of sand-clay mixture and then flooding it. Typically, since space has always been at a premium, the buildings of Venice stand literally on the edge of the canals, creating the city's unique appearance.

Platforms of larch boards were laid on the tops of piles, supporting foundation courses of water-resistant Istrian stone. The superstructure of the buildings was usually brick, sometimes stuccoed or (for greater prestige) faced with decorative marbles and architectural moldings. Each island had its *campo* (field), an open space too small to be dignified with *piazza*. The campo had a communal reservoir, fed with rainwater from the surrounding buildings, and (usually) a church, sometimes with a freestanding bell tower called a *campanile*. These open spaces were the center of community life, the location for markets, shops, and warehouses in the ground floors of the surrounding larger houses. The parts of the island remote from the campo were reached through unpaved streets and alleys. From the beginning of the twelfth century, narrow thoroughfares and the corners of canals and bridges were provided with street lighting—the first in any European city.

Venice was divided into *siestieri*, or sixths, one of which—the labyrinthine Santa Croce—was eventually merged with two others, Dorsoduro and San Polo, the city's commercial core since the eleventh century. The others were Cannaregio, Castello, and San Marco, which has been the seat of political power since the age of *La Serrenissima*—the Serene Republic of Venice.

The glorious and sometimes bloody history of that republic is beyond our present scope. Suffice it to say that mostly through canny business skills and judicious conflicts, by the end of the first millennium A.D., Venice had secured the northern end of the Adriatic and soon after that established herself as a key maritime trade center, not only in the Mediterranean but also across the world to distant China. During the Crusades and after 1204, her territories were extended to the Aegean islands, Crete, southern Greece, and even part of Constantinople. Competition with other Italian seafaring states, especially Genoa, simply served to increase her commercial dominance, and in the fifteenth century she expanded on the Italian Peninsula, claiming (among other cities) Treviso, Padua, Vicenza, and Verona. The fall of Constantinople to the Ottoman Turks in 1453, and the discovery of the New World in 1492, heralded her commercial and political demise. At Sapienza in 1499 the Venetian navy was defeated by the Turks, who took control of the Adriatic. At that moment, Vasco da Gama returned to Lisbon with news of a faster route to the Orient. Venice was forced to relinquish her long-held trade supremacy to the Portuguese, Dutch, and English. In 1797 the Treaty of Campoformio gave Venice to Austria; she next came under Napoleonic rule (1805–1814), and after several revolutions and wars of independence, in 1866 she was absorbed into the kingdom of Italy.

Venice is again in danger. The enemies are both natural and man-induced: eustacy (variation in sea levels due to global climate changes); seasonal high tides and water surges as well as subsidence, caused largely by mismanagement of subterranean water sources; and pollution. The combined result of the three means that, in effect, the city in the sea is drowning. In the twentieth century it sank about 10 inches (25 centimeters), about twice the average rate of the previous fourteen centuries. Only half of that was due to uncontrollable changes in sea level. Pollution is of several kinds: Venice has no drains; vast quantities of human and industrial waste of all sorts flow into the lagoon, and its self-cleansing capacity has long been overtaxed. Although authorities recognize the need to address these problems, there is a paradox: the resident population has been displaced by millions of tourists, changing the city's economic profile. Although a series of defensive measures has been planned since 1994, the municipality of Venice finds it increasingly difficult to meet the cost of maintaining its precious monuments. That is despite an April 1973 resolution of the Italian central government, which declared "the safeguarding of Venice and her Lagoon … to be a question of preeminent national interest" and guaranteed to protect her "landscape, historical, archaeology and artistic heritage" as well as to "preserve its environment from atmospheric and water pollution and to guarantee her social and economic vitality within the general framework of regional land development."

Further reading

Concina, Ennio. 1998. *A History of Venetian Architecture.* Cambridge, UK: Cambridge University Press.

Goy, Richard. 1997. *Venice, the City and Its Architecture.* London: Phaidon.

Romanelli, Giandomenico, and Mark E. Smith. *Portrait of Venice.* New York: Rizzoli.

Villa Savoye
Poissy, France

The Villa Savoye at 82 rue de Villiers, Poissy, has been described as "a house so important that architects travel from all over the world to experience its presence" and "an icon of European Modernism." Designed in 1928–1929 by the Swiss architect Charles Edouard Jeanneret (1887–1965), best known as Le Corbusier, it was completed by 1931. It is an architectural milestone, much copied (although often without an understanding of the design principles that underpinned it) and the fullest expression of the modernist belief, first expressed by Le Corbusier in 1923, that a house is "a machine for living in."

By that, he simply meant that a house should be designed in the same way as any contemporary machine—he cited automobiles, airplanes, and ocean liners—by clearly defining the associated problems and solving them as thoroughly as possible. At the beginning of the twentieth century that was a different way of making domestic architecture, whose forms were usually dictated by adherence to largely esthetic formulas. Yet Le Corbusier's houses were not altogether independent of geometry; he admitted that he was "possessed with the color white, the cube, the sphere, the cylinder and the pyramid." His standardized Citrohan House (1922) exhibited the five elements that he believed constituted modern architecture: the structure was raised on columns *(pilotis)*, an open floor plan; strip windows expressing the independence of the walls from the structural frame, a rejection of applied ornament, and a roof terrace. There followed the La Roche house (1923); the Michael Stein villa at Garches, France; and two houses in the Deutscher Werkbund's Weissenhof residential suburb in Stuttgart, Germany, both in 1927. All were functionally planned and starkly austere.

The Villa Savoye in semirural Poissy, designed a year later, was the climax of those experiments, and

Villa Savoye, Poissy, France; Le Corbusier, architect, 1928–1931. Detail of external ramp leading to the rooftop and solarium.

it also incorporated what Le Corbusier called the architectural promenade, an "itinerary" that offers "prospects which are constantly changing and unexpected, even astonishing" (Le Corbusier and Jeanneret 1964, 24). Because it was freestanding, built in a field surrounded by orchards, the architect was able to apply his five-point program without the limitations imposed by tighter urban sites. It is an unadorned horizontal white box supported on *pilotis* and authoritatively imposed upon the natural surroundings. It has been concisely described by others: "The dominant element is the square single-storied box, a pure, sleek, geometric envelope lifted buoyantly above slender *pilotis,* its taut skin slit for narrow ribbon windows that run unbroken from corner to corner" (Trachtenberg and Hyman 2001, 530).

The plan is almost square, organized on a 15.5-foot-square (4.75-meter) structural grid of posts that support the floors and roofs and carry the loads. The curved facade of the first floor allows an automobile to travel through a sort of undercroft. When family members alighted at the main door, the chauffeur drove on to the three-car garage. The radius of the curve is based on the turning circle of the 1927 Citroën. The servants' quarters are on the first floor. Family members and visitors entered the brightly day-lit foyer and proceeded to the second floor via a curved stair, or the ramp that forms the spine of the plan. The second-floor rooms—small, irregular bedrooms, a kitchen, internal bathrooms (lit by skylights), and a spacious salon that opens to the outside through sliding doors—were surrounded by bands of sliding metal windows and grouped around three sides of a roof terrace. Le Corbusier designed ranges of built-in aluminum cupboards along the walls and concrete slab benches for most rooms, artificial lighting was provided by suspended troughs of fluorescent tubes, and there were open-hearth fires in the salon and Madame Savoye's boudoir. The interior organization is partly independent of the structural grid, and there are one or two columns in seemingly inconvenient places. The second stage of the stair and ramp, now outside, lead to the roof, which has a solarium screened by curved walls. As radical as its architecture was, the Villa Savoye would hardly have been comfortable.

The house has had a troubled life. Madame Savoye did not like it and used it only as a summer retreat. During the Nazi occupation in the early 1940s it was used as a hay shed. By the end of the 1950s it was in ruinous disrepair and barely saved from demolition by André Malraux, then the French minister of culture.

Listed as a historic building in 1964 and the following year purchased by the government, the Villa Savoye now depends on the Ministry of Culture and Fondation Le Corbusier for its maintenance and conservation.

See also Weissenhofsiedlung

Further reading

Le Corbusier and Pierre Jeanneret. 1964. *Oeuvre Compléte, 1929–1934*. Zurich: Éditions d'architecture.

Meier, Richard, and Yukio Futagawa. 1972. *Villa Savoye, Poissy, France, 1929–1931*. Tokyo: A.D.A. Edita.

Trachtenberg, Marvin, and Isabelle Hyman. 2001. *Architecture, from Prehistory to Post-Modernism: The Western Tradition*. New York: Harry N. Abrams.

W

Washington Monument
Washington, D.C.

The largest freestanding stone structure in the world is the obelisk built in honor of George Washington that stands about halfway between the Capitol and the Lincoln Memorial in Washington, D.C. By legislation, it will remain the tallest structure in the U.S. capital. The 91,000-ton (82,700-tonne) monument is 555 feet, 5 inches (166.7 meters) high and 55 feet, 5 inches (16.67 meters) square at the base. Its load-bearing granite walls are 15 feet (4.5 meters) thick at the bottom and 18 inches (45 centimeters) thick at the top, reflecting the 10:1 proportion of the overall dimensions. The granite structure is faced with white marble; because it came from different quarries—first from Maryland and later Massachusetts—there is a perceptible variation in color at about one-third of the height. Around the internal stair, 200 memorial stone plaques are set, presented by individuals, societies, cities, states, and foreign countries.

At first, Washington acceded to the Congress's 1783 proposal to erect an equestrian statue of him in the planned federal capital. Faced with the problem of raising funds to build the city, he soon changed his mind. He died in 1799 and the following year, by agreement with his widow, Martha, Congress contemplated interring his remains in a marble pyramid beneath the dome of the Capitol Building, started six years earlier. Without money, the project was post-poned until 1832, the centenary year of Washington's birth. When his executors decided that his body should remain on his Mount Vernon property, the idea was abandoned.

Possibly reacting to official indecision, a group of influential Washington citizens established the Washington National Monument Society in 1833; Chief Justice John Maxwell was its president. Publicizing its intention in the press and by direct appeal to churches, societies, and individuals, the society set about fund-raising. All U.S. citizens were invited to contribute $1, for which a certificate would be issued, but it was not until 1836 that enough money had been collected to finance a design competition for American architects. That resulted in a stylistic potpourri of ideas, including a (larger) variation on the pyramid theme and at least a couple of Gothic Revival proposals. Meanwhile, the fund was growing while the society waited for the government to fix a location, which it did in 1848.

Robert Mills, said to be the first U.S.-born qualified architect, won the competition. He had been in government service for some years, designing among other public buildings the Patent Office and the Treasury in Washington, D.C. And about twenty years earlier he had produced a more modest Washington monument for Baltimore. His extravagant proposal for the national monument comprised a 500-foot (150-meter) obelisk, whose flattish pyramidal peak was adorned with a star; it rose from the center of a circular 110-foot-tall (33-meter) classical temple, between whose thirty-two Doric columns he proposed statues of America's founding fathers. Above a central portico an enormous toga-draped figure of George Washington held the reins of a four-horse chariot.

Construction began on Mills's obelisk in the middle of 1848. On 4 July the 12-ton (11-tonne) cornerstone of Maryland marble was laid according to Freemasonic ritual by the District of Columbia Grand Lodge, launching a long association with the brotherhood of which George Washington had been a member. The society actively solicited contributions to the building fund from Masonic lodges throughout the nation, an appeal it repeated in 1853. It also asked other fraternities for money, but even including the sponsorship of the states the fund was almost depleted by 1854. Work slowed to a crawl. Worse came to worst. In 1854 the anti-Catholic Know-Nothing Party seized control of the society's records and elected its own members to office. The takeover was occasioned by Pope Pius IX's gift of a block of stone from the Temple of Concord in Rome that was stolen and destroyed by party members. Under the two-year Know-Nothing regime, the stream of private gifts, already reduced to a trickle, dried up completely. The obelisk rose just a few feet, poor work at that, before it stopped altogether. A more serious hiatus followed, caused by the Civil War; for more than 20 years, the Washington Monument stood unfinished at a height of about 156 feet (47 meters).

In 1874, society secretary John Carrol Brent again pursued Masonic and other groups, this time with resounding, immediate success. Congress was less responsive, but the occasion of the American Centennial in 1876 raised national sentiment and funds were set aside. In August President Ulysses S. Grant authorized the government to complete the monument and to persuade the society to donate it to the American people. Public interest had waned by then, and Mills's design was challenged. The temple was omitted, and there was strong criticism of the entire proposal. For example, *American Architect and Building News* described it as a "monstrous obelisk, so cheap to design but so costly to execute, so poor in thought but so ostentatious in size" and called for it to be abandoned.

Plenty of alternative schemes were forthcoming. Some, including an equestrian statue over the entrance to a hollow Romanesque column and a Renaissance tower with a statue, incorporated a portrait of Washington. Others wanted to embellish Mills's obelisk: an observation platform with a statue, Gothic encrustation to the stump of the column, a decorated obelisk above a Mayan pyramid—so they went on. But (thankfully) it was decided to stay with the obelisk, but without the classical temple. The foundation was reinforced, and work recommenced on 17 August 1878 under the supervision of the U.S. Army Corps of Engineers, led by Lieutenant-Colonel Thomas Casey.

The proposed height was increased to 555 feet, 5 inches (166.7 meters), ostensibly to achieve a proportion of 10:1—claimed to be the standard for Egyptian obelisks—in relation to the base. That case was put on the advice of the U.S. minister to Italy, George Marsh. The reasoning was both specious and inaccurate, since that ratio would have yielded a height of 554 feet, 7 inches. And an analysis of the Egyptian models shows that there *was* no standard; they varied between 8.9:1 and 11.85:1. It is more likely that the dimensions represent some arcane Freemasonic numerology.

The Washington National Monument was completed on 6 December 1884, when the 3,300-pound (1500-kilogram) marble capstone was placed. At the pinnacle there was a 9-inch (22.9-centimeter) pyramid of cast aluminum, then a precious metal. The monument was dedicated on 1 February 1885 and officially opened to the public in October 1888. The cost was nearly $1.19 million.

A major restoration project, costing more than $9 million and jointly sponsored by Congress, a large retail chain, and several corporate partners, was announced in October 1997. Work commenced the following January. It was necessary to seal a number of exterior and interior stone cracks, repoint the 12 miles (19 kilometers) of external joints and 0.75 mile (1.2 kilometers) of interior joints, repair chipped stone, and clean more than an acre (0.4 hectare) of interior surface. The scaffolding was an architectural feat in its own right. Conceived by the postmodern architect Michael Graves and jointly designed by engineer Alan Shalders and James Madison Cutts Consulting Structural Engineers, the subtly sloping aluminum framework—all 37 miles (60 kilometers) of it—rose parallel to the taper of the monument. Concrete footings were built under the surrounding pavement to bear its weight, and an ingenious support and bracing system meant that it only lightly touched the great obelisk. The frame was sheathed in transparent fabric designed by Graves; its pattern of blue horizontal and vertical lines reflected the masonry beneath. The Washington National Monument reopened to the public in late spring 2000.

Further reading

Freidel, Frank, and Lonnelle Aikman. 1988. *George Washington: Man and Monument*. Washington, DC: Washington National Monument Association.

Olszewski, George J. 1971. *A History of the Washington Monument, 1844–1968*. Washington, DC: Office of History and Historic Architecture.

Watts Towers
Los Angeles, California

The Watts Towers comprise a group of imaginative structures at 1765 East 107th Street in south-central Los Angeles. Once threatened with demolition, they are now listed on the National Register of Historic Places and enjoy the dual status of a State of California Historic Park and Historic-Cultural Monument and a National Historic Landmark (a distinction bestowed in 1990). Someone has described them as "a unique monument to the human spirit and the persistence of a singular vision."

They were built between 1921 and 1954 by Sabato (Simon) Rodia (1879–1965). Rodia was born in a rural village named Ribottoli, near Naples, Italy, where he worked on the family farm and received no formal education. He emigrated to the United States when he was about fifteen years old (some sources say he was sent at the age of twelve). At first employed on the Pennsylvania coalfields, he later moved to the San Francisco Bay Area, where he married and had two children. Some time around 1918 he found work in the construction industry in southern California. In 1921, he purchased a triangular lot alongside a railroad track on dead-end East 107th Street in the Los Angeles multicultural working-class suburb of Watts. Immediately he began work on his fantastic creation, naming it Nuestro Pueblo (Our Town), which became America's best-known folk-art sculpture. Or is it architecture? Whichever, the official name of the work is "The Watts Towers of Simon Rodia."

For thirty years Rodia worked alone, without power tools, scaffolding, bolts, rivets, or welds—his only equipment was his tile-setter's tools and a window washer's safety harness—building a series of towers redolent of the fantastic work of the Catalan architect Antoni Gaudí in Barcelona. But there is no evidence that Rodia had even heard of Gaudí. The nine major structures include three towers, approximately circular in section and impressive for their individuality, purposefulness, and order. They are 99.5, 97.8, and 55 feet (31, 30, and 17 meters) tall, respectively. The slender main legs are assembled from steel pipes and rods, stiffened with circular hoops connected by spokes; all of the sections are tied together with wire, wrapped in wire mesh and then covered by hand-packed cement mortar. Quite intuitively, Rodia produced the *ferro-cimento* made famous by his compatriot, the engineer Pier Luigi Nervi, and his towers gained remarkably high structural strength.

But the towers are only part of the wonder Simon Rodia created on his small slice of America. The north and south perimeter fences of the property, almost 300 feet (93 meters) long, are of similar construction to the towers. Mailboxes emblazoned with his initials and the address flank the entrance. A fire destroyed the house in 1955 and only the front facade and a fireplace and chimney remain. East of the house is the patio with its decorated floor, from which the sculptures rise. To the north is a gazebo, with its 38-foot (11.6-meter) spire, three birdbaths, a center column, and a circular bench. There is also a "Ship of Marco Polo," with a 28-foot (8.6-meter) spire. Everywhere in the mortar that forms and molds every surface are embedded all manner of carefully selected and juxtaposed cast-off materials—bottle caps, broken glass (mostly green, some blue, but

Watts Towers, Los Angeles, California; Sabato (Simon) Rodia, architect, engineer, and builder, 1921–1954.

never clear), seashells, shards of cheap crockery, multicolored ceramic tile, mirrors, pebbles, and fragments of marble—that create a riot of color and texture. Like the work of Gaudí, Rodia's busy, beautiful buildings are beyond categorization and simply beggar description. Indeed, the artist could not provide a rationale for his own work: "I think if I hire a man he don't know what to do. A million times I don't know what to do myself." His motivation? "I wanted to do something for the United States because there are nice people in this country."

After World War II, confronted by major social changes in Watts, Rodia became increasingly withdrawn. In 1954, when he was approaching 75, he gave his property to a Latino neighbor and retired to Martinez, northern California. He never saw the towers again. In 1965 he was hospitalized after suffer-

ing a stroke and died shortly afterward, happy at having heard of public praise for his work. A memorial service was held at the base of his towers.

After fire ravaged Rodia's house in 1955, the Los Angeles Department of Building and Safety ordered the property demolished. But the Committee for Simon Rodia's Towers in Watts successfully fought to preserve it. Professional engineering tests conducted in 1959 proved the towers' structural strength and safety in earthquakes. The committee continued to preserve and maintain the structures for sixteen years, until in 1975 it turned over the towers and the adjoining Arts Center building to the city of Los Angeles for operation and maintenance. Three years later they became the property of the state of California, and a restoration program was initiated for the three tall towers. In 1985 responsibility for the towers and the Watts Towers Arts Center passed to the City of Los Angeles Cultural Affairs Department, where it remains. Both the Los Angeles County Museum of Art and the Getty Conservation Institute have shared in the conservation program since 1986. The towers were closed to the public after the 1994 Northridge earthquake, and a $1.9 million restoration project was launched in August 2000; the work was completed in 2001. In the meantime, the Environmental Affairs Department and the Los Angeles Conservation Corps planted over eighty new trees in the adjacent park.

See also Sagrada Familia (Church of the Holy Family)

Further reading

Goldstone, Bud, and Arloa Raquin Goldstone. 1997. *The Los Angeles Watts Towers.* Los Angeles: Oxford University Press.

Kaplan, Sam Hall. 1987. *LA Lost and Found: An Architectural History of Los Angeles*. New York: Crown Publishers.

Ward, Daniel Franklin. 1986. *Simon Rodia and His Towers in Watts: An Introduction and a Bibliography*. Monticello, IL: Vance Bibliographies.

Weissenhofsiedlung
Stuttgart, Germany

An acute accommodation shortage after World War I led many European cities to develop low-cost public

Weissenhofsiedlung, Stuttgart, Germany; Ludwig Mies van der Rohe and others, architects, 1927. This contemporary photograph shows how the white boxes contrasted with their conventional neighbors.

housing programs. In Stuttgart, Germany, the Württemberg group of the Deutscher Werkbund organized an exhibition, *Die Wohnung unserer Zeit* (The Housing of Our Times), in collaboration with the municipality. In 1927, twenty-one white stuccoed, flat-roofed examples of the "modern" house were built on city-owned land on the southwest slopes of Killesberg, a few kilometers from downtown Stuttgart. The dwellings and the philosophy were seminal. For the first time, architects were building not for an elite clientele but for families with limited means. There was overt social intent—the emphasis was on health and well-being, cleanliness, and efficiency—and a marked turning toward a new architecture that had the ordinary house as its main theme.

The Deutscher Werkbund, a coalition of architects, artists, and industrialists, had been founded in 1907 in a (largely successful) attempt to refine German "workmanship and [to enhance] the quality of production"—in short, to make Germany into Europe's top nation in design. Its goals may be exemplified by reference to the career of the architect Peter Behrens (1868–1940), who became artistic adviser to the Allgemeine Elektrizitats-Gesellschaft (General Electric Company, AEG), with a ubiquitous brief: besides designing AEG's industrial buildings, he produced coffeepots, fans, clocks, and radiators, as well as stationery, catalogs, packaging, and advertising material. Thus the Werkbund created a bond across the arts, and between art and industry.

The Weissenhofsiedlung project was strenuously promoted by the gallery owner Gustaf Stotz, and in 1925 the Berlin architect Ludwig Mies van der Rohe joined the board of the Deutscher Werkbund. The following year he became its first vice president and then was charged with producing the development plan of the *siedlung*. His first proposal was for houses grouped in terraced blocks that followed the site contours. He envisaged, with socialistic ardor, uneven rows of interlocked buildings, reached by pedestrian streets winding through interconnected public spaces and green areas—a true community. But because the Stuttgart municipality wanted to sell or rent the houses after the show closed, in the event a more conventional layout was chosen, with mostly single-family, freestanding units.

After much indecision and repeated revision of lists of participants, sixteen architects from five European countries were invited to build in the Weissenhofsiedlung. The houses were designed by Mies van der Rohe himself; the other Germans: Walter Gropius, Bruno Taut, Hans Poelzig, Richard Decker, Adolf Schneck, Adolph Rading, Hans Scharoun, Max Taut, and Ludwig Hilberseimer; Behrens; the Hollanders J. J. P. Oud and Mart Stam, both of whom designed row houses; the Austrian Josef Frank; the Belgian Victor Bourgeois; and the Swiss Le Corbusier. The dwellings were archetypes of the Modern Movement, what may well be the most significant group of buildings of twentieth-century architecture, based on the exploitation of the structural possibilities of new technologies. The homogeneous nature of the ensemble was certainly recognized beyond Europe, and in 1932 the New York Museum of Modern Art mounted an exhibition to show what became known as the International Style. The curators Henry-Russell Hitchcock and Philip Johnson identified the new "style" by a number of characteristics: cubic volumes, open plans, a skeleton structure that allowed the exteriors to be treated as skins rather than bearers of loads (often white painted concrete or stucco), and the decorative use of color and expressed structural detail rather than applied ornament.

The Weissenhofsiedlung was brushed aside by reactionary architects and critics, whose political preferences are betrayed by one contemporary epithet: "collective anthills of vicious Central African termites." It was only to be expected that the Nazi Party, coming to power in 1933, would also disapprove of the houses on political and artistic grounds, and five years later the "shameful blot on the face of Stuttgart" was publicly condemned as "degenerate art." Plans to have the entire suburb demolished and replaced by military buildings never eventuated. However, most of the Weissenhofsiedlung houses subsequently suffered war

damage; some were completely destroyed. After World War II the surviving buildings were neglected, unsympathetic changes were made, and infill construction took little account of the architectural context provided by the surviving icons of Modernism. The Society of Friends of the Weissenhofsiedlung was founded in 1977, and in 1981 the Federal Republic of Germany's minister for housing, regional, and urban planning authorized the "rehabilitation and restoration of the Weissenhofsiedlung." In 1987 the surviving buildings, designed by Behrens, Bourgeois, Frank, Le Corbusier, Mies van der Rohe, Oud, Scharoun, Schneck, and Stam, were restored.

See also Villa Savoye

Further reading

Irace, Fulvio. 1984. "Stuttgart: Weissenhof Case Study." *Domus* (April): 2–13.

Joedicke, Jürgen. 1989. *The Weissenhofsiedlung*. Stuttgart: Krämer.

Kirsch, Karin. 1994. *The Weissenhofsiedlung. Experimental Housing Built for the Deutscher Werkbund, Stuttgart 1927*. Cologne: Taschen.

World Trade Center Towers
New York City

On the morning of 11 September 2001, terrorists targeted the World Trade Center in Manhattan, first crashing a hijacked commercial jetliner into the upper levels of One World Trade Center, one of its twin 110-story iconic skyscrapers. A few minutes later a second hijacked aircraft sliced through the middle levels of Two World Trade Center, the other tower. (A third airliner crashed into the Pentagon outside Washington, D.C., while a fourth crash landed in a field in Pennsylvania, its intended target undetermined.) The effects were predictably devastating: both buildings burned fiercely before totally collapsing in clouds of dust and rubble that darkened the sky above Manhattan. A third building in the complex, a forty-seven-story office block (Seven World Trade Center), damaged by flying debris, followed soon after.

In an earlier raid in February 1993 Arab terrorists exploded a 1,200-pound (550-kilogram) truck

One and Two World Trade Center, New York City; Minoru Yamasaki and Associates and Emery Roth and Sons, architects, 1964–1970.

bomb in the Center's parking garage, creating a 150-foot diameter (46-meter) crater. Six people died, and over 1,000 were injured. Floors were destroyed for three levels below the point of detonation, but because of the load-bearing exterior walls, the structural stability of the building was largely unaffected. Tenants returned to their offices by the end of March. The cost of repairs was $250 million. The World Trade Center occupied a 16-acre (6.5 hectare) site a few blocks from Wall Street at the southwestern tip of Manhattan Island, near the bank of the Hudson River. Designed by Minoru Yamasaki and Associates and supervised by Emery Roth and Sons, it was the core of an urban renewal scheme sponsored by the Port Authority of New York and New Jersey to attract international firms to downtown Manhattan.

The surviving parts of the complex are a twenty-two story, 818-room hotel (Three World Trade Center); two nine-story office buildings (Four and Five World Trade Center); and an eight-story Customs House (Six World Trade Center). With the destroyed buildings, they were grouped around the 5-acre (2 hectare) landscaped Austin J. Tobin Plaza. Beneath it is The Mall, with about sixty specialty shops, banks, restaurants, and function spaces. Before the tragedy, about 500 international companies were located in the center, employing 50,000 people. It had its own subway stations and its own zip code. In March 1999 U.S. construction executives named the World Trade Center among the top ten construction achievements of the twentieth century.

For a short while the One and Two World Trade Center towers, at around 1,353 feet (411 meters), were the world's tallest buildings, but they were superseded in 1974 by the 1,442-foot (450-meter) Sears Tower in Chicago. In 1998 the Petronas Twin Towers in Kuala Lumpur, Malaysia, reached 1,483 feet (452 meters). Even higher buildings have been projected: for example, the Taipei Financial Center, to be completed in August 2002, will stand 1,660 feet (508 meters) tall and Hong Kong's Kowloon MTR Tower will be 1,903 feet (580 meters). As technically demanding as it is, great height does not qualify a building as an architectural feat. It was their structural system and the consequent creation of usable space that made the New York World Trade Center's towers remarkable. Ironically, it also was a contributor to their collapse.

Yamasaki's team was selected over a dozen other American architects. During the preliminary design phase, more than 100 proposals were reviewed, ranging from a single 150-story tower (its scale was far too large) to a series of lower towers (which "looked too much like a housing project"). Following Yamasaki's recommendation, the Port Authority decided on the twin towers as the focus for the complex. The final plan was produced, subject to some minor changes, in 1964 and construction began in 1966. The project, administered by the Tishman Realty and Construction Co., entailed over 700 contracts.

The site was first isolated by building a 3-foot (0.9-meter) thick, 70-foot-high (21.5-meter) wall, keyed into the bedrock. A six-level basement was built in the foundation excavation, and the spoil was used to create over 23 acres (9.3 hectares) of new land in adjacent Battery Park. The towers, each 208 feet (63.6 meters) square in plan, started to rise in March 1969.

The structural engineering firm of Worthington, Skilling, Helle and Jackson designed the world's highest load-bearing walls as vertical cantilevered steel tubes. Fourteen-inch square (0.36-meter), aluminum-clad box columns at 39-inch (1-meter) centers around the perimeter of the towers had spandrels welded to them at each floor, effectively making them into huge trusses. The facades were in effect steel lattices, a light, economical structure that provided efficient wind bracing at the outside surface of the building. Less than a third of the towers' surface area was glass; the slot-like windows were narrower than the columns between them. Thirty-three-inch- deep (0.84-meter) prefabricated steel trusses, spanning 60

feet (18.4 meters) to the towers' central service cores, carried the floors. Installation of 200,000 structural steel components—they were prefabricated in Seattle, St. Louis, and Los Angeles—presented a major logistical problem because of the tight urban building site. Delivery and fixing was managed by a computer-programed control system, and eight "kangaroo" cranes built in Australia were used to hoist the elements into place.

The towers' interiors were column-free, providing 4.48 million square feet (418,600 square meters) of net rentable space—about 75 percent of the total area (the average for contemporary high-rise buildings in the U.S. was only two-thirds of that). The vertical transportation system, which required fewer elevator shafts, also contributed to that efficiency. Each tower had ninety-nine elevators grouped in three vertical "zones". Express elevators served skylobbies at the forty-first and seventy-fourth floors; from the skylobbies and the plaza, four banks of local elevators carried passengers to each zone.

The World Trade Center complex opened in December 1970 and was dedicated in April 1973. During the course of construction the cost rose from the estimated $350 million to $800 million.

In the September 2001 attack, more than 5,000 people were killed and thousands more were injured. In late September, the last standing piece of the center, a seven-story wall from the south tower, was torn down and saved for possible use in a future memorial. It is unlikely that the elegant towers will rise again.

Further reading

Darton, Eric. 1999. Divided We Stand: A Biography of New York City's World Trade Center. New York: Basic Books.

Gillespie, Angus Kress. 1999. Twin Towers: The Life of New York City's World Trade Center. New Brunswick, NJ: Rutgers University Press.

Heyer, Paul. 1978. Architects on Architecture: New Directions in America. New York: Walker.

Glossary

Abutment   In bridge construction, the landward approach to the bridge.

Adobe   Sun-dried brick, made of clay, water and a bonding material (e.g., straw).

Aggregate   Broken stone, gravel or sand that is mixed with cement and water to form concrete.

Agora   The open space in a Greek town used for markets and public gatherings; equivalent to the Roman forum.

Aisle   In churches, the space between the nave and the outer wall.

Almonry   In monasteries, the room in which alms are distributed.

Ambulatory   In churches, the continuation of the aisle around an apse; originally used for liturgical purposes.

Apse   In churches, the (usually) semicircular termination at the east end that often houses the altar.

Arcade   A free-standing row of arches supported on piers or columns.

Archimedes' screw   A device comprising a spiral passage within an inclined closed cylinder, used for raising water; also Archimedean screw.

Ashlar   Squared blocks of smooth stone neatly trimmed to shape.

Atrium   An inner courtyard open to the sky or covered with a skylight.

Attic story   In classical architecture, a story above the main entablature.

Bailey   The courtyard inside castle walls; also known as a ward.

Barbican   The fortified gateway defending the drawbridge or entrance of a castle.

Barrel vault   The simplest form of vault, consisting of a series of semicircular arches extended prismatically.

Bas-Relief   Low-relief sculpture or carving, often applied as architectural decoration.

Bastion   A projecting tower at the angle of a fortification.

Battered wall   A wall with an inclined face.

Berm   An earthen mound or wall.

Béton brut   Concrete left unfinished except for the impression of the formwork on its surface; also known as "off-form" concrete.

Bevel   A right-angled corner cut off asymmetrically (i.e., at other than 45 degrees). *See* Chamfer.

Breccia   A type of stone consisting of sharp fragments embedded in a fine-grained matrix.

Brise-Soleil   A sun protection device; a shading screen.

Buttress   A masonry support built against an exterior wall of a building to absorb lateral thrusts from roof vaults.

Caisson   A watertight chamber used in underwater construction work or as a foundation.

Cantilever   A horizontal projection from a building (e.g., a balcony, beam or canopy) that is without external bracing and appears to be self-supporting.

Capital   The head of a column.

Caravansary   An inn for the lodging of caravans; also caravanserai.

Caryatid   In classical Greek architecture, a sculptured draped female figure supporting an entablature.

Cast-iron   A hard, brittle and nonmalleable alloy of iron, carbon, and silicon cast in a mold

Castellation   In castles, a battlement with alternating indentations and raised portions; also called crenellation.

Catenary — The natural curve taken up by a uniform filament hanging freely from two fixed points.

Cathedra — The throne of a bishop, that gives its name to cathedral churches.

Cella — The interior space of a classical temple.

Centering — The temporary formwork, usually timber, used to support arches or domes until the keystone is placed.

Chamfer — A right-angled corner cut off symmetrically (i.e., at 45 degrees).

Chancel — In churches, the space at the east end surrounding the high altar and containing seats for the clergy and choir.

Chevet — In churches (especially French medieval churches), the east end comprising the choir and apse, often with radiating chapels.

Choir — In churches, the part east of the crossing, where the services were sung.

Clerestory — The upper part of the wall of a church (above the aisle roofs) having windows in it; also, an upper part of any wall whose windows allow light into the center of a space.

Cloister — In monasteries, a covered passage, usually around three sides of a garden (or garth), and connecting the residential parts of the convent with the church.

Cob — Unburned clay mixed with straw, gravel and sand and used in wall construction.

Coffering — Decorative pattern on the underside of a ceiling, dome or vault, consisting of sunken square or polygonal ornamental panels. It reduces the weight of the ceiling without structurally weakening it.

Colonnade — A row of columns supporting an entablature or arches. *See* Arcade.

Coping — A covering of stones or other material used to protect the top of a wall from water penetration.

Corbel — A projecting block of stone built into a wall during construction; also, step-wise construction, as in an arch or dome.

Crepidoma — The stepped base of a Greek temple.

Crossing — In churches, the space formed by the intersection of the nave and the transepts.

Cupola — A dome, especially a small dome, on a circular or polygonal base.

Curtain wall — In castles, a connecting wall between two towers, surrounding the bailey.

Dado — The lower part of an interior wall when specially decorated or faced.

Deformation — The change in shape (usually quite small) that occurs in a structural member when loads are applied to it.

Downpipe — A vertical or near-vertical pipe that conveys rainwater to the ground from the upper parts of buildings.

Dressings — Masonry or moldings around openings or at the corners of buildings, of better quality than other facing work.

Drystone — Walls built without mortar.

Embrasure — In castles, a small opening in the wall or parapet, usually splayed on the inside.

Entablature — In classical buildings, the part between the tops of the columns and the roof, comprising the architrave (lower horizontal section that connects the columns); the frieze and the cornice, that projects to support the edge of the roof.

Faience — Twice-fired glazed terracotta.

Fan vault — Vault, largely confined to late English Gothic architecture, consisting of decorated solid concave-sided half-cones meeting at the apex of the vault.

Fanlight — A semicircular window above a door, of the same width as the door.

Flitch — A timber with a cross section exceeding 4 by 12 inches (10 by 30 centimeters).

Flying Buttress — An arch or half-arch that transfers the thrust of a vault or roof to an outer buttress and thus to the ground.

Formwork — A set of forms in place to hold wet concrete until it sets; also known as shuttering.

Frieze — The part of an entablature between the architrave and the cornice.

Galilee Porch — In churches, a vestibule at the west end.

Gothic — Relating to a style of architecture developed in northern France and western Europe from the middle of the twelfth century to the early sixteenth century.

Groin vault — Vault formed by two identical barrel vaults intersecting at right angles.

Keep — In castles, the main stone tower, the last line of defense.

Kiva — A Pueblo Indian ceremonial structure, usually round and partly underground.

Lantern — A small open-sided structure crowning a dome or roof, to admit light or air into the enclosed space below.

Lierne vault — A vault with tertiary (lierne) ribs spanning between the perimeter and diagonal ribs.

Lintel — A heavy beam over an opening in a wall, that supports the weight above it.

Loggia — A roofed open gallery overlooking an open court.

Machicolation — In castles, an opening between the corbels of a projecting parapet, usually above a portal, for discharging missiles upon attackers.

Megalithic — Constructed of large stones; the term is usually applied to cultures who built in this way.

Megaron — In pre-classical Greek palaces, a large rectangular hall.

Minaret	In Islamic architecture, a tall, slender tower with projecting balconies from which the *muezzin* calls the people to prayer.
Monolithic	An architectural element comprised of a single block of stone.
Mullion	A vertical member dividing units of a window or opening.
Naos	In Greek temples, the room in which the cult statue was placed.
Narthex	In churches, the enclosed porch or vestibule between the atrium and the nave.
Nave	In churches, the central principal space, extending from the narthex to the chancel.
Oculus	A circular or oval window, or a circular opening at the top of a dome.
Oratory	A place for prayer; often a private chapel.
Pagoda	A Buddhist temple, usually in the form of a cylindrical or polygonal tower with a series of roofs of diminishing size.
Parterre	A geometrically laid out ornamental garden (of French origin) with paths between the beds.
Pediment	In classical architecture, a triangular gable over a portico.
Pendentive	A concave triangular element, part of a sphere. Of Byzantine origin, it forms a transition between a square compartment and the drum or springing of a dome.
Piazza	An open square, especially in an Italian town. The Spanish equivalent is Plaza; German, Platz; English and French, Place.
Pilaster	An attached rectangular column (not necessarily structural) projecting slightly from a wall surface.
Pile	A long slender column (timber, steel, or reinforced concrete) driven into the ground to carry a vertical load.
Piloti	(French) Stilts that raise a building so that the ground floor is left open.
Polder	An area of low land (as in Holland) reclaimed from the sea or some other body of water, and surrounded by dikes.
Portcullis	In castles, a heavy timber or metal grille that protected the entrance and could be raised or lowered from within.
Porte cochere	A porch large enough for wheeled vehicles to pass through.
Portico	A porch or covered walk consisting of a roof supported by columns.
Pozzolana	A volcanic earth found in many places in southern Europe; also known as pozzoluana.
Propylaea	In Greek architecture, a ceremonial gateway.
Proskenium	In an ancient Greek theater, the "picture-frame" surrounding the stage.
Puddled Iron	Iron that has been converted from pig-iron to wrought iron by stirring it with an oxidizing agent in a reverberatory furnace.
Quoin	The dressed stones at the angle of a building.
Radome	Dome built to house large radar antennae in U.S. military installations; some are pneumatic, others geodesic.
Refectory	In monasteries, the dining hall.
Reveal	The side of a wall opening between the frame and the outer surface of the wall.
Rib & panel vault	A framework of diagonal arches (ribs) that carry thinner masonry panels spanning between them.
Rubble	Rough stones of irregular shapes and sizes; generally not laid in courses and often used as infill between ashlar facing walls.
Sanctuary	In churches, the space in which the altar is placed.
Scriptorium	In monasteries, the room in which monks copied and transcribed books.
Spandrel	The space between the right or left exterior curve of an arch and an enclosing right angle.
Stainless steel	A rust- and corrosion-resistant steel alloy containing chromium, and sometimes nickel or molybdenum.
Stucco	A material consisting of cement, sand, and a little lime, applied as a hard covering to exterior walls.
Stupa	A building type largely limited to Buddhism: a square base surmounted by an inverted circular bowl and spire.
Swag	In classical architecture, a sculpted garland of flowers or fruit hanging in a curve between two points.
Tabernacle	In churches, an ornamented receptacle in which the bread of the sacrament is reserved
Talus	In castles, a smooth sloping base at the external foot of a curtain wall; also known as a glacis.
Tesserae	Small pieces, usually cuboid, of marble, glass or metal used in mosaic work.
Tracery	In Gothic architecture, the ornamental ribwork in the upper part of window.
Transept	Transverse arm of cruciform plan church, intersecting the nave at a right angle.
Tribune	The apse of a basilican church.
Triforium	In churches, the arcaded wall-passage above the aisle, opening into the upper level of the nave.
Trilithon	Literally, three stones; a horizontal stone resting on two upright ones.
Truss	An assemblage of structural members, usually by triangulation, to form a rigid framework.
Tufa	A low density stone, made up of fine volcanic detritus fused together by heat; also known as tuff.

Tumulus	An artificial hillock or mound.	**Wrought-iron**	A tough, malleable and relatively soft commercial form of iron, with a very small carbon content, that also contains slag in mixture.
Vault	An arched structure of masonry usually forming a ceiling or roof.		
Vernacular	In architecture, relating to the common building style of a culture.	**Ziggurat**	In ancient Mesopotamia, a stepped pyramid with outside stairways, surmounted by a shrine or temple.
Voussoir	A wedge-shaped brick or stone that goes to form an arch or vault.		

Index

Note: page references in **bold** refer to main encyclopedia entries on the topic; references in *italic* refer to illustrations.

About the Authors

DR. DONALD LANGMEAD is Professor of Architectural History at the University of South Australia. He has written numerous works on architectural history, including *Willem Dudok: A Dutch Modernist* and *J. J. P. Oud and the International Style*.

DR. CHRISTINE GARNAUT is a Research Associate in the Louis Laybourne Smith School of Architecture and Design at the University of South Australia. She has written *Colonel Light Gardens: Model Garden Suburb* and has published widely on Australia's planning history.

Illustration Credits

Illustration Credits

Page 238	Francis G. Mayer/Corbis	Page 300	Sean Sexton Collection/Corbis
Page 241	Michael Freeman/Corbis	Page 305	Wolfgang Kaehler/Corbis
Page 243	Corbis	Page 307	Eye Ubiquitous/Corbis
Page 246	Hemera Studio	Page 309	Farrell Grehan/Corbis
Page 249	Bill Ross/Corbis	Page 311	Bettmann/Corbis
Page 252	Neil Beer/Corbis	Page 322	The Art Archive/Dagli Orti
Page 255	The Art Archive/Dagli Orti	Page 326	L. Clarke/Corbis
Page 257	Michael Freeman/Corbis	Page 330	Hemera Studio
Page 258	Courtesy Michael Stewart	Page 332	Michael Nicholson/Corbis
Page 261	How-Man Wong/Corbis	Page 334	Bill Ross/Corbis
Page 265	MIT Collection/Corbis	Page 336	Angelo Hornak/Corbis
Page 270	MIT Collection/Corbis	Page 339	Vanni Archive/Corbis
Page 272	AFP/Corbis	Page 341	Keren Su/Corbis
Page 276	AFP/Corbis	Page 343	Roger Wood/Corbis
Page 279	Hulton-Deutsch Collection/Corbis	Page 345	Hulton Getty/Archive Photos
Page 281	Hulton Getty/Archive Photos	Page 346	Vanni Archive/Corbis
Page 284	Patrick Ward/Corbis	Page 354	Bettmann/Corbis
Page 285	Robert Holmes/Corbis	Page 358	MIT Collection/Corbis
Page 289	Corbis	Page 364	Bettmann/Corbis
Page 291	Eric Crichton/Corbis	Page 365	Hulton Getty/Archive Photos
Page 293	Hulton-Deutsch Collection/Corbis	Page 367	Michael S. Yamashita/Corbis
Page 295	The Art Archive/Jarrold Publishing